# HANDBOOK OF WIRELESS
# LOCAL AREA NETWORKS

# INTERNET and COMMUNICATIONS

This new book series presents the latest research and technological developments in the field of Internet and multimedia systems and applications. We remain committed to publishing high-quality reference and technical books written by experts in the field.

If you are interested in writing, editing, or contributing to a volume in this series, or if you have suggestions for needed books, please contact Dr. Borko Furht at the following address:

**Borko Furht, Ph.D.**
**Department Chairman and Professor**
**Computer Science and Engineering**
**Florida Atlantic University**
**777 Glades Road**
**Boca Raton, FL 33431 U.S.A.**

**E-mail: borko@cse.fau.edu**

# HANDBOOK OF WIRELESS LOCAL AREA NETWORKS

Applications, Technology, Security, and Standards

Edited by
Mohammad Ilyas and Syed Ahson

## Taylor & Francis
Taylor & Francis Group

Boca Raton   London   New York   Singapore

A CRC title, part of the Taylor & Francis imprint, a member of the
Taylor & Francis Group, the academic division of T&F Informa plc.

Published in 2005 by
CRC Press
Taylor & Francis Group
6000 Broken Sound Parkway NW, Suite 300
Boca Raton, FL 33487-2742

International Standard Book Number-10: 0-8493-2323-1 (Hardcover)
International Standard Book Number-13: 978-0-8493-2323-2 (Hardcover)
Library of Congress Card Number 2004062380

---

### Library of Congress Cataloging-in-Publication Data

Handbook of wireless local area networks : applications, technology, security, and
    standards / edited by Mohammad Ilyas, Syed Ahson.
        p. cm.
    Includes bibliographical references and index.
    ISBN 0-8493-2323-1 (alk. paper)
    1. Wireless LANs--Handbooks, manuals, etc.  I. Ilyas, Mohammad, 1953- II. Ahson,
Syed.

    TK5105.78.H38 2004
    004.6'8--dc22                                2004062380

---

Taylor & Francis Group
is the Academic Division of T&F Informa plc.

Visit the Taylor & Francis Web site at
http://www.taylorandfrancis.com

and the CRC Press Web site at
http://www.crcpress.com

# CONTENTS

## SECTION VI: STANDARDS

# ABOUT THE EDITORS

**Mohammad Ilyas, Ph.D.**

Dr. Mohammad Ilyas received his B.Sc. degree in electrical engineering from the University of Engineering and Technology, Lahore, Pakistan, in 1976. From March 1977 to September 1978, he worked for the national Water and Power Development Authority in Pakistan. In 1978, he was awarded a scholarship for his graduate studies and he completed his M.S. degree in electrical and electronic engineering in June 1980 at Shiraz University, Shiraz, Iran. In September 1980, he joined the doctoral program at Queen's University in Kingston, Ontario, Canada. He completed his Ph.D. degree in 1983. His doctoral research was about switching and flow control techniques in computer communication networks. Since September 1983, he has been with the College of Engineering, Florida Atlantic University, Boca Raton, where he is currently associate dean for graduate studies and research. He is also interim associate vice president, Division of Research and Graduate Studies. From 1994 to 2000, he was chairman of the department. During the 1993–94 academic year, he was on sabbatical leave with the Department of Computer Engineering, King Saud University, Riyadh, Saudi Arabia.

Dr. Ilyas has conducted successful research in various areas including traffic management and congestion control in broadband and high-speed communication networks, traffic characterization, wireless communication networks, performance modeling, and simulation. He has published one book, four handbooks, and over 150 research articles. He has supervised 10 doctoral dissertations and more than 35 master's theses to completion. He has been a consultant to several national and international organizations. Dr. Ilyas is an active participant in several IEEE technical committees and activities, and a senior member of IEEE.

**Syed A. Ahson**

Syed Ahson received his B.Sc. degree in electrical engineering from Aligarh Muslim University, Aligarh, India, in 1995. He received his master's degree in computer engineering in July 1998 at Florida Atlantic University, Boca Raton. From September 1997 to September 1999, he was with NetSpeak Corporation (now part of Ne2Phone), a leading provider of Voice-over-IP solutions as a software developer. Since September 1999, he has been with iDEN Subscriber Division of Motorola, Inc., where he is a senior software engineer.

Syed Ahson has extensive knowledge of wireless data protocols (WAP 2.0 MMS, Wireless Email, Wireless Profiled TCP, Wireless Profiled HTTP) and has also worked extensively with Voice-over-IP Protocols (H.323, RTSP, RTP). He has published several research articles and taught undergraduate engineering courses as an adjunct professor at Florida Atlantic University.

# PREFACE

Wireless communications networks have been in existence for many years. Recent advances in microelectronics and transmission techniques, and the needs of users have led to the development of many applications of wireless communication networks. There is potential for many more wireless applications. In particular, there has been a tremendous growth in wireless local area networks (WLANs). Applications of WLANs include many important fields such as academia, healthcare, defense, home, search and rescue, and emergency management. There are several technical challenges related to WLANs which need to be addressed.

The *Handbook of Wireless LANs* provides technical information about all aspects of wireless local area networks. The areas covered range from basic concepts to research grade material, including future directions.

It reviews the current state of wireless local area networks and serves as a source of comprehensive reference material on this subject. It contains 25 chapters written by experts from around the world and is organized in the following six sections:

1. Introduction
2. Technology
3. Applications
4. Security I
5. Security II
6. Standards

The targeted audience for the handbook includes professionals who are designers or planners for wireless local area networks, researchers (faculty members and graduate students), and those who would like to learn about this field. It has these objectives:

■ To serve as a single comprehensive source of information and as reference material on wireless local area networks
■ To deal with important and timely topics of emerging communication technology
■ To present accurate, up-to-date information on a broad range of topics related to wireless local area networks
■ To present material authored by experts in the field
■ To present its information in an organized and well-structured manner

Although the handbook is not precisely a textbook, it can certainly be used as a textbook for graduate courses and research-oriented courses that deal with wireless local area networks. Any comments from the readers will be highly appreciated.

Many people have contributed to this handbook in their unique ways. The first and the foremost group that deserves immense gratitude is the group of highly talented and skilled researchers who have contributed 25 chapters to this handbook. All of them have been extremely cooperative and professional. It has also been a pleasure to work with Rich O'Hanley, Claire Miller, David Grubbs, and Andrea Demby of Auerbach Publications, and we are extremely gratified by their support and professionalism. Our families have extended their unconditional love and strong support throughout this project, and they all deserve very special thanks.

**Mohammad Ilyas**
*Boca Raton, Florida*

**Syed Ahson**
*Plantation, Florida*

# I

# INTRODUCTION

# 1

# IP MOBILITY

*Tauseef ur Rehman*

## INTRODUCTION

Wireless networks have recently received much attention. Support for mobility in Internet access is gaining significant interest as wireless/mobile communications and networking are proliferating, especially boosted by the widespread use of laptops and handheld devices.[1] Wireless LANs can be classified into two broad categories — infrastructured and infrastructureless (ad hoc). In certain wireless networks, one hop is needed to reach the mobile terminal. Ad hoc networks normally require a multi-hop wireless path from the source to the destination. Interest in such networks is increasing day by day.

Wireless networks can be broadly classified into two types:

1. Fixed wireless networks
2. Mobile wireless networks

*Fixed wireless networks* are mostly point-to-point and they do not support mobility (e.g., microwave networks, geostationary satellite networks). *Mobile wireless networks*, on the other hand, support mobility.[7] They can be further classified into two types — infrastructured (cellular) and infrastructureless (ad hoc) (Figure 1.1). This classification is not very strict. A mobile ad hoc network can also be infrastructured: several mobile nodes can form an ad hoc network and can also be connected to a wired gateway for Internet access.

Due to the increase in both computing power of the mobile device and wireless link bandwidth, a user can use his or her mobile device to perform certain tasks that were earlier not possible. In multi-hop mobile

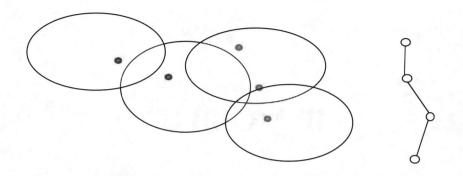

**Figure 1.1  Infrastructureless mobile ad hoc network.**

ad hoc networks, the focus will be on different type of routing protocols and medium access control techniques.

Several issues need to be taken into account while considering a wireless network. Quality of service (QoS) needs to be provided. If the delay produced in packet-switched networks can be removed to match them with the QoS of circuit-switched networks, then packet-switched networks would definitely be the choice for customers. This is not a simple problem. It actually means integrating delay-sensitive applications such as voice, audio, and video along with delay-insensitive applications such as e-mail, fax, file transfer, etc. The network must be able to differentiate between these two kinds of traffic and process both of them efficiently and on time. When the packet-switched network is a wireless LAN, then the situation becomes even more challenging. The TCP/IP stack was not designed with mobility in mind. Some solutions to this problem have been proposed, such as Mobile IP, etc.

On the wireless network side, the key characteristics are the following:

■ Mobility of the users
■ Bit errors in the wireless channels
■ Scarce wireless resources

On the IP network side, there are two key problems:

■ Lack of QoS support
■ Lack of data synchronization

Statistic from http://www.zakon.org/robert/internet/timeline/.

An important trend is the ongoing work and trials on a mobile device with multiple wireless communications interfaces. The rationale behind these efforts is that although there are diverse wireless/mobile communications technologies with their own characteristics (e.g., bandwidth and coverage), no single wireless/mobile communications standard is likely to be the norm in providing access to the Internet. Rather, these technologies have more or less complementary features. For example, the advantage of cellular technology is global coverage, whereas its weakness lies in the bandwidth capacity (currently tens of kilobits per second for the data traffic) and the operational cost.[4] In contrast to this, IEEE 802.11 wireless LAN technology has a bandwidth capacity on the order of megabits per second with little operational cost but has a relatively short range of coverage. With this perception, there have been approaches to equip a mobile node (MN) with multiple wireless communications interfaces. The most promising scenario so far is to equip the MN with a cellular network interface (e.g., GPRS, CDMA) and a wireless LAN interface (e.g., IEEE 802.11, HIPERLAN). With this configuration, the MN can connect to the Internet through the cellular network interface when outdoors and also through the 802.11 interface when indoors, if there is an available 802.11 wireless LAN access point.[4] In this era of wireless technologies convergence, we should note that diverse wireless coverage of technologies (e.g., cellular, wireless LAN, and Bluetooth) overlap each other, and there are frequent vertical handoffs between different kinds of wireless networks. To support seamless mobility to an MN with ongoing Internet connections, the most critical issue is handoff.[7] There are mainly three kinds of handoffs, depending on the situation:

1. Link-level
2. Intradomain (subnet-level)
3. Interdomain

Here intradomain handoff and interdomain handoff are also called microlevel mobility and macrolevel mobility, respectively. The mobility management framework should deal with handoff in all of these cases, seeking to minimize disruption. Especially, the interdomain handoff is likely to occur frequently in wireless technology convergence, and it will accompany considerable signaling traffic load and delay with the current solutions, to be detailed later on. Although there is a consensus that Mobile IP (described later) will be used to provide macrolevel mobility management in wireless/mobile Internet access, there have been a number of proposals for the microlevel mobility issue. Here, micro mobility is the case in which the MN is moving and thereby changing the point of attachment between subnets in the same wireless network (the same administrative domain).

## NEED FOR IP ADDRESSES

Mobile networks have seen a tremendous growth in the last decade of the 20th century and the beginning of the 21st. Various mobile devices such as personal digital assistants (PDAs), wireless laptops, and cell phones have sprung up.[14] Users of these devices want the same services that are available to their wired counterparts (e.g., WWW service, e-mail service, file-sharing service, etc.). These services are provided through the widely used TCP/IP stack. To provide the same services to the mobile devices, it is necessary that individual IP addresses be assigned to each such mobile device.[14] This is a very challenging task due to the dynamic nature of such mobile networks, and their full integration with the Internet will take some time. Packet switching is needed to perform this task as circuit-switched technology is too expensive for bursty data traffic typically present on wireless LANs.

Using another addressing scheme than IP will make these networks incompatible with the Internet. Also thousands of applications using the TCP/IP stack will be unusable. Therefore, to interconnect these networks with the Internet, IP addresses the need to be assigned to the nodes of the mobile wireless network. Most of the users now are mobile users as well as Internet users. They demand Internet access while they are on the move.

## TRAFFIC ISSUES

The original Internet normally used one traffic type for all applications, also known as *best-effort traffic*. However, for the integration of mobile networks with the mostly wired Internet, using one traffic type was not feasible. Moving MNs have different requirements from static nodes. Different schemes may be employed to support QoS in mobile networks. QoS support is especially important in wireless IP networks, where resources are scarce and should not be wasted. However, multiple traffic classes via the Type of Service (TOS) field in the IPv4 header format and via the Differentiated Services (DS) field in the IPv6 header format are supported.

In a wireless IP network, there would simultaneously exist different traffic types such as voice, audio, video, multimedia, and data. The applications may be classified into real-time (voice service) and non-real-time (e.g., e-mail and Web browsing). Therefore, different traffic types have different QoS demands that should be satisfied accordingly. Different nodes have links with different bandwidths ranging from kilobits per second to gigabits per second. Similarly, different applications are also heterogeneous. Some are real-time (e.g., VoIP, audio and video streaming), whereas some are not (e.g., Web browsing and e-mail).

## ROUTING PROBLEMS

A routing algorithm should have the following characteristics:

- Robustness
- Low computational complexity
- Convergence
- Optimality
- Fairness

Routing can be a challenging task in mobile wireless networks. The conditions of such a network are as follows:

- Fast-changing network topology
- A destination wireless node that may be multiple hops away from the source wireless node
- Nodes that may switch off/on, move in/out of the network anytime
- The presence of sleeping nodes that may receive traffic just for themselves and not forward traffic to others

These conditions do not allow the traditional dynamic algorithms implemented on wired networks to be implemented on wireless networks.

## MOBILITY CLASSES

There are four mobility classes that support IP mobility:

1. *Pico mobility* is the movement of an MN within the same base station (BS). The area extends to about 10 m in all directions, enveloping the user (for an example of pico mobility, see Figure 1.2).

10 m approx

**Figure 1.2  An example of pico mobility.**

2. *Micro mobility* is the movement of an MN across different BS with fast speed. This is supported by using link layer support.
3. *Macro mobility* is the movement of an MN across different subnets within a single domain or region. This is typically handled by Mobile IP.
4. *Global mobility* is the movement of an MN among different administrative domains or geographical regions. This is also handled by techniques such as Mobile IP (layer 3).

The goal is to provide uninterruptible connectivity during micro and macro mobility.

## FRAMEWORKS FOR SUPPORTING MOBILITY

There are several frameworks for implementing mobility in mobile wireless networks. The IETF has standardized two of them, Mobile IP and Session Initiation Protocol (SIP):

1. Mobile IP provides application-layer-transparent IP mobility. All the traditional applications using TCP/IP stack can continue to function in a mobile wireless network. The drawback of Mobile IP is that it does not provide additional features such as authentication and billing.[10]
2. SIP is an application layer protocol that can establish, modify, and terminate multimedia sessions or calls. The drawback of SIP is that it does not support TCP connections and is also not a solution for micro or macro mobility. These drawbacks disallow the use of SIP as the protocol for supporting mobility.

## MOBILITY MANAGEMENT

Mobility in wireless networks is of two types — user mobility and terminal mobility.[1] *User mobility* refers to the ability of end users to access calls and services from any terminal or any location. *Terminal mobility* may be referred to as the ability of a mobile terminal to access services from any location. A user may carry the terminal to any location and access the services provided by the wireless network. It is estimated that the number of mobile subscribers will reach about 60 million by 2005. Cellular networks are of many types: GSM, ADC, PDC, CT2, DECT, PACS, etc.[6,13] Location tracking or mobility management is the set of mechanisms by which location information is updated in response to endpoint mobility. The address and identifier of the endpoint need to be maintained for this purpose. In most cellular networks, precomputed routes are used, whereas in infrastructureless mobile ad hoc networks, packet switching is used. Every node computes its own routing table. In cellular wireless networks, usually central entities perform the function of coordination and control, whereas in mobile ad hoc networks, no such central entity exists.

Related work on IP-level mobility management can be broadly classified into three categories:

1. Work in the first category is focused on microlevel mobility (in short, micro mobility), mostly in the cellular network domain. As it takes considerable time to exchange the registration message between the foreign agent (FA) and the home agent (HA), most proposals in the first category have considered a special agent node in each administrative domain, which accommodates the local handoff within the administrative domain without contacting the HA of the MN.
2. Work in the second category mainly seeks to reduce disruption and packet loss in handoff. The most-proposed schemes suggest a cooperating scenario between the old FA and the new FA.
3. The third category is concerned with the authentication, authorization, and accounting (AAA) issue in regard to mobility in Internet service.

## MOBILE IP

*Mobile IP* is the method devised to provide seamless Internet access to the mobile wireless network. The idea of Mobile IP first emerged in 1995. It is still under development. The basic idea of Mobile IP is as follows: if a mobile node has an IP address of 120.54.32.1 and if it suddenly moves out of its subnet, then there are two solutions — either the mobile node may be assigned a new IP address, or through some mechanism it may keep on using its old IP address (e.g., 120.54.32.1). If the moving mobile node is involved in a file transfer with a Web server, receiving a multimedia transmission and downloading e-mail, then giving it a new IP address according to the first solution will result in termination of all such connections based on TCP, which is simply not acceptable. Research has proved that the second solution is much better. This second solution is known as Mobile IP.

In Mobile IP, the home address of an MN remains unchanged, and all TCP/IP connections can be made to it. When an MN moves to a network different from its home network, then a new IP address known as care-of address (COA) is assigned to it. The MN notifies the HA at its home network of its new position. The HA stores this information in a table and uses it to send traffic to the MN. Therefore, other hosts on the Internet do not need to know the COA of the Mobile IP. They can simply send their traffic to the home address of the mobile agent, where the HA will

**Table 1.1    The Network Architecture for a Wireless LAN**

Mobile IP/ Nomadic router
Network layer: VC support
Network layer: routing (DSDV, AODV, etc.)
Link layer (Acks, Priority, etc.)
MAC layer (MAC, MACA/PR, 802.3, etc.)
Clustering (cluster TDMA)
Connectivity
Radio channel

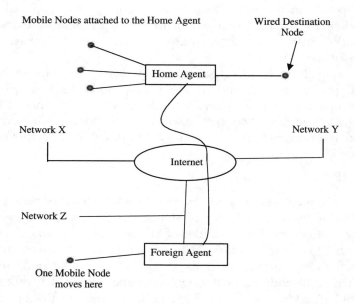

**Figure 1.3    Mobile IP architecture.**

send it to the COA of the MN. There is also a FA responsible for transmitting this traffic to the MN that is currently attached to that foreign network.[2] The HA is mostly implemented on a router in the home network. The network architecture for a wireless LAN is depicted in Table 1.1. See Figure 1.3 for Mobile IP architecture.

In Mobile IP, three major functions are used:

■ Dynamic COA allocation for the MN
■ Registration of the COA with the HA
■ Delivery of packets from the HA to the MN

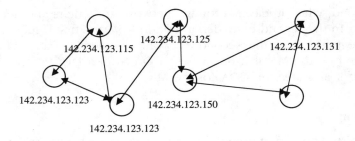

142.234.123.115    142.234.123.125    142.234.123.131

142.234.123.123    142.234.123.150

142.234.123.123

**Figure 1.4   Mobile nodes with IP addresses.**

## Mobile IPv4

As is obvious from its name, Mobile IPv4 uses the normally used 32-bit IP addresses (see Figure 1.4). The HAs and the FAs broadcast advertisements at regular intervals to announce their availability. As soon as an MN receives such an advertisement, it knows that its network has changed. These advertisement messages are an extended form of the Internet Control Message Protocol (ICMP). Router discovery messages are also known as *agent advertisements.*

The agent advertisements perform the following functions:

- Detect HAs and FAs
- Provide COAs
- Make available special features provided on foreign networks to the MN
- Let the MN know its current location

### Registration

The registration messages are of two types — registration request and registration reply. They are sent to User Datagram Protocol (UDP) port 434. UDP provides better performance than TCP, in wireless environments. During registration, the following may be performed by an MN:

- Contacting the HA and telling about its current location
- Temporarily routing services like Web browsing by using FA
- Renewing registration
- Deregistering from a foreign network after returning to the home network

There are two cases in which deregistration may be required: first, when an MN leaves a foreign network and returns to its home network and second, when it leaves a foreign network and enters another foreign network. In the second case, deregistration from the first foreign network and reregistration at the second network is essential. This function is performed by the MN if it receives its COA from a DHCP server and by the FA if it allocates the COA for the MN.

### Tunneling

After registration is performed, the packets destined for the MN will arrive at its HA, which will send them to the COA of the HA through the process of tunneling. This is performed through encapsulation. The original IP packet is encapsulated inside another packet by IP-within-IP encapsulation, as explained in RFC 2003. A special tunnel IP header is used with a value 4 in the outer protocol field. Except for the TTL field, the inner IP packet header is not modified. There are also other encapsulation techniques such as Minimal Encapsulation protocol and the Generic Routing Encapsulation (GRE) protocol.

The Minimal Encapsulation protocol combines the information from the tunnel header with the information from the inner Minimal Encapsulation header to reconstruct the original IP header.

The GRE protocol works in the following way. A source route entry (SRE) is provided in the tunnel header. By using the SRE, an IP source route, which includes the intermediate destinations, can be specified.

### ARP by HA

HA uses a proxy Address Resolution Protocol (ARP) to enable nodes located in the home network to communicate with the mobile. This is accomplished by sending gratuitous ARP messages. After that, every node will have updated information about the mobile node. Then the HA will use a proxy ARP to intercept node messages for the MN and will tunnel them to the COA of the MN.

### Mobile IPv6

Work on IPv6 was started in 1994 to address various problems with IPv4. IPv6 permits 128-bit addresses as compared to 32-bit addresses of IPv4. This is an increase of the address space by a factor of $2^{96}$. This allows on the order of $6 \times 10^{23}$ unique addresses per square meter of the surface of the Earth. This seems to solve the address allocation problem once and for all.

Major features of IPv6 are as follows:

■ Route optimization
■ Transparent use of the COA in the packet

- Use of COA as the source address
- Neighbor discovery and address auto configuration, hence eliminating the use of FAs
- Extensive use of IPSec for all security requirements
- Use of IPv6 routing header instead of encapsulation by a node to directly send traffic to the MN
- Use of anycast mechanism by mobile IPv6

## Extensions to Mobile IP

Several extensions are proposed for Mobile IP. Significant overhead is imposed by sending every packet destined for the MN through the mobile agent and using regular routing while sending the packet back from the MN. Therefore, certain extensions are proposed for Mobile IP so that a node may send traffic to the MN directly. The HA notifies all the related nodes about the COA of the MN through the binding update message. This poses a problem. If a node sends some packets to an MN on its COA and simultaneously the MN leaves that network, then those packets would be lost. Another extension to Mobile IP solves this problem also. By this extension, packets sent to a moved MN are directly forwarded to the new COA of the MN. Steps performed to achieve smooth data transfer between a network node and wireless nodes are as follows:

- A binding warning control message is sent by the corresponding node to the HA which tells the HA that the corresponding node is unaware of the new COA of the mobile node.
- A binding request message is sent by the corresponding node to the HA to request for the refreshment of its binding.
- An authenticated binding update message is sent by the HA to the corresponding nodes. This contains the current COA of the MN.
- The MN performs handoff with the corresponding node.

When an MN leaves a foreign network, it can either deregister itself or just perform handoff. When it reaches a new foreign network, it gets the address of the FA and sends a registration request to its HA by using the address of the FA as a COA. The HA, after processing, sends a registration reply.

## Multicasting Using Mobile IP

Multicasting using Mobile IP is also possible. There are many protocols for this purpose, such as MoM (Mobile Multicast) protocol, MMA (Multicast by Multicast agent), MNG (Mobile Network Gateway), etc. The current

IETF Mobile IP multicast requires remote subscription in which the mobile host (MH) is required to resubscribe to the multicast group on each FA by using a co-located COA. This has the advantage of providing the most efficient delivery of multicast datagrams, but it has a high price. The HA must also be a multicast router. Subscriptions must be done through the HA. For MoM, the HA need not forward a separate copy for each MH that it serves but only one copy for each foreign network at which its MH group members reside. The HA and FA maintain tables.

## *MoM Protocol Details*

- MH arrives at foreign network.
- MH returns to its home network.
- MH times out at a foreign network.
- A unicast packet for MH arrives at MH's FA.
- A multicast packet for group G arrives at HA.
- A tunneled packet arrives at FA from HA.

Case 1. MH arrives at foreign network.

Register with FA.
1. Create Visitor Table entry for MH.
2. Insert host name, HA information, and set timer.
3. Notify FA of MH's current group memberships.
4. For each multicast group that MH is in:
    a. Make entry in GroupInfoTable if needed.
    b. Add MH to group membership list for G.
    c. If this is the first MH from that HA, then add the MH's HA to the HA list for group G or else increment the host count for the MH's FA.
    d. Select a designated multicast service provider (DMSP) for this group from HA list. (The FA selects one HA as the DMSP if an MH is the first one to request subscription to group G at the foreign network.)
    e. If the chosen DMSP differs from the old DMSP, then perform DMSP handoff.

Register with HA.
1. Create or update Away Table entry for MH.
2. Record old FA, if any.
3. Insert host name, FA information, and set timer.
4. Notify HA of MH's group memberships.

5. For each multicast group G that MH is in:
   a. Make entry in GroupInfoTable if needed.
   b. Add MH to G's membership list if needed.
   c. If this is the first MH from this HA, then add the MH's FA to the FA list for group G, or else increment the host count for the MH's FA.
   d. If the MH's new FA differs from oldFA then decrement host count for old FA, discarding oldFA from list if count is zero.
   e. Record/update DMSP status (YES/NO) of HA for group G at FA (and at oldFA if needed).

Case 2. MH returns to its home network.

1. Notify the HA.
   a. Delete Away Table entry for MH, noting oldFA.
   b. For each multicast group G that MH is in:
      i. Delete MH from the membership list for G.
      ii. Decrement the host count for MH's oldFA, discarding oldFA from FA list if count is zero, and deleting oldFA from DMSP list if needed.
      iii. Select a DMSP from HA list for this group.
      iv. If chosen DMSP differs from the old DMSP, then perform DMSP handoff.

Case 3. MH times out at a foreign network.

1. Delete MH's entry from visitors list, noting HA.
2. For each multicast group G that MH is in:
   a. Delete MH from the membership list for G.
   b. Decrement the host count for MH's FA, discarding the HA from HA list if count is zero, and deleting the HA from DMSP list if needed.
   c. Select a DMSP from HA list for this group.
   d. If chosen DMSP differs from the old DMSP, then perform DMSP handoff.

Case 4. Unicast packet for MH arrives at MH's FA.

1. Look up FA information for MH in away table.
2. Encapsulate packet and tunnel it to the FA.

Case 5. A multicast packet for group G arrives at HA.

1. Forward multicast packet to local members.
2. Look up membership information for the away members of that group.

3. Encapsulate packet and forward to each FA for which the FA is the DMSP for group G. This could be done using a separate mobile IP unicast channel to each such FA or as a multicast tunnel to the set of FAs for which the HA is the DMSP of group G.

Case 6. A tunneled packet arrives at FA from HA.

1. Decapsulate the packet.
2. If the packet is a unicast packet for an MH, then forward to that host.
3. If the packet is a multicast packet for group G, then check for local members and forward using link-level multicast if local members are found

### Problems and Issues of MoM

1. The tunnel convergence problem.
2. The duplication problem.
3. Disruptions of Multicast Service.
4. Packets that are sent and received by MHs must always traverse the home network, making routing nonoptimal.
5. Multiple unicasts are used by the HA to tunnel multicast packets to FAs of MHs that are group members.

## Deficiencies in Mobile IP

Mobile IP has been proved to be slow. Recent research has proposed that IP should take support from underlying wireless network architecture to achieve good performance. Data-link layer may be used to provide support to Mobile IP. This is also known as *IP micro mobility* and *paging protocols.* Following are the issues of Mobile IPv4 related to macro mobility:

■ Asymmetric routing
■ Inefficient direct routing
■ Inefficient HA notification
■ Inefficient binding deregistration

Mobile IP does not provide capabilities for QoS provisioning, which is necessary to provide real-time support on mobile networks. Security is also an important issue for Mobile IP. Mobile IPv4 does not provide reliable authentication. Ingress filtering and location privacy are also required. Some of these issues have been resolved in Mobile IPv6, such as reducing overhead and enhancing security.

## SECURITY NEEDS

Modification of IP addresses for malicious purposes is very easy in a wireless ad hoc network, and this has to be taken care of. The IP security architecture is standardized by IETF and is also known as *IPSec*. With IPSec, security is directly applied to IP packets. The concept of security association (SA) is used. SA consists of three parameters: destination address, cryptography protocol, and an identifier that separates multiple SAs with the same destination host but using the same cryptography protocol. SAs are unidirectional. Two mechanisms are used to protect IP packets:

- The authentication header (AH) is used to protect the authenticity and integrity of an IP packet with a keyed cryptographic hash value.
- The encapsulating security payload (ESP) that transports encrypted IP packets, ensuring confidentiality and authenticity (optional), as well as integrity of the packet payload.

## CELLULAR IP

*Cellular IP* is defined as an extension to Mobile IP. It works locally in a cellular access network. Cellular IP can work with Mobile IP to support wide-area mobility. Cellular IP optimizes the cellular network for fast handovers.[11] It also provides integrated mobility control and location management functions at the wireless access points. As Mobile IP manages macro mobility, Cellular IP manages micro mobility.

### Architecture

Cellular IP networks are connected to the Internet via gateway routers. Mobile terminals are identified to the network by using the IP address of the BS (access router) as a COA. Because Cellular IP assumes that Mobile IP manages macro mobility, the HA tunnels the IP packets to the gateway router of the Cellular IP network. Within the network, packets are routed upon the home address of the mobile terminal. In the reverse direction, packets from mobile terminal are routed to the gateway hop-by-hop. After reaching the gateway router, packets are routed through Mobile IP.

### Routing

In a Cellular IP network, the gateway router periodically sends a beacon packet to the BSs in the wireless access network. BSs record the interface through which they last received this beacon and use it to route packets toward the gateway. Furthermore, BSs forward the beacon to mobile terminals. Each BS maintains a routing cache. Packets that are transmitted by MNs are routed to the gateway using standard hop-by-hop routing.

Each node in the Cellular IP network that lies in the path of these packets should use them to create and update routing-cache mappings. This way, routing-cache chain mappings are created, which can then be used to route the packets addressed to the MN along the reverse path.

As long as the MN is regularly sending data packets, nodes along the path between the MN's actual location and the gateway maintain valid routing entries. Information in the routing cache, which includes the IP address of the mobile and the interface from which the packets arrive, disappears after a certain time, called *route time-out*.

Every consecutive packet refreshes the routing information stored at the network nodes. Also, a mobile terminal may prevent a time-out from occurring by sending route-update packets at regular intervals, called *route-update time*. These are empty data packets. They do not leave the Cellular IP networks (i.e., they are discarded at the gateways).

## Location Management

Cellular IP uses two caches at each node in the access network. One is the routing cache (already discussed in preceding text). The other one is a paging cache, which is optionally implemented at the BSs. Although routing cache is primarily used to keep routing information for the ongoing connections, the paging cache is primarily used for idle users. Cellular IP defines an idle MH as one that has not received data packets for a system-specific time, called *active-state time-out*. MNs that are not regularly transmitting or receiving data (i.e., idle nodes) periodically transmit paging-update packets to maintain the paging caches, which may be used to route IP packets (when routing-cache mapping for that node is expired).

Paging-update packets are empty packets addressed to the gateway and are distinguished from a route-update packet by their IP-type parameter. These updates are sent to the base station that offers the best signal quality. Similar to data and route-update packets, paging-update packets are routed on a hop-by-hop basis to the gateway. So, maintaining the paging caches is accomplished similarly to the routing caches, except for two differences. First, any packet sent by the mobile updates paging-cache mappings, whereas paging-update packets do not update routing-cache mappings. Second, paging caches have a longer time-out than routing caches. Therefore, idle MHs have mappings in paging caches but not in routing caches. In addition, active MHs will have mappings in both types of cache. All update packets are discarded by the gateway, to isolate Cellular IP-specific operations from the Internet. After the paging time-out, paging mappings are cleared from the cache (e.g., when the mobile terminal is turned off).

Mappings always exist in the paging cache when the MN is attached to the network. If routing-cache mappings do not exist, incoming packets may be routed by the paging cache. However, paging caches are not necessarily maintained in all nodes.

## Handovers

In Cellular IP networks, the MN initiates a handover. MHs listen to beacons transmitted by BSs and initiate handover based on signal strength measurements.[8] To perform a handover, an MN has to tune its radio to the new BS and transmit a route-update packet. These update packets create routing-cache mappings and thus configure the downlink route from the gateway to the new BS.[11]

During the handover, the MN redirects its data packets from the old to the new BS. At the handover, for a time equal to the routing-cache time-out, packets addressed to the MN will be delivered to both the old and new BSs. If the wireless access technology allows listening to two different logical channels simultaneously, then the handover is soft. If the MN can listen to only one BS at a time, then the handover is hard (in this case, performances of the handover will be more dependent on the radio interface).

The routing-cache mappings will be automatically cleared the moment the time-out occurs. Two parameters define the handover performances: handover delay (i.e., latency) and packet loss. Handover delay is decomposed into rendezvous and protocol time. *Rendezvous time* refers to the time needed for an MN to attach to a new BS after it leaves the old BS. This time is closely related to wireless link characteristics (i.e., the rate of beacons transmitted by the BSs). *Protocol time* refers to the time spent to restore the connection once the MH has received a beacon from the new BS. Usually, rendezvous time is small and we may approximate handover delay to protocol time. The second parameter is packet loss during the handover.

Packet losses occur as follows: Packets are routed through the old BS until the arrival of the first packets through the new route. In a hard handover, some packets may be lost in this time interval. These losses are proportional to the *handover loop time*, which is defined as the transmission time from the crossover node to the old location of the MN, as well as the transmission time from the new location to the crossover node, which is the gateway in the worst case. Although IP packets may be lost during handover, Cellular IP has lower handover delay than Mobile IP. This is due to the local management of the handover (i.e., only local network nodes should be notified at the intradomain handover).

There is no need for communication with a HA that may be located far away from the MN's current network. To reduce packet losses during the handover in a Cellular IP network, a possible solution is semisoft handover. In this case, the routing-cache mappings are created before the actual handover takes place. So, before the handover to a new BS, the MN sends a semisoft packet to the new BS and immediately returns to listen to the old BS.

The idea with semisoft packets is to establish the new route between the gateway and the new BS before the handover execution. During this time, the MN is still connected to the old BS. After a time period called *semisoft delay* (e.g., 100 ms), the MN performs a regular handover. The semisoft approach, however, does not ensure a smooth handover. In reality, the transmit time from 80.

## Open Issues in Cellular IP

Cellular IP is a protocol and concept that integrates location management functions and fast handovers, which are usually found in today's mobile systems, with typical Internet routing and addressing mechanisms. Cellular IP solves micromobility, while Mobile IP handles the macromobility. However, there are several open issues.

First, the handover mechanism assures the local management of intra-domain handovers (i.e., micro mobility), but it is not impervious to packet losses. Losses disrupt typical Internet traffic, such as TCP flows. Semisoft handover reduces the losses, but still it does not guarantee zero loss. Second, Cellular IP does not provide mechanisms for QoS support, which is very important for some applications (e.g., real-time services). The protocol is basically proposed for the best-effort service, which is the dominant type of traffic on the Internet today. To be able to support multiple traffic classes with different QoS demands, we should integrate Cellular IP with some of the QoS mechanisms.

## Handover Mechanisms for Cellular Wireless Packet Networks

Besides Cellular IP, there are several other proposed solutions to micro mobility as an extension to Mobile IP. We refer to some of them, such as the multicast-based Mobile IPv4 algorithm and IP micro mobility support using Handover-Aware Wireless Access Internet Infrastructure (HAWAII). There are other micro mobility proposals such as vertical handoffs in wireless overlay networks, hierarchical FAs, as well as recent Internet drafts such as fast handovers for Mobile IPv6 and low-latency handovers in Mobile IPv4.

## QOS SPECIFICS OF WIRELESS NETWORKS

Wireless networks differ from wired networks in terms of access technology and in the characteristics of the transmission medium. In this section, we point to some important characteristics of the wireless medium that have influence on the communication quality.

### Cellular Topology

One of the main problems for wireless networks is the limited frequency spectrum. Therefore, the number of simultaneous connections over a particular geographical area is bounded by the capacity of the specific wireless access system. On the other hand, the capacity of wired (fixed) networks is not an issue because if we need a greater capacity, we may invest in additional infrastructure (e.g., by adding more twisted pairs or fiber).

To allow a greater number of users for a specific wireless technology, we need to use a cellular principle. Thus, a wireless network consists of wireless access points, the BSs, where each BS covers a particular geographical area. Due to fading (the power of radio waves decreases with distance), we may reuse the same frequencies by using appropriate frequency planning. For better frequency reuse, we group the available frequency carriers or bands into groups. The number of cells within a group defines the reuse factor. For example, in the time division multiple access (TDMA)-based Global System for Mobile Communications (GSM) system we have different frequency reuse patterns such as 3/9, 4/12, and 7/21. The notation x/y refers to all available frequency carriers (or bands) being divided into groups of y frequencies each, which are distributed in x different cells, a pattern that is repeated throughout the network.

In a dense area (with a large number of mobile users), smaller cells must be used due to frequency reuse and capacity requirements.

## ROUTING PROTOCOLS FOR IMPLEMENTING IP MOBILITY

Many routing protocols, i.e., Ad hoc On-Demand Distance Vector (AODV), Temporally-Ordered Routing Algorithm (TORA), Dynamic Source Routing Protocol (DSR), Destination-Sequenced Distance-Vector (DSDV) are proposed in the literature to provide routing services for the wireless network. These protocols act as providing the necessary service for the wireless network. DSDV can be implemented on a wireless LAN to provide IP mobility. DSDV is a distance vector–routing protocol. DSDV allows for freedom-loop guarantee. But it also has periodic broadcast overhead because it is proactive. DSR is another Mobile Ad-hoc Networks (MANET) routing protocol that has no periodic broadcast overhead, but the packet size is larger due to piggybacking of route information.

AODV shares the advantages of DSR and distance vector–routing protocol. Route discovery in AODV is similar to that of DSR. AODV maintains a routing table that contains the destination IP, destination sequence number, hop count, next hop, and lifetime. The route information in the routing table is invalid if a RERR packet is received or the route lifetime is expired.

Although TORA minimizes the reaction due to changes in network topology, the overhead is large due to IMEP (Internet MANET Encapsulation Protocol).

## ALL-IP MOBILE NETWORK

Once it is possible to assign IP addresses to mobile devices as well as fixed devices, we would have an all-IP network. All the Internet services such as WWW, FTP, TCP, VoIP, etc., would then be available to all these devices.

The main advantage of such an all-IP network is efficiency (due to statistical multiplexing and the use of packet switching instead of circuit switching), low operational costs, and the transparency of IP technology to different kinds of services. One service will cater to the demands of different types of users, e.g., a stock exchange update applet will show stock updates on personal computers, laptops, PDAs, and cell phones equally effectively. This will result in some very exciting possibilities.

## TCP OVER IP

TCP is the dominant standard for reliable data delivery on the Internet. Services such as WWW, FTP, and TCP traffic cover almost 95 percent of all bytes, 85 to 95 percent of all packets, and 75 to 85 percent of all flows. For real-time communication, mostly UDP is used. TCP was originally created for wired packet networks; hence TCP does not show good performance over wireless networks. Congestion is the norm rather than the exception on wireless networks. Multiple packet losses inside a TCP segment means that TCP Selective Acknowledgment (SACK) may be appropriate for wireless networks.

## POTENTIAL APPLICATIONS

There are profound advantages of IP mobility in wireless networks. We will consider two scenarios here:

1. Consider the scenario of a massive battlefield. An enormous wireless network is deployed with thousands of soldiers, hundreds of tanks, and other vehicles acting as network nodes (routers). Complete information regarding the battlefield is constantly transmitted to the headquarters, where it is analyzed and processed. Immediately, new instructions are sent back to the battleground, where they are received by the MNs, which adjust themselves accordingly.
2. Another scenario is that of an earthquake disaster area. With widespread destruction, wired access to the earthquake area may not be possible. In such a situation, a wireless ad hoc network can quickly be deployed.

Other applications include man-made disasters, relief operations, military applications, car-based networks, sensor networks, the provision of wireless connectivity in remote areas, collaborative computing, and video conferences.

## FUTURE

Future-generation mobile networks are going to be all-IP networks; thus, all types of information will be carried using IP packets. Although the basic infrastructure is in place, extensive research in the field of IP mobility and location management will be needed for full-scale deployment. Issues still outstanding are mobility management, security concerns, efficiency, etc.

## REFERENCES

1. R. Prasad, *Technology Trends in Wireless Communications*, Artech House, 2003.
2. Janevski, *Traffic Analysis and Design of Wireless IP Networks*, Artech House, 2003.
3. Mukherjee, Bandyopadhyay, Saha, *Location Management and Routing in Mobile Wireless Networks*, Artech House, 2003.
4. http://www.gsmworld.com.
5. http://www.3gpp.org.
6. K. Uehara et al., Design of Software Radio for Cellular Communication Systems and Wireless LANs, *IEEE PIMRC 2000*, pp. 474–478, 2000.
7. N. Nikolaou et al., Wireless Technologies Convergence: Results and Experience, *IEEE WCNC 2000*, pp. 566–571, 2000.
8. S. Helal et al., An Architecture for Wireless LAN/WAN Integration, *IEEE WCNC 2000*, pp. 1035–1041, 2000.
9. C. Perkins, IP Mobility Support, *IETF RFC 2002*, 1996.
10. C. Perkins and D. Johnson, Route Optimization in Mobile IP, Internet draft, draft-ietf-mobileip-optim-11.txt (work in progress), September 2001.

11. C. Perkins and K. Wang, Optimized Smooth Handoff in Mobile IP, *ISCC '99*, 1999.
12. E. Gustafesson, A. Jonsson, and C. Perkins, Mobile IP Regional Registration, Internet draft, draft-ietf-mobileip-regtunnel-03.txt (work in progress), July 2000.
13. K. El Malki et al., Low Latency Handoffs in Mobile IPv4, Internet draft, draft-ietf-mobileip-lowlatency-handoffs-v4-02.txt (work in progress), October 2001.
14. P. Calhoun et al., Foreign Agent Assisted Hand-off, Internet draft, draft-calhoun-mobileip-proactive-fa-03.txt, November 2000.
15. P. Calhoun and C. Perkins, Diameter Mobile IPv4 Application, Internet Draft, draft-ietf-aaa-diameter-mobileip-08.txt (work in progress), November 2001.
16. A. Hess and G. Shafer, Performance Evaluation of AAA/Mobile IP Authentication, technical report, Technical University, Berlin, 2001.
17. Helen J. Wang et al., Policy-Enabled Handoffs Across Heterogeneous Wireless Networks, 2nd IEEE Workshops on Mobile Computing and Applications (WMCSA '99), New Orleans, February 1999.
18. C. Perkins, Mobile IP Joins Forces with AAA, *IEEE Personal Communications Magazine*, pp. 59–61, August 2000.
19. C. Perkins and P. Calhoun, AAA Registration Keys for Mobile IP, Internet draft, draft-ietf-mobileip-aaa-key-08.txt (work in progress), July 2001.
20. K. Pahlevan et al., Handoff in Hybrid Wireless Data Networks, *IEEE Personal Communications Magazine*, pp. 34–47, April 2000.

# 2

# WLAN PERFORMANCE

*Shakil Akhtar*

## INTRODUCTION

The use of wireless LAN (WLAN) systems is rapidly growing in the communications industry. As the demand for WLAN products is increasing, it is becoming vital to understand the relevant performance aspects. WLANs can be used either to replace wired LANs or as an extension of the wired LAN infrastructure. The basic topology of the network consists of two or more wireless nodes or stations that have recognized each other and established communications. In the most basic form, stations communicate directly with each other on a peer-to-peer level, sharing a given cell coverage area. This type of network is often formed on a temporary basis and is commonly referred to as an ad hoc network or independent basic service set (IBSS).

In most cases, the basic service set (BSS) contains an access point (AP). The main function of an AP is to form a bridge between wireless and wired LANs. The AP is analogous to the base station used in cellular phone networks. When an AP is present, stations do not communicate on a peer-to-peer basis. All communications between stations or between a station and a wired network client go through the AP. APs are not mobile and form part of the wired network infrastructure. A BSS in this configuration is said to be operating in the infrastructure mode.

The popular physical (PHY) layer and medium access control (MAC) sublayer standard for WLAN are provided by the Institute of Electrical and Electronics Engineers (IEEE) in the form of a family of 802.11 protocols.[1] The basic access method for 802.11 is the Distributed Coordination Function (DCF), which uses Carrier Sense Multiple Access/Collision Avoidance

(CSMA/CA). This requires each station to listen for other users. If the channel is idle, the station may transmit. However, if it is busy, each station waits until transmission stops and then executes a random backoff procedure. This prevents multiple stations from seizing the medium immediately after completion of the preceding transmission.

Packet reception in DCF requires acknowledgment. The period between completion of packet transmission and start of the ACK frame is one short inter frame space (SIFS). ACK frames have a higher priority than other traffic. Transmissions other than ACKs must wait at least one DCF inter frame space (DIFS) before transmitting data. If a transmitter senses a busy medium, it determines a random backoff period by setting an internal timer to an integer number of slot times. Upon expiration of a DIFS, the timer begins to decrement. If the timer reaches zero, the station may begin transmission. However, if the channel is seized by another station before the timer reaches zero, the timer setting is retained at the decremented value for subsequent transmission. As mentioned earlier, DCF is the basic media access control method for 802.11, and it is mandatory for all stations. The Point Coordination Function (PCF) is an optional extension to DCF, which provides the capability to accommodate time-bounded, connection-oriented services such as cordless telephony. In 802.11, orthogonal frequency division multiplexing (OFDM) is the primary modulation technique, in which multiple carriers are simultaneously transmitted at equal power, each containing its own modulation. By using FFT and inverse FFT technology, the carriers are manipulated, and blocks of data or symbols are obtained. The multiple carriers of OFDM are modulated with a variety of techniques from BPSK through 16 QAM, and with various degrees of forward error correction coding. This creates multiple data rates up to 56 Mbps.

Although we present some of the related WLAN concepts, the main focus of this chapter is the performance of WLANs. First, a general overview of network performance is provided, followed by the performance issues of WLAN systems in light of the current technological developments. Next, three performance models are provided for wireless systems. The chapter ends with some concluding remarks about WLAN performance.

## NETWORK PERFORMANCE ISSUES

One of the basic requirements of a computer network is connectivity. Without a connected network, the computers may not communicate with one another. This requirement leads to different network topologies and interconnections. Another feature of a network is its capability to share resources. A network allows the sharing of disk space, printers, and other hardware and software. In addition to being cost effective, resource sharing allows the incremental growth of the network while providing leverage for the current investment. The network designer can effectively increase the network size without discarding the currently utilized resources.

The study of network performance allows a better understanding of network connectivity and resource sharing. For instance, a comparison may be made between the delays to run an application on a workstation versus running it on a server, which may lead to a decision of buying the required number of application licenses. Similarly, one LAN topology may be preferred over another, based upon performance. Currently, businesses and organizations debate the advantages and flexibility of deploying a wireless network versus a wired network. A requirements analysis that addresses the performance issues may help in taking this decision.

Performance of computer networks may be evaluated using three main techniques: analytical modeling, simulation, and direct measurements. Because these techniques complement each other and may have their own limitations, often their combination is used to predict the system performance under certain conditions. Often, the results from one approach confirm the results obtained from another if the models are built correctly. On the other hand, the problematic models may be fixed, the error often arising due to an incorrect assumption, incorrect use of a variable, or incorrect flow of data/control in the model.

Network mathematical analysis is often performed for a specific scenario with limited and very rigid conditions. Although simulation models usually provide a better understanding for design and development of networks, it requires software expertise with a deep understanding of network operation. Direct measurement allows thorough network troubleshooting, but network expansion and planning may be limited.

Network emulation refers to the ability to introduce the simulator into a live network. Special objects within the simulator are capable of introducing live traffic into the simulator and injecting traffic from the simulator into the live network. In this approach, live traffic may be passed through the simulator and the effect on it noted by objects within the simulation or by other traffic on the live network. In another approach, the simulator can include traffic sources or sinks that communicate with real-world entities. Using this combination of simulation and direct measurements is often very helpful in making performance predictions under "what-if" scenarios.

## WLAN PERFORMANCE

With higher bandwidths supported by the standards, WLANs are deployed for larger numbers of users and involve applications of e-mail, Web browsing, and database access. The current highest data rate supported by 802.11a is 54 Mbps for each of 12 (maximum) nonoverlapping channels with freedom from most potential RF interference. However, the need for higher data rates and techniques to improve performance of WLANs is crucial for many reasons, including the rise of multimedia and MPEG traffic in videoconferencing and mobile 3G and 4G applications.

The WLAN standard is drawn up by the IEEE 802.11 committee.[1] It consists of a family of specifications, with major definitions being 802.11, 802.11a, 802.11b, 802.11e, and 802.11g. All use Ethernet running CSMA/CA (Collision Avoidance as opposed to Collision Detection on Ethernet). The original 802.11 provides 1- or 2-Mbps transmission in the 2.4-GHz band, using either frequency hopping spread spectrum (FHSS) or direct sequence spread spectrum (DSSS). The 802.11a specification is an extension to 802.11 and provides up to 54 Mbps in the 5-GHz band. It uses an OFDM-encoding scheme rather than FHSS or DSSS. The actual available data rates for 802.11a are 6, 9, 12, 18, 24, 36, 48, and 54 Mbps, where the support of transmitting and receiving at data rates of 6, 12, and 24 Mbps is mandatory.

The 802.11b[2] specification (also referred to as 802.11 High Rate or Wi-Fi[8]) is also an extension to 802.11 that provides an 11-Mbps transmission (with a fallback to 5.5, 2, and 1 Mbps) in the 2.4-GHz band. The 802.11b specification uses only DSSS. The 802.11g[3] specification offers wireless transmission over relatively short distances at 54 Mbps in the 2.4-GHz band. The 802.11g specification also uses the OFDM-encoding scheme. The 802.11e specification is the newest addition to the 802.11 family, which is based upon 802.11a and includes quality-of-service (QoS) issues.

Although 802.11b provides 11-Mbps data rates, with only three non-overlapping channels, it can be extended to 802.11g to have 54-Mbps operation. However, the three nonoverlapping channels limitation still exists. With currently available technologies, 802.11a provides maximum performance, but more APs are needed because of the weaker range it has compared with 802.11b. Hence, the need to develop these standards still remains important.

It is noted that WLAN performance suffers severely when operated in a typical office environment (around 200-ft diameter). The degradation in performance is observed to be due to excessive packet loss and error rates, and results in much lower throughput than expected. Although most of the performance degradation is blamed on radio interference, other factors may cause limitations due to physical and MAC layers of the protocol. In general, these factors are modulation techniques and standards, the hardware used, quality of radio signals, processing speed of the stations, environmental effects such as path loss and echoes, radio interference, software design, and interfacing with high-level protocols.

One major factor affecting WLAN performance is adequate RF coverage. If APs are too far apart, then some users will be associating with the WLAN at less than the maximum data rate. For example, users close to an 802.11b AP may be operating at 11 Mbps, whereas a user at a greater distance may only have 2-Mbps capability. Performance may be maximized by ensuring that RF coverage is as spread out as possible in all user areas, especially the locations where the bulk of users reside.

Another factor affecting WLAN performance is RF interference. Mobile phones and other nearby WLANs can interfere with signals, degrading the operation of an 802.11b WLAN. These external sources of RF energy may block users and APs from accessing the shared air medium. As a result, the performance of 802.11b-based WLAN suffers when RF interference is present. However, deploying 802.11a networks using 5 Ghz may control this problem.

WLAN performance is heavily dependent upon the performance of underlying physical layer technologies. The access method relies on the physical carrier sensing. The underlying assumption is that every station can hear all other stations. This is not always the case, as one of the models presented in the next section illustrates. A problem occurs when the AP is within range of one station, but is out of range of another downstream station. In this case, the two stations would not be able to detect transmissions from each other, and the probability of collision is greatly increased. This is known as the *hidden-node problem*.

To combat this problem, a second carrier sense mechanism is available, which enables a station to reserve the medium for a specified period of time through the use of request-to-send or clear-to-send (RTS or CTS) frames. In the case described in the preceding text, a distant station sends an RTS frame to the AP. The RTS will not be heard by the second station. The RTS frame contains a duration or ID field that specifies the period of time for which the medium is reserved for a subsequent transmission. Upon receipt of the RTS, the AP responds with a CTS frame, which also contains a duration or ID field. Although the nearer station cannot detect the RTS, it will detect the CTS and adjust accordingly to avoid collision, even though some nodes are hidden from other stations.

The hidden terminals may cause performance degradation in WLANs, which can be controlled by using the RTS/CTS mechanism, as mentioned in the preceding text. Optimal WLAN performance may be obtained by adjusting the RTS/CTS threshold while monitoring the impact on throughput. The 802.11 standard requires that a station must refrain from sending a data frame until it completes a RTS/CTS handshake with an AP.

The "listen before talk," or CSMA/CA access method used in 802.11, guarantees equal duration of channel access to all devices irrespective of their needs. In other words, when a device with a low bit rate captures the channel, it penalizes other devices using a higher rate by degrading the speed of their connections. For example, when one wireless device connects to a WLAN at a lower bit rate than other devices due to being too far from the AP, performance of other network devices degrades. A possible remedy is 802.11e, which defines QoS mechanisms to support bandwidth-sensitive applications such as voice and video.

WLAN performance may also be improved by using a fragmentation option that divides the 802.11 data frames into smaller pieces sent separately to the destination. Each fragment consists of a MAC layer header, FCS (frame check sequence), and a fragment number indicating its position within the frame. With properly set values, fragmentation can reduce the amount of data that needs retransmission because the station only needs to retransmit the fragment containing the bit errors.

## PERFORMANCE MODELS

There are three models presented in this section. The first illustrates the use of TCP congestion control in a wireless environment. The second deals with the signal-to-noise ratio (SNR) estimate of 802.11g, and the third is about the performance enhancement of WLAN systems using RTS/CTS and fragmented frames.

### TCP Congestion Control Model

As an example of a network performance study, consider the TCP congestion control algorithm over a wireless link in which a host station A sends data to another host station B. Due to transmission interference and high incidence of errors in wireless communication, assume that every $n$th transmission from host A is lost (or corrupted). For example, if $n$ is 4 and A transmits sequences 1, 2, 3, 4, 5, 6, 7, 8, and 9, then the transmission numbers 4 and 8 will be dropped by the data-link layer. A limits the amount of data it sends, using the TCP slow start and congestion avoidance mechanisms. However, it does not implement fast retransmit or fast recovery.

We make the following assumptions in our simulation:

1. Headers and ACKs are of size 0.
2. The round trip time (RTT), which is set by the user, is preset to 1 s.
3. The retransmission time-out is set to RTT + 0.01 s.
4. The data in each frame is 1 KB.
5. The link bandwidth is 100 KB/s.
6. There are no other losses besides the ones mentioned above (however, every $n$th frame is lost irrespective of whether it is the original transmission or a retransmission).

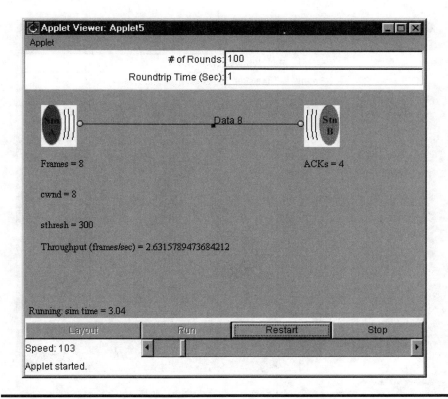

**Figure 2.1    Initial screen for TCP congestion control simulation.**

7. B sends an ACK for every frame it receives but may send cumulative ACKs for previously buffered frames.
8. B has buffer space to advertise infinite receiver window.
9. The congestion window (cwnd) at A does not increase when a duplicate ACK is received.

Figure 2.1 to Figure 2.4 display screen snapshots of the simulation run. The simulation window displays the following items:

1. An input window that includes the number of rounds intended and the RTT.
2. Running values of simulation time, frames sent or acknowledged, cwnd, sthresh, and throughput as number of frames successfully transmitted per unit time.
3. A sliding bar to control the speed of animation; sliding to the left increases the simulation speed.
4. A simulation trace that displays an animation of frames/ACK movement over the transmission path.

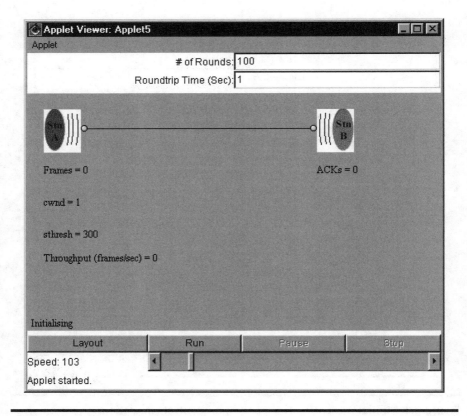

**Figure 2.2 Screen snapshot at 3.04 s displaying data frame 8 being transmitted to station B.**

A "simjava" simulation toolkit is used for building the model.[4–5] It is based on a discrete event simulation kernel and includes facilities for representing simulation objects as animated icons on screen. The model contains a number of entities, each of which runs in parallel in its own thread. An entity's behavior is encoded in Java, using its body() method. Entities have access to a small number of simulation primitives that can be used effectively to schedule events, wait for events, hold the entity, and create animation and traces. Simulation details are presented in Reference 6.

Using the model, traffic properties of wireless networks are modeled, and an error scenario using TCP congestion control mechanism is shown. The throughput for the simulated traffic was observed to be around 4.5 frames/s during the simulated time of about 4 s when cwnd = 8 and sthresh = 300.

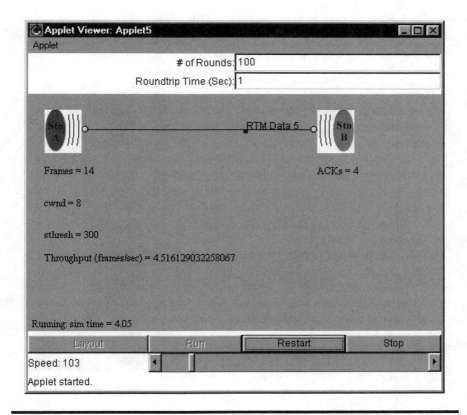

**Figure 2.3 Transmission loss of frame 5 is indicated by a duplicate ACK 4 by station B.**

## The SNR Estimate Model of 802.11g WLAN

The new IEEE 802.11g standard[1,3] for WLANs extends the data rate of IEEE 802.11b to 54 Mbps from its current level of 11 Mbps. This standard is based upon the 802.11 WLAN standard already being deployed. The new standard will also create data rate parity at 2.4 GHz, extending 802.11a. Due to efficient operation, high demand is expected for this new standard although 802.11a also allows for a 54 Mbps but at 5-GHz frequency. Also, a quiet environment is offered by 802.11a because there is no major source of interference at this frequency as there is at 2.4 GHz.

The IEEE 802.11g standard boosts WLAN speed to 54 Mbps by using OFDM technology. The IEEE 802.11g specification is backward compatible with the widely deployed IEEE 802.11b standard. The use of OFDM allows better efficiency and enables mixed network operation. This mixed operation

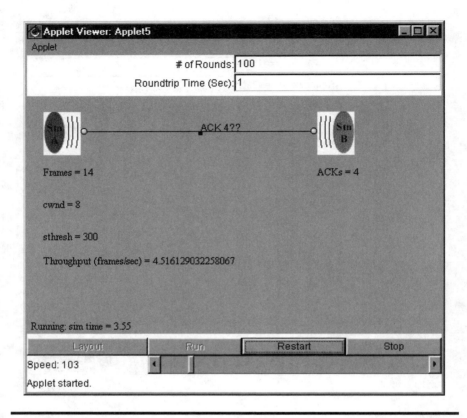

**Figure 2.4   Retransmission of frame 4 as the result of previous error in frame 5.**

allows legacy 802.11b devices to operate at 11 Mbps, while new 802.11g devices may operate at 54 Mbps on the same network. It is possible because the IEEE 802.11g standard maintains the spectral mask and carrier frequencies of the IEEE 802.11b standard. This extension is expected to improve access to fixed-network LANs and internetwork infrastructures. The added transmission speed gives wireless networks based on IEEE 802.11b (often called WiFi[11]) the ability to serve up to four to five times more users than they do now.

The WLAN standards are relatively new. However, some experimental and simulation results of these systems are available. For instance, 802.11 systems have been analyzed in Reference 7, where a multichannel (channel in which characteristics change with time) system has been modeled, allowing a performance comparison of 802.11, 802.11b, and 802.11a system designs. Bit error rate (BER) curves for various delay spreads have been obtained.

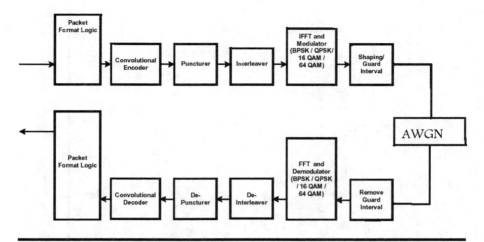

**Figure 2.5 The 802.11g implementation block diagram.**

Communication system reliability between the transmitter and receiver may be obtained in terms of SNR. Measuring SNR directly may not be always possible, and hence there are methods to estimate it. In adaptive system design, SNR estimation is commonly used for measuring the quality of the channel. Then, the system parameters are changed adaptively, based on this measurement. There are several SNR measurement techniques for communication systems, and some are specific for OFDM[8] and QPSK systems.[9]

A performance simulation model of the newly proposed 802.11g standard (an addition to the popular 802.11 standard), presented in Reference 10, uses communications blockset of Matlab/Simulink. Performance characteristics of the underlying modem are obtained in terms of BER versus SNR. We summarize here the simulation results for the physical layer of WLAN 802.11g corresponding to 54 Mbps.

By implementing and modeling characteristics of the physical 802.11g layer, we aim to understand the underlying OFDM modem and its interaction with the 802.11 PHY and MAC layers. The performance of the PHY layer through BER versus SNR is presented. The 802.11g block diagram and the Matlab implementations are given in Figure 2.5 and Figure 2.6, respectively. The major blocks are Encoder, Interleaver, Puncturer, and Modulator. The Convolutional Encoder block encodes a sequence of binary input vectors to produce a sequence of binary output vectors. This block can process multiple symbols at a time.

If the encoder takes $k$ input bit streams (that is, it can receive $2^k$ possible input symbols), then this block's input vector length is $L \times k$ for

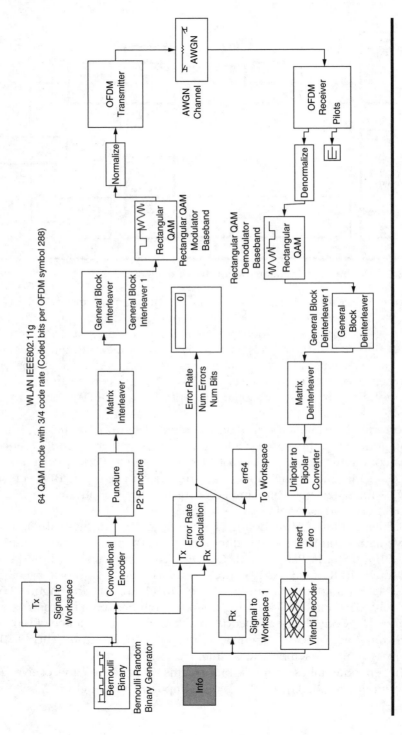

**Figure 2.6  Matlab/Simulink Simulation Block Diagram**

some positive integer $L$. Similarly, if the encoder produces $n$ output bit streams (that is, it can produce $2^n$ possible output symbols), then this block's output vector length is $L \times n$. The input can be a sample-based vector with $L = 1$ or a frame-based column vector with any positive integer for $L$.

Interleaver is the next important block. Interleaving is a key component of many digital communication systems involving forward error correction (FEC) coding. It is a standard DSP function used in many communications systems. Applications that store or transmit digital data require error correction to reduce the effect of spurious noise that can corrupt data. Digital communications systems designers can choose many types of error correction codes (EECs) to reduce the effect of errors in stored or transmitted data, depending upon the type of modulation desired.

Symbol interleavers and deinterleavers can mitigate the effects of burst noise. Typically, these functions are needed for transport channels that require a BER on the order of $10^{-6}$. Interleaving improves the efficiency of encoders and decoders by spreading burst errors.

OFDM is a special case of the multicarrier technique, in which frequency-shifted pulse shaping filters are used in transmitter and receiver. These filters structure the input to approximate the impulse shaping. To overcome the effect of multipath propagation, a short guard interval is introduced. With the guard interval, the OFDM system needs only one multiplication on each subcarrier as equalization.

Figure 2.7 shows the simulation results. The dashed curve shows the BER versus SNR, using 16-QAM modulation, which corresponds to 34 Mbps. The second curve is for 64-QAM + OFDM modulation, which corresponds to 54 Mbps. The result shows that 64-Modulation + OFDM is better than 16-QAM + OFDM in terms of BER versus SNR. We notice that the BER decreases when we increase the SNR, which is normal because the signal becomes stronger than the noises. We also notice that we can send data with small errors when SNR is greater than 35 dB.

## Performance Model of an 802.11 System with Fragmented Frames

Consider a three-node WLAN based upon 802.11 running at 1 Mbps. Two of the three nodes (nodes A and B) each transmit at an average rate of 80 kbps with packets of average size 1000 bytes and average interarrival time of 0.1 s (all following an exponential distribution). The third node acts only as a receiver and does not send any data (see Figure 2.8). We present the system performance model and study the effect on transmission as node A follows a trajectory, as shown in Figure 2.9. This study is based upon the simulation of an Opnet-provided wireless module.[12] It is

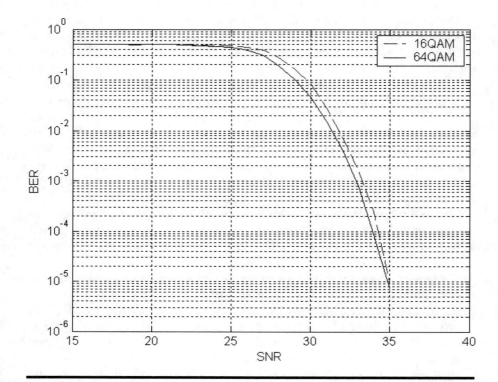

**Figure 2.7    Simulation results (BER versus SNR).**

noted in this model that the simulated nodes generate the traffic at lower layers, and the higher layers are replaced by a bursty source and a sink process within the node model. There is a MAC interface provided in which operation is independent of the higher layers.

We study the effect of frame fragmentation under a hidden-node scenario in the given WLAN. Using fragmentation for large-sized packets improves the reliability of data exchange between the stations. We also notice in this study that fragmentation may control the retransmissions and, in turn, the hidden-node problem, to some extent. However, because every data fragment requires an acknowledgment, the overall frame exchange per frame is higher than it would be without fragmentation. Fragmentation may also impact the total channel reservation time per data frame exchange.

The modeled WLAN system exhibits a hidden-node problem. Note that the radio coverage of nodes A and B coincide at the beginning of the simulation. However, as node A follows a left trajectory and moves out

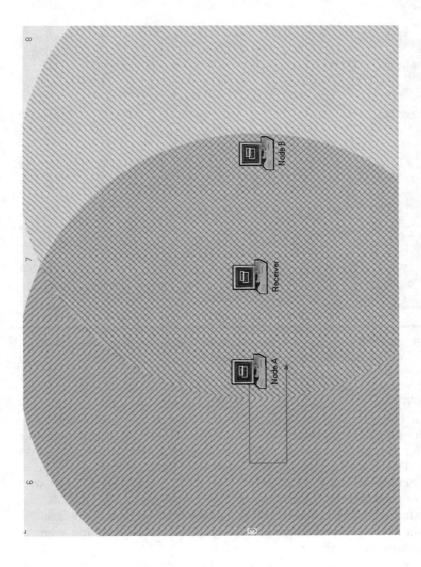

**Figure 2.8   A three-station WLAN with a defined trajectory for node A causing a hidden-node problem.**

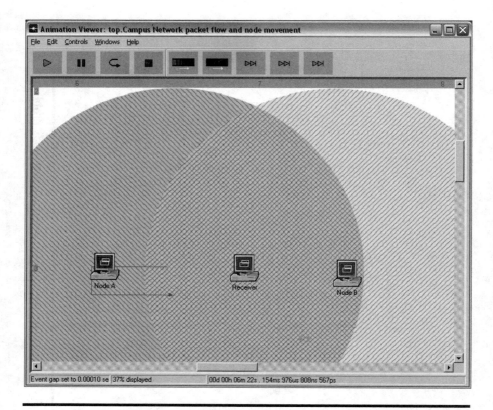

**Figure 2.9    Node B becoming a hidden node for node A (and vice versa) as node A moves along its trajectory between 350 and 650 s.**

of range of node B, the two nodes become hidden from one another during 350 to 650 s of simulation time. This results in a loss of received signal for both the nodes. In addition, their transmissions become very slow. The effect on transmission is shown in Figure 2.10, in which the two nodes cannot sense each other's transmission (received data drops to zero). This causes an increase in their sent traffic, which results in a sudden increase of delay, as shown in Figure 2.12, and also an excessive number of retransmission attempts, as shown in Figure 2.13. This adverse effect on traffic is noted because the amount of traffic sent is not controlled despite the collision avoidance mechanism in operation. The time average traffic is shown in Figure 2.11.

The four sets of results in Figure 2.10 to Figure 2.13 are for four different values of fragment thresholds for frames. Note that typical LAN protocols use packets several hundred bytes long (the longest Ethernet packet could be up to 1518 bytes long). There are several reasons why it is preferable to use smaller packets in a WLAN environment. First, due to the higher BER of a radio link, the probability of a packet getting corrupted increases with bigger packets. Second, in case of packet corruption (either due to collision or noise), the smaller the packet, the less the overhead required to retransmit it. Third, on a Frequency Hopping system, the medium is interrupted periodically for hopping, so the smaller the packet, the smaller the chance that the transmission will be postponed to after some time. The WLAN fragmentation or reassembly mechanism operates at the MAC Layer. Under this mechanism, the transmitting station is not allowed to transmit a new fragment until one of the following happens:

1. Node receives an ACK for the said fragment.
2. Node decides that the fragment was retransmitted too many times and drops the whole frame.

The station sends the frame called MAC Service Data Unit (MSDU), which can be up to 2 KB unless divided into fragments called MAC Protocol Data Units (MPDUs). The standard specifies a threshold to decide if an MSDU received from the higher layer needs fragmentation before transmission. The number of fragments to be transmitted is calculated based on the MSDU size and the fragmentation threshold. The destination station receives these fragments and stores them in the reassembly buffer until all fragments have been received.

## CONCLUSIONS

WLAN performance issues and three sample models are presented in this chapter. We note that currently the performance of WLAN systems largely relies on the PHY layer performance. RF coverage and proper placement of APs play an important role in improving performance. Often, RF interference and poorly designed network setup may cause reduced access speeds, sometimes due to problems caused by hidden downstream terminals that are not detected by a far station. However, the hidden-terminal problem may largely be controlled by the RTS/CTS mechanism or the available fragmentation option.

Also, results from three models dealing with TCP performance in a wireless setup, SNR measurements for 802.11g, and throughput enhancement by fragmentation have been presented. We note that the problem of maximizing throughput via more and better protocols and securing the network remains a hot research topic in WLAN performance.

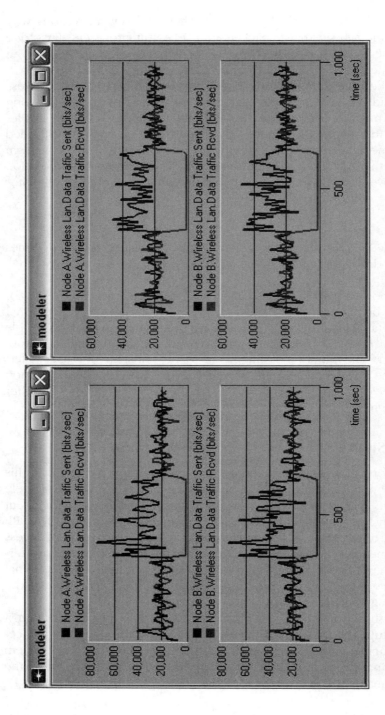

**Figure 2.10** WLAN instantaneous traffic for nodes A and B with hidden-node effect (when fragmentation thresholds are 2000, 1000, 500, and 100 bytes, respectively).

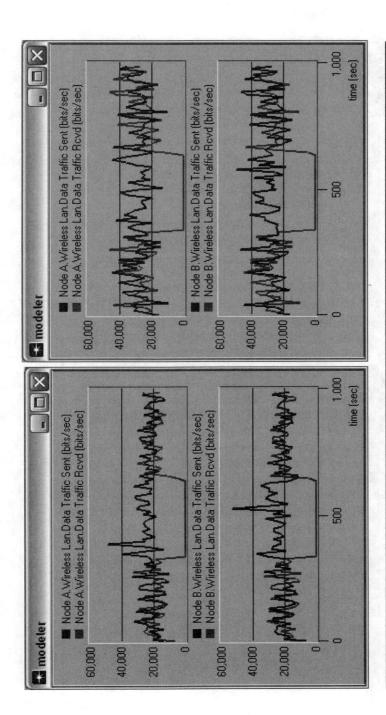

**Figure 2.10 (Continued)**

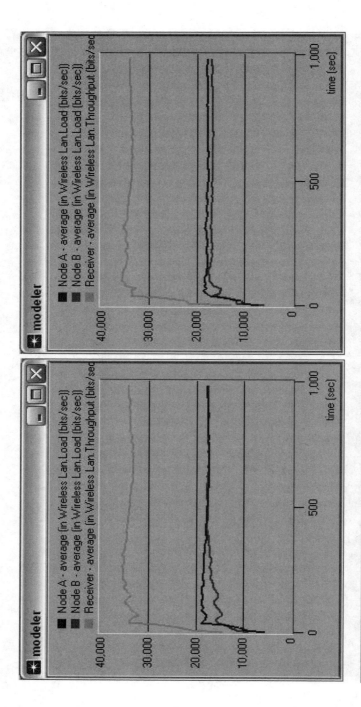

**Figure 2.11** WLAN average traffic for nodes A, B, and receiver with hidden-node effect (when fragmentation thresholds are 2000, 1000, 500, and 100 bytes, respectively).

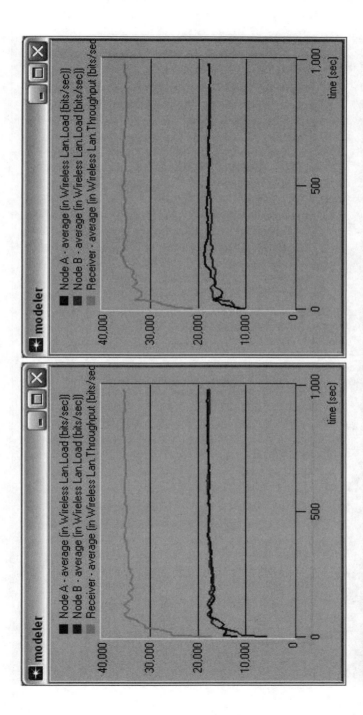

Figure 2.11 (Continued)

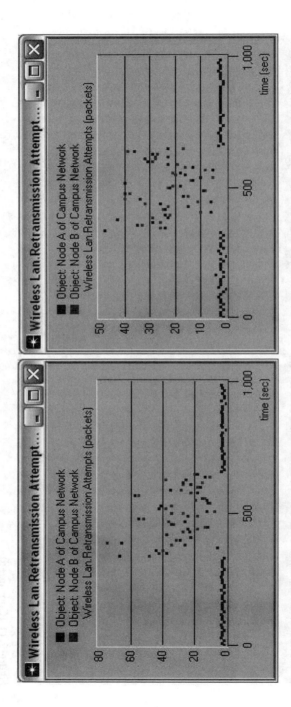

Figure 2.12 WLAN retransmission attempts for nodes A and B with hidden-node effect (when fragmentation thresholds are 2000, 1000, 500, and 100 bytes, respectively).

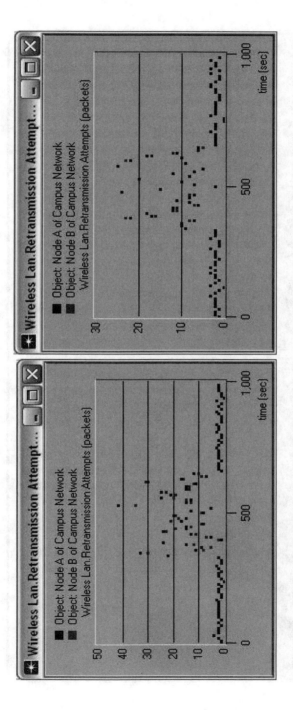

Figure 2.12 (Continued)

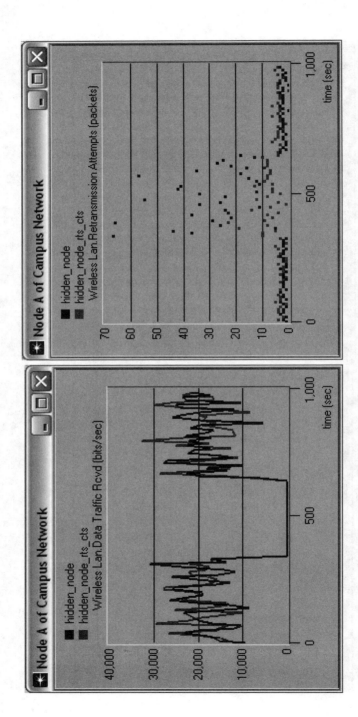

**Figure 2.13** Use of RTS/CTS to minimize the retransmission attempts under the hidden-node effect.

# REFERENCES

1. IEEE 802.11 Standard, Wireless LAN Medium Access Control (MAC) and Physical Layer (PHY) Specifications, 1999.
2. Wireless LAN Medium Access Control (MAC) and Physical Layer (PHY) Specifications: Higher Speed Physical Layer (PHY) Extension to IEEE 802.11b, 2003.
3. W. Carney, IEEE 802.11g: New Draft Standard Clarifies Future of Wireless LAN, white paper by Texas Instruments, 2002.
4. Ross McNab and F.W. Howell, Using Java for discrete event simulation, in *Proceedings Of the Twelfth U.K. Computer and Telecommunications Performance Engineering Workshop (UKPEW)*, University of Edinburgh, pp. 219–228.
5. Fred Howell and Ross McNab, Simjava: a discrete event simulation package for Java with applications in computer systems modeling, in *Proceedings of the First International Conference on Web-based Modeling and Simulation*, San Diego, CA, Society for Computer Simulation, January 1998.
6. S. Akhtar, Performance of mobile multimedia communication in hybrid networks, in *Proceedings of the Advanced Simulation Technologies Conference*, Washington, D.C., pp. 35–40, April 2000.
7. S.M. Nabritt, Modeling Multipath in 802.11 Systems, Comms Design online, http://www.commsdesign.com/story/OEG20021008S0001.
8. H. Arslan and S. Reddy, Noise power and SNR estimation for OFDM-based wireless communication systems, in *Proceedings of 3rd IASTED International Conference on Wireless and Optical Communications* (WOC), Banff, Alberta, Canada, 2003.
9. D.-J. Shin, W. Sung, and I.-K. Kim, Simple SNR estimation methods for QPSK-modulated short bursts, in *Proceedings of IEEE Global Telecommunications Conference*, Globecom 2001, San Antonio, TX, Vol. 6, pp. 3644–3647, 2001.
10. M. Boulmalf, A. Sobh, and S. Akhtar, Physical layer performance of 802.11g WLAN, in *Proceedings of Applied Telecommunications Symposium*, Washington, D.C., pp. 175–178, April 18–22, 2004.
11. Wi-Fi Alliance, URL: http://www.weca.net/.
12. Opnet Technologies, Wireless LAN Model Description, 2000.

# 3

# AN OVERVIEW
# OF THE SECURITY
# OF WIRELESS NETWORKS

*Eduardo B. Fernandez, Imad Jawhar,*
*María M. Larrondo Petrie, and Michael VanHilst*

## INTRODUCTION

More and more applications are being accessed through wireless systems, including commerce, medical, manufacturing, and others [Var00]. Wireless devices have become an extension of corporate databases and individuals. Their security compromises are as serious as any attack to the corporate database and may have damaging effects on the privacy of individuals and the protection of assets of an enterprise. Wireless devices include cellular phones, two-way radios, PDAs, laptop computers, and similar. These are normally portable devices with limitations of weight, size, memory, and power. The increase in functions in cellular devices creates new possibilities for attacks. Standard attacks against the Internet may now take new forms. Lists of vulnerabilities are already available, showing flaws in many existing products [Iss, Mit].

Communicating in the wireless environment has its own issues and challenges. It is characterized by relatively low bandwidth and data rates, as well as higher error rates, and the need for low power consumption (for mobile devices). The mobility of the nodes in cases such as ad hoc networks adds another significant layer of complexity and unpredictability.

There exist many different forms of wireless communications and networking [Sta02]. Some popular forms of wireless communications include:

*Satellite communication:* It uses microwave links, and provides global connection of many network infrastructures. There are three basic classes of satellites: GEO (Geostationary Earth Orbit), MEO (Medium Earth Orbit), and LEO (Low Earth Orbit).

*Cellular networks:* These are currently among the most widely used types of networks. The geographic area is divided into *cells*. Each cell is serviced by a *base station* (BS) and several base stations are served by a Mobile Telecommunications Switching Office (MTSO) or a similar structure. The latter provides connection to the wired telephone infrastructure. The new generation of cellular networks uses digital traffic channels, encryption, error detection/correction, and allows channel access to be dynamically shared by all users. *Global System for Mobile communication* (GSM) standard is widely used. The configuration of a typical cellular network is shown in Figure 3.1.

*Cordless systems:* They are used inside homes and buildings, and provide wireless communications between a cordless device such as a telephone and a base station. Typically, TDMA (Time Division Multiple Access) and TDD (Time Division Duplex) communication protocols are used in such systems.

*Wireless Local Loops (WLL):* They are used to provide last mile connections from the end user to the local switching telephone center. They have an advantage over their wired counterparts in low cost and relative ease of installation which can be done selectively and on demand.

*Mobile Internet Protocol (Mobile IP):* It provides nomadic access from different access points (APs) allowing the user to maintain connectivity as he or she moves from one access point to another. Mobile IP includes processes of registration, move detection, agent solicitation, and tunneling of data messages.

*Wireless Local Area Networks (WLANs):* They have increased popularity due to their characteristics of mobility, convenience, rapid deployment, and cost effectiveness, in addition to the small size, and increased power and speed of wireless devices. Two standards are typically used: IEEE 802.11 (Wi-Fi) and Bluetooth.

There are four types of WLANs:

*LAN extensions:* They allow connection between mobile wireless devices and a wired network. Some example applications are manufacturing, stock exchange, and warehouses.

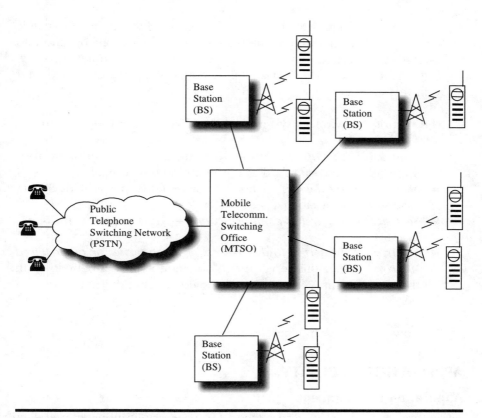

**Figure 3.1 Configuration of a cellular network.**

*Cross-building interconnects:* They allow fast wireless connections between buildings. Microwave communications with dish-shaped antennas are used. This type is a link more than it is a LAN.

*Nomadic access:* It is used to allow communication between mobile devices such as laptops, and PDAs to existing fixed wired networks. For example, applications can use such systems to transfer data from wireless devices to the home, office, or campus network.

*Mobile ad hoc networks (MANETs):* As mobile wireless computers and devices become increasingly smart, small, portable, and powerful, the need to interconnect these devices increases. MANETs allow such computing devices to establish networks on the fly without any pre-existing infrastructure. Numerous applications exist using MANETs such as disaster recovery, military missions, classrooms, and conferences. Multi-hop routing is used to provide communication between nodes (e.g., laptops or computers inside moving vehicles) that are out of range of each other. Each host provides routing capabilities to the mobile network. MANETs have dynamic topologies as nodes are allowed to move from one location to another, as well as join and leave the network at any time. Typically, these networks use Wi-Fi and Bluetooth, which are discussed later in this chapter.

The security of wireless systems can be divided into four sections:

■ *Security of the application* — this means the security of user applications and standard applications such as e-mail.
■ *Security of the devices* — how to protect the physical device in case it is lost or stolen.
■ *Security of the wireless communication* — how to protect messages in transit.
■ *Security of the server that connects to the Internet or other wired network* — after this server the information goes to a network with the usual security problems of a wired network (not discussed here).

We now look at each aspect in turn.

## APPLICATION SECURITY

### Application Development

There are two common approaches for user applications in wireless devices: WAP (Wireless Application Protocol), and applications based on the two standard component approaches, J2ME and .NET. The latter include standard object-oriented applications or applications using Web services. Middleware software supports wireless applications at both the client and server sides [Vau04]. Devices using Bluetooth can use Java [jav] or .NET.

### *WAP*

WAP is a thin-client (micro browser) development protocol specifically designed for development of user applications. WAP uses WML (Wireless Markup Language) and WMLScript to develop applications that can be interpreted at the browser and accessed at the server using HTTP (HyperText Transfer Protocol). WAP requires a gateway to the wired Internet, and cannot store and process data locally.

WAP uses WTLS (Wireless Transport Layer Security) [Ash01]. This protocol provides confidentiality, integrity, and authentication and uses RSA cryptography, but can also use Elliptic Curve Cryptography. It is based on the IETF SSL/TLS (Internet Engineering Task Force Secure Socket Layer and Transport Layer Security) protocols. WTLS provides security for communications between the WAP wireless device and the WAP gateway (discussed later). Current WAP devices use Class 2 WTLS, which enforces server-side authentication using public key certificates similar to the SSL/TLS protocols. Future Class 3 devices will also allow client-side authentication using certificates. This level will use a WAP Identity Module (WIM), with mandatory support for RSA public keys and optional support for ECC.

## Web Services

A Web service is a component or set of functions accessible through the Web that can be incorporated into an application. Web services expose an XML (eXtensive Markup Language) interface, can be registered and located through a registry, communicate using XML messages using standard Web protocols, and support loosely coupled connections between systems. Web services represent the latest approach to distribution and are considered an important technology for business integration and collaboration.

Wireless devices can access Web services using SOAP (Simple Object Access Protocol). Web services are still not widely used in portable devices [Gra04]. The limited processing power of portable devices and the lack of network reliability are serious obstacles for a full implementation. Using appropriate gateway middleware, it is however possible for portable devices to access Web services.

There are several toolkits that simplify the process of building applications using Web services. For example, Java-based client systems can use Sun ONE and kSOAP [Yua02a], while server-side systems can be built with Sun or IBM toolkits [ibm]. There are similar tools for .NET-based systems. In addition to the specific designs used, security also depends on the security of these component platforms [Fer04a].

The richness of Web services brings along a new set of security problems [Fer02]. All the attacks that are possible in wired systems are also possible in wireless systems using Web services, e.g., viruses, buffer overflow attacks, message interception, denial of service, etc. Web services introduce several extra layers in the system architecture and we have to consider the unique security problems of these layers. Since these are layers that run on top of the platform layers, the security of the platforms is still fundamental for the security of the complete system. Wireless systems using Web services have to face, in addition, the general vulnerabilities of wireless networks and may also add new security problems to these networks although this aspect has not been explored in detail [IM02]. There is also a variety of standards for Web services security and a designer of wireless devices should follow at least the most important ones to be able to have a credibly secure system. On the other hand, the extra layers bring more flexibility and fineness for security; for example, encryption can be applied at the XML element level and authorization can be applied to specific operations in a Web service interface. This greater security precision allows applying policies in a finer and more flexible way. WAP applications have fewer security risks compared to Web services. On the other hand, their functionality is considerably lower.

## Personalized Information

An important mobile application aspect is the delivery of personalized information to subscribers [Dog02]. Using specialized interfaces, users are able to select services offered by some companies; for example, lists of stores who have sales, stock market alerts, etc. Some of these services may be location-dependent, e.g., lists of nearby restaurants. Clearly, the companies that provide these services need to control access to their customer information, which in addition to the usual information about credit and Social Security Numbers now includes a privacy aspect (the company is able to track the client movements).

## Access Control to Sensitive Information in or through the Device

The portable device may contain files that need to be restricted in access and it is the function of its operating system to perform this control. Control of types of access is important; for example, a user may play a song, but she may not copy it. This type of control can complement other types of digital rights management.

When portable devices need to access corporate databases, some type of Role-Based Access Control (RBAC) is necessary where users can access specific data related to applications such as banking, shopping, health, navigation, and surveillance. Management and enforcement of application and institution constraints can be performed following PMI (Privilege Management Infrastructure) [Cha01]. PMI is a standard of International Telecommunication Union ITU X.509.

## Viruses and Other Malware

With increase in functions, the typical problems of larger systems are also appearing in portable devices. One of these problems is attacks by viruses [Fol01]. The first portable virus to appear was Liberty, followed shortly by Phage. The WML script language used by WAP can also be a source of possible attacks [Gho01]. The devices do not distinguish between script code from the phone or downloaded from potentially insecure sites, all of it executes with the same rights. An infected device can be used to launch denial of service attacks on other devices or the network. Similarly to wired systems, up-to-date antivirus programs are needed. Companies such as Symantec, McAfee, and Trend Micro have specialized products for handheld devices.

## Downloaded Contents

Downloaded contents may include malicious software Another issue is the control of unauthorized copying of downloaded contents, such as music, wallpaper, and games. This is a problem of digital rights management.

## Location Detection

Location detection is a problem unique to mobile devices. The actual location of the device should be kept hidden in some cases for privacy or for strategic reasons.

It is possible to control access to VLANs (Virtual Local Area Networks) by associating users with access points. There are products that can keep track of users and access points and use this information for network administration [Cox04].

## Operating System Security

Portable devices have evolved from having ad hoc supervisors to standard operating systems. Some systems use the Java run-time system as supervisor. High-end cell phones run complete operating systems such as Palm OS or Microsoft Windows CE, and provide IP networking capabilities for Web browsing, e-mail, and instant messaging. Some typical security features include:

- *A unique device identifier* — this can be provided and can be accessed by an application.
- *A kernel configuration with enhanced protection* — this allows using the protected kernel mode, instead of the full-kernel mode, while running threads to prevent accessing certain physical memory.
- *Digital authentication in the dial-up boot loader* — the dial-up boot loader is a program in ROM used to upgrade the OS image file (NK.bin) using flash memory or a remote server. The OS image file should be signed using digital encryption to verify its integrity before it is downloaded.

Smartphone [mic04] is a Windows CE-based cellular phone that comes bundled with a set of applications, such as address book, e-mail, and calendar. The provider that sells the Smartphone can limit the devices' ability to load and run programs. A locked cell phone either restricts unsigned applications or does not run them at all. Depending on the provider, an encryption key may be needed to run the application [Alf00], but cell phone codes have been successfully cracked [Rob99].

It is clear that, similar to larger systems, the operating system is fundamental for security. Because many of the security flaws of Microsoft's operating systems come from their general approach to systems design [Fer04], one should watch out for similar problems in their small OSs. The utilities of the OS are the main culprit in the attacks that have happened in wired systems and it is important to have utilities with strong security. For example, some products attempt to improve the security of e-mail systems [Kno04].

## DEVICE SECURITY

If a portable device is lost or stolen, user authentication can prevent somebody from gaining access to confidential information. Possibilities include PINs, pass-phrases, and biometrics. Some portable devices already have fingerprint readers [Rie00]. In networks that only authenticate the device instead of the user, losing a device is more serious than in networks that authenticate users or roles.

## COMMUNICATIONS SECURITY

An obvious problem of wireless communications is that they are very easy to intercept. This implies that some form of encryption is a must for the confidentiality of messages. The available approaches depend on the standard used. Cellular networks use GSM, while WLANs use two standard protocols:

- IEEE 802.11a (Wi-Fi) can reach up to 1800 feet (550 meters). Devices connect to APs that have unique identifiers, Basic Service Set identifiers (BSS IDs). APs are basically transceivers that take the radio signals to the WLAN switch, which performs all the required network management. WLAN switches support 802.11 at layer 2 and IP traffic at layer 3. The wireless network has a SSID (Service Set Identifier). It is also possible to set up Peer-to-Peer (P2P) networks.
- Bluetooth. A protocol for short-range (up to 100 meters) wireless networks. Bluetooth devices are typically structured into ad-hoc networks.

We describe these protocols in more detail and discuss their security in the following.

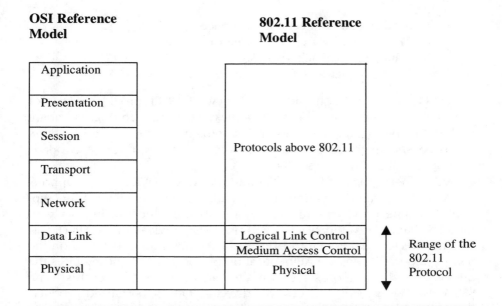

**OSI Reference Model**

**802.11 Reference Model**

| OSI Reference Model | 802.11 Reference Model |
|---|---|
| Application | |
| Presentation | Protocols above 802.11 |
| Session | |
| Transport | |
| Network | |
| Data Link | Logical Link Control |
| | Medium Access Control |
| Physical | Physical |

Range of the 802.11 Protocol

**Figure 3.2   Correspondence of the IEEE 802.11 layers to the layers of the OSI reference model.**

## IEEE 802.11 Wireless LAN Standard

It is the most widely used communications protocol for wireless LANs. The protocol resides in the physical and data link layers of the OSI (Open System Interconnection) model. It defines functions and specifications for the physical and MAC (Medium Access Control) layers. The MAC layer covers three functional areas: reliable data delivery, access control, and security. The protocol defines different building blocks such as BSS (Basic Service Set) and ESS (Extended Service Set). Each ESS consists of one or more BSS. Stations in a BSS compete for access to the shared wireless medium. Most ad hoc network routing protocols are designed and tested on top of the IEEE 802.11 protocol. Figure 3.2 shows the scope of the IEEE 802.11 standard in reference to the layers of the OSI model. It shows how the data link layer is actually divided into the MAC and LLC (Logical Link Layer). The latter is responsible for providing the upper layers with three types of services, which are:

1. Unacknowledged connectionless service
   a. No flow and error control support
   b. No guarantee of data delivery
2. Connection-mode service
   a. Logical connection is set up between two users
   b. Flow and error control are provided

3. Acknowledged connectionless service
   a. This service is a cross between previous two
   b. Datagrams are acknowledged
   c. No prior logical set-up required

802.11 uses Wired Equivalent Privacy (WEP). WEP provides device or access point authentication as well as message secrecy through a variant of the RC4 cryptographic algorithm. The implementation of this algorithm has been shown to be flawed [isa]. Access to the wireless network is controlled using a static key.

WEP is being replaced by Wi-Fi Protected Access (WPA). WPA supports the AES (Advanced Encryption Standard, also known as Rijndael) encryption algorithm, provides effective key distribution, and can interact with RADIUS (Remote Authentication Dial-In User Service) or LDAP (Lightweight Directory Access Protocol) servers. Authentication is based on the 802.1X and the Extensible Authentication protocol (EAP) and requires the use of an authentication server. An alternative (or complement) is using SSL VPNs (Virtual Private Networks) [Ave04]. Other specialized products detect unauthorized access points and users.

WLAN switches apply security controls, including authentication (a comparison of some of them is in [Che04]). Authentication can be provided locally or by connecting to a RADIUS or LDAP server.

Because the RSA algorithm is rather inefficient in its use of key length, *elliptic curve cryptography* (ECC) algorithms have been proposed [cer]. For example, an elliptic curve algorithm with a key length of 150 bits takes $3.8 \times 10^{10}$ MIPS-years to be broken by brute force, while the RSA with a key length of 512 takes only $3 \times 10^4$ years. However, this approach requires that all ECC users agree on a common set of parameters, otherwise the extra information needed effectively extends the key [Fer99].

## Bluetooth

Bluetooth is a wireless communications protocol, originated by Ericsson, that quickly was adopted by many companies. It is intended to work in a close proximity environment, such as homes, offices, classrooms, hospitals, airports, etc. Connections are established using designated master and slave nodes.

Bluetooth uses application profiles for different devices, synchronous connection-oriented (SCO) for data, and asynchronous connectionless (ACL) links for voice, which are multiplexed on the same (Radio Frequency) RF link. Frequency-hopping spread spectrum with a high 1600 hops/sec rate is used to reduce interference, and provide low power, low cost radio communications. It operates in the Industry Scientific and Medical (ISM) band at 2.45 GHz with a transmission power of 1 to 100 mW, a range of

10 to 100 meters, a maximum bit rate of 1 Mbps, and an effective data transfer rate of 721 Kbps.

Bluetooth provides authentication and message 128-bit encryption using hierarchical keys. Devices can be discoverable or invisible. In discovery mode a device is visible to any other device within range, which can make it vulnerable to attacks from those devices.

## GSM

GSM implements authentication based on a cryptographic challenge response protocol. For encryption it uses the A5 algorithm [a5]. It also includes Anonymity using temporary identifiers. Details of GSM security can be found in [hac02].

## GATEWAY SERVER SECURITY

Gateways are devices that control the flow of traffic into or out of a network. Although definitions differ, for this context a gateway can be thought of as a device that passes packets between subnets (real or virtual), and performs operations above OSI layer 3 (session, flow control, protocol conversion, and application specific). Gateways can also be the source of vulnerabilities [Juu01a]. Gateways are important to wireless networks and mobile wireless devices for several reasons:

■ Wireless networks do not afford the same physical levels of security as wired networks. Due to resource constraints, mobile wireless devices are themselves often less secure than wired devices. *Wireless security gateways* can protect a wired network from untrusted wireless hosts. Unlike firewalls, for which hosts are either "inside the firewall" or "outside the firewall," the distinction between inside and outside is somewhat blurred for mobile wireless devices. A company's trusted workers may need "inside" kinds of connectivity while using wireless devices. Conversely, visitors may need "outside" kinds of connectivity while connecting to the company's wired network through an access point inside the corporate firewall. Wireless security gateways address these issues by performing two-way authentication and limiting access privileges on a per-device basis.

■ Mobile wireless devices often have limited resources that cannot support the same protocols as wired devices. They may therefore use resource-sharing protocols which must be translated in a protocol gateway to enable interaction with standard Internet protocol services. For example, a *WAP gateway* translates protocols in the WAP suite, including WML (HTML), WML Script (CGI), WBMP

(BMP), WBXML (XML), WSP (HTTP), WTP (TCP/IP), WTLS (SSL), and WDP (UDP). These kinds of translation pose security issues both because the wireless protocols are often less secure than the corresponding wired protocols and because, in translation, encrypted data takes an unencrypted form inside the gateway.

■ Wireless devices often exist on subnets that do not support the full Internet addressing scheme. For example devices may use IP addresses [ipv] reserved for local access only, or otherwise not support all of the capabilities needed for WAN access. Gateways can provide a bridge between these local subnets and a broader WAN, (i.e., Internet). Common small office home office (SOHO) wireless switches provide NAT to allow local devices to all access the Internet using a single IP address. Similarly, a Personal Mobile Gateway with WAN connectivity like GSM or GPRS (General Packet Radio Service) can allow Bluetooth, 802.11, or 802.15 devices on a PAN (Personal Area Network) to have full Internet connectivity. The fact that devices behind a NAT gateway do not have unique IP addresses has implications for some security strategies (i.e., IPSEC-AH [Sta99]).

■ Mobile wireless devices may be involved in various sorts of commerce, such as M-commerce and downloading multimedia streams with digital rights. Depending on how you look at it, where conflicting privacy and ownership interests come into play, "trusted gateways" can bridge the no man's land, or encapsulate the overlap as a trusted third party. This space is an area of active research and is, as yet, not as well defined as the other gateway functions. Issues here are closely tied to digital rights management. See for example the Shibboleth project [shi].

The Internet was built on "transparency" and the "end-to-end principle." Roughly stated, transparency "refers to the original Internet concept of a single universal logical addressing scheme, and the mechanisms by which packets may flow from source to destination essentially unaltered." [rfc2775] The end-to-end principle holds that functions of data transmission other than transport, such as data integrity and security, are best left to the transmission endpoints, themselves. This allows applications to be ignorant of the transport mechanisms, and transport systems to be ignorant of the data being transported. Gateways, by their nature, violate one or both of these principles.

## Gateway Deployment Strategies

At the basic network level, gateways are viewed as servers or end-systems. But gateways create their own overlay networks and may be involved in

ISO level 2 and level 3 routing. The use of gateways can greatly complicate problems of network management. Their deployment should be carefully considered within a comprehensive network coverage and security strategy.

The main reason for using a wireless security gateway is that intruders may gain access through an insecure wireless access point and mount an attack on the internal network. As indicated earlier, 802.11b, Bluetooth, and WAP are all potentially insecure. Access points with stronger security are possible using Cisco or 802.1x protocols. Typically, a large site or campus, will need many access points for good coverage. The cost of numerous high-end access points and the problem of managing them, especially when they are not all from the same vendor, is a major concern. A common strategy is to use simple ("thin") access points and put one or more security gateways between all wireless access points and the wired network. Then, even if anyone can establish a connection to an access point, they will be challenged at the gateway. The gateway might use IPSec, VPN, or LDAP encryption and authentication. Cisco also has LEAP (Lightweight Extensible Authentication Protocol) which they are pushing as PEAP (Protected Extensible Authentication Protocol) for a standard. There are several products that include SSL VPNs and gateways [Ave04].

Several strategies are available to ensure that access points connect only to a gateway. Access points could be physically wired on a separate subnet, where gateways provide the only bridge to the main wired network. Over a large area, the need to maintain two wired networks, one for access points, may be impractical. Multiple smaller networks can be used, each with its own gateway. Multiple gateways can share a common, central management tool — like CA or HP OpenView. They may also be arranged in master/slave relationships, i.e., for configuration and fail-over. Another alternative is to use access points that VPN tunnel to a single gateway, using the regular wired network as the transport medium.

Gateways can grant different users different levels of trust. The easiest way to set this up is to differentiate users by their IP address, and grant different levels of service (i.e., bandwidth) and different kinds of access (i.e., specific protocols like FTP [File Transfer Protocol] and HTTP, and specific destination hosts) using ISO level 2 (IP address) and level 3 (protocol type) filtering. Access classes can be grouped by role, and identified by predefined ranges of IP addresses.

By grouping IP addresses, the IP address can also be used to distinguish between wired and wireless clients, e.g., to deliver content appropriate to small or large screens, or to put a WAP service behind the gateway or firewall. Other parameters, such as signal strength will be harder to expose.

Basing access privilege on statically assigned IP addresses makes systems difficult to manage and upgrade. Imagine having to change thousands of statically assigned IP addresses to accommodate a new access policy. A better approach uses DHCP (Dynamic Host Configuration Protocol) and MAC addresses. The DHCP servers are configured with fixed MAC to IP address mappings which are much easier to maintain and can be upgraded as needed. The dynamically assigned IP address serves as a kind of token to gain specific levels of access. To hide these IP addresses from snoops, use one of the newer (or evolving) standards for level 2 encryption in the client and access point (i.e., Tunneled Transport Layer Security).

## GATEWAY SERVICES

Any system granting access to clients should include a separate method for authenticating the user. MAC addresses can be spoofed. The gateway may provide its own authentication service, or act as a proxy for a remote authentication service available elsewhere on the network. Various authentication services can serve this function, including RADIUS and Windows Active Directory. Using an underlying operating system's authentication may allow the user to log in to both the network and a machine with a single sign-in. 802.1x proposes this approach. A "captive portal" directs every http request from a not yet authenticated user to the authentication service (and blocks all other types of requests).

There are situations where wireless clients are not capable of performing a standard authentication behavior. Sensors on a shop floor or in a wireless automotive network might be examples. In these cases, with very limited privileges, statically assigned access may be justified. But the security implications must be carefully considered and strong encryption should be used.

Roaming is another issue that gateways can address. Roaming users may move out of range of their current access point and into range of several alternative access points. Handover delays may affect streaming applications like Voice-over-IP (VoIP) and video. Secure access points might require the user to be re-authenticated, while gateways offer other options. The 802.11 Fast Roaming Study Group and 802.21 working group are looking for standard ways to address roaming, as is a partnership among Proxim, Avaya, and Motorola.

WAP devices use WTLS instead of SSL, due to the assumed WAP client's resource constraints. The basic WAP configuration involves a WAP gateway that translates between the various WAP protocols and the corresponding Internet protocols. The WAP gateway translates between WTLS and SSL by decrypting the message as it comes in and then re-encrypting it in the other protocol before passing it on. Decrypting the message in the WAP gateway is only one of many WTLS vulnerabilities [Juu01b, Saa99]. Better security can be achieved by using an encryption protocol in the layer above WTLS/SSL that works directly between the client and server end-points.

PKI-based encryption is the logical candidate for end-to-end encryption, e.g., for M-Commerce applications. But PKI (Public Key Infrastructure) is resource intensive. The special processing could be handled by a SIM (Subscriber Identification Module) or WIM (WAP Identity Module) smartcard, but smartcards add cost to small devices. Research is currently underway to use a remote server to perform the heavy processing part of the RSA/ECC algorithm implementation, while still holding all key parameters in secrecy by the client [eti].

Resource overhead for even basic internet connectivity can be an issue for very small devices, such as those imagined for wearable and ubiquitous computing. A special class of gateway, called personal mobile gateway (PMG), has WAN capability (e.g., GSM/GPRS) and shares it with other little devices with PAN connectivity (i.e., Bluetooth, 802.11, 802.15). The delegation can be general, or specific to the type of applications needed (SMS [Smart Messaging System], voice, digital photos, video, etc.) Security issues at this level are beyond the scope of this discussion.

Government wireless installations are required to meet the National Institute of Standards and Technology Federal Information Processing Standard NIST FIPS 140-2 standard for cryptographic modules [nis]. RADIUS does not meet this standard. For such applications a FIPS 140-2 compliant gateway and corresponding authentication server software must be used. The physical vulnerability of gateways in unattended locations may also need to be addressed. By encasing the gateway's circuitry in a special hardened plastic security potting resin, any attempt at physical tampering will be easily recognized.

In any discussion of security and gateways the limitations of gateways must be emphasized. Gateways form part of a perimeter defense for wired networks. They do not solve the vulnerability of any network to insiders with malicious intent. In addition, while the gateway strategy addresses the threat to the network from malicious wireless devices, it doesn't protect wireless devices from malicious access points.

## THE FUTURE

There is serious concern about the vulnerabilities of wireless systems. The easy access to the medium by attackers is a negative aspect compounded by the design errors in the early protocols [Arb03, Juu01b, Saa99]. The US Department of Defense recently issued Directive 8100.2 that requires encrypting all information sent in their networks according to the rules of the Federal Information Processing (FIP) standard [dod]. The provision also calls for antivirus software. It is interesting to observe that their concern is again mostly about transmission and they don't seem to be worried about the other aspects of security.

On the other side, Ashley et al. arrived at the conclusion that WAP provides excellent security [Ash01]. It is true that Wi-Fi is becoming more secure and Bluetooth appears reasonably secure, but they (and WAP) cover only some of the security layers. A basic security principle indicates that security is an all-layer problem, one or more secure layers is not enough [Fer04a]. While some of the layers are still insecure, it is not possible to have true security.

Third generation systems will have voice quality that is comparable to public switched telephone networks. VoIP will bring its own set of security problems. The new systems will have also higher data rates, symmetrical and asymmetrical data transmission rates, support for both packet and circuit switched data services, adaptive interface to the Internet to reflect common asymmetry between inbound and outbound traffic, more efficient use of available spectrum, support for wide variety of mobile equipment, and more flexibility. All of these are the potential sources of new security problems.

As mentioned earlier, Web services are not delivered directly to portable devices but transformed in the gateway. Most of the use of Web services for mobile systems is now between servers [Gra04]. However, this situation is changing and predictions indicate that Web services in cell phones will be arriving soon. In fact, Nokia just announced a Service-Oriented Architecture for smart mobile phones [Yua04]. Security will be an important issue for this generation of smart and complex devices.

Security patterns are a promising area to help designers build secure systems [Fer04b]. Several patterns have been found in the Bluetooth architecture, including versions of the Broker, Layers, Lookup, and Bridge patterns [Gam95]. However, no security patterns for wireless systems have been found yet. This is an area to explore.

# REFERENCES

[a5] Cracking the A5 algorithm, http://jya.com/crack-a5.htm

[Alf00] M. Alforque, "Enhancing the Security of a Windows CE Device," Microsoft Developers Network, http://msdn.microsoft.com/library/enus/dnce30/html/winsecurity.asp

[Arb03] W. Arbaugh, "Wireless security is different," *Computer,* IEEE, August 2003, 99-102.

[Ash01] P. Ashley, H. Hinton, and M. Vandenwauver, "Wired versus wireless security: The Internet, WAP, and iMode for e-commerce," *Proceedings of the. 17th Annual Computer Security Applications Conference,* 2001.

[Ave04] Aventail, "Practical solutions for securing your wireless network," White paper, 2004. http://www.aventail.com

[cer] Certicom, http://www.certicom.com

[Cha01] D. W. Chadwick, "An X509 role-based PMI," 2001, http://www.permis.org/files/article1_chadwick.pdf

[Che04] B. Chee and O. Rist, "The Wi-Fi security challenge," *InfoWorld,* May 17, 2004. http://www.infoworld.com

[Cox04] J. Cox, "Vendors offer tools to control, secure WLANs," Network World, June 7, 2004, http://www.nwfusion.com

[dod] http://www.dtic.mil/whs/directives/corres/html/81002.htm

[Dog02] A. Dogac and A. Tumer, "Issues in mobile electronic commerce," *Journal of Database Management*, January-March 2002, 36-42.

[eti] http://www.eti.hku.hk/eti/web/p_s_wiress.html

[Far01] R. Farrow, "Wireless security: A contradiction in terms?," Network Magazine, Dec. 5, 2001. Also in: http://www.networkmagazine.comarticle/ NMG20011203S0008

[Fer99] A. D. Fernandes, "Elliptic-Curve Cryptography," *Dr. Dobbs Journal*, December 1999, 56-63.

[Fer02] E. B. Fernandez, "Web services security," in *Web Services Business Strategies and Architectures,* P. Fletcher and M. Waterhouse (Eds.), Expert Press, UK, 2002, 290-302.

[Fer04a] E. B. Fernandez, M. Thomsen, and M. H. Fernandez, "Comparing the security architectures of Sun ONE and Microsoft .NET", in *"Information security policies and actions in modern integrated systems,"* C. Bellettini and M.G. Fugini (Eds.), Idea Group Publishing, Hershey, PA, 2004.

[Fer04b] E. B. Fernandez, "A methodology for secure software design," *Proceedings of the 2004 Int. Conf. on Software Engineering Research and Practice (SERP'04)*, 21–24, June, 2004, Las Vegas, NV, CSREA Press, Vol. 1, 130–136.

[Fol01] S. Foley and R. Dumigan, "Are handheld viruses a significant threat?," *Comm. of the ACM,* vol. 44, No 1, January 2000, 105-107.

[Gam95] E. Gamma, R. Helm, R. Johnson and J. Vlissides, *Design patterns –Elements of reusable object-oriented software*, Addison-Wesley, 1995.

[Gho01] A. K. Ghosh and T. M. Swaminatha, "Software security and privacy risks in mobile e-commerce," *Comm. of the ACM*, vol. 44, No 2, February 2001, 51-57.

[Gra04] P. Gralla, "Mobile Web services: Theory vs. reality," http://SearchWebServices.com, 10 February, 2004.

[hac02] "The GSM security technical whitepaper for 2002," http://www.hackcanada.com/blackcrawl/cell/gsm/gsm_security.html

[ibm] IBM Corp., AlphaWorks Web Services Toolkit, http://www.alphaworks.ibm.com/tech/webservicestoolkit

[IM02] "Security in a Web Services World: A Proposed Architecture and Roadmap: A Joint White Paper from IBM Corporation and Microsoft Corporation," Microsoft Developers Network, 7 April 2002, http://msdn.microsoft.com/library/en-us/dnwssecur/html/securitywhitepaper.asp

[ipv] IPv6 home page, http://www.ipv6.org

[isa] "Security of the WEP algorithm," http://www.isaac.cs.berkeley.edu/isaac/wep-faq.html

[Iss] Internet Security Systems, X-Force Database, http://xforce.iss.net/xforce

[jav] "Java APIs for Bluetooth Wireless Technology (JSR-82) Specification," http://jcp.org/en/jsr/detail?id=82

[Juu01a] N. C. Juul and N. Jorgensen, "Security limitations in the WAP architecture," Workshop at OOPSLA 2001, http://www.dnafinland.fi/oopsla/wap.pdf

[Juu01b] N. C. Juul and N. Jorgensen, "WAP May stumble over the gateway (security in WAP-based mobile commerce)," http://www.dat.ruc.dk/~nielsj/research/papers/wap-ssgrr.pdf

[Mit] The Mitre Corporation, Common Vulnerabilities and Exposures List, http://cve.mitre.org

[nis] http://csrc.nist.gov/publications/fips/fips140-2/fips1402.pdf

[rfc2775] http://www.faqs.org/rfcs/rfc2775.html

[Rie00] M. J. Riezenman, "Cellular security: better, but foes still lurk," *IEEE Spectrum*, June 2000, 39-42.

[Rob99] S. Robinson, "Researchers crack code in cell phones', *The New York Times*, Dec. 7, 1999.

[Saa99] M.-J. Saarinen, "Attacks against the WAP WTLS protocol," *Proceedings of Communications and Multimedia Security '99*, http://www.jyu.fi/~mjos/wtls.pdf

[shi] http://shibboleth.internet2.edu/

[Sta99] W. Stallings, *Cryptography and network security: Principles and practice* (2nd Edition), Prentice-Hall, 1999.

[Sta02] W. Stallings, *Wireless Communications and Networks,* Prentice-Hall, 2002.

[Var00] U. Varshney, R. J. Vetter, and R. Kalakota, "Mobile commerce: A new frontier," *Computer*, IEEE, October 2000, 32-38.

[Vau04] S. M. Vaughan-Nichols, "Wireless middleware: Glue for the mobile infrastructure," *Computer*, IEEE, May 2004, 18-20.

[Yua02a] M. J. Yuan, "Access Web services from wireless devices," *Java World*, August 2002, http://www.javaworld.com/javaworld/jw-08-2002/jw-0823-wireless.html

[Yua02b] M. J. Yuan, "Securing wireless J2ME: Security challenges and solutions for mobile commerce applications," *IBM DeveloperWorks*, (1 June 2002), IBM. http://www-106.ibm.com/developerworks/wireless/library/wi-secj2me.html

[Yua04] M. J. Yuan, "SOA and Web services go mobile, Nokia-style," July 6, 2004, http://www.sys-con.com/story/?storyid=45531&DE=1

# 4

## WLAN TECHNOLOGIES

*Theodore Zahariadis and Christos Douligeris*

### INTRODUCTION

Wireless local area networks (WLANs) have changed traditional networking practices by providing tremendous flexibility and liberating users from network cables. The WLAN market has grown rapidly because wireless technology has evolved to meet the fundamental needs of businesses and technology consumers. Major application areas of WLAN include the following:

- Enterprise and corporate departments deploy WLANs to instantly set up meetings and conference calls, support nomadic or part-time employees, reduce the cost of cabling and recabling the physical plant, and provide rapid response to changes in demand.
- Small office/home office (SOHO) users utilize WLANs to share Internet connections, printers, and peripherals, and to create backup connectivity solutions while remaining highly scalable and flexible, without the costs of cabling.
- Mobile workers and executives use public-access WLANs in "hot spots" (e.g., cafes, airports, trains, lobbies) and remote corporate sites to instantly connect to the Internet or the corporate network.
- Home users deploy in-home WLANs to share a broadband Internet connection among multiple PCs or home-networking appliances without drilling holes and installing cables throughout the home.

## WLAN Classification

There are many WLAN technologies. Based on the physical (PHY) layer, they may be categorized into two groups:

Infrared (IR) — IR technology requires that the transmitter and the receiver be in a direct line (line-of-sight) and that no object be positioned between them. This requirement has restricted the wide deployment of IR in data networks, even though it assures security from eavesdroppers.

Radio Frequency (RF) — RF technology is more flexible and does not require line-of-sight positioning. RF technologies can be further categorized as narrowband and spread spectrum. Narrowband technologies include microwave transmission, which is not very suitable for in-door WLAN networks, as offices and SOHO environments are rich in reflections and mirroring effects. Alternatively, spread spectrum technologies distribute the signal over a number of frequencies, thus avoiding signal interception in most cases. The fact that RF technologies suffer from eavesdropping makes the integration of security features and encryption techniques a normal feature of RF networking products. Spread-spectrum technologies are further categorized into direct sequence spread spectrum (DSSS) and frequency hopping spread spectrum (FHSS) technologies. The DSSS systems transmit a signal over multiple frequencies simultaneously, which allows the complete use of the allocated frequency band, and makes the communication highly resistant to interference. A receiver retrieves the original data by match filtering. The FHSS systems hop over entire bands of frequencies in a particular sequence, which is carefully selected to minimize interference on the in-door network. FHSS allows for a simpler radio design compared to DSSS but leads to greater hopping overhead.

## Frequency Bands

To simplify the operation and expand the deployment base of WLANs, the vast majority of standardized WLAN technologies operate in unlicensed frequency bands, characterized as Industry, Scientific, and Medical (ISM) bands. Among the various ISM bands, the 2.4-GHz and the 5.8-GHz bands are the most widely used frequencies for WLAN communications. Especially widespread is the use of the 2.4-GHz band, the worldwide spectrum recognized by international regulatory agencies such as the Federal Communications Commission (FCC), the United States, the European Telecommunications Standards Institute (ETSI), in Europe, and the Radio Equipment Inspection and Certification Institute (MKK), in Japan, for unlicensed radio operations, as summarized in Table 4.1.

Table 4.1   Global Spectrum Allocation at 2.4 GHz

| Region | Allocated Spectrum |
| --- | --- |
| United States | 2.4000–2.4835 GHz |
| Europe | 2.4000–2.4835 GHz |
| Japan | 2.471–2.497 GHz |
| France | 2.4465–2.4835 GHz |
| Spain | 2.445–2.475 GHz |

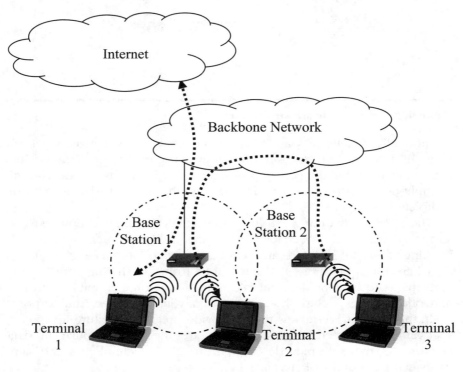

**Figure 4.1   Structured network architecture.**

## WLAN Network Architecture

WLAN communication of terminals is supported via two general network architecture topologies — the structured mode and the ad hoc mode.

In the structured mode (Figure 4.1), the terminals communicate with the backbone network via a base station, which aggregates access for multiple wireless stations onto the wired network. Multiple base stations

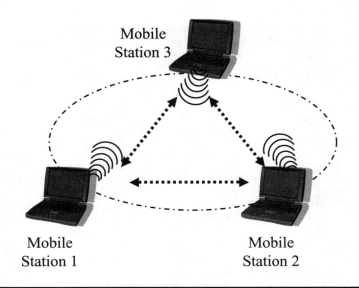

**Figure 4.2    Ad hoc network architecture.**

are normally connected via a fixed, wired network, which usually is an Ethernet network. The structured topology is useful for providing wireless convergence of buildings, campuses, or hot-spot areas by developing multiple access points, with radio coverage that may overlap to provide complete coverage.

The ad hoc networking topology allows direct communication between terminals without the need of a base station (Figure 4.2). Ad hoc networking is useful for fast and easy setting up of wireless networks anywhere for applications that may range from file sharing between the participants to battlefield communications. Common examples of ad hoc network locations include hotel rooms, convention centers, and airports.

In this chapter, we review the most widespread and mature RF spread-spectrum WLAN technologies, as they are the most popular in data networks. A more detailed description on WLAN technologies and standards can be found in Reference 1.

## BLUETOOTH

Bluetooth is intended to serve as a universal, low-cost, air interface that aims to replace the plethora of proprietary interconnecting cables between a variety of personal devices.[2] The technology was named "Bluetooth" to honor the Danish king Harald Blåtand (loosely translated to "Bluetooth"), who united Demark and Norway in the 10th century, because wireless technology unites people.[3]

Bluetooth is an FHSS wireless system operating in the unlicensed 2.4-GHz ISM band. Bluetooth 1.0 is a short-range (10 cm to 10 m), rather narrowband technology that can transfer up to 720 kbps. Bluetooth 2.0 is expected to be able to transfer up to 20 Mbps in ranges of up to 50 m.

## Network Architecture

Bluetooth communication is based on ad hoc networking, which means that there is no base station or access point, but there exist mechanisms and messages for Bluetooth devices to discover each other and establish communication links. A Bluetooth ad hoc network may be "open," in which a number of people are around a conference table and information is shared between them, or "closed," when a Bluetooth mobile phone, laptop, and PDA of the same person communicate, but no other devices within range are allowed to take part in the communication by any means.

Nevertheless, after the initial communication and without the need for base stations, the Bluetooth standard defines a structured network architecture based on a star network topology called "piconet." When a Bluetooth device initiates a communication, it defines a new piconet cell having this device as the center of the cell, called master device of the cell. All the other devices in this cell are considered slave devices. During the lifetime of a piconet, the master device may be changed dynamically. Piconet cells may be combined to form a scatternet. In a scatternet one or more Bluetooth devices may be members of more than one piconet, and may act as master in one cell and slave in another.

In Figure 4.3, a scatternet consists of two piconets; piconet A has one master and four slave devices, whereas piconet B has one master and three slave devices. It is worth noting that the master device of piconet B operates as an additional slave device to piconet A.

## Bluetooth Packet Format

Bluetooth, as mentioned in the previous section, is an FHSS system that operates in the unlicensed frequency range of 2.4 to 2.4835 GHz. It defines 79 channels with a channel spacing of 1 MHz and a hopping rate of 1600 hops/s. Within each piconet, the master device defines the hopping sequence and events, and all the slaves of the piconet have to synchronize with the master device. Apart from the frequency hopping, the master's clock is used for the time division slot and the encryption. Bluetooth uses a slotted time division duplex (TDD) mechanism. The TDD slot structure is shown in Figure 4.4. Each packet is transmitted on a different frequency, normally within a single time slot, but it can be extended to cover up to five slots. If more than one slot is used for a single packet transmission,

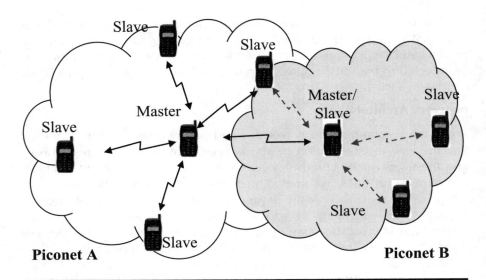

**Figure 4.3   Bluetooth scatternet network architecture.**

the frequency does not hop until the entire packet has been transmitted. Each time slot has a duration of 625 $\mu s$, and a 220 $\mu s$ guard time between the end of reception of a packet and the start of transmission of the next packet guarantees the minimal number of packet collisions.

## Bluetooth Protocol Stack

The Bluetooth protocol stack (Figure 4.5) supports a physical separation between the link manager layer and the higher layers at the Host Control Interface (HCI), which is common in most Bluetooth implementations. The baseband layer provides the lower-layer functionality, which is required for air interface packet framing, establishment and maintenance of piconets, and link control.

The link manager layer is responsible for link set up and control and supports a number of procedures including authentication, encryption control, physical parameters control, and master-to-slave switching. The HCI provides a mechanism whereby the higher layers of the protocol stack can delegate the decision on whether to accept connections to the link manager and whether to switch on filters at the link manager. The HCI supports a bidirectional transmission of asynchronous connectionless (ACL) data and synchronous connection oriented (SCO) streams such as audio across it.

The Logical Link Control Adaptation Layer Protocol (L2CAP) provides connection-oriented and connectionless data services to higher-layer protocols. It supports multiplexing of higher-layer protocols, establishment and removal of logical connections for SCO services, and segmentation

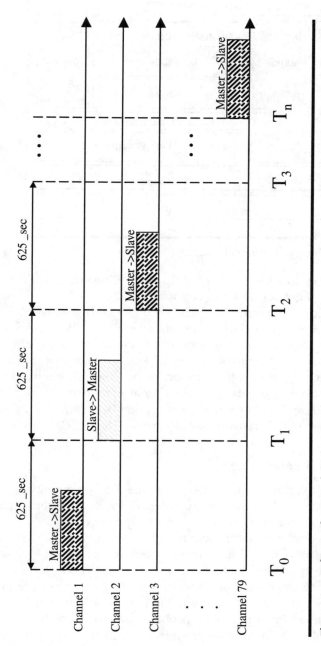

**Figure 4.4  Bluetooth TDD slot structure.**

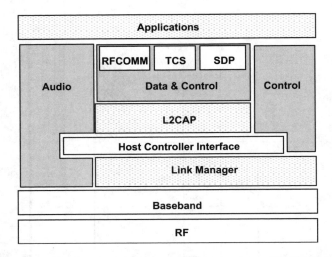

**Figure 4.5   Bluetooth protocol stack.**

and reassembly to allow packets of up to 64 KB to be transported. It also provides for negotiation of the connection's QoS.

The Service Discovery Protocol (SDP) is a higher-layer Bluetooth-specific protocol, which allows Bluetooth devices to discover what services are available on a device. SDP follows the client/server paradigm. It uses the service discovery database server to store and retrieve service records and attributes. The RFCOMM component provides an emulation of serial ports. RFCOMM may emulate up to 64 serial ports between two Bluetooth devices. The Telephony Control Specification (TCS) provides an adaptation layer that enables Q.931 call-control services over L2CAP.

## Evolution of Bluetooth

In Bluetooth 1.0, throughput, coverage range, interoperability, and security were not considered as major requirements. However, during the maturing phase of the standard, new requirements appeared.

The first one was interoperation. For security reasons, during the establishment of a new link, Bluetooth devices exchange public keys to confirm their identities.[5] The Bluetooth 1.0 specification leaves important details open to interpretation. As a result, devices from different manufacturers may fail to negotiate the initial link establishment. Bluetooth 1.1 solved this initial interoperability problem by requiring the slave devices to acknowledge the master device and confirm their roles as slaves.

Another problem is related to the protocol frame structure. Bluetooth 1.0 optionally supports up to five slots per packet, but not all Bluetooth devices support the five-slot format option. In Bluetooth 1.1, the slave is able to acknowledge the master device, inform it about the packet sizes that it can accept, and control the communication flow.

Finally, Bluetooth 1.0 had to assure worldwide operation. However, the frequency plans of some countries, including Japan, France, and Spain, utilize segments of the 2.4-GHz frequency for noncommercial purposes (e.g., military communications). To extend coverage in these countries, Bluetooth 1.0 specified a different hopping pattern with 23 hops, which does not use the specified frequencies of the 2.4-GHz spectrum. However, Bluetooth devices that use the 79-hop pattern are incompatible with those that follow the 23-hop pattern; thus they cannot be used in these countries. To overcome this interoperability issue, the Bluetooth SIG managed to gain worldwide permission, and Bluetooth 1.1–compliant devices use only the 79-hop pattern to communicate within the 2.4-GHz frequency.

Market requirements for longer coverage range and broader communications have driven the evolution of Bluetooth 1.1 into a newer standard Bluetooth 2.0. Bluetooth 2.0 is expected to achieve transfer rates of up to 20 Mbps in ranges of up to 50 m. Moreover, the direct integration of Bluetooth 2.0 chips into mobile terminals (e.g., PDAs and cell phones), will offer the users the capability to use interchangeably the local Bluetooth connections wherever available, or roam to the third-generation (3G) and 2.5G mobile networks. This will enable greater flexibility in introducing the wide availability of personal access networks (PAN).

## IEEE 802.11

IEEE 802.11 has been one of the pioneering and most important efforts to specify and develop a data-oriented WLAN standard.[6] The widespread acceptance of IEEE 802.11 (and especially its major evolution 802.11b) derives from industry standardization, which has ensured product compatibility and reliability among the various manufacturers. IEEE 802.11 has been so widely accepted throughout the world that the general term WLAN refers primarily to the IEEE 802.11 standard.

The development of IEEE 802.11 started in 1990, and the first version was almost finalized by mid-1994. In 1997, the Institute of Electrical and Electronics Engineers (IEEE) validated the initial IEEE 802.11 specifications as the standard for WLANs. The first version of IEEE 802.11 provided for 1 to 2 Mbps data transfer rates and a set of fundamental signaling methods and services. Extensions to IEEE 802.11 provide up to 54 Mbps.

### Network Architecture

IEEE 802.11 supports communication of terminals via both structured and ad hoc network architectures. In the structured mode (Figure 4.6), the

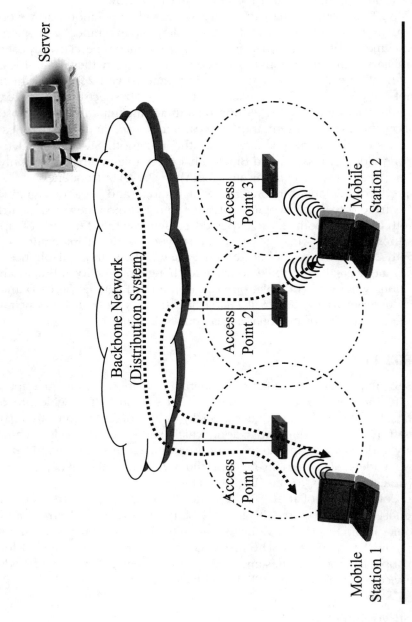

**Figure 4.6  IEEE 802.11 structured network architecture.**

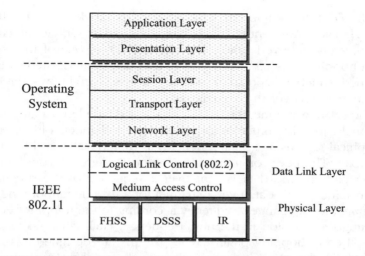

**Figure 4.7    IEEE 802.11 protocol stack.**

access point aggregates traffic from multiple wireless stations onto the wired network. The configuration that consists of at least one access point connected to the wired network infrastructure and a set of wireless end stations is called *basic service set* (BSS). Wireless connections between the access points are supported via a special frame format that effectively tunnels original frames over the 802.11 wireless networks. The set of two or more BSSs is called *extended service set* (ESS).

## Protocol Stack

As all IEEE 802.x protocols, IEEE 802.11 covers the lower layers of the OSI model, and specifies the physical and the Medium Access Control (MAC) layers.[7] Moreover, the IEEE 802.2 logical link control (LLC), 48-bit addressing, and the upper layers of the protocol stack remain unchanged as in other 802 LANs, allowing for very simple bridging between wireless and wired networks that follow the IEEE standards (Figure 4.7).

### 802.11 Physical Layer

Similar to Bluetooth, the initial IEEE 802.11 standard operates within the unlicensed 2.4-GHz ISM band. The IEEE 802.11 standard supports several WLAN technologies in the unlicensed bands of 2.4 GHz, and shares the same MAC over two PHY layer specifications: DSSS and FHSS technologies. Infrared technology is also supported, but it has not been adopted by any manufacturer.

In the DSSS technique, the 2.4-GHz band is divided into 14 channels of 22 MHz each. Adjacent channels partially overlap. Only three channels are completely nonoverlapping. Data is sent across one of the channels without hopping to other channels. The user data is modulated by a single predefined wideband-spreading signal. The receiver knows this signal, and is able to recover the original data.

In the FHSS technique, the 2.4-GHz band is divided into a large number of channels. Just like Bluetooth, the number of channels differs between geographical regions, i.e., 79 frequencies in United States and Europe and 23 in Japan. The peer communication endpoints agree on the frequency-hopping pattern, and data is sent over a sequence of the subchannels. The transmitter sends data over a channel for a fixed length of time, called *dwell time*; then it changes frequency according to the hopping sequence, continuing transmission in the new frequency. Each conversation occurs over a different hopping pattern. The patterns are designed to minimize the possibility that two senders simultaneously use the same subchannel. As the dwell time is rather long, the transmitter can send multiple consecutive symbols at the same frequency. FHSS techniques allow for a relatively simple radio design, but are limited to speeds no higher than 2 Mbps. This limitation is driven primarily by FCC regulations that restrict subchannel bandwidth to 1 MHz, leading to high amount of hopping overhead.

## IEEE 802.11 MAC Layer

The IEEE 802.11 specifies a single MAC protocol for all 802.11 physical layers. This design decision is quite important because it greatly simplifies interoperability, while enabling chip vendors to achieve higher production volumes and keep prices low.

Like the IEEE 802.3 Ethernet MAC, the 802.11 MAC has to support multiple users on a shared medium by having the sender sense the medium before sending any signals and by handling collisions that occur when two or more terminals try to communicate simultaneously. In the Carrier Sense Multiple Access with Collision Detection (CSMA/CD) protocol, used in wired Ethernet, a terminal must be able to transmit and listen at the same time. In radio communications, the terminal is not able to transmit and receive simultaneously, thus it is not able to detect a collision. Thus, 802.11 uses a modified protocol, called Carrier Sense Multiple Access with Collision Avoidance (CSMA/CA) or Distributed Coordination Function (DCF). CSMA/CA attempts to avoid collisions by using explicit packet acknowledgments (ACKs), which means that an ACK packet is sent by the receiving station to confirm the integrity of the data packet that arrived.

The algorithm works as follows. Initially, the terminal senses the medium. If the medium is sensed idle, the terminal may transmit. If the medium is sensed busy, the terminal has to wait. After the medium is sensed idle, the terminal has to wait two additional time periods. The first period depends on the packet to be transmitted. If it is an ACK, this period is one short interframe space (SIFS); otherwise it is a DCF interframe space (DIFS). The second is a random backoff period, which prevents multiple terminals from seizing the medium immediately after completion of the preceding transmission. After both periods have passed and the channel has not been seized, the terminal may start transmission. Otherwise, the whole process restarts. Figure 4.8 gives a graphical presentation of the algorithm.

To ensure robustness, the 802.11 MAC layer provides cyclic redundancy check (CRC) and packet fragmentation. Packet fragmentation splits large packets to smaller units, because the larger packets have a higher probability to be corrupted. This technique reduces the number of retransmissions or requires retransmission of shorter messages, thus improving the overall wireless network performance.

## IEEE 802.11b

The major drawback of IEEE 802.11 was its limited throughput. In 1999, recognizing the critical need for higher data transmission rates in most business applications, IEEE enhanced the initial 802.11 standard to the 802.11b standard, which is able to support transmissions of up to 11 Mbps. The IEEE 802.11b enables WLANs to achieve performance and throughput comparable to wired 10-Mbps Ethernet.

In parallel to the standards bodies, to ensure interoperability and compatibility across all market segments, IEEE 802.11 product manufacturers agreed on a compliance procedure called Wi-Fi (Wireless Fidelity standard). Moreover, a Wireless Ethernet Compatibility Alliance (WECA) was formed to certify Wi-Fi interoperability of new products, to certify cross-vendor interoperability and compatibility of IEEE 802.11b wireless networking products, and to promote IEEE 802.11b for business and home applications.[8] Members of WECA include WLAN semiconductor manufacturers, WLAN providers, computer system vendors, and software makers.

### IEEE 802.11b PHY Layer Enhancements

The main enhancement of 802.11b was the standardization of a physical layer able to support higher speeds of 5.5 and 11 Mbps. As 802.11 FHSS systems cannot support higher speeds without violating current FCC regulations, the DSSS technique was selected as the exclusive physical layer technique. In this way, 802.11b is backwards compatible and can interoperate at 1 Mbps and 2 Mbps only with the 802.11 DSSS systems and not with FHSS systems.

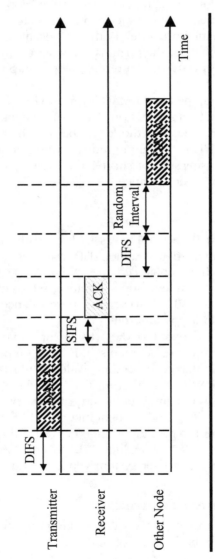

**Figure 4.8 IEEE 802.11 CSMA/CA algorithm.**

**Table 4.2   802.11 Physical Layers Differences**

| Physical Layer | Data Rate | Bits/Symbol | Code | Modulation | Symbol Rate |
|---|---|---|---|---|---|
| 802.11 | 1 Mbps | 1 | Barker sequence | BPSK | 1 Msps |
| 802.11 | 2 Mbps | 2 | Barker sequence | QPSK | 1 Msps |
| 802.11b | 5.5 Mbps | 4 | CCK | QPSK | 1.375 Msps |
| 802.11b | 11 Mbps | 8 | CCK | QPSK | 1.375 Msps |

To increase the data rate, 802.11b specifies an advanced coding technique called complementary code keying (CCK). CCK consists of a set of 64 code words, each of 8 bits length, which can be distinguished by the receiver even in the presence of noise or interference. CCK encodes 4 bits per carrier to achieve a data rate of 5.5 Mbps, and 8 bits per carrier to achieve 11 Mbps. Both speeds use QPSK modulation and a 1.375-Msps symbol rate. The differences between the 802.11 physical layers are shown in Table 4.2.

The IEEE 802.11b may operate in distances of up to 300 to 400 m in outdoor environments and 30 to 50 m in indoor environments with low noise. To support noisy environments as well as extended range, 802.11b uses dynamic rate degradation. When the terminal moves beyond the optimal range or if substantial interference occurs, 802.11b degrades transmission at lower speeds, falling back to 5.5, 2, and finally 1 Mbps. Vice versa, if the terminal returns within the optimal range or the source of interference disappears, the connection automatically increases the transmission speed.

The major drawback of IEEE 802.11b is the lack of QoS and synchronous channels for voice and video communications. Therefore, a number of IEEE 802.11 extensions have been initiated, aiming at enhancing the bandwidth, security, and QoS issues of 802.11. Some of these extensions are reviewed in the next paragraph.

## IEEE 802.11a

IEEE 802.11a is one of the most promising evolutions of IEEE 802.11b; hence, we will study it in more detail. The 802.11a standard is similar to 802.11b, but provides wireless data speeds of up to 54 Mbps and uses the 5.8-GHz spectrum range, which has less interference than the 2.4 GHz spectrum.

## PHY Layer

The PHY layer of IEEE 802.11a is a multicarrier system based on orthogonal frequency-division multiplexing (OFDM). It uses 52 carriers; 48 among these are data carriers, which carry user data traffic, and 4 are pilot carriers, which are used for synchronization and system control purposes. Each carrier is 300 kHz wide. Inclusion of the 52 carriers along with the channel spacing covers the 20-MHz bandwidth.

IEEE 802.11a uses various modulation schemes, namely binary phase shift keying (BPSK), quadrature PSK (QPSK), 16-quadrature amplitude modulation (QAM), and 64-QAM with $\int$ or $\Omega$ error-correcting code overhead. According to the modulation, each one of the 48 data carriers may transmit raw data at rates ranging from 125 kbps to 1.5 Mbps; hence, the total raw bandwidth may vary from 6 to 72 Mbps. Assuming that 64-QAM is used for maximum bandwidth (72 Mbps) reduced by $\Omega$ error-correction code overhead, IEEE 802.11a may achieve up to 54 Mbps of useful traffic.

## MAC Layer

IEEE 802.11a uses a MAC protocol almost identical to IEEE 802.11b, based on CSMA/CA. An IEEE 802.11a terminal must initially sense the medium for a specific time interval, and if the medium is idle, it can start transmitting the packet. If the medium is not idle, the terminal begins a backoff process and waits for a certain time interval (minimum 34 $\mu$s). When the backoff time has expired, the terminal can try to access the medium again. As collisions in a wireless environment cannot be detected, a positive acknowledgement is used to notify that a frame has been successfully received.

Figure 4.9 shows the format of a packet data unit in 802.11a, including the preamble, header, and physical layer service data unit (PSDU) or payload. The preamble is mainly used for frame synchronization. The header is mapped on one OFDM symbol and coded with the most robust modulation and error correction scheme (BPSK $\int$). It contains information about the type of modulation and the coding rate used in the rest of the packet (rate field), the length of the payload (length field), and the parity bit. The tail field is used to reset the encoder and provide information to the decoder. The service field is currently used to initialize the descrambler (7 bits), whereas 9 bits are reserved for future use. The pad field is used for mapping the Physical Payload Data Unit (PPDU) into an integer number of OFDM symbols.

IEEE 802.11a products are already available. However, there are certain barriers to be overcome before IEEE 802.11a is widely accepted. First of all, the coverage range is very short. The distance of 50 m is barely

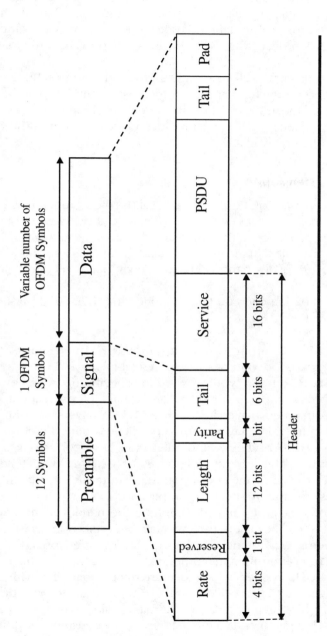

**Figure 4.9  IEEE 802.11a frame and packet format.**

adequate for in-office or home usage. Another reason is the selection of the 5-GHz frequency band. This band is not available worldwide. Japan, for example, permits the use of a smaller band that contains half the channels. In Europe, the standard does not comply with various EU requirements. Moreover, IEEE 802.11a does not provide any QoS mechanisms.

All these constraints still restrain the large adoption of IEEE 802.11a, limit the volume production, and keep the costs high. On the other hand, a new standard, the IEEE 802.11g, has been developed that operates in the 2.4-GHz frequency band and provides up to 54 Mbps. IEEE 802.11g products are already available.

## IEEE 802.11 Extensions

The wide acceptance of IEEE 802.11/802.11b, the competition, and the users' requirements for broader, more secure, and guaranteed QoS communications necessitated IEEE to create a number of task groups aimed at enhancing the initial standard. Each task group has specific objectives related to the protocol's bandwidth, frequency band, security, QoS, and interoperability.

Apart from 802.11a and 802.11g, the following IEEE 802.11 task groups are currently active:

- **802.11d** task group works towards 802.11b versions at other frequencies, for countries where the 2.4-GHz band is not available.
- **802.11e** task group works toward the specification of a new 802.11 MAC protocol to accommodate additional QoS provision and security requirements over legacy 802.11 PHY layers. It replaces the Ethernet-like MAC layer with a coordinated time division multiple access (TDMA) scheme, and adds extra error correction to important traffic. The key part of the 802.11e is the 802.1p standard, which provides a method to differentiate traffic streams in priority classes to support streaming applications.
- **802.11f** task group aims to improve the handover mechanism in 802.11 such that users can maintain a connection while roaming between access points attached to different networks.
- **802.11g** task group aims to define a new standard, which will use the 2.4-GHz frequency band, feature full compatibility with already-deployed 802.11b devices, and yet provide up to 54 Mbps. IEEE 802.11g has already been in the market since 2003.
- **802.11h** aims to enhance the control over transmission power and radio channel selection to 802.11a, to be acceptable to the European regulators.

- **802.11i** aims to enhance 802.11 security. Instead of the Wired Equivalent Privacy (WEP), a new authentication/encryption algorithm based on the Advanced Encryption Standard (AES) will be proposed.
- **802.11j** will propose the issue of 802.11a and HIPERLAN/2 interworking.

# HIPERLAN

High-Performance Radio LAN (HIPERLAN) is the European answer to the IEEE 802.11 standard.[9,10] In 1991, ETSI organized a committee to specify the new high-performance wireless protocol. The major difference between IEEE 802.11 and HIPERLAN was that the latter started from scratch as a new specification with specific objectives. The first draft version of the HIPERLAN standard was published in 1995. The standard allowed bandwidth of 23.529 Mbps with support for multi-hop routing, both asynchronous and time-bounded communication, prearranged or ad hoc network topology, and power saving.

## Network Architecture

Just like IEEE 802.11, HIPERLAN supports both structured network architecture and ad hoc peer-to-peer networking. Moreover, HIPERLAN optionally supports packet forwarding. In this case, a node operates as an intermediate reflector that forwards packets to other nodes (*forwarder*). A forwarder node broadcasts packets to all HIPERLAN nodes or unicasts packets to a specific node. To achieve the latter, the reflector node has to maintain and dynamically update a database, where routing and addressing information are stored.

## Software Architecture

Similar to most WLAN standards, HIPERLAN specifies only the two lower layers of the OSI protocol stack, whereas the upper layers remain unchanged for interoperability reasons (See Figure 4.10).

The requirements for high bit rate and low power transmission, for safety and autonomy reasons, limited the standard to distances of up to 100 m, although the bandwidth and the multichannel transmission required a large spectrum of 150 MHz. Thus, the Conference of European Posts and Telecommunications Administration (CEPT) selected the 5-GHz band, which was divided into 5 channels. The lower 3 channels are available throughout Europe, whereas the upper 2 channels are available only in some countries.

The HIPERLAN PHY layer uses the Gaussian minimum shift keying (GMSK) and the frequency shift keying (FSK) technologies for delivering

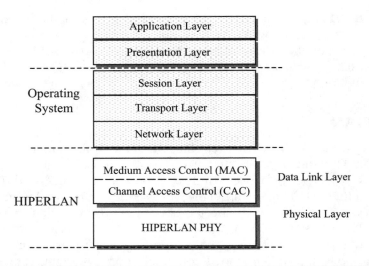

**Figure 4.10  HIPERLAN protocol stack.**

data with speeds up to 23 Mbps. The MAC layer is based on carrier sensing, but it differs from the IEEE 802.11 standard because it introduces packets prioritization via a nonpreemptive priority multiple access (NPMA) mechanism.

The HIPERLAN channel access control sublayer supports time-bounded communication through the following algorithm. In case the terminal senses that the air interface is idle for a sufficient time interval, immediate transmission is allowed. If it is busy, channel access is granted in three phases: the prioritization, the elimination, and the transmission phases. The purpose of the prioritization phase is to exclude nodes with lower priority packets from further channel access contention. It consists of 1 to 5 slots, one per priority category. Following a descending-priority per-slot strategy, each node transmits a data burst, if it has not sensed a data burst of higher priority.

## HIPERLAN/2

HIPERLAN/2 is the European proposition for a broadband WLAN operating with data rates of up to 54 Mbps at the PHY layer on the 5-GHz frequency band. HIPERLAN/2 is supported by the ETSI, and developed by the Broadband Radio Access Networks (BRAN) group.

HIPERLAN/2 is a flexible radio-LAN standard designed to provide high-speed access to a variety of networks including 3G mobile core networks, ATM networks, and IP-based networks, as well as for private use as a WLAN system. HIPERLAN/2 is a connection-oriented Time Division Multiplexed

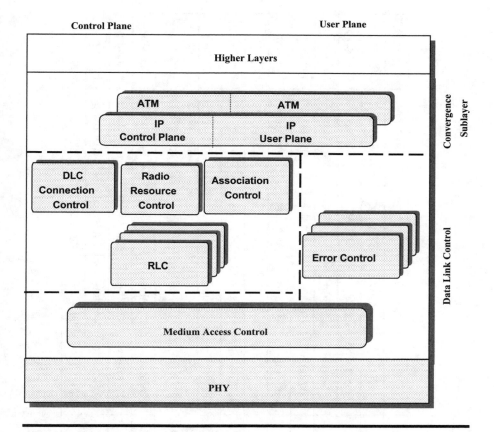

**Figure 4.11 HIPERLAN/2 protocol stack.**

(TDM) protocol. Data is transmitted on connections that have been established prior to the transmission, using signaling functions of the HIPERLAN/2 control plane. This makes it straightforward to implement support for QoS. Each connection can be assigned a specific QoS; for instance, in terms of bandwidth, delay, jitter, bit-error rate, etc. It is also possible to use a more simplistic approach when each connection can be assigned a priority level relative to other connections. This QoS support, in combination with the high transmission rate, facilitates the simultaneous transmission of many different types of data streams, e.g., video, voice, and data.

### Software Architecture

The HIPERLAN/2 protocol stack is shown in Figure 4.11. At the physical layer, HIPERLAN/2 uses OFDM to transmit the analog signals in a 20-MHz bandwidth.[11] Just like IEEE 802.11a, it uses BPSK, QPSK, 16-QAM, and

**Table 4.3 Wireless Technologies Summary**

| | Bluetooth 1.0/1.1 | Bluetooth 2.0 | 802.11 | 802.11b | 802.11a | 802.11g | HiperLAN |
|---|---|---|---|---|---|---|---|
| Frequency Band | 2.4 GHz | | 2.4 GHz | 2.4 GHz | 5.8 GHz | 2.4 GHz | 5.8 GHz |
| Physical | FHSS | | FHSS, DSSS | DSSS | OFDM | OFDM | GMSK/FSK |
| Access | Master-slave, polling | | CSMA/CA | CSMA/CA | CSMA/CA | CSMA/CA | NPMA |
| Max Range | 10 cm–10 m | | 250–300 m | 250–300 m | 50–100 m | 50–100 m | 100 m |
| Power | Very low | | Medium | Medium | Medium | Medium | Medium |
| Complexity | 1x | | 2x | 2x | 3x | 2.5x | 3x |
| QoS | Yes | | Not inherited | | | | Yes |
| Security | Medium | | Medium | | | | Strong |
| **Throughput** | | | | | | | |
| Physical | 1 Mbps | 10 Mbps | 2 Mbps | 11 Mbps | 54 Mbps | 54 Mbps | 23.5 Mbps |
| Effective | ≤0.7 Mbps | ≤8 Mbps | ≤1.5 Mbps | ≤7 Mbps | ≤30 Mbps | ≤22 Mbps | ≤20 Mbps |
| Region | Worldwide | | Worldwide | Worldwide | United States | Worldwide | Europe |
| Promoters | 2000+ | | 500+ | 500+ | 100+ | 500+ | <20 |

64-QAM modulations and $\int$, and $\Omega$ error-correcting overhead. Moreover, it uses 52 OFDM carriers, 48 for data and 4 pilots, achieving 12, 24, 48, or 72 Mbps coded bit rates and up to 54 Mbps data rate.

The difference between the HIPERLAN/2 and the IEEE protocols lies on the data-link control layer (DLC).[12] In the HIPERLAN/2 case, the DLC contains the MAC layer, error control, and DLC control modules and convergence sublayers. The MAC protocol is built from scratch, implementing a type of dynamic TDMA/TDD scheme with centralized control.

## SUMMARY

A summary of the presented WLAN technologies is shown in Table 4.3. Until recently, IEEE 802.11 technology was too expensive as compared to wired technologies. However, advances in VLSI, volume production and competition have significantly reduced the cost of the IEEE 802.11 b/g implementation. Comparing the coverage area, Bluetooth is rather limited to "room-distances," while it is the best solution for implementation of Personal Area Networks (PANs).

## REFERENCES

1  Zahariadis, Th., *Home Networking Technologies and Standards*, Artech House, Boston, MA, October 2003.
2. Bisdikian, C., An Overview of Bluetooth Wireless Technology, IEEE Communications Magazine, Vol. 39, No. 12, pp. 86–94, December 2001.
3. Miller, B., Bisdikian, C., *Bluetooth Revealed: The Insider's Guide to an Open Specification for Global Wireless Communications*, Prentice Hall, Upper Saddle River, NJ, 2001.
4. Shepherd, R., Bluetooth wireless technology in the home, *IEE Electronics and Communications Engineering Journal*, Vol. 13, No. 5, pp. 195–203, October 2001.
5. Zahariadis, Th., Pramataris, K., Zeros, N., A comparison of competing broadband in-home technologies, *IEE Electronics and Communications Engineering Journal*, Vol. 14. No. 4, pp. 195–203, August 2002.
6. Zyren, J., Petrick, A., Brief Tutorial on IEEE 802.11 Wireless LANs, Intersil, February 1999, http://www.intersil.com/an9829.pdf.
7. ISO/IEC, IEEE 802.11 Local and Metropolitan Area Networks: Wireless LAN Medium Access Control (MAC) and Physical (PHY) Specifications, ISO/IEC 8802–11:1999(E).
8. Brewer, B. et al., Security of the WEP algorithm, http://www.isaac.cs.berkeley.edu/isaac/wep-faq.html.
9. LaMaire, R.O., Krishna, A., Panian, J., Bhagwat, P., Wireless LANs and Mobile Networking: Standards and Future Directions, IEEE Personal Communications, August 1996, pp. 86–94.
10. ETSI, TC-RES, Radio Equipment and Systems (RES; High Performance Radio Local Area Network (HiperLAN); Functional Specification, ETSI 06921, Sophia Antipolis Cedex–France, July 1995, Draft ETS 300 652.

11. ETSI, Broadband Radio Access Networks (BRAN); HiperLAN Type 2 Functional Specification; Data Link Control (DLC) Layer; Part 1: Basic Data Transport Function, ETSI report TR 101 761–1, Ver. 1.1.1, April 2000.

12. ETSI, Broadband Radio Access Networks (BRAN); HiperLAN Type 2; Data Link Control (DLC) Layer; Part 2: Radio Link Control (RLC) Sublayer, ETSI report TR 101 761–762, Ver. 1.1.1, April 2000.

# 5

# WIRELESS SECURITY PROTOCOLS

*Saeed Rajput*

## INTRODUCTION

Wireless networks[1] can be broadly classified into four categories: (1) satellite networks,[2] (2) cellular networks,[3] (3) wireless LANs,[4] and (4) personal networks.[5] Several fundamentally different communication protocols are used in cellular networks.[6] Consequently, the security issues are addressed differently. Wireless LANs, on the other hand, are all flavors of the same basic standard[7] — IEEE 802.11 — and therefore a single security protocol addresses security issues uniformly. Satellite networks are highly proprietary and are therefore considered inherently secure. Although some would disagree with this hypothesis of implicit security, we will leave out satellite networks from our discussion in this chapter. The personal networks are either 802.11 or Bluetooth. Bluetooth is now standardized as IEEE 802.15.[5]

Whereas the securing of traditional voice traffic (also known as "Plain Old Telephone Service" or POTS) is limited by the capabilities of the POTS handsets, data traffic is not affected by such limitations. Consequently, several options for securing data traffic are available. However, the protection of voice traffic on cellular networks is still limited to securing the air link between the Mobile Station (MS) and the Base Station (BS), due to legacy voice networks. Although the same type of link security is also available for data traffic in almost every wireless protocol that supports data, there are other options available as well. Therefore, architects responsible for comprehensive security of wireless data traffic must understand the network layers and the security options available (Figure 5.1).

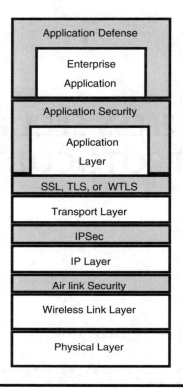

**Figure 5.1 TCP/IP network layers where security can be provided in wireless-based networks.**

| Link Header | IP Header | Transport Header | Application Message | Link Trailer |
|---|---|---|---|---|

**Figure 5.2 Extent of protection available at the link level.**

All link layer security protocols protect information on a physical channel between two endpoints that are directly connected by a physical link (copper or fiber) or over the air (RF or laser). The advantages of protecting the link layer are that it protects the complete data load of a frame. In other words, the application data as well as all headers introduced by the protocols are protected, as shown in Figure 5.2. The disadvantages include weak cryptographic algorithms due to hardware limitations. We will see that weaknesses have been discovered in almost

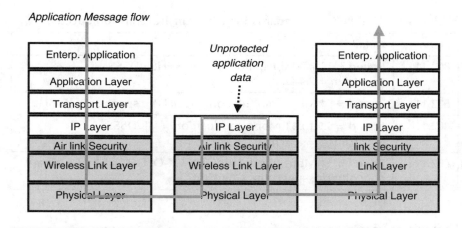

**Figure 5.3 Application data is exposed at every node if only link security is used.**

all the existing link layer wireless protocols. All next-generation wireless standards promise improvements to their security protocols.

Stronger cryptographic algorithms in hardware are now more feasible than better hardware technologies. However, other disadvantages of link layer security make combining link layer security with other high-level security protocols desirable for organizations. These disadvantages include the obvious vulnerability introduced due to decryption of protected data at every node (network hop). In wireless networks, this node is usually the BS, where data is extracted from the wireless protocol and moved to regular data networks, as shown in Figure 5.3. A malicious BS can do significant harm, because it is capable of looking at and modifying every frame. It is also infeasible to use public-key cryptography for per-frame non-repudiation, due to performance reasons and the fact that two end devices on a physical channel do not necessarily represent the end user. Thus, if non-repudiation is required for some high-level business application (e.g., stock trades), wireless-level link security is not adequate.

Because the protocols used for protection of links in wireless communications are highly varied and are often used, we will discuss them in separate sections. In the following section, we provide an overview of the 802.11 protocol, discuss the WEP and its weaknesses, and consider the TKIP and 802.11i protocols. Subsequent sections discuss Bluetooth security, provide a treatment of the security aspects of GSM and CDMA cellular networks, discuss the higher level security protocols that could be of interest to wireless security architects, and provide a brief overview of future improvements planned for cellular security.

**Table 5.1  IEEE 802.11 Standards for Different Transmission Speeds**

| Standard | Details |
| --- | --- |
| 802.11 | Speed = 1 or 2 Mbps, modulation = FHSS/DSSS, spectrum = 2.4 GHz |
| 802.11a | Speed = 54 Mbps, modulation = COFDM, spectrum = 5 GHz |
| 802.11b | Speed = 5.5 to 11 Mbps, modulation = HR/DSSS, spectrum = 2.4 GHz |
| 802.11g | Speed = 54 Mbps, modulation = CCK-OFDM, coding = binary convolution, spectrum = 2.4 GHz |

**Table 5.2  IEEE 802.11 Standards that Specify Functionality**

| Standard | Function |
| --- | --- |
| 802.11e | Support for multimedia and QoS. MAC layer specification (works with all; a, b, and g) |
| 802.11c | Started but later abandoned. Was about bridging layer that was later included in 802.11d |
| 802.11d | Physical layer requirements for international regulatory compliance |
| 802.11f | Recommended practice (to improve handover) |
| 802.11h | Better control over transmission power to 802.11a for Europe (spectrum managed 802.a) |

## OVERVIEW OF 802.11 PROTOCOL

IEEE 802.11 is a series of standards.[7] We summarize and classify them in Table 5.1 and Table 5.2. These tables do not include IEEE 802.11i, which is currently being standardized to fix and enhance the security of the original specification. The range of 802.11 networks is often a few hundred meters. We will first provide an overview of some of the important concepts that are necessary to understand security issues, then discuss the existing security protocols, and finally discuss what is being done to fix the current security problems permanently.

### IEEE 802.11 Overview

The overview of the IEEE802.11[8] protocol frame is shown in Figure 5.4. The preamble of the frame starts with an 80-bit synchronization sequence,

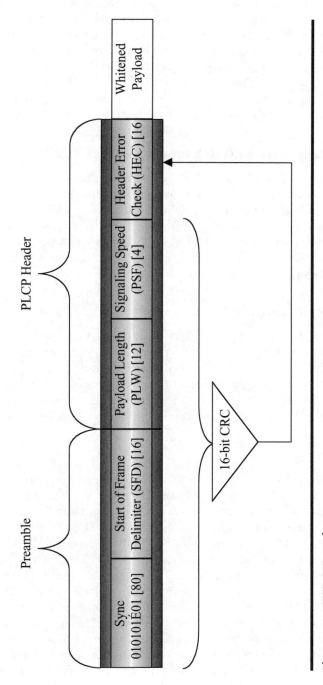

**Figure 5.4  IEEE 802.11 frame.**

which helps the receiver lock on to the transmitter clock. This is followed by a Start of Frame Delimiter (SFD), which marks the beginning of the frame and also identifies it as an IEEE 802.11 frame. The preamble is followed by the Physical Layer Convergence Protocol (PLCP) header. The first field in the PLCP header is a 12-bit PSDU Length Word (PLW) that stores the length of the payload carried by this frame. The PLCP Signaling Field (PSF) indicates the speed of the communication. The last field of the PLCP header is the 16-bit Cyclic Redundancy Check (CRC) calculated over the header.

## Wired Equivalent Privacy (WEP) Protocol of IEEE 802.11

This is an option in IEEE 802.11 for providing security that was, at the time of specification, thought to provide the same level of security as conventional wired networks. However, since its publication, several vulnerabilities have been discovered, making some simple attacks feasible.[9] We first present the original standard as specified, and then we discuss the weakness in the standard. Finally, we will discuss the efforts that are underway to fix the standard.

### WEP: Security Objectives

WEP is designed to encrypt the frame body and to compute a checksum before transmission. It uses the Rivest Cipher 4 (RC4) stream cipher (provided by RSA Security, Inc.[10]) for encryption. The receiver performs decryption. Thus, WEP is a link-encryption protocol that encrypts data over the wireless link between 802.11 stations. Once the frame enters the wired side of the network (shown in Figure 5.5), protection no longer applies.

### WEP: Encryption

Most implementations of WEP use RC4 128-bit key encryption (although the original standard specifies[2] weaker encryption). The 128-bit key is used as a seed to an RC4 random sequence generator that produces a keystream. A keystream's length is equal to the length of the frame's payload together with a 32-bit Integrity Check Value (ICV). The actual frame is encrypted by taking the eXclusive OR (XOR) of each payload bit with the corresponding bit of the keystream, as shown in Figure 5.6. In mathematical notation, $\{p\} \oplus \{r\} = \{c\}$, where $\oplus$ indicates the XOR operation. We can also use the notation $E_K[\{p\}] = \{c\}$ to emphasize that the plaintext sequence is encrypted with the key K. The ICV is a checksum that the receiver recalculates and compares with the one received with the frame. A mismatch between the two indicates a corrupted payload. Then the receiver can reject the frame and optionally flag the user. Please

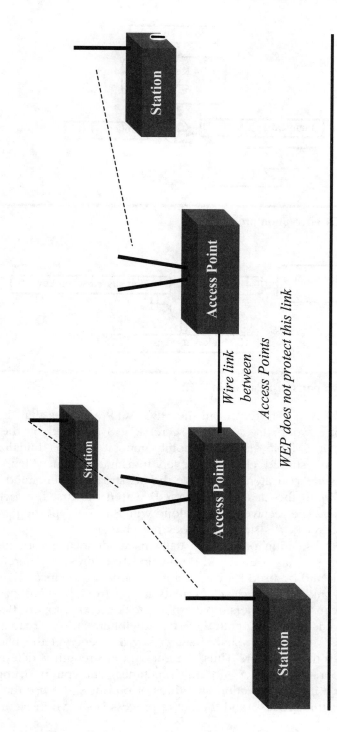

**Figure 5.5   Wired link between APs.**

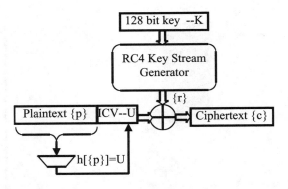

**Figure 5.6  RC4 encryption process.**

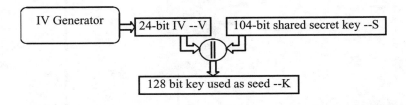

**Figure 5.7  Two parts of RC4 key in WEP.**

note that the ICV is different from the HEC. WEP specifies that the key K be split into two parts: a shared secret key S supplied by the user (usually kept at 104 bits) and a randomly generated 24-bit Initialization Vector (IV) V that should[3] change for every transmitted frame (Figure 5.7). The two parts are concatenated before the keystream is generated. WEP appends the IV in the clear (unencrypted) within the first few bytes of the frame body. The receiver uses V along with S to decrypt the payload. The shared secret key is static and does not change.

Different transmitting payloads usually have identical information at the beginning, such as source and destination IP addresses. Therefore, a different V[4] should be used for each frame; otherwise, identical plaintexts will be encrypted to identical ciphertexts in the front portions of these frames. This can offer crackers a pattern to break the encryption. Because the plaintext V is sent along with the payload ciphertext, we say that WEP encryption uses the 104-bit shared secret key S to encrypt the plaintext {p}, rather than the 128-bit K. Thus, the encryption strength is determined by the number of bits in S, not K. Sometimes, encryption strength is misrepresented in the marketing literatures of products. We use the notation $E_s[\{p\}]$ for this process, and the entire process is shown in Figure 5.8.

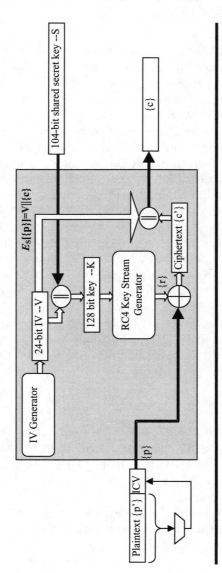

**Figure 5.8 Encryption process of WEP.**

| Init. Vector [24] | Pad [6] | Key ID [2] | Data 8xn; n ≥ 1 | ICV [32] |
|---|---|---|---|---|

**Figure 5.9  Encapsulated WEP frame.**

WEP allows all stations in a network to use the same key, called the *default key*, for encryption. It also allows each MS to have a different key that is shared with the access point (AP). This key is called a pairwise key. When devices are programmed to use the default key, all traffic in the system is encrypted with the default key. When a station is programmed for a pairwise key, the point–point traffic is encrypted with it. However, group messages and the broadcast messages are still encrypted with the default key. Therefore, a station using a pairwise key still needs the default key. WEP uses a key-backup mechanism to allow for convenient updating of both default and broadcast keys, by allowing the stations to retain two keys of each kind at once. When the keys are being reprogrammed over the network, one key can be the old key and the other the new key. Therefore, at any given time, an MS may have as many as four keys: (1) an old default key, (2) a new default key, (3) an old pairwise key, and (4) a new pairwise key. An AP, on the other hand, must therefore retain two default keys and two pairwise keys for each MS. Different keys are identified by a unique 2-bit key ID that is transmitted with each frame. The receiver uses this ID to locate the proper key for decryption from its database. The WEP frame format is shown in Figure 5.9. The encryption adds 4 bytes in front and 4 bytes for IV at the end of the data. The portion of the frame that is encrypted is the shaded area.

## WEP Authentication

WEP also attempts to provide a secure authentication method intended to prove to a legitimate AP that the station has the shared secret key. However, due to the serious vulnerability of this algorithm as discussed in the following subsection, it is strongly recommended that deployments disable this feature, because it not only is easy to compromise the authentication process, but it also leaves the entire WEP protocol vulnerable. We now describe the original authentication mechanism proposed in the IEEE 802.11 WEP protocol.

Figure 5.10 shows the messages exchanged between the station and the AP. It shows that once a station requests authentication from the AP, the AP sends out a challenge, which is just a random sequence {p}. The station uses the shared secret key S to encrypt that challenge and returns

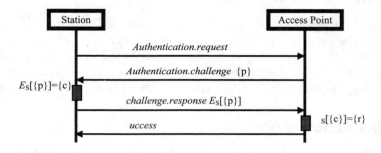

**Figure 5.10   WEP authentication message exchanges.**

the ciphertext. To verify that the station has the correct secret key S, the AP decrypts the ciphertext received to retrieve the original challenge.

## WEP Authentication Vulnerability

The encryption algorithm described in the section titled "WEP Authentication" is an example of a poorly thought-out protocol. An attacker who passively observes this exchange will have observed {r}, {c}, as well as V. The attacker can use these to deduce the keystream {r}, using the fact that RC4 simply XORs the plaintext with the keystream.

As shown in Figure 5.6, the ciphertext is produced by taking the XOR of the plaintext with the random sequence produced by the RC4 algorithm. In mathematical notation, we can express this as: $\{p\} \oplus \{r\} = \{c\}$ .

$$\{\{p\} \oplus \{r\}\} \oplus \{p\} = \{c\} \oplus \{p\}$$

$$\text{L.H.S} = \{r\} \oplus \{p\} \oplus \{p\}$$

$$= \{r\} \Rightarrow$$

$$\{c\} \oplus \{p\} = \{r\}$$

This indicates that once the plaintext {p} (which in this case is the random challenge sent out by the server) is known, it is simple to obtain the RC4 keystream {r}. Ironically, the V used with this keystream is also available to the attacker. All the attacker has to do is collect all samples of IVs along with the corresponding keystreams to fully decrypt any information that is exchanged between the station and the AP.

## WEP Encryption Vulnerabilities

The main causes of other encryption vulnerabilities in WEP are the relatively short IVs, the static keys, poor selection of the ICV algorithm, and, to a lesser degree, the strength of the RC4 encryption algorithm itself. Some of the vulnerabilities result from improper implementations. In the following discussion, we will point out some improper implementations.

### Inadequate Size of IV

Changing V per packet was intended to prevent keystream deduction. However, it does not achieve this goal. Because the size of V is only 24 bits, in the best possible implementation, its values will eventually repeat after 16,777,215 frames, when all possible 24-bit combinations will have been used. Assuming an average frame size of 1,500 bytes and a channel speed of 11 Mbps, all possible IV values can be consumed in less than 6 hr on a busy channel. Thus, the stations will eventually have no choice but to repeat the same IV for different data packets. For a 54-Mbps channel, the repetition will occur in less than an hour. For the frames that are encrypted with keystreams having the same shared key and same IV, the keystream used will be the same. The attacker can utilize the well-known vulnerability of stream ciphers, which is that encrypting two messages with the same key can reveal information about both messages. Thus, if

$$\{c_1\} = \{p_1\} \oplus \{r\}$$
$$\{c_2\} = \{p_2\} \oplus \{r\}$$

then,

$$\{c_1\} \oplus \{c_2\} = \{p_1\} \oplus \{r\} \oplus \{p_2\} \oplus \{r\}$$
$$= \{p_1\} \oplus \{p_2\}.$$

Note that if the plaintext of one stream is known, then that of the other stream can be obtained as well. There are many situations in wireless communications in which the plaintext can be predicted. One example is the frequent use of known IP addresses. More general techniques are discussed in Reference 11. If we have N ciphertexts that all use the same keystream $\{r\}$, the cryptanalysis problem becomes easier as N increases because the pairwise XORs of every pair of plaintexts can be calculated.[12] The attacker can therefore collect enough frames that are encrypted with the same IV, and eventually he or she will be able to deduce the keystream or even the shared secret key. From that point on, the attacker can decrypt any subsequent 802.11 frame.

To make matters worse, the WEP standard recommends — but does not require — changing the IV for every frame. Further, it does not say anything about IV selection. Many implementations select IVs poorly. For instance, some implementations initialize their IVs to zero every time the system is power-reset and then increment it for every frame. This can result in reuse of IVs even before all possible IVs are exhausted. Selecting random IVs makes matters worse due to the birthday paradox, and collisions can be expected as soon as 500 packets are generated.

Finally, we want to point out that if a malicious receiver uses not-so-old IV values, there is no protection provided by WEP to detect it.

### Static Keys

802.11 does not provide automatic key renewal among stations. As a result, users may use the same keys for months, and even years. This leaves the shared secret keys unchanged for extended periods, and makes the problem mentioned in the preceding subsection worse. The attacker will have enough time to collect samples of IVs and ciphertext for successful cracking of the key.

### Poor Choice of ICV Algorithm

The 4-byte ICV checksum is concatenated at the end of the input plaintext. This is a CRC-32 linear checksum, not a cryptographic checksum like MD5 or SHA. This leads to further vulnerability, as initially discovered in Reference 13. Due to linearity, we have $h[\{p_1\} \oplus \{p_2\}] = h[\{p_1\}] \oplus h[\{p_2\}]$. This linearity property, coupled with the fact that each RC4 bit is encrypted independently (and therefore is also linear), makes it is possible for the attacker to launch an active attack. The attacker can introduce bitwise modifications in the plaintext and compute the new encrypted checksum without knowing the keystream, while making the receiver believe that he or she has received an uncorrupted message.

Let us assume that the attacker captures ciphertext $V \mid \mid \{c\}$, where $\{c\} = E_s[\{p\} \mid \mid h[\{p\}]]$, and that the plaintext $\{c\}$ is not known to the attacker. We also assume that although the attacker does not know the plaintext, he or she knows which bits are to be modified in the message. To do so, the attacker creates a message $\{e\}$ of the same size as $\{p\}$, with the bits to be modified set to 1 and all other bits reset to 0. Now the modified message can be expressed as $\{p'\} = \{p\} \oplus \{e\}$. He or she now has to construct a ciphertext $\{c'\}$ that decrypts to $\{p'\}$.

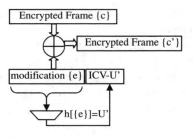

**Figure 5.11   Bit-flipping attack on WEP.**

$$\{c'\} = E_s[\{p'\} \mid\mid b[\{p'\}]]$$

$$= \{r\} \oplus (\{p'\} \mid\mid b[\{p'\}])$$

$$= \{r\} \oplus (\{p\} \oplus \{e\} \mid\mid b[\{p\}] \oplus b[\{e\}])$$

$$= \{r\} \oplus (\{p\} \mid\mid b[\{p\}]) \oplus (\{e\} \mid\mid b[\{e\}])$$

$$= (\{r\} \oplus (\{p\} \mid\mid b[\{p\}])) \oplus (\{e\} \mid\mid b[\{e\}])$$

$$= \{c\} \oplus (\{e\} \mid\mid b[\{e\}])$$

This shows how simple it is to generate the modified encrypted frame. Using this bit-flipping attack, several attacks can be launched. One example is launching a known plaintext attack. For instance, consider the scenario where the attacker first flips a few bits in a frame, as shown in Figure 5.11. The IEEE802.11 receiver believes that the frame is original. This packet eventually reaches an upper layer such as the TCP layer on another station. This layer eventually detects the corrupted message through its own error-handling mechanisms. It returns a high-level error message to the sender that is highly predictable and can be captured by the attacker to deduce the keystream. This attack is shown in Figure 5.12.

### Message Insertion Attacks

The reuse of old IVs or keystreams does not raise any alarms, nor is there any protection against the premature use of IV values. As a result, if an attacker ever gets hold of the entire plaintext of any frame, he or she can inject arbitrary traffic into the network. This can be accomplished by first recovering the keystream $\{p\} \oplus \{c\} = \{p\} \oplus (\{p\} \oplus \{r\}) = \{r\}$ and then constructing new messages $\{p'\} \oplus \{r\} = \{c'\}$. The process can be repeated indefinitely.

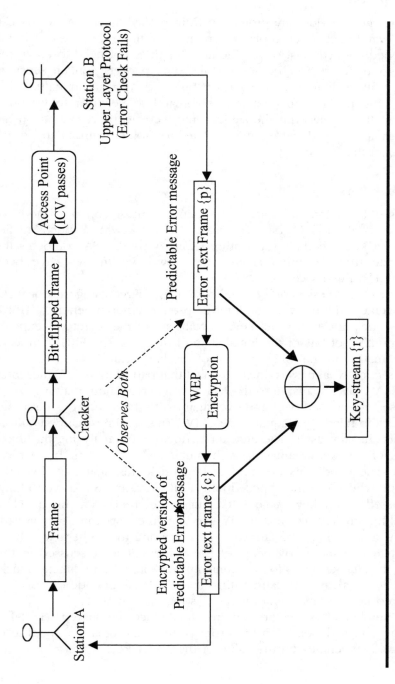

**Figure 5.12  Deducing the keystream by inserting an error.**

## Replay Attacks

WEP does not provide any protection from replay attacks. Attackers can record a valid packet and replay it a number of times, hoping that it will be decoded correctly just once. The attacker does not need to know the shared secret for this to be successful. Some packets contain sensitive data that fits well in just one packet but is dangerous in nature. For instance, the packet can contain a command to delete a folder on a computer. If a folder with the same name were to be recreated on the computer, a replayed packet may delete that folder without the user's approval in the future.

## RC4 Weak Keys

RC4 has a large internal state, which, in the most common mode of operation, is about 1700 bits. To make it to work with a much smaller key size, a key scheduling algorithm (KSA) is used. KSA computes the initial state from an input smaller key. Two weaknesses have been discovered in this process.[14]

The first weakness is due to the existence of large categories of weak keys, together with the way the internal pseudo random generator (PRG) of RC4 works. For each weak key, a small part of the secret key exposes a large number of bits of the initial state of the RC4. The PRG then leaks this information in the initial portion of the keystream {r}.

The second weakness is due to the fact that part of the key V presented to the KSA is also available to the attacker, whereas the remaining part S is kept secret but used over and over again with the same V, as discussed in the section titled "Inadequate Size of IV." This weakness can be utilized to extract the key itself. The attacker deduces S by analyzing the initial portions of the keystreams $\{r\}_V$ with much less work than brute force analysis. The conditions are perfect to launch this attack in the WEP algorithm, where V is the exposed part of the key, and S is a (very) long-term shared secret key. As pointed out in Reference 14, this attack is practical for any key size and any IV size in a time that grows only linearly with key size, and can be launched by observing the ciphertext only.

To take maximum advantage of the vulnerabilities discussed in this section, the attacker waits for a potentially weak key K and directly attacks it. Actually, it is easier to launch this attack if V is appended in front of the shared key S, which unfortunately is the case in WEP.

The number of IV vectors required to deduce the shared secret key using this attack depends on the order of the concatenation, the size of the IV, and sometimes on the value of the secret key.

## Other Weaknesses

There are other lackluster weaknesses in WEP, making it inadequate for protecting networks containing sensitive information. WEP only provides a method for authenticating stations to the APs and does not enable the MS to authenticate the AP. As a result, a hacker can reroute data through an alternative unauthorized path that avoids other security mechanisms. Instead of one-way authentication, wireless LANs need to implement mutual authentication to avoid this problem.

## Temporal Key Integrity Protocol (TKIP) and the Wi-Fi Protected Access (WPA) Interim Solution

To address the weaknesses discussed earlier, IEEE is working to define a new appendix to 802.11. This standard is called IEEE802.11i. It defines a Robust Security Network (RSN). RSN is a complicated standard, and while it was being worked out, the manufacturers decided to implement a subset of this standard called WPA. WPA in turns relies on TKIP.

The attacks described above mostly rely on the relatively static key, poor ICV algorithm, and short IV. For encryption to be effective, it must minimize the reuse of the shared secret key S. This can be done by:

1. Renewing shared secret keys often, possibly for every frame transmission.
2. Using a cryptographic hash function for the ICV algorithm that is also a function of a shared secret between the station and the AP.
3. Letting the RC4 algorithm run for some time before extracting the keystream {r}.
4. Using a larger IV value, which reduces the possibility of repetition.

This decreases the time available for a hacker to break into the network and makes it much harder, if not impossible, to compromise the security of the network.

IEEE 802.11i provides two more privacy methods: TKIP and CCMP. TKIP is just an enhancement of the encryption algorithm, which fixes the problems noted in the preceding text. Thus, both WEP and TKIP are based on the RC4 algorithm, but TKIP works by improving the utilization of RC4. To reduce the attack probability, it periodically renews the encryption keys. However, the CCMP is based on the Advanced Encryption Standard (AES).

TKIP is designed so that existing WEP-based APs and stations can be upgraded through relatively simple firmware patches, and new or

← Extension IV →

| TSC1 [8] | WEP Seed [8] | TSC0 [8] | Pad [5] | Ext. IV [1] | Key ID [2] | TSC2 [8] | TSC3 [8] | TSC4 [8] | TSC5 [8] | Data 8xn; n ≥ 1 | MIC [64] | ICV [32] |
|---|---|---|---|---|---|---|---|---|---|---|---|---|

**Figure 5.13  TKIP frame format.**

upgraded equipment can still interoperate with TKIP-enabled devices. TKIP only fixes the confidentiality and integrity services; it does not provide authentication service and should only be considered as an interim measure.

TKIP uses a 128-bit temporal key that is shared between the station and the AP. In addition to the ICV, TKIP specifies a keyed cryptographic Message Integrity Code (MIC). The transmitter computes and appends this code to the frame before the ICV (Figure 5.13). The receiver verifies the MIC after decryption. Because this MIC is keyed by a key only known to the transmitter and receiver, message integrity code protects against forgery attacks.

To defend against replay attacks, a TKIP transmitter adds a TKP Sequence Counter (TSC) T, so that the receiver can drop frames that are received out of order. This 48-bit number also serves as an IV.

The temporal key is combined with the station's MAC address and the TSC, using a special cryptographic message integrity code called *Michael*, as shown in Figure 5.14. The receiver recovers the TSC from the received frame (the frame format is shown in Figure 5.13) using the same MIC function, and computes the key needed to correctly decrypt the frame. The key-mixing function is designed to counter weak-key attacks.

Effectively, TKIP uses a much larger IV to produce the data encryption key. This ensures that each station has a different keystream even when the temporal key is the same. The temporal keys are changed after every 10,000 packets. New keys are distributed dynamically. TKIP reuses the existing WEP format shown in Figure 5.9. It specifies another 12 bytes — 4 bytes are specified for an Extension IV (EIV) immediately after the IV field and 8 bytes for the new MIC just before the ICV. The MIC is also encrypted.

The WEP seed field is set to a value that is a function of TSC1 and TSC0, and it is designed to avoid weak keys. The EIV bit indicates the presence or absence of an extended IV. This bit is set for TKIP and is clear for regular WEP frames. This allows for backward compatibility with WEP equipment. TSC5 is the most significant byte of the TSC, whereas TSC0 is the least significant byte. The TSC T can be considered as a concatenation of two parts; we will use the notation $T_H \mid\mid T_L$ to indicate this. TSC1 $\mid\mid$ TSC0 forms the lower IV sequence number ($T_L$), which is used with the TKIP phase-2 key mixing, and TSC5 $\mid\mid$ TSC4 $\mid\mid$ TSC3 is

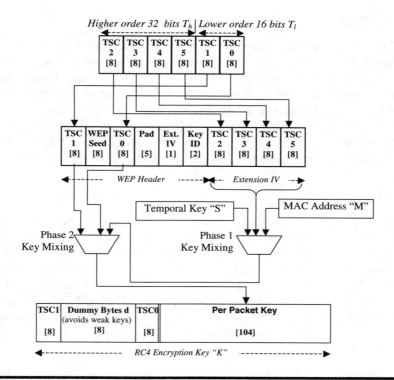

**Figure 5.14   Relationship between TCS, MAC, session key, and the RC4 encryption key.**

the upper IV sequence ($T_H$), which is also used with TKIP phase-1 key hashing. When the lower 16-bit sequence in TSC1 || TSC0 rolls over, a new data encryption key is calculated, and the upper 32 bits in TSC5 || TCS4 || TCS3 are incremented by 1. When this number rolls over, a new phase 1 value is calculated.

## Keys Used in TKIP

Three layers of keys are used in TKIP. The devices are programmed with a long-term shared or Pairwise Master Key (PMK) M, or it is established using a high-level key-agreement algorithm. From this key, the temporal encryption key, temporal receiver MIC key, and temporal transmitter MIC key are derived. These temporal keys are updated every 10,000 packets. However, the temporal encryption key is not directly used for encryption. It is only used as an input to the phase-1 key-mixing stage. For every packet, a new one-time-use key called the per-packet key is computed as shown in Figure 5.14.

## MIC in TKIP

TKIP's MIC algorithm, Michael, offers only weak defenses against fake messages. It is designed to work with the majority of existing hardware, and is the best that could have been achieved with this constraint. To provide protection from weaknesses, TKIP defines some countermeasures. It uses a temporal key that is different from the encryption key. To protect against more sophisticated attacks, TKIP uses different MIC keys for the transmitter and receiver. TKIP computes the MIC on the payload before it is fragmented or encrypted. Michael generates a 64-bit MIC.

## 802.11i

The 802.11i standard can be considered to have three main components stacked in two layers. On the lower level are improved encryption algorithms. Two options are specified in addition to WEP: one is TKIP and the other is the stronger Counter mode with the CBC-MAC Protocol (CCMP). TKIP is intended for legacy equipment and CCMP for future WLAN equipment. The upper layer specifies IEEE 802.1X — a generic standard for access control. IEEE 802.1X is a general-purpose standard. 802.11i uses 802.1X to authenticate users and manage encryption keys. We have already described the TKIP standard.

## CCMP

CCMP is an encryption method that uses AES.[15] AES is a symmetric block encryption algorithm that can be used in a number of different modes or algorithms. 802.11i specifies that it be used in the Counter mode with CBC-MAC (CCM). For encryption, the data is divided into 128-bit blocks, and each data block is passed through multiple encryption rounds. AES provides multiple key-size options for varying security needs; however, 802.11i specifies a key length of 128 bits. CCMP is mandatory for any device claiming to be 802.11i-compliant. Figure 5.15 shows the format of an AES-encrypted frame. 16 bytes are appended to the original data. The AES-encrypted frame is identical to a TKIP frame, the only difference being that the legacy WEP ICV is not included any more.

CCMP also uses a 48-bit IV, but it is called a packet number (PN). This number is used together with other information to initialize MIC calculation and frame encryption. The CCMP algorithm is designed in such a way that the AES encryption blocks are used for both MIC calculation and packet encryption. The two processes proceed in parallel and also use the same temporal key K, which is derived from the master key. The master key can be established as part of the 802.1X exchange (discussed later), or it can be configured manually.

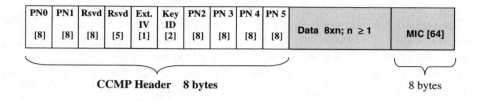

| PN0 [8] | PN1 [8] | Rsvd [8] | Rsvd [5] | Ext. IV [1] | Key ID [2] | PN2 [8] | PN 3 [8] | PN 4 [8] | PN 5 [8] | Data 8xn; n ≥ 1 | MIC [64] |
|---|---|---|---|---|---|---|---|---|---|---|---|

CCMP Header    8 bytes                                                      8 bytes

**Figure 5.15    CCMP-encrypted frame format.**

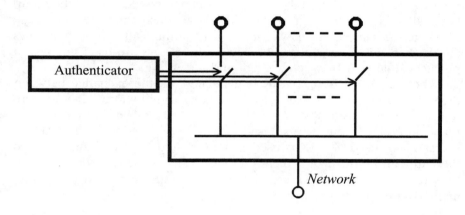

Authenticator

Network

**Figure 5.16    Authenticator blocks unauthenticated supplicants from access network.**

## Overview of 802.1X

IEEE 802.1X is a standard for both wired and wireless networks. It restricts the access of a station to the network until the user has been authenticated by the network. 802.1X does not reinvent the wheel, and is designed to work with a number of other existing authentication protocols that verify credentials and generate encryption keys. It also defines an infrastructure with three roles: (1) an authenticator to enforce authentication, (2) a supplicant (usually the station) to request access to the network, and (3) an Authentication Server (AS) to actually authenticate the credentials of the supplicant. The physical point at which the supplicant connects to the network is called a *port*. Each port is connected to only one supplicant and vice versa. A port is also associated with an authenticator that controls its state. The state of the port is either open or closed. As shown in Figure 5.16, the authenticator changes the port state to open only if the supplicant has been authenticated.

In the simplest implementations, functionality AS also resides in the authenticator. In moderate and large-sized networks, AS is a separate entity on the wired side of the AP. 802.1X adapts IETF's Extensible Authentication Protocol (EAP)[16] for authentication so that it can be used over LANs. This adaptation is called *EAPoL*. However, we use the acronym EAP for simplicity. EAP may rely on protocols such as Kerberos, RADIUS, and PKI for actual authentication. These protocols are used by regular ASs that already exist on the enterprise network. 802.1X does not bind the organization to use any specific AS.

As in WEP, two sets of keys are generated in 802.1X-compatible systems: the session keys (also referred to as pairwise keys) and the group keys (also referred to as groupwise keys). Stations connected to a single AP share the same group keys for multicast traffic. Different session keys are used for different stations, however. These keys are used to create a private virtual port between a client and the AP.

When an AS is used, a PMK is generated via the exchange between the station and the AS. Other lower-level keys used by MAC layer encryption are derived from the PMK. This is done to ensure that the PMK is not compromised due to extensive use. When an AS such as RADIUS server is not used, 802.1X is used in a preshared key configuration, and the PMK can be manually entered into each station. This model is similar to WEP. The session keys are still used, and the improved encryption methods are still fully supported.

### *Authentication Protocols*

Upper-layer authentication protocols are not specified in the 802.11i standard. However, they are an important consideration in any significant WLAN deployment. Some popular authentication protocols include EAP with Transport Layer Security (EAP-TLS),[17] Protected EAP (PEAP), EAP with Tunneled TLS (EAP-TTLS), and Lightweight EAP (LEAP). Tunneled Transport Layer Security (TTLS), Protected EAP (PEAP), and EAP-Transport Layer Security (EAP-TLS) are all TLS-based methods. LEAP is an exception, but it was one of the first authentication protocols for wireless networks developed by Cisco, and so it is quite popular. LEAP is based on a challenge-password hash exchange, in which the AS issues a challenge to the client after which the client returns the password to the AS after first hashing it with the challenge text.

EAP-TLS is a certificate-based authentication protocol and is easy to use with servers that already support Web-based certificate-authentication systems. The network administrator has to program the initial certificates on the station and AS, and no further intervention is required. To establish the shared PMK, the station and AS take part in an EAP-TLS handshake.

The station and AS exchange credentials and random data that is used by them to compute PMK. The AS then sends the encryption keys to the AP through a secure channel. Finally, the AP exchanges messages with the station and both derive temporal encryption keys to be used for lower layers.

PEAP and EAP-TTLS are IETF drafts and are not commonly used in products. They also use TLS but make it possible for stations to authenticate without certificates.

## BLUETOOTH

The company L.M. Ericsson originally conceptualized the Bluetooth protocol to replace cables connecting devices over very small distances (such a network is sometimes called a Personal Area Network [PAN]). Now it is expanding in scope to compete with small-distance applications of 802.11. IBM, Intel, Nokia, and Toshiba were the first supporters, but now IEEE is using it as the basis for its 802.15.1 standard.

### Protocol Overview

A fundamental unit of networking in Bluetooth is the piconet, with one master node that controls up to 255 slave nodes within a distance of 10 m. Up to seven slaves can be active simultaneously. The rest are kept in the low-power parked state in which the devices can respond to beacon signals and activation commands only. Multiple piconets can be connected with a bridge slave node to form a scatternet. Piconet is a TDM system in which the master controls the clock, and slaves are allocated time slots to communicate with the master. Slaves cannot communicate directly with each other.

From a high-level application development perspective, Bluetooth specifies 13 profiles, with more planned. The generic profile is the simplest, and other profiles are built on top of it. It establishes, maintains, and secures links. The other relatively simple profile is the service discovery profile that finds out what services other devices are offering. Generic and service discovery profiles are mandatory for all Bluetooth devices. Of the other profiles, the serial profile is the most interesting because it defines a transport protocol and emulates a serial line. Many legacy applications that expect a serial interface use this profile. Other important profiles are the file transfer and the headset profiles.

From a communications engineering perspective, Bluetooth can be regarded as several layers of protocols. Unlike the TCP/IP stack or ISU model, these protocols are not stacked in a well-defined manner. However, for understanding the security provided by Bluetooth, it is important to

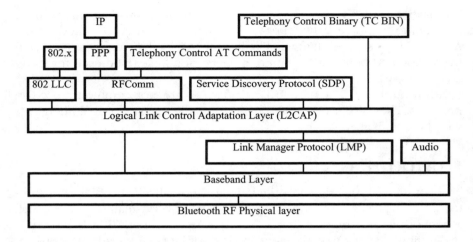

**Figure 5.17    Bluetooth protocol stack.**

understand the approximate relationship between the layers depicted in Figure 5.17.

The lowest layer deals with the radio transmission, modulation, and bit-error issues. It uses a low-power signal (1 mW). To avoid interference from other devices operating in the same spectrum (2.4 GHz), it uses fast frequency hopping. The band is divided into 79 channels of 1 MHz each, with FSK modulation giving 1 bit/Hz. A significant portion of this capacity is consumed by the overhead, giving a maximum data throughput of 780 kbps. All nodes in a piconet hop synchronously. The hopping rate is 1600 hops/s and is based on a sequence dictated by the master. When a node joins a piconet, the master provides a key for the hopping sequence so that node can synchronize.

The baseband layer supports inquiry, paging, and synchronization between different Bluetooth devices. The master's baseband layer controls the time slots, which are numbered from 0 to $2^{27}$ 1. The master usually transmits in the even numbered and slaves in the odd numbered time slots. These slots are grouped into frames that are composed of either one, three, or five slots. Time slots are 625-ms long, during which at most 625 bits can be transmitted. Consecutive time slots belonging to different frames need a 160-ms settling time between them. The simplest frame has only one time slot. In such a frame, 259 bits are unusable due to the settling time, and only 366 bits are usable. Further, the frame inserts a 126-bit access code and header overhead. The remaining 240 bits can be used for data. At the other extreme, the longest frame has 3125 (5 × 625) total possible bits, of which many are lost due to the settling time and other physical layer over heads and only 2781 bits are available.

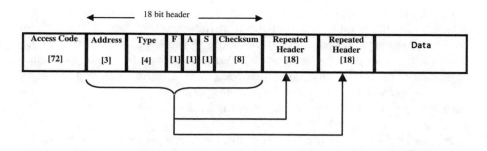

**Figure 5.18    Bluetooth baseband frame format.**

Figure 5.18 shows the structure of the baseband frame. The address field identifies one of the eight devices that are currently active in the piconet (one master and seven active slaves). The type field specifies, among other things, the length of the frame (1, 3, or 5). Figure 5.18 also shows an eight-bit checksum, which is computed over header bits only.

Each frame is sent over a link that is established by the LMP as a logical channel. The LMP also handles power management, quality of service (QoS), negotiation of baseband packet size, and security issues including authentication, encryption, and generation, exchange, and verification of keys. Security is either enabled at the time of link establishment, or it can be enabled by the upper-layer application any time after link establishment. In either case, LMP handles the details of securing the link. Four types of frames can be defined in the type field of Figure 5.18: Asynchronous Connectionless Link (ACL), Synchronous Connection Oriented (SCO), poll, and null. We will discuss only two, ACL and SCO. SCO frames carry application data within ACL and SCO links, as discussed in the following paragraph. The type field also specifies the size of frame and type of error correction used.

Two types of links are used by the upper layers: the SCO link and the ACL. SCO is used for real-time data such as voice connections in which timely delivery is preferred over accuracy. The SCO frames are allocated fixed time slots in each direction. There are no retransmissions even when frames are detected to be corrupt. Error correction codes can still be used to reduce bit-error probability. The audio layer uses SCO links, and a slave can have up to three SCO links to the master.

ACL is used for packet-oriented data where timely delivery is less important than accuracy. Frames on this link are handled in a manner similar to IP packets, where every effort is made to deliver the packets to the destination but the delivery is not guaranteed.

The L2CAP uses ACL channels. The L2CAP layer in a slave device can establish only one ACL channel with the master. LC2CAP hides the upper layers from details of the lower layers by accepting 64-KB packets and fragmenting them into frames at the transmitter, while reassembling them at the receiver to present them back to the upper layer as packets. L2CAP also routes the packets so that they are delivered to the appropriate destination protocol. This layer provides flow control by using some of the flags carried in the frame header (Figure 5.18). L2CAP uses a stop-and-wait protocol, in which the transmitter sends a frame and waits for an ACK flag to be set to 1 in the next frame — which acts as a piggyback acknowledgment — before sending the next. It also looks at the flow bit F, which, if set to 1, means that the receiver's buffer is full and the transmitter should stop sending more frames to that device until it "sees" F reset to 0 in a returned frame from the same device. The sequence bit S is used as a frame sequence number. In the stop-and-wait protocol, a single-bit sequence number is enough to detect retransmissions. Hence, S is used for this purpose. The access code field in the frame header (Figure 5.18) identifies the link; frames that are exchanged on a link of a piconet all have the same channel access code.

SDP is used to find services and their characteristics on the network. Applications use it when they are setting up communications with another Bluetooth device. The 802 LLC layer was added by IEEE in 802.15.1 to provide compatibility with other 802-based networks. The Cable Replacement Protocol (RFCOMM) emulates a serial line for legacy applications that were designed to use the serial interface. The most interesting applications are the ones that use PPP and the regular modem AT command sets. The Telephony Control — Binary (TCS BIN) is used to establish speech calls between devices and to control mobile phones.

## Security

Security is of lesser concern in Bluetooth compared with WEP, because the attacker has to get close to the master, increasing the chances of detection. However, a miniature device that attacks and transmits critical information on an alternative channel is a possibility. Keeping this in view, Bluetooth defines a sophisticated security architecture. Bluetooth applications can use one of three security modes. Security mode 1 provides no security at all and is also called *discoverable mode*. Here, the devices do respond to inquiries from not-yet-trusted devices. One use of this mode is to allow new devices to join a piconet group. In security mode 2, security procedures are initiated after channel establishment at the L2CAP level is complete. This mode allows each application to have a distinct security policy, and is more flexible than mode 3. It is enforced at the service level. In security mode 3, a device initiates security procedures before the link setup on the LPM level is completed. This is the most secure mode.

As is clear from the previous paragraph, security in Bluetooth is available at multiple layers. Some are of the opinion that this starts at the physical layer, where the signal is spread over many frequencies (frequency hopping), using a modulating sequence. However, the device obtains this sequence over the channel from the master, and so it is not a secret.

## Link Security Overview

Each Bluetooth device is assigned a unique 48-bit IEEE address that is public but unique for each unit. This number is used in many security procedures discussed here. To set up a secure channel, the master and slave each check if the other has the shared secret key. This is done with an authentication protocol, without ever exchanging the key. After successful authentication, they negotiate the security services: authentication, encryption, or both. Then a 128-bit link key is established.

To simplify description, we define a *Bluetooth session* as the period during which a device remains a member of a particular piconet. The session terminates when the device disconnects from the piconet. The security in a session is based on the current link key for that session.

For encryption, Bluetooth uses the $E_0$ stream cipher, whereas SAFER+ is used for integrity control. To encrypt, the plaintext is XORed with the keystream produced by a stream cipher, just as in the WEP protocol. Weaknesses in $E_0$ have been identified, which are summarized in Reference 18.

## Link Keys

Security in Bluetooth is completely controlled by 128-bit symmetric keys called *link keys*. A link key is directly used for authentication, and is also used as one of the inputs when computing the connection encryption key.

From the life-cycle point of view, there can be two types of link keys, semipermanent and temporary. A semipermanent link key is used for multiple sessions and is stored in the device. The temporary link key, on the other hand, is discarded after a session and is renegotiated for every session. The current link key being used in a session can be either the semipermanent key or the temporary key. To accommodate different needs of various applications, four types of link keys are specified in version 1.2 of the standard: the combination key $K_{AB}$, unit key $K_A$, $K_{master}$, and $K_{init}$.

The unit key $K_A$ is the simplest to understand. It is dependent on a single device and could be hard-coded into the device, or it can be generated during the very first installation of the device. In either case, it is long term and rarely changes. This makes security derived from the unit key implicitly weak. Devices with limited resources and memory can use unit keys so that they have to store only one link key. Devices that are shared by a large number of other devices may also opt for this; or else they will have to store a combination key for every other device. The combination key $K_{AB}$ is derived from information in both the devices A and B. It is used in the same way as $K_A$, but it provides better security.

In a point-to-multipoint piconet, some information must be distributed securely to several slaves. In such situations, a link key called master key ($K_{master}$) is used as a replacement for the current link keys in all the devices. When the master wishes to broadcast, it generates the $K_{master}$ and sends it to all the slaves, with a command asking them all to make $K_{master}$ their current link key. This key then remains valid until the termination of the current connection.

The initialization key — $K_{init}$ — is used to protect a link during the initialization process when no other keys are available or when a link key has been lost. This key is derived from a random number, a PIN code, and the 48-bit IEEE device address (BD_ADDR). The PIN is either a fixed number hard-coded into the device that has no user interface, or it can be entered by the user at the time of initialization of both devices. If no PIN is available, a default value of zero can be used. Entering a PIN in both devices provides the most secure means of initialization and protects against some of the attacks mentioned in Reference 18. The PIN code can also come from applications that could generate it based on other application-level key-agreement protocols such as the Diffie–Hellman algorithm.

### Initialization

When no link key is available between two communicating devices, the LMP automatically starts initialization. Each slave device in the piconet is initialized separately. The initialization consists of the four steps:

1. Generating the initialization key $K_{init}$
2. Generating a link key (either $K_A$ or $K_{AB}$) and exchanging these keys
3. Authentication
4. If the encryption is enabled, generating the encryption key from the current link key in each device

Please note that the initialization is also possible without security, which facilitates operation in mode 1 as discussed in the section titled "Security." The generation of $K_{init}$ is a function of BD_ADDR, the user-provided or hard-coded PIN, and a random number IN_RAND sent by one of the devices. The PIN is the only protection against an attacker who is capable of observing the initialization messages.

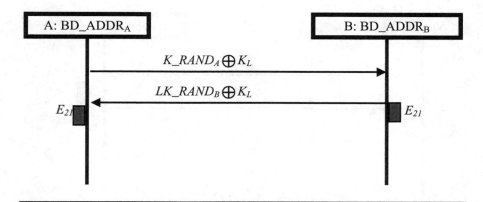

**Figure 5.19 Exchanging random numbers for computing combination key.**

After the initialization, whenever a new session is established, the already-established link key is used for authentication. However, the encryption key $K_C$ is always renewed.

### Unit Key $K_A$

The unit key $K_A$ is either hard-coded or generated when the device is used for the first time, and not during the initialization of a device in the piconet. When it is not hard-coded, the key is generated locally as a function of BD_ADDR and a random number. This function is the Bluetooth's E21 algorithm. The generated key is then sent to the other device by XORing it with $K_{init}$. The application decides which of the two devices' unit key will be used as the link key. If the unit key has to be changed, all devices using its key as the link key must be reinitialized.

### Combination Key $K_{AB}$

To compute the combination key, each device first generates a random number and sends it to the other device securely by XORing it with an existing link key (which could be $K_{init}$), as shown in Figure 5.19. All message exchanges shown in this chapter are LMP data units. Once the random numbers have been exchanged, the combination key is computed as follows using Bluetooth's $E_{21}$ algorithm:

$$LK \_ K_A = E_{21}(LK \_ RAND_A, BD \_ ADDR_A)$$

$$LK \_ K_B = E_{21}(LK \_ RAND_B, BD \_ ADDR_B)$$

$$K_{AB} = LK \_ K_A \oplus LK \_ K_B$$

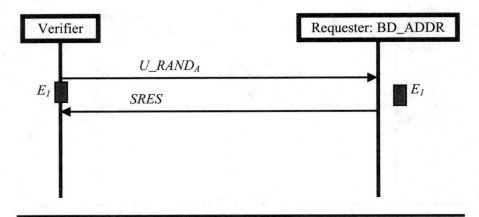

**Figure 5.20 Bluetooth link authentication handshake.**

## Encryption Keys

An encryption key $K_C$ is used for providing the confidentiality service. When encryption is enabled by the LMP, a new encryption key is generated, the size of which is kept flexible. However, unlike the authentication key, the encryption key cannot be controlled by the application. This lack of control over the encryption key is built into the design to meet legal regulations that may apply in different parts of the world.

$K_C$ is computed using Bluetooth's $E_3$ algorithm in one of two possible ways. If the current link key is not a master key ($K_{master}$), $K_C$ is computed as a function of the authenticated ciphering offset (ACO) (see Figure 5.21), the current link key, and a 128-bit random number EN_RAND. Otherwise, $K_C$ is computed as a function of $K_{master}$, a 96-bit number that is BD_ADDR repeated twice (BD_ADDR || BD_ADDR), and EN_RAND.

$$K_C = \begin{cases} E_3(K_L, ACO, EN\_RAND); & K_L \neq K_{master} \\ E_3(K_{master}, BD\_ADDR \,||\, BD\_ADDR, EN\_RAND); & K_L = K_{master} \end{cases}$$

## Authentication

Figure 5.20 shows the authentication protocol. Both the verifier and requester know the 48-bit address of the requester. They must have a shared link key. If that is not the case, $K_{init}$ is established as described in the section titled "Initialization." The master sends a challenge AU_RAND$_A$ to the requester. The requester computes a signed response (SRES) from AU_RAND$_A$, BU_ADDR$_B$, and the link key $K_L$, using the algorithm $E_1$ shown in Figure 5.21. The SRES is sent to the verifier. $E_1$ also produces an ACO that can be used as input to the encryption key generation process. Once

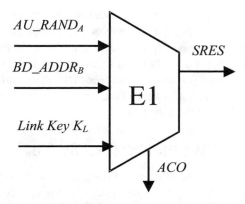

**Figure 5.21  Bluetooth's E1 algorithm used for authentication.**

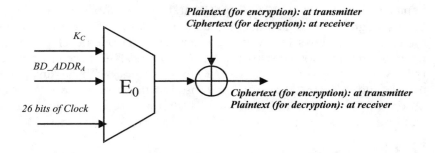

**Figure 5.22  Encryption process.**

the verifier receives the SRES, it can be recomputed locally and compared with the one received. If the two match, the requester has proven that it has the link key (without ever sending it).

Authentication can be invoked by the LMP layer during connection establishment when operating in security mode 3. Authentication can also be explicitly executed any time after a connection has been established. This would be the case with operation in security mode 2, in which the application has more flexibility of security.

## Encryption

Each frame is encrypted separately after the CRC bits are appended but before the error correction code's redundant bits are appended. Header and access code bits are not encrypted either. The encryption algorithm $E_0$ uses the device address (BD_ADDR), 26 bits of the master real-time clock, and the encryption key $K_C$ as input, as shown in Figure 5.22.

As with any other stream cipher, the encryption algorithm generates a binary keystream $\{k_{cipher}\}$. However, to protect against obvious attacks, before using the key $K_C$ for actual keystream generation, the $E_0$ algorithm modifies the encryption key, generating another key $K_C$. A new $K_C$ is generated for every frame (it is different for every frame because it depends on the clock). The maximum effective size of this key is between 1 and 16 bytes, and it is set at the factory.

## CELLULAR NETWORK PROTOCOLS: GSM

GSM was originally introduced in Europe, but is now being widely deployed all over the world. If we consider end-to-end cellular communications security, we need to discuss the entire cellular network architecture, which includes the cell phone (also called the Mobile Station–MS), the Base Transmission Station (BTS), Base Station Controller (BSC), the Mobile Switching Center (MSC), Gateway MSC (GMSC), and the backbone public network. Even though there are attempts to define mobile-specific security protocols for each of the links between these entities, it is our belief that generic network security protocols such as IPSec will prevail in those environments. Here, we will only concern ourselves with the security of the air link between the MS and the BTS because it is exposed to the widest possible range of observers. Other links are established within a service provider's network, and one can assume that they can be trusted to a certain extent.

### Overview of GSM

GSM provides GPRS service, which provides Internet and other data services to GSM subscribers. Compared with earlier cellular standards, GSM is capable of serving many more subscribers with the same bandwidth and power, using the spectrum efficiently with a combination of Frequency Division Multiple Access (FDMA), Time Division Multiple Access (TDMA), efficient speech coding, and modulation schemes. The GSM standard specifies frequency bands near 900 MHz, 1800 MHZ, or 1900 MHz, depending on the flavor of the standard being used. The basic channel at a particular frequency is subdivided into 577-ms time slots, with 8 time slots constituting a TDMA frame of 4.6 ms. Multiple frames are grouped into multi-frames in a hierarchy, with the hyperframe as the largest structure. Within a hyperframe, each TDMA frame is identified by a 22-bit sequence number.

### GSM Security

The built-in security of GSM is designed for call privacy and integrity of billing data. The security involves three entities, the Subscriber Identity

Module (SIM), MS, and the network. The SIM is a smart card with an onboard processor that is capable of performing cryptographic computations. The subscriber's anonymity is ensured by temporary identification numbers. The confidentiality of the communication itself on the radio link is ensured by encryption algorithms. The MS is identified with a temporary Mobile Subscriber Identity (TMSI) issued by the network, and it changes during handoffs for additional security. The owner of the MS is uniquely identified by the international Mobile Subscriber Identity (IMSI).

IMSI and a subscriber authentication key $K_i$ are maintained in the SIM, and are used for authentication based on challenge–response in such a way that IMSI or $K_i$ are never exposed on the channel. Traffic over air is encrypted with a temporary randomly generated key $K_C$. The SIM also stores the PIN of the owner.

GSM uses three security-related algorithms, which are named A3, A5, and A8. Together they provide authentication, confidentiality, and integrity services. A3 is used during authentication, A5 is a stream cipher that is used for encryption, and A8 is the key-agreement algorithm. Algorithms A3 and A8 are not specified in the specification document; only the interfaces are specified. It is left to the operators to select a particular algorithm. However, the vast majority of operators use the officially unpublished COMP128 algorithm. This algorithm has been leaked and cryptanalysis performed[19] and the GSM industry has attempted to repair it, releasing (still unpublished) COMP128-2. Two versions of A5 (A5/1 and A5/2, with A5/2 being the weaker version for export) were part of the original GSM specification, but the algorithm was never made public. These algorithms have been reverse engineered as well as attacked. Another algorithm A5/3 has been standardized and is based on the KASUMI algorithm.[20]

The GSM network contains an Authentication Center (AC) that stores the IMSI, TMSI, and $K_i$ for each subscriber. It also maintains a current location identifier for the subscriber Location Area Identity (LAI). After authentication has taken place (as discussed in the following subsection), i.e., the MS is trusted by the network, optional encryption (discussed in the subsection titled "Encryption") is invoked, and the network generates the TMSI and sends it to the MS over the secured channel. The TMSI remains valid in the location area (identified by LAI) in which it was issued. When the MS moves outside this area, LAI has to be used in addition to the TMSI.

## Authentication

For challenge–response authentication, the network generates and sends a 128-bit random number (RAND) and sends it to the MS. The MS sends this RAND to the SIM, which uses it along with $K_i$ as input to the A3 algorithm to compute a 32-bit number SRES, i.e., $SRES = A3(RAND, K_i)$. The MS returns this SRES to the network as a response to the challenge. The network repeats the calculation after obtaining $K_i$ from the AC to verify the identity of the subscriber. The standard does not specify that the RAND should not to be repeated by the network. The MS does not verify if it has seen the same RAND before, which introduces a weakness into the system.

### Encryption

SIM computes a 64-bit encryption key, using $K_i$ and RAND (previous subsection) as inputs to the algorithm A8, i.e., $K_C = A8(RAND, K_i)$. $K_C$ is used to encrypt and decrypt the data between the MS and BS. $K_C$ can be changed at regular intervals determined by the network's security policy. Actual encryption of voice and data traffic is done with A5 algorithm. The network can invoke the encryption by sending a command to the MS.

### GPRS Security

GPRS uses the same security model as GSM. However, a greater choice of algorithms is provided. Seven GPRS Encryption Algorithms (GEA) are allowed. The algorithm is negotiated at the start of the call.

## CELLULAR NETWORK PROTOCOLS: CDMA

CDMA uses a 42-bit code to seed a shift register that generates a pseudo random sequence of length $2^{42}$ 1. This sequence is used to spread the signal over a wide band (hence the name spread spectrum), making it look like random noise. This spreading of the signal also effectively scrambles the transmissions. The receiver must have the correct code to extract the data in transmission. Thus, CDMA makes eavesdropping inherently very difficult by design.

### Security

CDMA uses a 64-bit A-key, the electronic serial number (ESN) of the MS, and a random binary number called RANDSSD, for authentication. The A-key is stored in the MS and AC of the network. Protecting the A-key from exposure is essential for security of the CDMA system. The A-key is also used as one of the inputs in the generation of two subkeys for encryption and data integrity. To do so, the A-key, the ESN, and the network-supplied challenge RANDSSD are provided to the Cellular

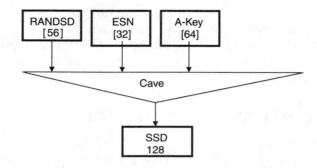

**Figure 5.23   Generating SSD with CAVE algorithm.**

Authentication and Voice Encryption (CAVE) algorithm to obtain a 128-bit Shared Secret Data (SSD), as shown in Figure 5.23. We will see that this CAVE algorithm is used for multiple purposes.

SSD is considered to have two 64-bit parts: SSD_A, which is used to create authentication signatures, and SSD_B, which is used to generate encryption keys. The network copies the SSD when a handoff occurs while the MS is roaming, so that the local network can continue to secure calls in the same way as the home network where the keys are first established. A fresh SSD is generated when the MS returns to the home network or when a handoff occurs to a completely different system.

## Authentication

For authentication, the MS applies the CAVE algorithm to the SSD_A and the challenge RAND sent to MS by the network, and generates an 18-bit signature AUTH_SIGNATURE. It sends this signature to the network. The operator can choose two methods for generating RAND. He or she can either send a global challenge to all mobile stations or send a unique challenge to each MS. The former method is faster.

The A-key is reprogrammed by updating both the MS and the network AC. One of the update methods is called OTASP (over-the-air service provisioning), and it utilizes a 512-bit Diffie–Hellman key-agreement algorithm. This feature can be used to initiate a new service. It can also be used to immediately stop a hacked MS.

## Encryption

The MS also uses the CAVE algorithm and the key SSD_B to generate a Private Long Code Mask (PLCM), which is used to change the long code (discussed earlier as the seed to a 42-bit random number generator).

Because the long code is used to scramble the voice signal, the spread sequence is further protected because the attacker cannot determine PLCM even if the long code is known. A separate data key is also produced by the CAVE algorithm, and an encryption algorithm ORYX is used to encrypt and decrypt data traffic.

## FUTURE PLANS FOR CELLULAR NETWORKS

Third-generation cellular technologies are planning to add more secure security protocols. These include 128-bit privacy and authentication keys. They are also planning to use more secure algorithms; for example, CDMA2000 networks are planning to use Secure Hashing Algorithm-1 (SHA-1) for hashing and integrity and the Advanced Encryption Standard (AES [Rijndael]) algorithm for encryption. More sophisticated key-agreement protocols are being developed so that the keys can be updated more frequently. KASUMI is another algorithm being considered for encryption and message integrity in some protocols.

## APPLICATION OF HIGHER-LEVEL SECURITY PROTOCOLS

In the introduction, it was mentioned that the link layer security may not address all security needs for most organizations. Therefore, it is important for a wireless security architect to understand the security service available in the higher-layer protocols. In this section, we discuss some of these protocols.

IPSec[21] is a network-level security protocol that provides machine-to-machine security. Among the wireless protocols,[22] it is most often used for 802.11. This protocol has the advantage that it is at a high enough level to avoid node-level decryption and low enough to support all protocols that run on top of IP. Usually, an IPSec client is required on every MS, whereas an IPSec gateway is needed on the network. When using IPSec, one has to deal with the issue of scalability because at least one tunnel is required for each MS. Issues have been raised regarding the key management provided by IPSec.[21]

Transport layer security protocols such as SSL, TLS, or WTLS work at layers further up in the protocol stack. They are usually embedded in browsers and sometimes in e-mail software. WTLS needs the support of the BS. TLS and SSL are usually built into Web servers and nothing needs to be done at the BS. WTLS is part of WAP.[23] WAP was originally designed because standard TCP performs suboptimally[24] over wireless networks.[25-28] Therefore, WAP defined its own stack of network layers that is different from the TCP/IP stack. WTLS[29] is just one of the layers that resides on top of the wireless transport layer (WTL) in WAP. The performance issues of WTLS on hardware devices with limitations in Reference 30. Even though WTLS is defined as a transport layer security protocol, it suffers some of the same vulnerabilities that are known to exist in link layer protocols; the network's gateway has to decrypt the message before it is reencrypted and sent to the actual Web server.[31] A solution is proposed in Reference 31, but it is not used in practice. Other security issues are analyzed in Reference 32. WTLS also uses public-key cryptography, thus introducing performance challenges, which are discussed in Reference 33.

Practical interoperability constraints have led people to use a variation of TCP (instead of WAP) wireless networks, which is discussed in Reference 34. This paper[34] also discusses techniques that make IPSec-type security possible in conjunction with these methods.

Modern enterprises are becoming increasingly dependent on wireless networks. Therefore, any comprehensive security policy of an enterprise should also consider securing wireless networks. Network components that implement link layer encryption, IPSec, and transport-level security are examples of conventional security components. Conventional security wisdom dictates implementation of other security components including firewalls, intrusion detection systems, honeypots, and ASs. Translation of high-level business rules into security policies of the components is very complex and does not scale well for larger network infrastructures. Lower-level Internet security issues,[35] ASs, firewalls, and classical access control models are well summarized by Ping Lin and Lin Lin.[36] We conclude that traditional components have limited utility when it comes to controlling business processes because they all operate at the network protocol level. Reference 36 captures the current trends of security in enterprises by advocating that Access Control Lists (ACLs) should be maintained within the system or resources they protect. We advocate reversing this trend. Some vendors have started to address this issue by embedding security in the high-level application software. Modifying each application to make it secure essentially implies rewriting all critical applications — a prohibitive undertaking. With an estimated 80 percent of all security breaches attributed to authorized users (insiders),[37] a revolutionary approach to securing the enterprise is needed. Lin and Lin[36] also identify a problem faced during enterprise integration, where each system recognizes its own users locally and assigns them its own set of IDs and security-related attributes. To address this issue, several single sign-on solutions have been advertised but most are specific to platforms or frameworks, as identified by Burt et al.[38] Thuraisingham et al.,[39] also point out that access controls must be embedded into application code, for the special case of Web and E-commerce applications. This is difficult if at all possible. The embedding of application code in application is the classical area of application security. Halliden[40] addresses this problem by using security APIs developed for business application developers. This approach to securing applications is hard to implement[41] even in brand new applications.

A new approach to securing the enterprise is emerging with which organizations can centralize access control, satisfy corporate governance requirements, and secure business processes. This is called application defense and is indicated in Figure 5.1. This approach separates the security code from the application code, making it possible to write well-tested security code once and to use it with all applications, with trivial modifications.

## CONCLUSION

We have looked at security issues of wireless networks, including important protocols for cellular networks, wireless LANS, and personal networks. We have reviewed link-level security in depth because it is fundamentally unique in all wireless networks. We have seen that wireless protocols have weaknesses, which are being addressed by the standardization bodies. However, even when these protocols are made fully secure, the need to decrypt the packets at the BSs and master nodes implies that some end-to-end security services cannot be provided by these protocols. Therefore, we also provided an overview of other higher-layer security options that can be used on top of wireless networks to provide more comprehensive security services.

## REFERENCES

1. Varshney, U., Networking support for mobile computing, *Communications of the AIS,* Jan. 1999.
2. Chitre, P. and Yegenoglu, F., "Next-generation satellite networks: architectures and implementations, *IEEE Communications Magazine*, Vol. 37, Issue 3, March 1999, pp. 30–36.
3. Sarikaya, B., Packet mode in wireless networks: overview of transition to third generation, *IEEE Communications Magazine*, Vol. 38, Issue 9, Sept. 2000, pp. 164–172.
4. Stallings, W., IEEE 802.11: moving closer to practical wireless LANs, *IT Professional,* Vol. 3, Issue 3, May–June 2001, pp. 17–23.
5. Gutierrez, J.A. et al. IEEE 802.15.4: a developing standard for low-power low-cost wireless personal area networks, *IEEE Network*, Vol. 15, Issue 5, Sept.–Oct. 2001, pp. 12–19.
6. Ojanpera, T. et al. Analysis of CDMA and TDMA for 3rd generation mobile radio systems, *Vehicular Technology Conference*, 1997 IEEE 47th, Vol. 2, 4–7 May 1997, pp. 840–844.
7. Varshney, U., The status and future of 802.11-based WLANs, *Computer,* Vol. 36, Issue 6, June 2003, pp. 102–105.
8. http://grouper.ieee.org/groups/802/11/index.html.

9. Petroni, N.L., Jr. and Arbaugh, W.A., The dangers of mitigating security design flaws: a wireless case study, *IEEE Security & Privacy Magazine*, Vol. 1, Issue 1, Jan.–Feb. 2003, pp. 28–36.

10. Rivest, R. The RC4 Encryption Algorithm, *RSA Data Security*, Mar. 1992.

11. Dawson, E. and Nielsen, L., Automatic cryptanalysis of XOR plaintext strings, *Cryptologia,*, Apr. 1996, pp. 165–181.

12. Singh, S. *The code book: the evolution of secrecy from Mary Queen of Scots, to quantum cryptography,* Doubleday, New York, 1999.

13. Borisov, N.I., Goldeberg and Wagner, D., Intercepting mobile communications: the insecurity of 802.11, *Proc. of 7th Annual Int. Conference on Mobile Computing and Networking*, ACM, Rome, Italy, 2001, pp. 180–188.

14. Fluhrer, S., Mantin, I. and Shamir, A., Weaknesses in the key scheduling algorithm of RC4, *8th Annual Workshop on Selected Areas in Cryptography*, 2001.

15. Ferguson, N., Housley R., and Whiting, D., AES mode choices OCB vs counter mode with CBCMAC, Presentation to 802.11, doc., IEEE 802.11-01 r1., Nov. 2001.

16. Blunk, L. and Vollbrecht, J., RFC 2284 - PPP Extensible Authentication Protocol (EAP), *IETF*, http://www.faqs.org/rfcs/rfc2284.html, March 1998.

17. Aboba, B. and Simon, D., RFC2716 - PPP EAP TLS Authentication Protocol, *IETF*, http://www.faqs.org/rfcs/rfc2716.html, Oct. 1999.

18. Jakobsson, M. and Wetzel, S., Security Weaknesses in Bluetooth, *Topics in Cryptology: CT-RSA 2001*, Berlin: Springer-Verlag, LNCS 2020, pp. 176–191, 2001.

19. Briceno, M., Wagner, D. and Goldberg, I., http://www.isaac.cs.berkeley.edu/isaac/gsm.html

20. 3rd Generation Partnership Project: Specification of the 3GPP Confidentiality and Integrity Algorithms Document 2: KASUMI Specification, *ETSI/SAGE Specification*, Ver. 1.0 1999.

21. Perlman, R. and Kaufman, C., Analysis of the IPSec key exchange standard, *Proc. 10th IEEE Intl. Workshops on Enabling Technologies: Infrastructure for Collaborative Enterprises, 2001*, 20–22 June 2001, pp. 150 – 156.

22. Wei Q. and Srinivas, S., IPSec-based secure wireless virtual private network, *Proc. MILCOM 2002*, Vol. 2, 7–10 Oct. 2002, pp. 1107–1112.

23. WAP Technical specifications 2.0 http://www.wapforum.org/what/technical.htm.

24. Stevens, W., TCP slow start, congestion avoidance, fast retransmit, and fast recovery algorithms, *IETF, RFC 2001*, 1997.

25. Caceres, R. and Iftode, L., Improving the performance of reliable transport protocols in mobile computing environments, *IEEE J. Select. Areas in Communications*, vol. 13, June 1994, pp. 850–857.

26. Partridge, C. and Shepard, T., TCP performance over satellite links, *IEEE Network*, vol. 11, Sept. 1997, pp. 44–49.

27. Mascolo, S. et al., TCP Westwood: Bandwidth estimation for enhanced transport over wireless links, *Proc. 7th Annual. Int. Conf. Mobile Computing Networking (MobiCom'01)*, 2001, pp. 287–297.

28. Holland G. and Vaidya, N. Analysis of TCP performance over mobile ad hoc networks, *Proc. 5th ACM Int. Conf. Mobile Computing Networking (MobiCom'99)*, Aug. 1999, pp. 219–230.

29. Heikkine, T., Wireless Application Protocol, *Tik-111-350 Multimediasemi-anaari.* http://www.tml.hut.fi/Opinnot/Tik-111.550/1999/Esitelmat/Wap/wap/WAP.html.

30. Kahraman, G. and Bilgen, S., Integrated transport layer security: end-to-end security model between WTLS and TLS, 2003 (ISCC 2003). *Proc. 8th IEEE International Symposium on Computers and Communication,* Vol. 2, 30 June-3 July 2003, pp. 1141–1146.

31. Eun-Kyeong, K, Yong-Gu, C. and Ki-Joon C., Integrated transport layer security: end-to-end security model between WTLS and TLS, *Proc.s. 15th Intl. Conference on Information Networking, 2001,* 31 Jan.-2 Feb. 2001, pp. 65–71.

32. Radhamani, G. and Ramasamy, K., Security issues in WAP WTLS protocol, *IEEE Int. Conference on Communications, Circuits and Systems and West Sino Expositions 2002,* Vol. 1 29 June-1 July 2002, pp. 483–487.

33. Levi, A. and Savas, E., Performance evaluation of public-key cryptosystem operations in WTLS protocol, (ISCC 2003). *Proc. 8^th IEEE International Symposium on Computers and Communication, 2003,* Vol. 2, 30 June-3 July 2003, pp. 1245–1250.

34. Zhang, Y., A Multilayer IP Security Protocol for TCP Performance Enhancement in Wireless Networks, *IEEE J. on Selected Areas in Communications,* Vol. 22, Issue 4, May 2004, pp. 767–776.

35. Oppliger, R., Security at the Internet layer, *Computer,* Vol. 31, Issue 9, Sept. 1998, pp. 43–47.

36. Lin, P. and Lin, L., Security in Enterprise Networking: A quick tour, *IEEE Communications Magazine,* Jan. 1996, pp. 56–61.

37. Gerwig, K., Business: The 8th layer: Shoring up security an imperfect art, *netWorker,* Vol. 4, Issue 2, June 2000, pp. 13–16.

38. Burt, C. C., Bryant, B. R., Raje, R. R., and Auguston, M., Model Driven Security: Unification of Authorization Models for Fine-Grain Access Control, *Proc. IEEE Int. Enterprise Distributed Object Computing Conference 2003.* 7th IEEE International, Sept. 2003, pp. 159–171.

39. Thuraisingham, B. et al., Directions for Web and e-commerce applications security, *Proc. 10th IEEE International Workshops on Enabling Technologies: Infrastructure for Collaborative Enterprises, 2001,* June 2001, pp. 200–204.

40. Halliden, P.W., Security for distributed applications, *European Convention on Security and Detection, 1995,* May 1995, pp. 156–160.

41. Whittaker, J., Why secure applications are difficult to write, *Security & Privacy Magazine,* IEEE, Vol. 1, Issue 2, Mar–Apr 2003, pp. 81–83.

# II

---

# TECHNOLOGY

# 6

# FROM WLANS TO AD HOC NETWORKS, A NEW CHALLENGE IN WIRELESS COMMUNICATIONS: PECULIARITIES, ISSUES, AND OPPORTUNITIES

*Salvatore Rotolo, Danilo Blasi, Vincenzo Cacace, and Luca Casone*

## INTRODUCTION TO AD HOC NETWORKS

Two main advantages, simple yet very valuable, lay the basis for the introduction and the widespread diffusion of wireless local area networks (WLANs) in application areas such as office automation and home networking. These are their ease of installation because of the absence of wires and their capability of supporting communications among movable terminals.

In most cases WLAN systems are based on *single-hop operation*; that is to say, a pair of terminals, whenever out of the reciprocal range of radio coverage, can connect to each other only through the use of an infrastructure providing access point devices and centralized control and management facilities.

Significant studies have been made recently of *multi-hopping operations*, generalizing the concept of peer-to-peer interconnection between terminals, out of immediate visibility. For example, let us consider two terminals, A and B, not directly capable of interconnection, accessing another terminal, C, to exchange information: this "third node" is able to reach both nodes A and B and relay all messages not addressed to it.

It is then easy to extend this relaying concept to all elements in the network, more appropriately defined as *network nodes* able to support any communication between the source and the destination through an arbitrary number of wireless intermediate steps, forming multi-hop paths. In case a fixed infrastructure and a centralized management are not in place, these networks, commonly called ad hoc networks (AHNs), should qualify as *self-configuring* and *self-organizing*, and should exploit their enhanced communication capabilities by letting nodes concur to implement the necessary networking functions for automatic operation and minimizing or even completely avoiding any manual setup. Such characteristics may lead both to very high levels of *scalability,* because network management relies on the ability of each node to use local resources only, and to strong *reliability,* which is dependent on the nodes' capacity to react to any anomalous event or failure by giving rise to automatic reconfiguration procedures. Both the above properties, combined with the introduction of node redundancy in AHNs, result in very strong robustness of the overall system.

At first glance, it could appear sufficient to reuse all already established WLAN technologies to implement AHNs just by segmenting paths in more single-hop connections, each one managed as a WLAN operating in ad hoc (AH) mode. On the other hand, as will appear clearer later in this section, this approach, fully based on *local management* of the transmission resources, disregards interaction between contiguous WLANs.

The main goal of this chapter is to give an overview of AHNs, neither exhaustive nor thoroughly detailed but trying to bring readers closer to the basic themes considered as most relevant for the academic and industrial research activities in the sector.

The rest of this chapter is organized as follows: The section titled "The Application Scenarios" describes some scenarios very well suited to AHNs. The section "Peculiarities of Ad Hoc Networks" illustrates the main peculiarities and issues of such networks. A design approach for AH networking, overcoming the limits of the ISO/OSI layered model, is presented in the section titled "The Protocol Framework," and the section "Some Relevant Issues" briefly overviews some related design aspects.

## THE APPLICATION SCENARIOS

AHNs originated in the early 1970s from the Packet Radio Network (PRNET) Project sponsored by the U.S. Department of Defense (DoD). In the early 1980s, concepts that evolved at PRNET were adopted by the Survivable Adaptive Radio Networks (SURAN) Project.

Both initiatives had as their goal laying the foundations for a packet-switched network (similar to the Internet), fully wireless and suitable for military applications such as communications among soldiers and fighting vehicles in hostile battlefield environments without the availability of any networking infrastructure.[1]

Only recently, following the massive diffusion of mobile user terminals (cellular phones, pagers, PDAs, etc.), the research community has started to look at civilian applications for AHNs, especially where such solutions could well complement the existing commercial systems. Important examples may be found in the projects MANET,[2] WINS,[3] and TERMINODES;[4] additional references can also be found in Reference 5. All these activities testify to the growing interest of academic and industrial researchers in AHN.

Among the many areas benefiting by AHN implementations, the field of *environmental control and monitoring* is worth mentioning. In such a case the network nodes are based on specialized sensors that are able to react to particular events, to make local computations, and to exchange data with other instrumentation or control machines (machine-to-machine interfacing). Ambient parameters can be monitored and measured, and results can then be supplied to the users (human-to-machine interfacing). Previous descriptions correspond to what is widely known as the Wireless Sensor Network (WSN).[6] The main factors suggesting the adoption of AH networking in such systems are their capability to establish infrastructure-less wireless communications in difficult or even inaccessible locations and their effectiveness to increase the robustness of the overall system in all cases of critical events by always having some running nodes able to perform networking functions instead of partially, or even fully, damaged or exhausted network members.

In a completely different scenario, a challenging possible application of the AH mode is its access to the Internet *anytime, anywhere*, in the sense of allowing users to connect through their own terminals to the worldwide network in total autonomy, without locational constraints. In the AH mode of operation, the *last-mile* connection might be implemented by a multi hop path to the nearest available access point or IP gateway. These are widely known as mesh-based mobile networks.[7] Valuable civilian applications may be foreseen in case of emergencies, disastrous events, rescue operations, and communications in Third World countries.

Additionally, a significant example of specialized scenarios is the car network,[8] devoted to traffic control — possibly in combination with a global positioning system (GPS) — or used for advanced intervehicle communications (i.e., multimedia communication, multiplayer gaming, etc.). Also, in this case, AH networking fits well the requirements of a collaborative system in which all (or many) elements have the twofold role of user and supplier of a number of services.

## PECULIARITIES OF AD HOC NETWORKS

AHNs imply a completely new approach to networking design and bring the architectures of nodes and transmission protocols very close to the specific field of application of each considered case study: scenario parameters such as the nodes' density, their different resources, their relative positions in the considered environment, the extension of the operating area, the type of established communications (e.g., one-to-one, one-to-many, etc.), and the specific types of traffic (e.g., real-time, best-effort, etc.), just to list a limited number of items, pose totally different requirements and may appreciably influence the choice of design. From this perspective, it appears even clearer that applications are very different from each other: the particular choice in the first context might not be as good in another. Keeping in mind this variety of scenarios and flexibility of use, a number of characteristics of such networks are usually considered in the design process.

Obviously, issues related to the radio channel are important. Several solutions foresee the use of the Industrial, Scientific, and Medical (ISM)[9] frequency band. On the one hand, this is advantageous because of free license usage and a range of frequencies common to many countries. On the other hand, the ISM band requires coexistence with many other standards (such as Bluetooth[10] or WaveLAN[11]), already allocated to the same radio bands, or even with other generic electronic appliances (such as cordless phones or microwave ovens), which are potential sources of interference. Moreover, transmissions will have to comply with precise and strict transmission rules (e.g., low maximum output power level and specific spread spectrum modulation schemes). In some cases, to effectively exploit the available frequency range, the entire bandwidth is sliced into small parts by creating a number of nonoverlapping subchannels whose management necessarily also involves the MAC sublayer. Also, the adopted transmission power levels assume great importance in AHNs, especially in the case of high-density and battery-powered nodes. Minimal transmission power, together with the implementation of smart strategies for energy saving, may considerably increase node life duration and channel utilization. However, using less transmission power may involve longer multi-hop paths with a major impact on routing protocols and overall topology management. Additionally, the use of variable power levels may give rise to unidirectional links; that is, a pair of nodes cannot communicate with each other because of the mismatching of covered radio range. The handling of such cases is cumbersome, and the MAC sublayer usually acts like a filter so that upper-layer protocols operate in bidirectional-link conditions.

Even when all nodes transmit at maximum power, particular topology constraints may require multi-hop transmissions to guarantee full network coverage. Considering also that no centralized control system is provided, new challenges for wireless AH MAC protocols, with respect to classical wired cases, are then posed. Moreover, effective usage of the medium requires that the MAC implement certain quality of service (QoS) levels whenever they are imposed by applications. The creation of subsets for nodes — such as those identified by network organization algorithms — can be exploited to implement alternative MAC coordination schemes by enabling some types of centralized transmission policies.

As stated before, node mobility is just another key feature of AHNs. It obviously impacts on their topology, often in a very unpredictable way, in space and in time: the setting up or the dropping of a link, an occasional event in wired as well as in single-hop wireless networks, becomes very frequent, and routing protocols have to react to these events and manage them promptly, implementing fast tracking of changing topologies and quick recalculations of the available paths. The optimized route calculation should be based on a plurality of parameters of the network; for instance, an algorithm could discard links showing a received power level under a specified threshold. A too-low level is considered the sign of a very weak link, prone to too many errors or even to termination. In a different manner, an algorithm could evaluate the relative mobility among the nodes and require that only those hosts showing a low relative mobility will act as relayers, aiming at potentially increasing the stability of routes. Alternatively, an algorithm can exploit the presence of a predetermined virtual infrastructure, assigning to it crucial routing functions such as *route discovery*; finally, an algorithm could determine paths that for certain reasons can guarantee, more than other paths, certain degrees of QoS. Generally speaking, any routing protocol has to guarantee the robustness of a path over time. This may be achieved by predicting and preallocating alternative paths to be used whenever the first choice fails (*backup paths*) or quickly recalculating paths on the basis of local information (*local recovery procedures*). In any case, the number and the repetition rate of control data exchanges has to be minimized to preserve bandwidth for data and to minimize energy consumption.

When considering transport layer issues, the high degree of distributed intelligence naturally present in an AHN should be taken into account. Many features, in fact, are usually provided by the system already, and an accurate analysis should concentrate on the provisioning of such complementary functions as are strictly necessary to manage the application's traffic flow. It would be possible, in this way, to avoid the replication of any useless functionality. A "tiny" transport protocol, for instance, could implement only the flow control feature and leave to the underlying network mechanisms the execution of congestion control and the notification of transmission errors. Many MAC protocols, for instance, already provide acknowledgment frames.

*Security* can pose several problems. Usually the *Admission Control* procedures are implemented by a central unit and are not provided in AHNs. In principle, this is a weakness that becomes even more noticeable when looking at all nodes actively participating in overall operations. That is why in AHNs there is the need to take a comprehensive approach, protecting a system already at very low networking levels. Also, considering both scenario and application characteristics, some precautions are required even at the MAC sublayer to keep a malicious node from receiving or transmitting sensitive data. But this is only one facet of the problem because a malicious node could even cause impaired operations, for instance, by communicating false control data to mix up routing schemes, with a major, even global, impact. For such reasons, communications, and consequently participation in networking activity, should be restricted to authorized nodes only. A continuous monitoring of each element is required in this regard with suitable trusting criteria to allow rapid isolation of all items exhibiting illicit behavior.

A final consideration should be *scalability*. An AHN may spread and grow over a territory with different levels of density: it follows a clear requirement for distributed algorithms. For reasons of simplicity, such algorithms should use local control data processing and involve the smallest part of neighboring nodes.

## THE PROTOCOL FRAMEWORK

Ad hoc networks, as their Latin name clearly highlights, are tailored to the precise requirements of scenarios and applications. Moreover, these requirements have to be met despite frequent and random variations in border conditions (radio channel, topology, data traffic, etc.). All this implies the rapid handling of a large number of parameters in the most efficient and dynamic way. From this perspective the rigid subdivision of networking protocols in precise ISO/OSI layers, actually the winning approach for the Internet, shows its limitations to new designers who have to cope with unprecedented challenges. For example, distributed network organization algorithms may be needed to create a virtual infrastructure that could be used for routing, access to channel, and other control functions. Also, application protocols might be required to work with lower-layer mechanisms to achieve end-to-end service quality. As far as security is concerned, integrated solutions should be adopted to prevent attacks in all layers.

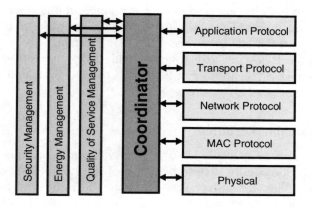

**Figure 6.1 Protocol Framework. Each protocol may, by means of a Coordinator, exchange info with any other.**

For a comprehensive view of all requirements and the exploitation of all potential synergies between protocols, it may be necessary to use the so-called *cross-layer design*[12] that allows handling and trimming of parameters even across layers not adjacent in the stack. Moreover, the cross-layer design often shows itself to be the only way to resolve interdependencies among layers: for instance, transmission power control can be steered by a routing algorithm able to optimize multi-hop behavior or by a MAC willing to shrink collision domains. Such demands could be contradictory but can be amalgamated by protocol coordination.

Figure 6.1 summarizes the layers controlling specific networking functions, such as security, energy, or QoS management. The coordinator for logical tasks is able to cross-correlate information and actions at different levels. Actually, how the functional block diagram presented in Figure 6.1 may be implemented in real-world cases is still an open point because of the extreme complexity involved, and is still largely to be discovered by researchers in this field.

## SOME RELEVANT ISSUES

From what has been discussed in the previous paragraphs, it follows that the design of protocols should take into account many varied factors. In some cases, it is possible to start from existing solutions — already studied for structure-based, wired, or wireless networks. In other cases, it becomes necessary to introduce new and more specific approaches.

Traditionally, attention to AHNs has been focused on mechanisms and protocols allowing efficient packet data exchange in multi-hopping mode, particularly with regard to network organization, route establishment and maintenance, and medium access control. Therefore, these are briefly reviewed in the following text, together with a brief section dedicated to energy consumption — another critical aspect for nodes usually compelled to operate using scarce resources.

## Energy: A Valuable Resource

Lack of resources, especially electric power, figures prominently in the use of AHNs. Considering the architecture of the nodes, two main blocks share responsibility for power consumption: the radio interface and the processing unit. The running energy cost of the first generally represents the dominant factor[13] and strongly depends on the operating node state: transmission, reception, idle, or sleep.[14] Two different techniques are proposed to minimize energy consumption: (1) *power-control* techniques to limit transmitted power; and (2) *energy-saving* techniques to economize on energy by turning off the radio interface when it is not in use.

In traditional networks, the presence of an access point guides decisions on what nodes should be put into the sleep state, the sleep duration, and what transmission power level should be used. Conversely, in AHNs, these tasks become more complicated and should be performed in a distributed and coordinated manner, also taking into account the multi-hopping mode of operation. For instance, a node, not being either a receiver or transmitter, cannot go into sleep state in cases where it is cooperating with others in network maintenance operations.

As far as power control is concerned, the application of suitable strategies, in addition to improving overall energy consumption, impacts on two key related items: the channel spatial reuse and the network topology. Regarding the first point, greater transmission power, corresponding to greater propagation distances and wider interference-prone areas, leads to greater collision domains and, consequently, to a lower number of simultaneous connections per unit of area[15] (see Figure 6.2). As a result, weaker transmission power may increase the overall channel utilization. With regard to the second point, diminished transmission power implies lower correct reception ranges, and therefore more hops will be required to reach the selected destination. This has a twofold effect. First, given that the power level of a signal propagating along the wireless channel is inversely proportional to at least the square of distance traveled, it should be clear that splitting a given transmission into two or more steps (i.e., multi-hop transmission) may reduce the required total energy.[16] Second, the probability of losing routed paths and the tendency for packet

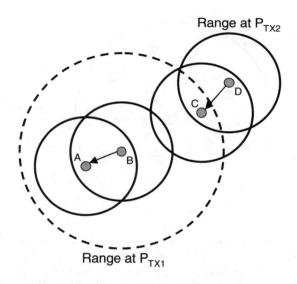

**Figure 6.2  Power control increases channel utilization. If B transmits to A using a power level $P_{TX1}$, its transmission would prevent C from receiving data from D. On the contrary, by using a power level $P_{TX2}<P_{TX1}$, both transmissions can occur simultaneously.**

delivery to be delayed increase correspondingly and a reduced overall throughput is obtained.

Several power control schemes are suggested by the literature (see Figure 6.3).[17] To avoid the eventuality of unidirectional links, all network nodes should be set at the same transmitting power level, but the correct setting to meet a target global topology property (for instance, network connectivity, number of links per path, maximum number of neighboring elements, etc.) is an issue. This setting must not only be evaluated in a distributed manner but has to be dynamically adapted to network characteristics such as traffic load, node density, and other relevant parameters.[18] Each node could have its emission level adjusted to be just strong enough to reach the destination (whenever possible) or its closest neighbor, the best relaying node. Alternatively, each node could have its transmission power adjusted to be just strong enough to reach the nearest neighbor on the forward direction. This behavior, however, could result in the formation of longer paths and unidirectional links.

When considering a radio range of about 100 m — something rather reasonable in most cases of WLANs — transmission state power consumption may not be much higher than in idle or receiving states:[14] power control, as described in the preceding text, may not be so effective. It is, therefore, preferable to try to prolong the sleep state for the greatest

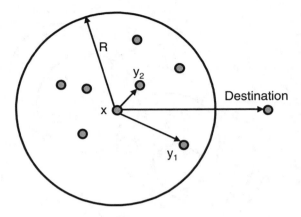

**Figure 6.3 Power control approaches. Node x may transmit using a fixed power level (R represents the coverage radius); alternatively, it may adapt its transmission power level to reach (i) the Destination directly (if possible), (ii) the nearest Destination's neighbor $y_1$, or (iii) the nearest neighbor $y_2$ towards the Destination.**

number of nodes by means of suitable energy-saving schemes. Having more sleeping nodes also means having less contention induced by nodes trying to catch the channel. A smart energy-saving protocol should allow nodes to go into sleep mode without adversely affecting connectivity and operations: for instance, it could provide a *caching procedure* for data flowing to sleeping nodes to prevent the loss of important information, or it could use an auxiliary, very-low-power channel, separated from the data channel, to wake up the sleeping nodes whenever there is data for them. In all these cases, the presence of a network organization scheme could help in making available nodes that can handle more tasks than others.

## Organizing an Ad Hoc Network

AHNs are anarchic by nature; they grow spontaneously, change rapidly, and include either few or many nodes in various topologies. Events may happen at any time in an unpredictable manner; new nodes may appear, old nodes disappear, nodes move and stop, and so forth. To maintain operations in such an infrastructureless environment, every member of the network should be notified about any kind of variation that might have occurred. Such information should flow with control signals and the dimension of overheads grows with the number of nodes and with the dynamics of topology/connection changes. The autonomous organization of the network becomes a critical issue.

A way to support both reliable communications and efficient network management is to put in place distributed protocols that select nodes or group them into subsets creating a hierarchy among them: this corresponds to a *virtual backbone infrastructure*. These distributed protocols should base their operation as much as possible on local information and implement intelligent ways to decide when and how to react to topology changes.

The most commonly used organization method is to group nodes into subsets, usually called *clusters*. In some schemes, one element inside each cluster is elected *cluster head* and takes the role of the group's leading node; sometimes, this principal node is used only as a cluster representative (e.g., the cluster head's identifier is used to identify the whole cluster);[19] in turn, it can act as the coordinator of intra- and intercluster communications.[20]

Some algorithms partition the network without assigning any particular role to nodes;[21] consequently all nodes decide, in a distributed manner, which sets have to be formed and what members are to be included. This can be done, for instance, on the basis of the preferred number of nodes per cluster[22] or by the estimation of the paths' duration.[23]

Clustering mechanisms can also be distinguished in accordance with the structure topology they create; some of them (e.g., see Reference 19, Reference 20, and Reference 24) provide nodes with affiliation only when they have direct wireless links with target cluster heads; others, (e.g., see Reference 25) conversely, create multi-hop paths between nodes and related cluster leaders. The latter approach tends to define fewer groups than the former, even if nodes are more numerous.

Another way to introduce a level's hierarchy into an AHN is to identify a virtual backbone, usually called a spine.[26] A spine is a self-organizing network structure, composed of a small number of nodes and links interconnecting them, which aggregates information associated with networking functions — typically, routing. Nonspine nodes are said to be dominated by spine nodes, in the sense that they work like hosts in traditional fixed networks, while spine nodes take care of the main networking tasks.

The cluster–spine combined approach may be an effective solution in big networks.[27] The network is subdivided into rather big clusters; a spine managing each cluster is then identified, and a node in each spine, namely the *root*, is dedicated to managing and controlling intercluster communications.

The need for having (and maintaining) a virtual infrastructure strongly depends on both the purposes and characteristics of the network to be implemented. For instance, small or low-varying topology networks would probably not gain appreciable performance improvements by the introduction of any kind of hierarchy. On the other hand, managing large networks comprising hundreds or even thousands of nodes, such as flat ones (i.e., no distinction among the nodes), can introduce too much overhead, thus causing the waste of precious system resources. Moreover, assuming node heterogeneity — a more realistic case than homogeneity — special leading roles could be assigned to nodes (e.g., a cluster head or spine member) exhibiting either more resources (like energy) or some valuable characteristics (such as being motionless).[24,27]

Furthermore, the type of AH infrastructure has to be chosen in accordance with the particular application: Clusters are good for supporting scalability and mobility; spines are more useful whenever nodes have differing movement rates (slower nodes can be selected to dominate faster ones).

The creation of a virtual infrastructure does not offer advantages only in terms of topology control features but may also offer effective support to solve many other typical issues.

An important aspect to consider first relates to the establishment and maintenance of paths between nodes: In a cluster-based structure, it may not be necessary to run these operations at a hop-by-hop level, but such functions can be managed at a more aggregate stage.[26] Second, the mechanism of access control relative to the transmission medium can benefit from a virtual infrastructure by offering a larger degree of freedom in choosing a MAC policy; for instance, for what may be considered suitable scheduling strategies. Climbing up the protocol stack to the top level, we may find the application layer benefiting from a virtual infrastructure. Leading nodes may collect and process information and offer various services much better than others.

Finally, in addition to their features, it must be emphasized that such distributed protocols should be extremely simple, calling for very little processing and very low levels of information exchange, even in the case of frequent and dynamic changes. This is not at all easy to achieve, but it is the only way that may guarantee a suitable trade-off between costs and benefits.

## MAC Protocols in Ad Hoc Networks

The effective use of the wireless medium for transmitting data packets among AH nodes poses a number of problems not present to the same degree in wired applications. The broadcast nature of the medium, the way transmitted power decreases while electromagnetic signals spread over a distance, and the fact that a network interface cannot simultaneously receive and transmit are just a few examples of the peculiarities of the wireless channel. The *hidden-terminal* problem[28] and the *exposed-terminal* problem[29] are probably the most well-known related issues, whose solutions have been studied for several years.[28,30] Usually, however, these solutions have been implemented in MAC protocols designed for wireless LANs. Even if they can be applied "as is" to AHNs — initially as a number of overlapped WLANs — obviously they cannot be fully optimized for such systems. Multi-hop transmissions, together with the absence of any centralized control system, in fact, lead to other issues, mainly the need to effectively coordinate, in a distributed manner, access to the medium attempted by hosts not in the same radio coverage range. This indicates why AHNs call for well-designed MAC protocols. In the following paragraph a brief overview of MAC protocols is given.

MAC protocols may be distinguished by the way they transmit data packets (i.e., if they are prone to collisions or not), thus classifying them as either contention-free or contention-based. A MAC protocol of the former type forces only one node to access the channel at a given time, thus avoiding many collisions. Usually, the access scheme consists of a *group access phase*, which permits a node to issue a *registration query*, possibly contending with other entering nodes, to acquire the right to use the channel, and a subsequent *transmission phase*, when all the registered nodes send their own packets in a predetermined order. Quite often, contention-free MAC protocols come with a slotted time structure: it turns to be natural, in fact, to group slots and assign them to nodes or phases dynamically over time. The problem of such a transmission approach in AHNs lies in the way the protocol propagates the assignment up to the receiver's one-hop neighbors or, in other words, within the sender's two-hops neighborhood: such nodes, in fact, must be kept from concurrently transmitting anything, otherwise the receiver will not correctly receive the incoming packet (recall the hidden-terminal problem). As a solution, such MAC schemes introduce various control packet exchange techniques and several transmission policies,[31,32] usually at the cost of a more complex slot/frame structure and higher overhead. There are, nevertheless, some other schemes (e.g., see Reference 33), which try to solve this issue in a different manner. Instead of using an access phase, they are based on a periodic exchange of information among neighbors in their one-hop neighborhoods, so that each host is able to collect useful data regarding its own two-hops neighborhood. This collected information is then processed by each station to determine whether it can transmit, always aiming at minimizing collisions and maximizing channel reuse.

Contention-based MAC protocols, on the other hand, try to organize channel usage in a multi-hop environment by exploiting asynchronous channel accesses. Their simplicity, coming from the absence of any explicit coordination among neighbors, requires more complex mechanisms for controlling, in a distributed way, how any node may access the channel.

In such approaches, the main issue is related to the fact that any host, before actually sending a data packet, has to make sure that all the potentially interfering one-hop and two-hops neighbors will not try to access the channel for a reasonable period of time. For that reason, various forms of *handshaking* procedures between the sender and the intended receiver that were meant to acquire the channel may be carried out through the exchange of control packets,[29,34–36] as well as by the adoption of signaling tones.[37,38]

Most of the contention-based MAC protocols use the well-known *backoff mechanism* to control asynchronous channel accesses and to guarantee a certain degree of fairness in the channel exploitation among competing nodes. Basically, backoff algorithms try to achieve these goals by spreading transmission attempts over time, taking into account a variable number of parameters. For the backoff window, in addition to the standard binary exponential backoff scheme (see, for instance, Reference 29), some other schemes[35,39] propose different stretching/shrinking processes aiming at improving the spreading of transmission attempts with respect to some metric. Other algorithms (e.g., the one in Reference 35) drive the backoff function on the basis of which traffic class the incoming packet belongs to, thus assigning to some classes a higher priority than others. In other cases (e.g., Reference 40) the backoff algorithm depends on scheduling information coming from queued packets.

Most MAC protocols transmit frames over a single channel shared among all the network nodes. Given that it is the simplest solution to be implemented, its greatest disadvantage is represented by the fact that each transmission may generate destructive interferences over large areas, thus lessening the effectiveness of spatial/temporal channel reuse. To overcome this problem, some protocols using multiple channels have been proposed. The overall available bandwidth is split into a number of smaller subchannels, which allow neighboring stations to transmit simultaneously. Because of the typical shortage of a great number of subchannels, the MAC sublayer is usually in charge of dynamically assigning subchannels to nodes whenever the need arises. To cope with the aforementioned issues related to an AH environment, the selection is carried out in such a way that every node has a channel that is unique within its entire two-hop neighborhood. After each node is given with its own channel, there are basically three ways[41,42] a channel may be actually selected for conveying an incoming transmission: transmitter-based, in which the channel assigned to the transmitting node is selected; receiver-based, in which the channel assigned to the receiving node is used; and pairwise-based, in which both sender and receiver agree about the channel they will share. It is important to note that the first scheme avoids collisions at the receiver side, but it needs the receiver's knowledge of the channel the sender is going to transmit on. The second scheme makes broadcasting impossible, but it does not need any coordination at the receiver side. The third scheme is similar to the first one, but it calls for much more coordination among nodes. Whenever coordination is needed, it is usually achieved by handshaking between senders and receivers.[42]

## Routing in AHNs

Routing in AHNs is much more critical and complex than in more classical, wired, or wireless packet-switched networks. There are two main reasons for this: first, the absence of a network infrastructure and a centralized control, a factor causing all nodes to be involved in the management of data exchange, and second, because of the high probability of quick and unpredictable changes in the overall topology, causing a rapid obsolescence of active routes. These factors have to be carefully taken into account when designing a routing protocol, in addition to the usual physical constraints (i.e., energy, transmission power, bandwidth, range, and obstacles). The Mobile Ad-hoc Networks (MANET) Working Group of the Internet Engineering Task Force (IETF), the authoritative international body, has provided lists of attributes for routing solutions in the AH field.[7] Several academic and industrial research centers have drawn inspiration for their activities on the basis of this preliminary work.

Depending on the application, a routing protocol will have to favor one or the other feature. Different types of taxonomy may be defined, as in Reference 43, for example, which shows an interesting way of grouping protocols according to their peculiarities.

A first distinction may be made by considering two main classes: the *single-channel* and the *multi-channel* schemes. The single-channel approach uses a contention-based MAC, whereas the multi-channel one bases the medium access control mechanism on time or frequency division. The greatest number of algorithms proposed in the literature belongs to the first class; significant examples of multi-channel protocols are described in Reference 44.

It is also possible to distinguish between *flat* and *hierarchical* schemes. The former are those having nodes with the same routing function; the latter, on the contrary, are formed by different kinds of nodes, some of which play specific roles, as far as networking functions are concerned (e.g., coordination of communications, data exchange, flow control, etc.). A typical example of a hierarchical scheme is based on node clusters, such as in Reference 45. In such situations the leading role is performed by a *cluster head*, a node representing the cluster it belongs to or through which it coordinates internal communications, or a *gateway*, a node located at the border between two (or more) clusters, usually controlling intercluster links. A different typology of hierarchical schemes supports the creation of a spine; Reference 26 provides an example of a spine-based approach, whereas an example of a hybrid cluster-spine scheme is presented in Reference 46.

The criteria being used to discover routes for data may lead to the definition of further classes. *Proactive* or *Pre-elaborated* schemes include all solutions using a periodic exchange of control data among nodes to find out paths for each source–destination pair: Reference 47 and Reference 48 describe protocols belonging to this category. *Reactive* or *On-demand* schemes tend to create a new route only when strictly necessary; that is, only when data to be delivered are present. Usually, this class of algorithms includes a *route discovery phase* during which the route is established and a *route maintenance phase* during which the validity of the choice is frequently checked. Typical examples are proposed in Reference 49 and Reference 50.

Routing protocols may also be subdivided by looking at the nature of routing control information. It is therefore possible to define *geographical* schemes, mainly based on node location;[51] *signal-strength* schemes, mainly based on the measurement of the intensity of received signals;[52] or, finally, *topology schemes* (this solution has already been widely adopted in classical networks), based on the exchange of *link state*–like topological information, as in Reference 53. An algorithm that uses a combination of these last two schemes is presented in Reference 54.

It is also worthwhile to mention *link-reversal solutions*, which emphasize local reactions to link disconnections,[55] and solutions based on the *zone concept*, which is a portion of the network defined on the basis of geographical[56] or topological[57] information, in which each node "knows" the entire zone topology.

All these types are specifically concerned with the algorithms forming point-to-point routes (*unicast*) and are *address-centric*, in the sense that greatest importance is given to node identification, regardless of physical or logical address.

However, as in classical systems, other kinds of solutions exist, oriented to point-to-multipoint (*multicast*) communications. These mainly support multimedia applications and suit very well the topology of *mesh networks*. A list of multicasting routing protocols may be found in Reference 58.

The area of *sensor networks* represents a typical field of application for *data-centric* solutions. Here, the concept of the precise identification of the nodes vanishes, and much emphasis and importance is given to the event detected by a node and to data representing the results of both gathering and processing operations performed by a distributed system. Further information can be found in Reference 59 and Reference 60.

Many other cases, often belonging to more than one of the groups already described, may be found in several references in the bibliography.

## CONCLUSIONS

This section on AHNs has attempted to show the relevance of alternative technologies in respect to traditional communication systems, either wired or wireless.

The advantages conferred by many important features — flexibility, scalability, robustness against failures, self-configuration and self-management in the absence of any networking infrastructure, and peer-to-peer communication — are, for sure, key factors suggesting applications in a large number of different scenarios.

Therefore, AHNs appear to be the enabling technology for transmitting "anything, anytime, anywhere." Nevertheless, just because of this, finding a correct design is a difficult and demanding job. A great number of variables must be considered simultaneously, often under the pressure of conflicting requirements.

Up to now, research activities have moved along the path of refining and adapting protocols that were designed for the classical schemes so as to function in completely new situations. The use of the ISO/OSI model undoubtedly underlies the great success of today's telecommunications field, but many signs already suggest new, revolutionary approaches.

The transition from moderately static to highly dynamic networks will shortly call for innovative architectural design. With networking protocols, much recent work is already aimed at strongly integrated solutions, providing cooperation across the full protocol stack and allowing shortcuts between noncontiguous layers. The so-called *cross-layer optimization* appears to be the new path toward a concrete realization of systems able to emerge from research laboratories and test-beds for pervasive application in daily life.

## BIBLIOGRAPHY

1. Freebersyser, J. and Leiner, B., A DoD perspective on mobile ad hoc networks, in *Ad Hoc Networking*, Perkins, C.E., Ed., Addison-Wesley, 2001, pp. 29–51.
2. Mobile Ad-hoc Networks (MANET) Charter, http://www.ietf.org/html.charters/manet-charter.html.
3. Wireless Integrated Network Sensors (WINS), http://www.janet.ucla.edu/WINS/.
4. Terminodes, http://www.hec.unil.ch/yp/Terminodes/.
5. Ad Hoc Wireless Networking Links, http://www.antd.nist.gov/wctg/manet/adhoclinks.html.

6. Estrin, D. et al., Instrumenting the world with wireless sensor networks, in *Proceedings of IEEE ICASSP 2001*, 4, 2033, 2001.

7. Corson, S. and Macker, J., Mobile Ad hoc Networking (MANET): routing protocol performance issues and evaluation considerations, http://www.ietf.org/rfc/rfc2501.txt.

8. Gerla, M., Xu, K., and Hong, X., Exploiting mobility in large scale ad hoc wireless networks, in *Proceedings of IEEE 18th CCW*, 1, 34, 2003.

9. Golmie, N., Interference in the 2.4 GHz ISM band: challenges and solutions, National Institute of Standards and Technology (NIST), white paper, June 2001.

10. The official Bluetooth wireless info site, http://www.bluetooth.com/.

11. Wireless LAN clients and chip sets, http://www.agere.com/client/wlan.html.

12. Goldsmith, A.J. and Wicker, S.B., Design challenges for energy-constrained ad hoc wireless networks, *IEEE Wireless Commun.*, 9(4), 8, 2002.

13. Li, Q., Aslam, J. and Rus, D., Online power-aware routing in wireless ad-hoc networks, in *Proceedings of 7th ACM/IEEE MobiCom*, 1, 97, 2001.

14. Jamieson, K., Implementation of a power-saving protocol for ad hoc wireless networks, Master's thesis, Massachusetts Institute of Technology, 2002.

15. Gupta, P. and Kumar, P., The capacity of wireless networks, *IEEE Trans. Inf. Theory*, 46(2), 388, 2000.

16. Rodoplu, V. and Meng, T.H., Minimum energy mobile wireless networks, *IEEE J. Sel. Areas Commun.*, 17(8), 1333, 1999.

17. Hou, T. and Li, V.O.K., Transmission range control in multihop packet radio networks, *IEEE Trans. Commun.*, COM-34, 38, 1986.

18. Park, S.J. and Sivakumar, R., Quantitative analysis of transmission power control in wireless ad-hoc networks, in *Proceedings of IEEE ICPPW 2002*, 1, 56, 2002.

19. Baker, D.J., Ephremides, A., and Flynn, J.A., The design and simulation of a mobile radio network with distributed control, *IEEE J. Sel. Areas Commun.*, Sac-2(1), 226, 1984.

20. Gerla, M. and Tsai, J.T., Multicluster, mobile, multimedia radio network, *ACM J. Wireless Networks*, 1(3), 255, 1995.

21. Lin, C.R. and Gerla, M., Adaptive clustering for mobile wireless networks, *IEEE J. Sel. Areas Commun.*, 15(7), 1265, 1997.

22. Ramanathan, R. and Steenstrup, M., Hierarchically-organized, multihop mobile wireless networks for quality-of-service support, *ACM J. Mobile Networks and Applications*, 3(1), 101, 1998.

23. McDonald, A.B. and Znati, T., A mobility-based framework for adaptive clustering in wireless ad-hoc networks, *IEEE J. Sel. Areas Commun.*, 17(8), 1466, 1999.

24. Basagni, S., Distributed and mobility-adaptive clustering for multimedia support in multi-hop wireless networks, in *Proceedings of 50th IEEE VTC–Fall*, 2, 889, 1999.

25. Amis, A.D. et al., Max-min D-cluster formation in wireless ad hoc networks, in *Proceedings of 19th IEEE INFOCOM*, 1, 32, 2000.

26. Sivakumar, R., Sinha, P., and Bharghavan, V., CEDAR: core extraction distributed ad hoc routing, *IEEE J. Sel. Areas Commun.*, 17(8), 1369, 1999.

27. Blasi, D. et al., Availability clustering: a spine-based clustering scheme to exploit nodes heterogeneity in ad hoc networks, in *2nd Med-Hoc Net Conf.*, 2003.

28. Tobagi, F.A. and Kleinrock, L., Packet switching in radio channels: part II — The hidden terminal problem in carrier sense multiple-access and the busy-tone solutions, *IEEE Trans. Commun.*, COM-23(12), 1417, 1975.

29. Karn, P., MACA — a new channel access method for packet radio, in *ARRL/CRRL Amateur Radio 9th Computer Networking Conference*, 1, 134, 1990.

30. Kleinrock, L. and Tobagi, F.A., Packet switching in radio channels: part I — carrier sense multiple-access modes and their throughput-delay characteristics, *IEEE Trans. Commun.*, COM-23(12), 1400, 1975.

31. Tang, Z. and Garcia-Luna-Aceves, J.J., Collision-avoidance transmission scheduling for ad-hoc networks, in *Proceedings of IEEE ICC 2000*, 3, 1788, 2000.

32. Zhu, C. and Corson, M.S., A five-phase reservation protocol (FPRP) for mobile ad hoc networks, in *Proceedings of 17th IEEE INFOCOM*, 1, 322, 1998.

33. Bao, L. and Garcia-Luna-Aceves, J.J., A new approach to channel access scheduling for ad hoc networks, in *Proceedings of 7th ACM/IEEE MobiCom*, 1, 210, 2001.

34. Fullmer, C.L. and Garcia-Luna-Aceves, J.J., Solutions to hidden terminal problems in wireless networks, in *Proceedings of ACM SIGCOMM 1997*, 1, 39, 1997.

35. Bharghavan, V. et al., MACAW: a media access protocol for wireless LANs, in *Proceedings of ACM SIGCOMM 1994*, 1, 212, 1994.

36. Garcia-Luna-Aceves, J.J. and Tzamaloukas, A., Reversing the collision-avoidance handshake in wireless networks, in *Proceedings of 5th ACM/IEEE MobiCom*, 1, 120, 1999.

37. Haas, Z.J. and Deng, J., Dual busy tone multiple access (DBTMA) — a multiple access control scheme for ad hoc networks, *IEEE Trans. Commun.*, 50(6), 975, 2002.

38. Sobrinho, J. and Krishnakumar, A.S., Quality-of-service in ad hoc carrier sense multiple access wireless networks, *IEEE J. Sel. Areas Commun.*, 17(8), 1353, 1999.

39. Aad, I. and Castelluccia, C., Differentiation mechanisms for IEEE 802.11, in *Proceedings of 20th IEEE INFOCOM*, 1, 209, 2001.

40. Luo, H., Lu, S., and Bharghavan, V., A new model for packet scheduling in multihop wireless networks, in *Proceedings of 6th ACM/IEEE MobiCom*, 1, 76, 2000.

41. Joa-Ng, M. and Lu, I., Spread spectrum medium access protocol with collision avoidance in mobile ad-hoc wireless networks, in *Proceedings of 18th IEEE INFOCOM*, 2, 776, 1999.

42. Jain, N., Das, S.R., and Nasipuri, A., A multichannel CSMA MAC protocol with receiver-based channel selection for multihop wireless networks, in *Proceedings of 10th IEEE IC3N*, 1, 432, 2001.

43. Feeney, L.M., A taxonomy for routing protocols in mobile ad hoc networks, SICS Technical Report T99/07, Oct. 1999.

44. Chiang, C.C. et al., Routing in clustered multihop mobile wireless networks with fading channel, in *Proceedings of 5th IEEE SICON*, 1, 197, 1997.

45. Jiang, M., Li, J., and Tay, Y.C., Cluster based routing protocol functional specification, http://www.ietf.org/internet-drafts/draft-ietf-manet-cbrp-spec-*.txt.

46. Das, B., Sivakumar, R., and Bharghavan, V., Routing in ad-hoc networks using a virtual backbone, in *Proceedings of 6th IEEE IC3N*, 1, 1, 1997.

47. Perkins, C.E. and Bhagwat, P., Highly dynamic destination-sequenced distance-vector (DSDV) routing for mobile computers, in *Proceedings of ACM SIGCOMM 1994*, 1, 234, 1994.

48. Clausen, T. and Jacquet, P., Optimized link state routing protocol (OLSR), http://www.ietf.org/rfc/rfc3626.txt.

49. Perkins, C., Belding-Royer, E., and Das, S., Ad hoc on-demand distance vector (AODV) routing, http://www.ietf.org/rfc/rfc3561.txt.

50. Johnson, D.B., Maltz, D.A., and Hu, Y.C., The dynamic source routing protocol for mobile ad hoc networks (DSR), http://www.ietf.org/internet-drafts/draft-ietf-manet-dsr-*.txt.

51. Basagni S. et al., A distance routing effect algorithm for mobility (DREAM), in *Proceedings of 4th ACM/IEEE MobiCom*, 1, 76, 1998.

52. Dube, R. et al., Signal stability-based adaptive routing (SSA) for ad hoc mobile networks, *IEEE Pers. Commun.*, 4(1), 36, 1997.

53. Chen, T.W. and Gerla, M., Global state routing: a new routing scheme for ad-hoc wireless networks, in *Proceedings of IEEE ICC 1998*, 1, 171, 1998.

54. Gerla, M., Hong, X., and Pei, G., Landmark routing protocol (LANMAR) for large scale ad hoc networks, http://www.ietf.org/internet-drafts/draft-ietf-manet-lanmar-*.txt.

55. Park, V. and Corson, S., Temporally-ordered routing algorithm (TORA) version 1 functional specification, http://www.ietf.org/internet-drafts/draft-ietf-manet-tora-spec-*.txt.

56. Joa-Ng, M. and Lu, I.T., A peer-to-peer zone-based two level link state routing for mobile ad hoc networks, *IEEE J. Sel. Areas Commun.*, 17(8), 1415, 1999.

57. Haas, Z.J., Pearlman, M.R., and Samar, P., The zone routing protocol (ZRP) for ad hoc networks, http://www.ietf.org/internet-drafts/draft-ietf-manet-zone-zrp-*.txt.

58. Ren, X. and Wang, H., A survey on multicast routing for wireless mobile ad hoc networks, in *Proceedings of ICWN 2003*, 1, 539, 2003.

59. Intanagonwiwat, C., Govindan, R., and Estrin, D., Directed diffusion: a scalable and robust communication paradigm for sensor networks, in *Proceedings of 6th ACM/IEEE MobiCom*, 1, 56, 2000.

60. Braginsky, D. and Estrin, D., Rumor routing algorithm for sensor networks, in *1st ACM WSNA*, 1, 22, 2002.

# 7

# P2PWNC: A PEER-TO-PEER APPROACH TO WIRELESS LAN ROAMING

*Elias C. Efstathiou and George C. Polyzos*

## INTRODUCTION

Deployment of 3G is slow and the promise of ubiquitous wireless Internet at low cost and high speed is still unfulfilled. We present here *P2PWNC* (Peer-to-Peer Wireless Network Confederation), an incrementally deployable networking infrastructure that could provide a *Peer-to-Peer* (P2P) solution to the problem of limited Internet access. Instead of relying on a few cellular and public WLAN operators, we propose to combine the resources of numerous smaller providers and present one unified access network to roaming users. In P2PWNC, providers use low-cost WLAN technology for their individual networks and the typical provider is a household with a broadband Internet link. The P2PWNC scheme relies on providers that allow nearby visitors to access these links over the WLAN. Such a scheme is becoming possible in urban areas where WLAN coverage is already on the rise due to the technology's low cost and the proliferation of home broadband connections.

Why should households open their WLANs for passersby and participate in P2PWNC? The idea is that participating householders earn the right to access other P2PWNC WLANs when they roam away from home, at no extra fee. Non-providing users are excluded and are therefore given the incentives to participate in the system in order to rip the benefits of free roaming. P2PWNC is therefore a P2P community where participating providers may consume from other providers, and where, additionally, the only currency is payment "in kind." Here we also compare P2PWNC to other WLAN-based public Internet access schemes and we present P2PWNC in the context of current P2P research.

It has been observed[1] that, without a mechanism for motivating contribution in P2P systems, rational and selfish participants tend to "free ride" on the contributions of others. In the P2PWNC context, free riders would try to benefit from the community when roaming while eliminating their participation cost by not setting up their own WLANs.

In order to encourage contribution, P2PWNC employs an accounting mechanism based on digitally signed statements that encode consumption actions. Roaming users sign these statements, or *receipts*, which are then stored by contributing providers and are used as evidence of a provider's good standing whenever requested. We will show that this mechanism enables participants to reliably detect and exclude free riders in a decentralized manner (i.e., without relying on any central server).

This brings us to the most important characteristic of P2PWNC. A P2PWNC community is completely self-organized. We assume no central authorities that certify identities, issue currency, or have the ability to punish or reward. Such authorities do not even exist as physically distributed structures formed by a subset of the participants. In P2PWNC all participants have equivalent roles and they alone control the execution of the accounting mechanism. P2PWNC's self-organization affects all our design decisions but it addresses an important requirement that can make P2PWNC socially acceptable and can contribute to its success. The P2PWNC protocols we propose could be implemented in WLAN access points and WLAN card drivers, and could fuel the deployment of a single, global, roaming system. Designing P2PWNC so that its accounting system is efficient, incentive-compatible, and secure against Sybil attacks[2] and collusion is our primary goal.

In the P2PWNC terminology, to join P2PWNC a participant only needs to set up a *P2PWNC server* that adheres to the *P2PWNC protocol*. We assume, however, that strategic or malicious participants can tamper with these servers and make them execute arbitrary variations of the protocol. Potential hosts of P2PWNC servers could include, for example, residential WLAN access points (access *routers*, to be more precise). These devices only require a few software modifications in order to function as P2PWNC servers.

To access these servers, roaming users would require a *P2PWNC client*. We assume clients are non-tamperproof also; clients could include, for example, WLAN-enabled mobile phones, PDAs, and laptops. Digital certificates stored in the clients would affiliate a client with its home server. Since P2PWNC users should also be operators of P2PWNC servers, such an affiliation should always exist. We use the term *administrative domain* (or simply *domain*) to describe the collection of components and users consisting of (1) a home server and its local WLAN, (2) the server's affiliated clients, and (3) the users of these clients, which we will also call the domain's *members*. A domain, such as a household, constitutes a single P2PWNC peer and P2PWNC can be seen as a P2P community of administrative domains.

Here we will consider only citywide P2PWNCs comprising, perhaps, a few thousand WLANs (enough WLANs to cover almost all areas of a densely populated metropolis). We limit ourselves to citywide P2PWNCs because, as we will show, the design of the accounting mechanism assumes a specific user mobility pattern that is appropriate for urban areas but cannot be applied *as-is* to a sparse (e.g. a nationwide) P2PWNC. Furthermore, a basic rule is that only participants who *can* actually provide service may rip the benefits of P2PWNC. People that live in areas where visitors are extremely rare would either need to become affiliated with another more successful WLAN or place their P2PWNC servers in more public locations. We do not expect these limitations to be significant in practice because P2PWNC would be advantageous mostly as an everyday alternative to expensive cellular broadband services, and not as a "24/7" reliable commercial service. It is reasonable to assume that the P2PWNC "anarchy" costs in efficiency compared to a more regulated and centralized model. However, our results indicate that P2PWNC could be practical, even if sub-optimal.

## DESIGN

Throughout the following analysis, the terms *normal* and *normally* will refer to the correct execution of the proposed P2PWNC protocol. We assume that strategic and/or malicious participants can always deviate arbitrarily from this protocol (but are limited to performing computationally feasible calculations). Before presenting P2PWNC's design, we restate a number of its required characteristics. First, P2PWNC is open to all. To join it, participants only need to setup a server and start accepting P2PWNC visitors. Second, P2PWNC is self-organizing. All its participants have equivalent roles and no authorities (distributed or centralized) are relied upon, not even when new participants join. Third, since the two P2PWNC software components (clients and servers) may run on relatively resource-constrained devices (mobile handsets and home access points respectively), device constraints must be taken into account.

## P2PWNC Identities, Entities, and Certificates

A *P2PWNC identity* is a unique public-private key pair. The private key is normally kept secret by the entity that wishes to adopt this identity and is used to sign digital statements. Assuming knowledge of the public key, anyone can verify that a statement was signed by this private key (i.e. this identity) and by extension, by the entity that is in possession of the private key. We will refer to a public key as the *name* of the corresponding entity from now on.

Normally, only two types of entities would adopt P2PWNC identities: P2PWNC servers (the providers in P2PWNC) and P2PWNC clients (the devices used by roaming users when consuming the resources of providers). The name of a P2PWNC server also acts as the name of the corresponding P2PWNC domain. An entity could, of course, create an arbitrary number of identities simply by generating appropriate key pairs. However, if entities guard their private keys, other entities cannot hijack their identities (i.e., cannot sign statements in their name).

There are two types of *identity certificates* in P2PWNC: server (or domain) certificates and client (or user) certificates. Server certificates are signed statements that bind a server name to the server's IP address. Because in P2PWNC there are no trusted third parties to certify this binding, server certificates are simply self-signed, i.e. the private key that corresponds to the name included in the certificate is the key that signed the certificate.

Client certificates, on the other hand, are statements signed by a server's private key that bind the server's name to a specific client name and to an expiration date. The client certificate encodes the member-home domain affiliation, i.e., the fact that the user using the client is a member of the domain that operates the home server. There could be many ways for domains to interpret membership, and we consider this problem out of our scope. However, we do assume that domains would in general be small and symmetric in their consumption and contribution ability.

P2PWNC clients, in addition to storing their own identity and certificate, also store a copy of their home server certificate. By presenting both certificates to foreign servers, P2PWNC clients declare (1) their name, (2) the name of their home server, and (3) their server's IP address.

Because P2PWNC clients may get lost or stolen (and in addition to the usual protections like PIN numbers and passwords), the client certificates contain an expiration date. After this date, visited servers would normally consider them invalid and would ignore the devices that presented them. Legitimate domain members, therefore, need to renew their client certificates with the help of their home server. An adversary cannot create client certificates that affiliate a client with a specific home server without access to the home server's private key. However, anyone can affiliate a client with a (fake) server identity created just for this purpose. Note that from the point of view of a visited server, a client certificate affiliates a device with a home server name and IP address, but nothing more. We have not yet established if this server name corresponds to a real contributing P2PWNC domain.

P2PWNC domains represent the peers of P2PWNC, as we mentioned before. A separate P2PWNC server resides in every domain. These servers operate as WLAN access points and IP routers, and provide services to both domain members and nearby visitors. Everyday server usage by local domain members is not considered contribution to the community and does not involve P2PWNC mechanisms. Only visitor sessions are considered P2PWNC transactions. In the P2P view of P2PWNC, the two transacting peers are always the visited domain and the home domain of the visitor, representing the provider and the consumer respectively.

When a server provides service to visiting clients, clients sign a *receipt* that servers then store locally to use as proof of contribution. Clients sign receipts with their private keys and this takes place in the background of a WLAN session at specific intervals determined by the server. Each receipt contains (1) a copy of the client certificate, (2) a copy of the client's home server certificate, (3) a copy of the visited server certificate, (4) a timestamp, and (5) session-specific information (more specifically, the amount of traffic in kilobytes forwarded for the client by the server).

By using the certificates the receipts contain, any interested party examining a receipt can check if the client private key that corresponds to the client name found on the receipt has indeed signed this receipt. They can also check if the home server private key that corresponds to the home server name found on the receipts has signed the client certificate. By following what is essentially a two-certificate chain, interested parties can verify the identity of both the consuming and the providing server involved in the transaction that a receipt encodes.

A P2PWNC transaction normally proceeds as follows. When a roaming user first requests service from a visited domain, the client presents its two certificates (the client certificate and the home server certificate, which are both stored in the client device) to the visited server. Clients normally detect the presence of servers through appropriate beacon messages that servers broadcast, which include the visited server certificate.

For the analysis below, assume for the sake of clarity that the visited server has somehow been satisfied with the client's credentials and authorizes use of the local WLAN. The P2PWNC transaction that follows is a series of *post-pay* exchanges. In the beginning, the server forwards only a limited amount of Internet traffic for the user. The server then asks the client to sign a receipt with appropriate parameters. If the client declines, no further resource consumption is allowed. To avoid persistently dishonest clients, the server may keep track of all possible client-identifying elements (P2PWNC identities, link-layer (MAC) addresses, and, if possible, information such as device location and distance) and attempt to increase the cost of dishonest behavior by delaying or denying suspect login requests. If the client does sign a receipt and the server confirms the signature via the client's public key, the receipt is stored in the server and the session is allowed to continue.

At intervals specified by the server, new receipts are requested from the client. In the session-specific field, server and client may account for any resource whose consumption they can both confirm (such as amount of traffic forwarded) so that the client agrees to sign it. In order to minimize the number of receipts that a server will eventually store, the server will ask from the client during the WLAN session to sign receipts that have the same timestamp. In such a case, and if the server accounts for volume of forwarded traffic, the volume across receipts would represent the total volume for the entire session up to that moment. Each such receipt would make the previous obsolete and the visited server would only store the latest one because other servers would not accept as proof of contribution more than one receipt signed by a specific client having the same timestamp.

## Attacks Against the Accounting Mechanism

In the interest of readability, we emphasize that the server and the affiliated clients that represent the two "faces" (provider and consumer) of a unique P2PWNC domain are both parts of the same P2PWNC peer. In what follows, when we refer to entities or nodes that "sign/store/present receipts," "are free-riding," "are visiting a server," or "are attacking the mechanism," our focus is always on P2PWNC peers even though the actual action may be performed by only one of the peer's two parts. We always imply that *clients* sign receipts, *servers* store and present these receipts to other servers, and clients visit servers. Also, if a domain/peer is labeled *attacker* or *free rider*, the label applies to both the home server and to all its affiliated clients, i.e. the entire domain/peer.

We define a directed *receipt graph* such that nodes represent domain names (i.e. server names) and edges represent receipts. Edges are directed from the consuming domain (the domain of the roaming visitor) to the providing domain (the domain of the visited server). This graph will play an important part in the following discussion. An edge exists if there is at least one receipt connecting the two domains, i.e. at least one client affiliated with the consuming domain has signed a receipt for resources consumed at the providing domain.

As we mentioned already, the primary reason for the existence of the accounting mechanism outlined above is to enable participants to detect free-riding visitors (more accurately, visitors from free-riding domains) in order to exclude them. However, as-is, the P2PWNC mechanism is exposed both to Sybil attacks[2] and to colluding participants. With Sybil attacks, an attacker can first create any number of server and client identities since creating such identities only requires the generation of zero-cost public-private key pairs. The attacker can then generate any number of client certificates that affiliate the client identities with any one of the server identities. Then, the attacker can use these client identities to sign any number of fake receipts that would show one of its server identities having provided service to an arbitrary number of clients from various home servers. Because each P2PWNC server is associated with an IP address that appears in its server certificate, and because other servers expect to find a server's stored receipts by querying this address, the attacker would have to control one or more IP addresses and include only these addresses in the various fake server certificates.

If the attacker visited a foreign server and presented all the receipts (visitors, or more accurately their home servers, "present" receipts by replying to receipt queries) that he fabricated as evidence of contribution, how would the visited server be able to detect that the attacker is really only a free rider? One possible approach would be for the visited server to examine the IP addresses of servers in the receipt graph (these addresses are contained in the server certificates which are, in turn, contained in the receipts), in order to search for suspicious patterns such as numerous addresses that belong to the same subnet, or multiple server names associated with the same IP address, or addresses that do not respond to network probes.

None of these possible defenses, however, is also *collusion-resistant*. Assume, for example, that the attacker has colluded with one or more owners of real P2PWNC servers (or random owners of Internet hosts). Then, the pattern of IP addresses contained in the graph can be made to look more realistic since any two colluding peers can always sign receipts for each other even if no service has actually been provided–this is an example of *false trading*. A fundamental problem with false trading is that if two or more peers collude like this there is no way for another server to detect it just by requesting a list of the stored receipts.

## The P2PWNC Authentication Algorithm

We propose a simple authentication algorithm for a prospective provider that is attempting to detect possible Sybil attacks and collusion. *The prospective provider should only authorize those visitors that can present*

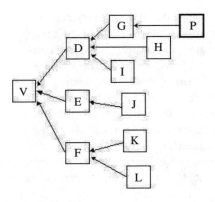

**Figure 7.1** A visitor from domain V is visiting domain P. The prospective provider P detects itself in the visitor tree (rooted at the visitor V). P can verify that the P → G receipt was indeed signed by one of its affiliated clients. The connecting path is P → G → D → V. Identities G, D, and V could be aliases of the same entity, or they could represent colluding domains. This does not change the fact that P did in fact receive service.

*receipts that form a tree, which contains the provider and is rooted at the visitor.*

This algorithm is based on the following assumptions: (1) P2PWNC domains only "trust" domains for which they have signed receipts, where by *trust* we mean "believe not to be a free rider," (2) P2PWNC domains can only accept as authentic the receipts that are signed by their own members, since any other receipt could be fabricated, and the certificates contained in it could be fake, (3) we assume that domains trust their members to sign only real receipts, (4) we assume that servers present their stored receipts whenever anyone requests them.

By detecting itself in such a *visitor receipt tree*, as seen in Figure 7.1, the prospective providing domain can trust that there exists at least a non-fake and non-free-riding domain in the receipt path that connects itself to the root (i.e. the visitor's home domain) because the provider trusts its roaming members to sign receipts only when resources are actually provided. The possibility that all the tree nodes could represent aliases of the visitor's home domain, or that they could represent domains that are in collusion with the visitor's home domain does not change the fact that the provider detected one of its prior consumption acts (i.e., consumption by one of its roaming members) through *this* tree. We will now see how the provider can use this information.

It is not uncommon[3] in open self-organizing P2P systems to use graphs such as the receipt graph above to establish trust between two previously unknown peers (or peer identities to be more precise). Here, we also make the following observation. For a P2P system like P2PWNC, its primary reason-of-being is to enable the consumption of resources provided by the P2PWNC community. As we already mentioned, our method for providing appropriate incentives is to require that peers "pay" the community in kind, in order to be able to consume community resources. If a prospective providing domain actually detects itself in a visitor tree as described above (possibly in more than one place in the receipt tree), the domain is "reminded" that it did in fact consume from the community.

Because, however, of the ease with which Sybil attacks and collusion can occur, a visitor tree where the above algorithm does not terminate successfully (i.e. the prospective provider does not detect itself in it) could very well represent a completely fabricated tree. If this were true, the provider would want to exclude the visitor in order to give him the correct incentives to participate in the community by contributing.

A fundamental P2PWNC guideline for designers of such contribution policies must be: "encourage visitors from domains that provably contribute to the community and discourage (by excluding) visitors from domains that cheat or free-ride." The above algorithm is a simple way to decide this. Although it can cause *false alarms* (see later), it is also an appropriate algorithm, considering the P2PWNC trust model.

Once they detect one of their receipts in the tree, peers should never consider this receipt in future authorization decisions again. This would protect against colluding peers that attempt to exploit a single prior consumption action of the prospective provider more than once, as seen in Figure 7.2. This rule applies only if edges have no weights.

If the edges of the receipt tree are somehow weighted (for example, if receipts contain the volume of traffic forwarded during the corresponding session) and if these weights are taken into account to calculate the amount of contribution to the requesting peer, the only weight worth considering is the weight of receipts signed by (members of) the prospective provider, since all the others receipts could have been fabricated. Also, a modified version of the previous rule applies, where the contributed amount of resources should now be subtracted from the weight before any future authorization decisions are made.

We claim that with the algorithm and the guidelines presented above, P2PWNC would give good chances to "good" peers to be accepted by the visited domains within which they happen to roam. This set of guidelines (or *strategy*), however, is quite general. In the process of evaluating them we will also propose a number of appropriate enhancements that will resolve some of the issues that will become evident.

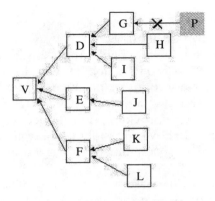

**Figure 7.2** **After authorizing usage of the WLAN by V, P should not take the connecting receipt (P → G) into account in authorization decisions again. If P did not do this, and if D were to visit P after V, a search conducted by P rooted at D would detect the same receipt. However, what happens if the D → V receipt is fake? (This could happen if D and V were both aliases of the same entity, or if D and V were colluding.) If P were to authorize D, the wrong incentives would be provided: colluding domains, for example, could serve P only once (corresponding to the P → G receipt) and then take advantage of this ad infinitum (the [G → D → V → …] path could be elongated indefinitely with Sybil attacks or collusion).**

## EVALUATION AND ENHANCEMENTS

### Expiring Receipts

Our first enhancement involves the timestamps found on the receipts. There are two reasons why these are important. First, since P2PWNC servers do not have infinite storage capacity they can only store a limited number of receipts that visitors from other domains sign. Is there a reason to discard only certain receipts and replace them with new ones, and if yes, which receipts should be discarded first? To answer this question we must refer to results from previous work on an analytic model of P2PWNC.[4] Those results indicate that there exists a simple contribution rule that can bring a system like P2PWNC near a good operating point from the perspective of total economic welfare: the *fixed-fee* rule in Reference 4 states that if all participants keep a fixed amount of resources *constantly* available for visitors, P2PWNC social welfare can approach the maximum achievable, as the number of participants becomes large, and if exclusions are possible.

This rule is impossible to enforce given the P2PWNC trust model: unless there is a trusted "police force" roaming from domain to domain, there is no other way to verify the availability of Internet bandwidth in a WLAN unless a trusted entity is physically present in the WLAN's coverage area. We can, however, "approximate" that fixed-fee rule. First, we note that the existence of stored (non-fake) receipts on a server indirectly proves that this server had resources available in the past. Second, we note that if most participants were to place greater "value" on receipts that are more recent, this would give incentives to other peers to seek recent receipts. In such a scheme, ideally, peers would want their contribution actions to occur just before their consumption actions. Assuming, however, that roaming clients wish to consume *whenever and wherever* they are (which is the reason they joined P2PWNC in the first place) and that clients have no way of contacting their home servers other than through P2PWNC itself, and assuming that their home servers are not being visited constantly by foreigners (i.e. visits are relatively rare), the need for recent receipts will drive home servers to keep their WLAN open, always waiting for new visitors, instead of just relying on visitor receipts they accumulated in the (perhaps distant) past. This would also naturally drive peers to discard old receipts in favor of new ones. In addition, this would allow new participants (or participants that changed their identity for some reason) to increase their standing in the community relatively fast, as old contribution actions (more specifically, the lack thereof) are not taken into account in any case.

## Simulation

How would discarding older receipts affect the authentication algorithm mentioned above? If older receipts were discarded, wouldn't paths in the receipt graph, connecting a prospective provider to a visiting consumer, be harder to find? How often would participants who do in fact keep a fixed amount of resources available for visitors be mistaken for free riders?

To measure the percentage of such *false alarms*, we programmed a custom simulation of a P2PWNC community that used a basic version of the authentication algorithm we described before: if, according to the algorithm, a connecting receipt path is found, the visitor is authorized to use the WLAN (and another new receipt is created as a result). If a connecting path is not found, the visitor will be denied access (and no new receipt will be created). In our simulations all authentications that fail to find a connecting path between prospective provider and visiting consumer represent a false alarm because none of the simulated participants is attempting to free ride: all simulated domains are always willing to contribute resources to visitors coming from domains that (provably) contribute to the community.

In order to choose an appropriate *user mobility model* for our simulations, we assumed that a metropolitan-sized P2PWNC would have some elements of a *small-world*.[5] What this means is that for relatively small communities (i.e. citywide P2PWNC communities) any two (active) participants would probably be connected via a short chain of receipts: a contributing peer would probably have consumed in the past either from the currently visiting consumer, or from a peer that consumed from the current consumer, and so on. Our first objective is to see how discarding older receipts affects the discovery of these paths.

Borrowing a small-world model from Reference 5, our set of nodes representing P2PWNC domains (WLANs) are identified with the set of lattice points in an $n$ x $n$ square. For a universal constant $l$ all domains have local relationships with domains within lattice distance $l$. Each domain also has exactly $d$ distant relationships with domains chosen uniformly at random during system initialization. These relationships may model social and organizational relationships that could cause interactions, and, by extension, P2PWNC transactions. The distant relationships in particular can also simply model, for example, specific WLANs that a user passes by on his way to work everyday. Note that the idea behind the local set of relationships in small-world graphs is that most of a domain's "local friends" are also "friendly" with each other.

For our set of simulations, all P2PWNC domains are *symmetric* (per our previous assumption) with only one affiliated client each. We used a slotted-time model and for a given set of *movement probabilities* encoded in the universal constants $p_l$, $p_d$, and $p_r$ (all in the range [0, 1]), all clients decide where to roam in every time slot, effectively spending $p_l$ fraction of their time accessing domains from the local set (choosing uniformly at random which one); $p_d$ fraction of their time accessing one of the $d$ domains in the distant set (choosing uniformly at random which one); and $p_r$ fraction of their time accessing any P2PWNC domain chosen uniformly at random. This last probability of movement acknowledges that completely random movement is also a possibility in such a setting. Note that the three probabilities do not necessarily add up to one, as there is always a fourth probability, which is that a client is either inactive, or home at the time. For our experiments, we generally vary these three probabilities from 0 to 0.05.

Given this model, the resulting receipt graph will have small-world properties, i.e. it will combine elements from random graphs due to the distant sets of relationships when $p_d > 0$ (and also because of random user movement when $p_r > 0$) and it would also contain connections caused by the local set of relationships when $p_l > 0$. These connections would give it a small average graph diameter.

In the simulations, to allow for the first receipts to be created, a percentage of the domains, $p_a$ in the range (0, 1), is altruistic at the beginning of time. Altruistic domains authorize visiting users without running the authentication algorithm. However, they do ask for signed receipts. We also apply the following heuristic in order to change an altruistic domain to a regular one (i.e. non-altruistic): every altruistic domain stops being altruistic when the domain's affiliated client successfully passes his first authentication in a non-altruistic domain. There are various local policies that domains may use to decide something similar. What is important is that new domains must start off by being altruistic in order to collect a number of receipts and that they should then eventually turn non-altruistic. Note that $p_a$ cannot equal one because in such a case the heuristic would not work and all domains would remain altruistic forever. It cannot equal zero either because no receipts would ever be created. For all our experiments we set $p_a$ equal to 0.99 and we observed how fast altruist numbers are reduced to zero. In these simulations we do not consider domains joining or leaving P2PWNC. In a real-world deployment, as we said, it would be necessary for new domains to remain altruistic for some time after they have joined the system in order to increase their good standing quickly, a behavior we approximate here with the $p_a$ parameter and the heuristic.

The receipts in our simulations do not contain any session-specific information. We assume that if a user is eventually authorized to use a visited domain, a new receipt connecting the consuming and the providing domain will be created. The offered resource that is accounted for is, in essence, a "WLAN session" of a standard duration.

The main result of our simulations is that for many sets of user mobility patterns there exists a certain value for what we call the receipt *TTL* (Time-to-Live) parameter, above which false alarms start to represent only a very small percentage of the total number of visitor authentications, and below which the system collapses (i.e. it eventually reaches 100% false alarms). The TTL represents the average duration that receipts should be kept by providers before being discarded. The higher the TTL, the greater, on average, the number of stored receipts a provider would have to keep in order to prove its good standing, but this way, good providers would be recognized as such for most of the time.

Note that in collapsing systems what happens is that, if a false alarm is signaled, this results in yet another receipt not being created. As time progresses, the total number of receipts in the system decreases (due to the discarding of older receipts) and, eventually, we return to the initial conditions (i.e. with no receipts in the system). However, without altruists anymore because of our assumed heuristic (which eventually turns all peers to non-altruistic), and since all visits result in false alarms, visitors are never allowed to access visited WLANs.

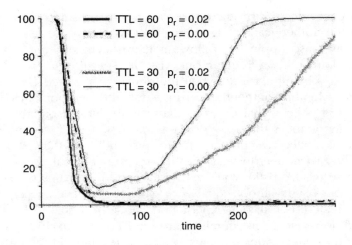

**Figure 7.3  The percentage of false alarms.**

## Results

We include here results from indicative simulation runs. We obtained similar results for a great number of P2PWNCs and user mobility patterns. Here, we simulate 1024 domains arranged in a 32x32 grid with $l = 1$, $d = 2$, $p_l = 0.05$, $p_d = 0.05$, and $p_a = 0.99$. In the following figures (Figures 7.3 and 7.4), we depict results from simulations with two different values for the remaining parameters. For the TTL, we chose 30 and 60 (note that time is measured in simulation steps). For $p_r$, we chose 0.00 and 0.02. For every step of the simulation, all users may visit foreign WLANs according to the preset relationships and probabilities of movement. With these parameters, users stay at home 90% or 88% of the time (if $p_r = 0.00$ or $p_r = 0.02$ respectively).

Figure 7.3 supports our main finding, namely that there exists a TTL value above which there would be insignificant gains in performance since false alarms are very near zero, and below which the system eventually collapses (false alarms reach 100% of authentications when, eventually, all old receipts are discarded and no new ones can be issued because altruists have disappeared). Note that random movement is not required to achieve low false alarm rates. Of course, in the case of system collapse, random movement can delay the collapse somewhat. In the real world, collapse can be avoided and false alarm percentages can be even lower if some domains remain altruistic or if domains employ an appropriate heuristic according to which they would switch from non-altruistic to altruistic and back again. Note that the percentage of false alarms is in

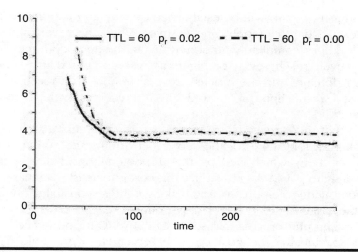

**Figure 7.4 The average tree depth at which a successful authentication test terminates**.

general near 1% (and can reach even smaller values). A failure percentage around 1% is near or below what we experience when, as cellular users in the process of making an outgoing call, a cellular base station cannot honor our request.

Figure 7.4 is a direct result of our small-world model and a characteristic of random graphs in general: in the two sustainable systems (those estimating TTL = 60) it takes only four steps on the average in the receipt graph to reach a domain from any other domain by following the receipt path that connects them. Also, it takes slightly less than four on the average if there is some element of completely random movement. A general result not shown here is that the more randomly connected the graph is (higher $d$ and $p_d$, or $p_r$) the easier it is to find connecting paths, even for small TTL estimations.

## Pre-computed Receipt Trees

Even though the authentication algorithm could work theoretically (i.e. it can detect all free riders and cause a very small percentage of false alarms), there is still the practical issue of efficiency. Additional simulation results not shown here suggest that even with only 1024 domains in P2PWNC, several hundreds of receipts need to be accessed in a breadth-first search before a connecting receipt path is established. The fact that the average tree depth computed in the previous section is close to four is misleading. In actuality what happens is that when consumers visit one of the domains in the local set (local, according to our small-world model), it is likely that a connecting path will be established quickly and that the search will terminate at a depth of only one or two. When, however, a consumer visits a distant or completely random domain, the tree depth may increase and may even reach seven or eight (as was computed in experiments with 1024 domains). In the process, several hundreds of receipts need to be accessed (results indicate around 100–200 receipts for the 1024-domain experiments).

Given that these receipts are stored across different P2PWNC servers, it could take a long time for the server performing the authentication to access all of them, which is a cost that the potential providing server may not be willing to pay. A way to reduce this cost is to assist potential providers by pre-computing receipt paths, and making it the responsibility of the visitor and of the visitor's home server. For example, each P2PWNC server can keep a background process that searches P2PWNC for new receipts.

More specifically, the goal of each server is to pre-build a tree of receipts of a certain depth and with a specific minimum number of distinct nodes, with the server in question at the root of this tree, and update it regularly to include new receipts that may have entered the system (and to discard older receipts according to the TTL estimation). A home server could then send latest versions of this tree to its roaming clients whenever the opportunity arose. This does not have to be very often, and it depends on the TTL. If an appropriate TTL is in the order of 5-10 days, then refreshing client trees at least once per day would allow clients to roughly keep up with the latest changes in the receipt graph–and in any case, "good" participants should have several redundant receipt paths connecting them to any other good participant. If visitor trees contain around 100-200 receipts signed using Elliptic Curve Cryptographic signatures, this would represent approximately 100-200 KBs of data, which is not a significant burden, even for (modern) mobile phones.

There is a way with which the cost of pre-computing trees can be further split between prospective consumers and prospective providers, further enhancing the efficiency of the scheme in an incentive-compatible way. Prospective providers can also maintain their *outgoing receipts* (sent to the servers by their roaming members, assuming that the members are actually trusted to do so). The purpose of the visitor tree and the provider's outgoing receipts is to serve as a subset of the P2PWNC receipt graph, and in particular, the subset that the provider will check in order to find a connecting path to the client. By relying only on these two structures, an online and expensive breadth-first search across many domains can be avoided. Instead, by merging the two structures, the provider attempts to establish a path in their union with only local information. Finding such a path is equivalent to saying that the two structures would have at least one node in common. Using standard probability computations we can show that the chances of this happening increase rapidly even with a modest increase in the number of distinct nodes in the visitor tree and in the number of outgoing receipts that a server maintains. This is true even as the number of P2PWNC domains becomes large. (This is similar to the birthday paradox.)

Essentially, the authentication algorithm overhead could be moved to a background process, which would search through P2PWNC servers and build trees of a pre-determined size, without a serious effect on the results of the authentication algorithm. The visitor authentication problem is thus reduced to combining and checking two receipt structures of two previously unknown parties. The visitor pre-computed trees are essentially credentials that can (perhaps) prove a visitor's good standing, and the provider outgoing receipts are needed for efficiency and in order to increase the probability of detecting connecting paths (without considering outgoing receipts visitors would have to maintain, ideally, visitor trees containing every possible destination).

## RELATED WORK

The P2PWNC scheme and an initial economic analysis were presented before in Reference 6. There, P2PWNC was examined from a business perspective as a low-cost P2P alternative to existing WLAN roaming schemes and several implementation issues were outlined. In Reference 4, an analytic model was used to derive a near-optimal contribution policy for a very similar system using incomplete information.

P2PWNC is not the only scheme proposed for a WLAN-based wide-area Internet access infrastructure. Because WLAN technology is already enjoying success in laptops and PDAs, many commercial providers are seeing that the economic value of these devices is greatly reduced because WLAN coverage is still not ubiquitous. The lack of universal WLAN roaming standards has resulted in a fragmented market of commercial public WLAN operators: no single subscription can offer roaming users a substantial coverage "footprint." Without roaming support, operators provide only limited coverage and this fact makes it difficult for them to attract new customers and to fund additional investments in infrastructure. On the non-commercial side, the majority of residential, business, and university WLANs are, unfortunately, closed to outsiders, and the various open *Community Wireless Networks*[7,8] that are available in several cities cover (relatively) small areas only, because they rely solely on altruistic participants.

Efforts with the same goal as P2PWNC (i.e. ubiquitous wireless Internet) are under way. WLAN roaming standards are being established that allow customers from one operator to access WLANs belonging to other operators. These efforts focus both on the technical[9,10] and on the roaming settlement issues.[11,12] In parallel, several hotspot aggregators[13,14] are attempting to unite smaller operators by coordinating AAA (Authentication, Authorization and Accounting) and roaming settlement. These efforts still result in relatively "spotty" coverage even for the largest of aggregators, and at substantial fees for the roaming customer. Also, these solutions focus on commercial WLAN operators and are inappropriate, for example, for residential WLANs owned by individual householders.

P2PWNC's differences from these schemes are: (1) P2PWNC does not require any central authorities or roaming brokers, (2) P2PWNC is a pure P2P system where the only currency is payment in kind, and (3) P2PWNC is open to all WLANs that wish to join it. Community wireless networks are perhaps more similar to P2PWNC because they too can be based on a decentralized architecture. P2PWNC, however, unlike community wireless networks, does not simply rely on participant altruism but uses an accounting mechanism to provide incentives for contribution and form a P2P community where consumers of the P2PWNC resource (wireless Internet connectivity) must also be providers of this resource.

There exist companies[15,16,17] that follow a business model that has similarities to the WLAN-sharing aspect of P2PWNC. These models also involve the sharing of WLAN connections among householders and nearby visitors. However, unlike P2PWNC, these models involve a third party (usually the vendor of the software/hardware that enables this connection sharing) that "partners" with the household in a revenue-sharing scheme. More importantly, these business models do not assume a P2P scheme but require visitors to have a subscription with the third party, or to pay per wireless session using their credit card or prepaid cards. As such, these models are more similar to the hotspot aggregation model, but focus instead on residential providers and not on public venue owners. In P2PWNC, we assume that successful vendors of P2PWNC software/hardware components will not employ this kind of revenue-sharing scheme but would stay true to the P2P nature of P2PWNC, respecting the independence of P2PWNC's participants, and making profit by selling P2PWNC equipment only. That is why the main objective of this work is to produce an open P2PWNC specification that can be implemented by various manufacturers.

On the P2P research front, there are many proposed incentive mechanisms for contribution. One of the most popular P2P applications, the file-sharing Kazaa system,[18] employs a simple ranking scheme whereby peer contribution is awarded with advancement in rank. During congestion, a prospective provider can select and serve only those of the requesting consumers that have high-enough ranking. This scheme is, however, exposed to very simple attacks: the ranking of each peer is stored on the peer's local software agent, which can be tampered with[19] and the ranking set to an arbitrary value. As a result, prospective providers cannot trust the rankings that remote peers declare.

Other designs propose to store this ranking across several remote peers. In Reference 20, a peer's *karma* (its "standing" in the community) is kept by a set of peers called its *bank-set* that increase its karma as resources are contributed and decrease it as they are consumed. This design is, however, exposed to Sybil attacks, and requires strong identifiers and central authorities (such as credit cards, as the authors propose) to provide security. The authors claim that their secure entry algorithm can limit the rate of Sybil attacks even in a completely decentralized setting. However, the proposed cryptographic puzzle that new participants need to solve (in order to include the solution in their generated node ID) is not a serious setback for attackers. Attackers can still generate as many IDs as they want, it only takes longer (computationally) for them to do so. This problem is characteristic of a larger class of designs for completely self-organized systems that do not address Sybil attacks directly. After an attacker controls a substantial percentage of system identities all system assumptions regarding rank-holders and malicious node percentages are invalidated.

Solutions for the incentive problem that feel more natural include micropayment schemes. In Reference 21 such a scheme is proposed. Unfortunately, this scheme, as is the case with most micropayment schemes, assumes the existence of a bank or broker from whom peers must purchase the initial "digital coins." The focus in Reference 21 is mainly on reducing broker load (i.e., is the broker required to monitor every transaction, or not?). There are logical extensions to these schemes that distribute the bank across a subset of peers and rely on threshold cryptography for generating group signatures without exposing the group's private key to a single peer. Distribution solves a lot of scalability problems, as well as the single-point-of-failure problem, but not the administrative problem of empowering a set of users to function as "the bank," and effectively making them more equal than the others. Such systems cannot be initialized in a purely spontaneous manner. Also, distributed cryptographic protocols are generally complex.

Such a distributed scheme is employed by COCA,[22] which is not a P2P resource-sharing community, but rather a decentralized certification authority. A scheme like COCA could have been used to certify identities in P2PWNC (and, thus, protect against Sybil attacks), but the problems mentioned above would remain. The core COCA servers would form the logical equivalent of a central authority, and it is not clear how participants would join the core group. If only trusted peers join it, the trust model is not appropriate for P2PWNC. If anyone can join it, the core group itself is exposed to Sybil attacks.

The schemes described in References 3 and 23 resemble P2PWNC in that they use completely self-organized trust-inference schemes. The purpose of each system is different. Reference 3 describes a platform for implementing cooperative applications over the Internet. The authors use trust graphs and trust paths to enable unknown users to establish a level of trust. P2PWNC shares a lot of ideas with this work, even though our focus is on implementing a very specific system that needs to be deployable in the real world. The work in Reference 23 describes a way to establish trust in self-organized ad hoc networks, using chains of digital certificates.

## CONCLUSIONS

Is the P2PWNC scheme realistic? The technology to implement it already exists. In addition, broadband connections to the Internet are proliferating and WLAN signals will soon pervade many cities. Most of these resources, however, may remain under-exploited: WLAN owners are always being advised about the "security issues" of their networks, and WLAN vendors are always aiming for products that are resistant to outside attacks. P2PWNC, on the other hand, promotes a sharing regime that affects legitimate network usage, but only in a controlled way.

We believe that universal roaming at the cost of a single broadband subscription constitutes an appealing proposal, even with limited guarantees. Since P2PWNC is incrementally deployable, its success or failure does not have to involve financial risks and it is something that can be experimented with. The main output of this work will eventually consist of a protocol specification for both P2PWNC clients and servers.

A distinctive characteristic of P2PWNC is that the P2PWNC identities are zero-cost by design. In P2PWNC, the possibility for Sybil attacks and collusion is acknowledged in the basic design, and all design decisions revolve around these possibilities.

In conclusion, the P2PWNC system is an open and self-organized Internet access infrastructure, built on top of a Wireless LAN sharing community that can increase its coverage over time with independent and small investment decisions from individual participants. P2PWNC is a simple substitute to other proposed WLAN roaming schemes and joining it can be as simple as joining a file-sharing network. The peers in P2PWNC are independent WLANs that provide access to each other's owners.

Here, we presented the design of the P2PWNC accounting mechanism, we discussed its incentive and security problems, we evaluated the design, and we compared P2PWNC to related work.

This research is supported by the project "Mobile Multimedia Communications" (EP-1212-13), funded by the research program "Herakleitos--Fellowships for Research in the Athens University of Economics and Business," which is co-financed by the Ministry of National Education and Religious Affairs of Greece and the European Union, through the program "EPEAEK II."

# REFERENCES

1. Adar E, Huberman B. 2000. Free riding on Gnutella. *First Monday*, 5(10).
2. Douceur J.R. 2002. The Sybil attack. IPTPS'02: First International Workshop on Peer-to-Peer Systems, Cambridge, MA.
3. Lee S, Sherwood R, Bhattacharjee B. 2003. Cooperative peer groups in NICE. INFOCOM'03, San Francisco, CA.
4. Courcoubetis C, Weber R. 2004. Asymptotics for provisioning problems of peering wireless LANs with a large number of participants. WiOpt'04: Modeling and Optimizations in Mobile, Ad Hoc and Wireless Networks, Cambridge, UK.
5. Kleinberg J. 2002. The small world phenomenon: an algorithmic perspective. STOC'00: 32nd ACM Symposium on the Theory of Computing, Portland, OR.
6. Efstathiou E, Polyzos G. 2003. A peer-to-peer approach to wireless LAN roaming. WMASH'03: First ACM International Workshop on Wireless Mobile Applications and Services on Wireless LAN Hotspots, San Diego, CA.
7. Seattle Wireless Community Network. http://www.seattlewireless.net.
8. Athens Wireless Metropolitan Network. http://www.awmn.gr.
9. GSM Association. PDR IR.61, WLAN roaming guidelines. http://www.gsmworld. org/documents/wlan/ir61.pdf.
10. 3GPP. TS 23.234, 3GPP system to WLAN interworking. http://www.3gpp.org/ ftp/Specs/html-info/23234.html.
11. Anton B, Bullock B, Short J. 2003. Best current practices for wireless internet service provider roaming. Wi-Fi Alliance Public Document. http://www.wi-fi. org/opensection/downloads/WISPr_V1.0.pdf.
12. Wireless Broadband Alliance. http://www.wirelessbroadbandalliance.com.
13. Boingo Wireless Inc. http://www.boingo.com.
14. iPass Inc. http://www.ipass.com.
15. Netshare by Speakeasy Inc. http://www.speakeasy.net/netshare/
16. Linspot by Biontrix b.v.b.a. http://www.linspot.com.
17. Pacific Wi-Fi. http://www.pacificwi-fi.com.
18. Kazaa by Sharman Networks Ltd. http://www.kazaa.com.
19. Kazaa Hack. http://www.khack.com.
20. Vishnumurthy V, Chandrakumar S, Gun Sirer E. 2003. KARMA: A secure economic framework for peer-to-peer resource sharing. IPTPS'02: First International Workshop on Peer-to-Peer Systems, Cambridge, MA
21. Yang B, Garcia-Molina H. 2003. PPay: micropayments for peer-to-peer systems. CCS'03: 10th ACM Conference on Computer and Communication Security, Washington D.C.
22. Zhou L, Schneider B, Renesse R. 2002. COCA: A secure distributed online certification authority. *ACM Transactions on Computer Systems*, vol. 20, Nov. 2002.
23. Capkun S, Buttyan L, Hubaux J-P. 2003. Self-organized public-key management for mobile ad hoc networks. *IEEE Transactions on Mobile Computing*, 2(1).

# 8

## SPACE–TIME PROCESSING FOR WLANS

*Noor Muhammad Sheikh and Asad Ch*

### INTRODUCTION

The focus of recent research in wireless systems, especially wireless LANs, is to find ways and means to increase throughput or to decrease bit error rate (BER). One of the conventional approaches can be to increase the signal strength and thereby increase the signal-to-noise ratio (SNR); however, this is not a practical approach in wireless LANs because increasing output power for one of the LAN cards will increase the interference for the rest of the cards, decreasing overall system throughput. Alternatively, more bandwidth can be allocated, which is also not a practical solution because bandwidth is costly. Thus, the question is, how to decrease BER and effectively increase throughput? This can be achieved by the use of extensive error correction schemes and increase of redundancy in the data. However, if the redundancy increases to a certain threshold, it cannot be used for real-time networks because it will result in much higher processing delay. This is also the case with interleaving; up to a certain level, an interleaver can improve the system BER performance, but after that it will result in unwanted latency.

A discussion of the factors responsible for BERs can begin with channel equalizers, which can help to nullify the effect of channel impurities up to a certain level. Usually, equalizers try to equalize the effect of multipath fading, considered to be an enemy of the signal. If seen in a positive way, multipath is an ally rather than an enemy to the signal, and if we can use this multipath property properly, we can improve our throughput and overall system performance.

Such use is found in maximal ratio combining (MRC) with multiple antennas. One of the innovations in this field is space–time coding, in which we process the signal both in time and space.

Figure 8.1 shows the conventional MRC with one transmitting and two receiving antenna.[1] At a particular instant, a symbol is transmitted. The signal reaches the two receivers by two different paths, namely $h_1$ and $h_2$. These two paths can be modeled by two complex quantities having a phase and a magnitude:

$$h_1 = \alpha_1 e^{j\theta_1}$$

$$h_2 = \alpha_2 e^{j\theta_2}$$

$\alpha_1$, $\alpha_2$ are fading magnitudes, and $\theta_1$, $\theta_2$ are phase values.

Noise is added at each receiver, so the received signal at each receiver can by given as:

$$y_1 = h_1 S + n_1$$

$$y_2 = h_2 S + n_2$$

And in matrix form

$$\begin{pmatrix} y_1 \\ y_2 \end{pmatrix} = S \begin{pmatrix} h_1 \\ h_2 \end{pmatrix} + \begin{pmatrix} n_1 \\ n_2 \end{pmatrix}$$

The receiver combines the two received signals $y_1$ and $y_2$ as:

$$\tilde{S} = h_1^* y_1 + h_2^* y_2$$

$$\tilde{S} = h_1^* \left( h_1 S_x + n_1 \right) + h_2^* \left( h_2 S + n_2 \right)$$

$$\tilde{S} = \left( \alpha_1^2 + \alpha_2^2 \right) S + h_1^* n_1 + h_2^* n_2$$

Expanding the above, we get
  Choose $S_i$ iff

$$\left( \alpha_1^2 + \alpha_2^2 \right) \left| S_i \right|^2 - \tilde{S} S_i^* - \tilde{S}^* S_i \leq \left( \alpha_1^2 + \alpha_2^2 \right) \left| S_k \right|^2 - \tilde{S} S_k^* - \tilde{S}^* S_k \, \forall i \neq k$$

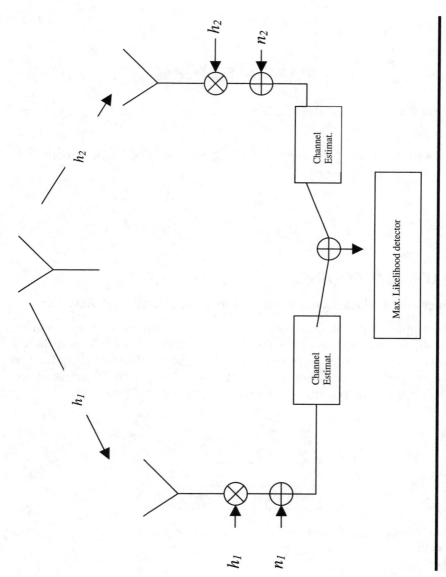

**Figure 8.1  Maximal ratio combining.**

Choose $S_i$ iff

$$\left(\alpha_1^2 + \alpha_2^2 + \alpha_3^2 + \alpha_4^2 - 1\right)\left|S_i\right|^2 + d^2\left(\tilde{S}_2, S_i\right) \le \left(\alpha_1^2 + \alpha_2^2 + \alpha_3^2 + \alpha_4^2 - 1\right)\left|S_k\right|^2 + d^2\left(\tilde{S}_2, S_k\right)$$

i.e., for $k$ choose $S_i$

$$d^2\left(\tilde{S}_1, S_i\right) \le \left(\tilde{S}_1, S_k\right) \forall i \ne k$$

Similarly, for $S_i$, choose $S_i$ iff

$$\left(\alpha_1^2 + \alpha_2^2 + \alpha_3^2 + \alpha_4^2 - 1\right)\left|S_i\right|^2 + d^2\left(\tilde{S}_2, S_i\right) \le \left(\alpha_1^2 + \alpha_2^2 + \alpha_3^2 + \alpha_4^2 - 1\right)\left|S_k\right|^2 + d^2\left(\tilde{S}_2, S_k\right)$$

or

$$d^2\left(\tilde{S}_2, S_i\right) \le d^2\left(\tilde{S}_2, S_k\right), \forall i \ne k$$

## SPACE–TIME CODING

### Space–Time Block Codes for Two Transmitters and One Receiver

The simplest form of space–time coding, mentioned by Alamouti, is given as follows: during the first symbol period, two symbols, $S_1$ and $S_2$, are transmitted simultaneously from $Tx_1$ and $Tx_2$, respectively. During the next symbol period, the transmitter $Tx_1$ transmits $-S_2^*$, and $Tx_2$ transmits $S_1^*$ Where (*) means complex conjugate,[1] as shown in Table 8.1:

**Table 8.1**

| Symbol Period | $Tx_1$ | $Tx_2$ |
|---|---|---|
| 1 | $S_1$ | $S_2$ |
| 2 | $-S_2^*$ | $S_1^*$ |

Assuming that channel response remains constant over the two symbol periods, the two paths $h_1$ and $h_2$ can be given as:

$$h_1\left(T = 1\right) = h_1\left(T = 2\right) = \alpha_1 e^{j\theta_1}$$

$$h_2\left(T = 1\right) = h_2\left(T = 2\right) = \alpha_2 e^{j\theta_2}$$

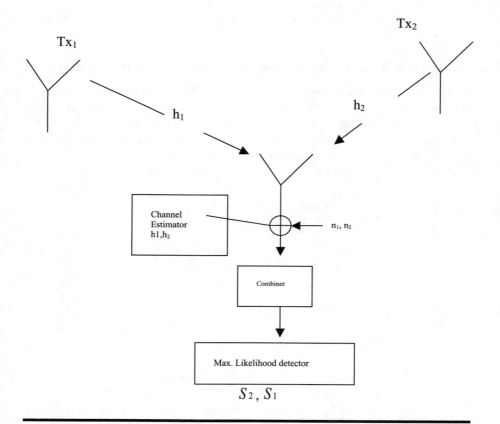

**Figure 8.2   Space–time block codes for Tx = 2 and Rx = 1.**

Now the received signal can be given as:

$$Y_1 = Y(T+1) = b_1 S_1 + b_2 S_2 + n_1$$
$$Y_2 = Y(T+2) = -b_1 S_2^* + b_2 S_1^* + n_2$$

The received signals are combined as follows to get $\tilde{S}_1$ and $\tilde{S}_2$:

$$\tilde{S}_1 = b_1^* Y_1 + b_2 Y_2^*$$
$$\tilde{S}_2 = b_2^* Y_1 - b_1 Y_2^*$$

$$\tilde{S}_1 = \left(\alpha_1^2 + \alpha_2^2\right)S_1 + b_1^* n_1 + b_2^* n_2$$

$$\tilde{S}_2 = \left(\alpha_1^2 + \alpha_2^2\right)S_2 - h_1 n_2^* + h_2^* n_1$$

## Space–Time Block Codes for Two Transmitters and Two Receivers

In the case of two transmitters and two receivers, the encoding and transmission scheme is the same as that for two transmitters and one receiver. The difference in this system is the number of channels, which have now been increased to four, i.e., now channel transfer functions are $h_{11}$, $h_{12}$, $h_{21}$, and $h_{22}$, as shown in Figure 8.3.[1]

Now we have four received signals, as shown in Table 8.2.

**Table 8.2**

| Symbol Period | Rx 1 | Rx 2 |
|:---:|:---:|:---:|
| 1 | $y_1$ | $y_3$ |
| 2 | $y_2$ | $y_4$ |

The four signals can be given as:

$$y_0 = h_{11}S_1 + h_{12}S_2 + n_1$$

$$y_1 = h_{11}S_2^* + h_{12}S_1^* + n_2$$

$$y_2 = h_{21}S_1 + h_{22}S_2 + n_3$$

$$y_3 = h_{21}S_2^* + h_{22}S_1^* + n_4$$

where $n_1$, $n_2$, $n_3$, and $n_4$ are complex random variables representing noise.

The output of the combiner to be sent to the maximum-likelihood detector is

$$\tilde{S}_1 = h_{11}y_1 + h_{12}y_2^* + h_{21}^*y_3 + h_{22}y_4^*$$

$$\tilde{S}_2 = h_{12}^*y_1 - h_{11}y_2^* + h_{22}^*y_3 - h_{21}y_4^*$$

Substituting values in the above equation, we get:

$$\tilde{S}_1 = \left(\alpha_1^2 + \alpha_2^2 + \alpha_3^2 + \alpha_4^2\right)S_1 + h_{11}^*n_1 + h_{12}n_2^* + h_{21}^*n_3 + h_{22}n_4^*$$

$$\tilde{S}_2 = \left(\alpha_1^2 + \alpha_2^2 + \alpha_3^2 + \alpha_4^2\right)S_2 + h_{11}n_2^* + h_{12}^*n_1 - h_{21}n_4^* + h_{22}^*n_3$$

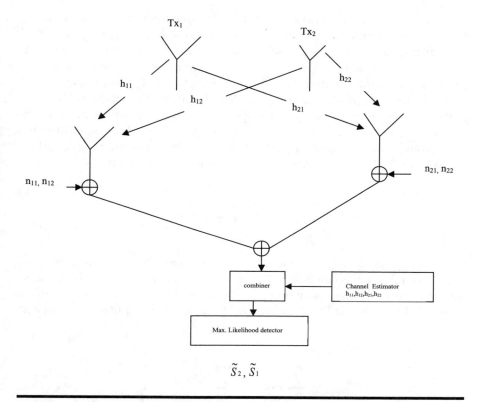

**Figure 8.3  Space–time block codes for Tx = Rx = 2.**

The results obtained by Alamouti's scheme for two transmitters and two receivers were similar to those obtained by four-branch maximal receiver ratio combining, but the power transmitted per antenna is halved as compared to the Maximal Ratio Combiner (MRC).

One of the most interesting results of Alamouti's scheme was that the combined signals from the two receiving antennas were obtained by the simple addition of the combined signals from each receiving antenna. The conclusion was that by using two transmitting and $m$ receiving antennas, with each of the $m$ antennas having its own combiner, and then simply adding the outputs of all the combiners, we obtain the same diversity order as $2m$ branch MRC.

## Generalization of Space–Time Block Codes for Any Number of Antennas

The extension of the space–time block codes was studied by Tarokh and colleagues, who presented a general technique for constructing space–time block codes for more than two Tx and Rx systems that they considered in a wireless communication system, in which the base station is equipped with $n$ and the remote terminal is equipped with $m$ antennas. At each time slot $T$, signals $S_T^i$ where $i = 1, 2, \ldots, n$ are transmitted simultaneously from the transmitting antennas. The coefficient $h_{i,j}$ is the path gain for the path between the $i$th transmitting and the $j$th receiving antenna. The path gains are modeled as sample independent complex Gaussian random. The wireless channel is assumed to be quasi-static so that the path gains are constant over a frame length and vary only from one frame to another.

At time $T$, the signal $y_T^j$ received by antenna $j$ is given as

$$y_T^j = \sum_1^n i h_{i,j} s_T^i + n_T^j$$

where $j = 1, 2, \ldots, m$ and where $n_T^j$ are independent samples of a zero-mean complex Gaussian noise. The average energy of the symbols transmitted from each antenna is normalized to be $1/n$. Assuming that perfect channel state information is available, the receiver computes the decision metric as

$$\sum_{T=1}^l \sum_{j=1}^m \left| y_T^j - \sum_{i=1}^n h_{ij} S_T^i \right|^2$$

over all code words

$$S_1^1 S_1^2 \cdots S_1^n S_2^1 S_2^2 \cdots S_2^n \cdots S_l^1 S_l^2 \cdots S_l^n$$

and decides in favor of the code word that minimizes this sum.

Given the perfect channel state information at the receiver, we may approximate the probability that the receiver decides erroneously in favor of a signal[2]

$$e = e_1^1 e_1^2 \cdots e_1^n e_2^1 \cdots e_2^n \cdots e_l^1 e_l^2 \cdots e_l^n$$

assuming that

$$S = S_1^1 S_1^2 \cdots S_1^n S_2^1 S_2^2 \cdots S_2^n \cdots S_l^1 S_l^2 \cdots S_l^n$$

was transmitted. This analysis leads to the following diversity criterion.

■ Diversity Criterion For Rayleigh Space–Time Code: To achieve the maximum diversity, the matrix

$$B(S,e) = \begin{bmatrix} e_1^1 - S_1^1 & e_2^1 - S_2^1 & \cdots\cdots & e_l^1 - S_l^1 \\ e_1^2 - S_1^2 & e_2^2 - S_2^2 & \cdots\cdots & e_l^2 - S_l^2 \\ e_1^3 - S_1^3 & e_2^3 - S_2^3 & \cdots\cdots & e_l^3 - S_l^3 \\ \vdots & \vdots & \cdots\cdots & \vdots \\ e_l^n - S_1^n & e_2^n - S_2^n & \cdots\cdots & e_l^n - S_l^n \end{bmatrix}$$

has to be full rank for any pair of distinct code words $S$ and $e$. If $B(S,e)$ has minimum rank $y$ over the set of pairs of distinct code words, then a diversity of $ym$ is achieved, in which $m$ is the number of receiving antennas.

These codes perform well in Rician environments in the absence of perfect channel state information and under a variety of mobility conditions and environmental effects.

## CODE DESIGN AND DECODING

### Code Design Basics (Hurwitz–Radon Problem)

A real orthogonal design of size $n$ is an $n \times n$ orthogonal matrix with entries $\pm S_1, \pm S_2, ..., \pm S_n$. The existence problem for orthogonal designs is known as the Hurwitz–Radon problem, and was completely settled by Radon. An orthogonal design exists only for n = 2, 4, or 8. Given an orthogonal design, one can negate certain columns to arrive at another orthogonal design in which all the entries of the first row have positive signs. Example of each of a 2 × 2, 4 × 4, and 8 × 8 orthogonal designs are given below:

$$\begin{pmatrix} S_1 & S_2 \\ -S_2 & S_1 \end{pmatrix}$$

$$\begin{bmatrix} S_1 & S_2 & S_3 & S_4 \\ -S_2 & S_1 & -S_4 & S_3 \\ -S_3 & S_1 & S_4 & -S_2 \\ -S_4 & -S_3 & S_2 & S_1 \end{bmatrix}$$

$$\begin{bmatrix} S_1 & S_2 & S_3 & S_4 & S_5 & S_6 & S_7 & S_8 \\ S_2 & S_1 & S_4 & -S_3 & S_6 & -S_5 & -S_8 & S_7 \\ -S_3 & -S_4 & S_1 & S_2 & S_7 & S_8 & -S_5 & -S_6 \\ -S_4 & S_3 & -S_2 & S_1 & S_8 & -S_7 & S_6 & -S_5 \\ -S_5 & -S_6 & -S_7 & -S_8 & S_1 & S_2 & S_3 & S_4 \\ -S_6 & S_5 & -S_8 & S_7 & -S_2 & S_1 & -S_4 & S_3 \\ -S_7 & S_8 & S_5 & -S_6 & -S_3 & S_4 & S_1 & -S_2 \\ -S_8 & -S_7 & S_6 & S_5 & -S_4 & -S_3 & S_2 & S_1 \end{bmatrix}$$

These matrices can be identified with the complex number $S_1 + S_2 j$ and the quaternionic number $S_1 + S_2 i + S_3 j + S_4 k$, respectively.

## The Coding Scheme

Tarokh and colleagues used the theory of orthogonal designs to construct generalized space–time block codes for any number of transmitter and receiver antennas. It is assumed that transmission at the baseband employs a real signal constellation with elements providing a diversity order of $nm$. The maximum transmission rate is $b$ bits per second per hertz (bits/s/Hz). At each time slot, $nb$ bits arrive at the encoder and select constellation signals, at each time slot $T = 1, 2, \ldots, n$, the entries $S_T^i, i = 1, 2, \ldots, n$ are transmitted simultaneously from transmit antennas $1$, $2, \ldots, n$.

## The Decoding Algorithm

The rows of the orthogonal design matrix are all permutations of the first row with possibly different signs. Let $\varepsilon_1, \varepsilon_2, \ldots, \varepsilon_n$ denote the permutations corresponding to these rows and let $\delta_k(i)$ denote the sign of $S_i$ in the $K$th row of the matrix. Then $\varepsilon_k(p) = q$ means that $S_p$ is up to a sign change of the $(K, q)$th element of the orthogonal design matrix. As the columns of the orthogonal design matrix are pairwise-orthogonal, it turns out that minimizing the metric of $\sum_{T=1}^{l} \sum_{j=1}^{m} \left| y_T^j - \sum_{i=1}^{n} b_{ij} S_T^i \right|^2$ amounts to minimizing $\sum_{i=1}^{n} S_i$,

where

$$S_i = \left( \left| \left[ \sum_{T=1}^{n} \sum_{j=1}^{m} y_T^j b_{ET(i),j}^* S_T(i) \right] - S_i \right|^2 + \left( -1 + \sum_{b,e} |b_{k,l}|^2 |S_i|^2 \right) \right)$$

and where $h^*_{\varepsilon T(i),j}$ denotes the complex conjugate of channel response. The value of $S_i$ depends only on the code symbol, the received symbols, the path coefficients, the structure of the orthogonal design, and the maximum-likelihood detection rule is to form the decision variables

$$Y_i = \sum_{T=1}^{n} \sum_{j=1}^{m} y_T^j h^*_{ET(i),j} S_T(i)$$

for all $i$ = 1, 2, ..., n and decide in favor $S_i$ of among all the constellation symbols if

$$S_i = org \, \underset{S \notin A}{\min} |Y_i - S|^2 + \left(-1 + \sum_{k,l} |h_{k,l}|^2 \right) |S|^2$$

## Complex Orthogonal Designs

A *complex orthogonal design* of size $n$ is defined as an orthogonal matrix with entries $\pm S_1, \pm S_2, ..., \pm S_n$, their conjugates $\pm S_1{}^*, \pm S_2{}^*, ..., \pm S_n{}^*$, or multiples of these entries by $\pm i$, $i = \sqrt{-1}$. The method of encoding is the same as described previously for real designs. The decoding metric again separates into decoding metrics for the individual symbols $S_1$, $S_2$, ..., $S_n$. An example of a complex orthogonal design is[2]

$$\begin{pmatrix} S_1 & S_2 \\ -S_2^* & S_1^* \end{pmatrix}$$

This design is the same as given by Alamouti. Maximum-likelihood detection amounts minimizing the decision statistic

$$\sum_{j=1}^{m} \left( \left| y_1^j - h_{1,j} S_1 - h_{2,j} S_2 \right|^2 + \left| y_2^j + h_{1,j} S_1^* - h_{2,j} S_1^* \right|^2 \right)$$

over all possible values of $S_1$ and $S_2$. The minimizing values are the receiver estimates $S_1$ and $S_2$, respectively. This is equivalent minimizing the decision statistic for detecting $S_1$

$$\left| \sum_{j=1}^{m} \left[ y_1^j - b_1^* + \left( y_2^j \right) b_{2,j} \right] - S_1 \right|^2 + \left( -1 + \sum_{j=1}^{m} \sum_{i=1}^{m} |b_{i,j}|^2 \right) |S_1|^2$$

and the decision statistic for detecting $S_2$

$$\sum_{j=1}^{m} \left| \left( y_1^j b_{2,j}^* - \left( y_2^j \right)^* b_{1,j} \right) - S_2 \right|^2 + \left( -1 + \sum_{j=1}^{m} \sum_{i=1}^{2} |b_{i,j}|^2 \right) |S_2|^2$$

## Sporadic Codes

For $n = 3$, 4, rate-¾ generalized complex linear processing orthogonal designs given by Tarokh and colleagues are:

For $n = 3$

$$\begin{bmatrix} S_1 & S_1 & \dfrac{S_1}{\sqrt{2}} \\[2ex] S_2^* & S_1 & \dfrac{S_1}{\sqrt{2}} \\[2ex] \dfrac{S_3^*}{\sqrt{2}} & \dfrac{S_3^*}{\sqrt{2}} & \dfrac{-S_1 - S_1^* + S_1 - S_2^*}{2} \\[2ex] \dfrac{S_3^*}{\sqrt{2}} & \dfrac{S_3^*}{\sqrt{2}} & \dfrac{S_2 + S_2^* + S_1 - S_1^*}{2} \end{bmatrix}$$

For n = 4

$$\begin{bmatrix} S_1 & S_2 & \dfrac{S_3}{\sqrt{2}} & \dfrac{S_3}{\sqrt{2}} \\[2ex] -S_2^* & S_1^* & \dfrac{S_3}{\sqrt{2}} & -\dfrac{S_3}{\sqrt{2}} \\[2ex] \dfrac{S_3^*}{\sqrt{2}} & \dfrac{S_3^*}{\sqrt{2}} & \dfrac{-S_1 - S_1^* + S_1 - S_2^*}{2} & \dfrac{-S_2 - S_2^* + S_1 - S_1^*}{2} \\[2ex] \dfrac{S_3^*}{\sqrt{2}} & \dfrac{S_3^*}{\sqrt{2}} & \dfrac{S_2 + S_2^* + S_1 - S_1^*}{2} & \dfrac{-\left( S_1 - S_1^* + S_2 - S_2^* \right)}{2} \end{bmatrix}$$

These codes are designed using the theory of amicable designs. Apart from the two preceding designs, Tarokh and colleagues did not find any other generalized designs in higher dimensions with rate greater than 0.5.

## Maximum-A-Posteriori Decoding

Using Bayes' rule of the probability of $x_1, x_2, \ldots, x_k$ being transmitted, given that $y_1, y_2, \ldots, y_{qn}$ are received, is given by

$$P\left(x_1,\ldots,x_k \mid y_{11},\ldots y_{qn}\right) = P\left(y_{11},\ldots,y_{qn} \mid x_1,\ldots,x_k\right)^o P\left(x_1,\ldots,x_k\right)$$

According to Bauch, for a nondispersive Rayleigh Fading channel, it can be given as:

$$P\left(y_{11},\ldots,y_{qn} \mid x_1,\ldots,x_k\right) = \frac{1}{\left(\sigma\sqrt{2\pi}\right)^{qn}} \exp\left\{-\frac{1}{2\sigma^2}\sum_{t=1}^{q}\sum_{i=1}^{n}\left|y_{ti} - \sum_{j=1}^{P}b_{ij}g_{ji}\right|^2\right\}$$

where $\sigma^2$ is the noise variance and $g_{ij}$ is the entry of transmission matrix. Also,

$$P\left(x_1 \mid y_{11},\ldots,y_{qn}\right) = P\left(y_{11},\ldots,y_{qn} \mid x_1\right)^o P\left(x_i\right)$$

where $i = 1, 2, \ldots, k$

The above system for the case of two transmitters and two receivers can be given here now as: number of transmitters = number of receivers = $k = n = p = 2$. And assuming $P(x_1, \ldots, x_k) = C$, where $C$ is a constant, now the a-priori information is given as:

$$P\left(x_1,\ldots,x_k \mid y_{11},\ldots,y_{qn}\right)$$

$$= C \gg \frac{1}{\left(\sigma\sqrt{2\pi}\right)^{qn}} \exp\left\{-\frac{1}{2\sigma^2}\sum_{t=1}^{q}\left|y_{t1} - \sum_{j=1}^{P}b_{ij}g_{j1}\right|^2 + \left|y_{t2} - \sum_{j=1}^{P}b_{ij}g_{j2}\right|^2\right\}$$

$$= C' \gg \exp\left\{-\frac{1}{2\sigma^2}\sum_{t=1}^{q}\left|y_{t1} - b_{t1}g_{11} - b_{t2}g_{21}\right|^2 + \left|y_{t2} - b_{t1}g_{12} - b_{t2}g_{22}\right|^2\right\}$$

$$= C' \gg \exp\left\{-\frac{1}{2\sigma^2} \sum_{t=1}^{q} \left|y_{t1} - b_{t1}x_1 - b_{t2}x_2\right|^2 + \left|y_{t2} - b_{t1}\bar{x}_2 - b_{t2}\bar{x}_1\right|^2\right\}$$

where

$$C' = C \gg \frac{1}{\left(\sigma\sqrt{2\pi}\right)^{qn}}$$

Also,

$$P\left(x_1 \mid y_{11}, \ldots, y_{q2}\right)$$

$$= C' \gg \exp\left\{-\frac{1}{2\sigma^2} \sum_{t=1}^{q} \left[\left|y_{t1} - b_{t1}x_1\right|^2 + \left|y_{t2} - b_{t1}\bar{x}_1\right|^2\right]\right\}$$

$$= C' \gg \exp\left\{-\frac{1}{2\sigma^2} \sum_{t=1}^{q} \begin{bmatrix} -b_{t1}x_1\bar{y}_{t1} - \bar{b}_{t1}\bar{x}_1 y_{t1} - b_{t2}\bar{x}_1\bar{y}_{t2} \\ -\bar{b}_{t2}x_1\bar{y}_{t2} - \bar{b}_{t2}x_1 y_{t2} + \left|x_1\right|^2 \sum_{i=1}^{2} \left|b_{ti}\right|^2 \end{bmatrix}\right\}$$

where $\left|y_{t1}\right|^2 \left|y_{t2}\right|^2$ are constants with respect to $x_1$ so included in $C$. Applying more mathematics, we can obtain the following results:

$$P\left(x_1 \mid y_{11}, \ldots, y_{q2}\right)$$

$$= C \gg \exp\left\{-\frac{1}{2\sigma^2}\left[\left|\left[\sum_{t=1}^{q}\left(\bar{b}_{t1}y_{t1} + b_{t2}\bar{y}_{t2}\right)\right] - x_1\right|^2 + \left(-1 + \sum_{l=1}^{q}\sum_{i=1}^{2}\left|b_{li}\right|^2\right)\left|x_1\right|\right]\right\}$$

$$P\left(x_2 \mid y_{11}, \ldots, y_{q2}\right)$$

$$= C \gg \exp\left\{-\frac{1}{2\sigma^2}\left[\left|\left[\sum_{t=1}^{q}(\bar{b}_{l2}y_{l1} + b_{l2}\bar{y}_{l2})\right] - x_2\right|^2 + \left(-1 + \sum_{l=1}^{q}\sum_{i=1}^{2}|b_{li}|^2\right)|x_2|\right]\right\}$$

# CHANNEL ESTIMATION

## Channels Estimation for MIMO Space–Time Systems

A key feature of all space–time techniques is their being open loop, i.e., channel knowledge is not required at the transmitter. Although several noncoherent space–time coding schemes that do not require channel information at the receiver as well have been developed, they suffer a significant performance penalty compared with coherent techniques. The noncoherent techniques are more suitable for rapidly fading channels that experience significant variation within the transmission block. For quasi-static or slowly varying fading channels, training-based channel estimation at the receiver is a very common practice. More specifically, current single-antenna wireless packet communication systems provide for a training sequence to be inserted in a packet to aid in channel estimation at the receiver end. This motivates the need to develop practical high-performance training-based channel estimation algorithms for multiple-antenna systems. This can easily be achieved for narrowband transmissions that encounter flat fading by using orthogonal pilot training sequences. For the multiple transmitting and single receiving antenna transmission scenarios, the receiver observes the superposition of training sequences transmitted through different channels. The training sequences that achieve the channel estimation minimum mean square error (MMSE) have an impulse-like autocorrelation sequence and zero cross correlation. This last property makes the channel estimation problem different for multiple-antenna systems from single-antenna systems, and has motivated research in this area. For the multiple-transmit-antenna scenario, a straightforward method to achieve zero cross correlation is to transmit training symbols only from one antenna at a time. This approach results in a high peak-to-average power ratio and, hence, is undesirable in practice. For implementation purposes (to avoid nonlinear amplifier distortion), it is desirable to use constant-amplitude training sequences, which can be classified into two main categories according to the training symbol alphabet size. The first approach constructs optimal sequences from a root-of-unity alphabet without constraining the alphabet size. Such sequences are the perfect roots-of-unity sequences (PRUS) or polyphase sequences that have been proposed in the literature for different applications. For any training sequence length, there exist optimal training sequences that belong to a root-of-unity alphabet. The training sequence length determines the smallest possible alphabet size. The second approach in the literature constrains the training sequence symbols to belong to a specific constellation, typically binary phase-shift keying (BPSK) or quaternary phase-shift keying (QPSK), to have a simpler transmitter/receiver implementation. In this case, optimal sequences do not exist for all training lengths. Instead, exhaustive searches can identify suboptimal sequences according to some performance criteria. The training sequence best suited to a particular application depends on the training sequence length (which for standardized systems is predetermined), the number of channel taps to estimate, and the signal constellation used. A PRUS of a predetermined length may not belong to a standard constellation, and exhaustive searches are in many cases computationally prohibitive. Restricting the training sequence alphabet size to BPSK would reduce the search space to sequences, which is still large and would increase the achievable MMSE. The paper by Christina Fragouli and colleagues proposed a method to easily identify training sequences for multiple transmit antennas that have the following attractive properties:

1. They belong to a standard constant-amplitude signal constellation of size $2^m$, m = 1, 2, ..., such as BPSK, QPSK, 8-PSK, etc.
2. They can be easily identified or constructed for an arbitrary training sequence length and an arbitrary number of unknown channel taps.
3. They result in negligible MSE increase from the lower bound.

The main idea is to reduce the problem of designing multiple training sequences with impulse-like auto correlation and zero cross correlation to designing a single training sequence with impulse-like auto correlation. This makes exhaustive searches more practical and, thus, facilitates the identification of good training sequences. In some cases, no search is necessary because optimal sequences are available from published results in the literature. Moreover, when optimal sequences do not exist, instead of exhaustive searches they proposed a method that identifies suboptimal sequences from a standard signal constellation with a small MSE increase from the respective lower bound.

## Channel Model and Optimal Training Sequences

Consider a system that employs two transmit and one receive antennas. The analysis can be generalized to multiple transmit/receive antennas. Two signals $S_1$ and $S_2$ are simultaneously transmitted over two frequency-selective channels $h_1$ and $h_2$. Each channel is modeled as a finite-impulse response (FIR) filter.[3] The received signal can be expressed as:

$$y(k) = \sum_{i=0}^{-1} h_1(i)S_1(k-i) + \sum_{i=0}^{-1} h_2(i)S_2(k-i) + n(k)$$

where $n(k)$ is assumed to be additive white Gaussian noise (AWGN).

The observed training sequence output that does not have interference from information or preamble symbols can be expressed as:

$$y = Sb + n = \left[ S_1(L, N_t) S_2(L, N_t) \right] \begin{bmatrix} h_1(L) \\ h_2(L) \end{bmatrix} + n$$

and

$$S_i(L, N_t) = \begin{bmatrix} S_i(L-1) & \cdots\cdots & S_i(0) \\ S_i(L) & \cdots\cdots & S_i(1) \\ \vdots & \cdots\cdots & \\ S_i(N_t-1) & \cdots\cdots & S_i(N_t-L) \end{bmatrix}$$

For $i = 1, 2$. The linear least-square channel estimates, assuming that $S$ has full column rank, can be calculated as:

$$\hat{b} = \begin{bmatrix} \hat{b}_1 \\ \hat{b}_1 \end{bmatrix} = \left( S^H S \right)^{-1} S^H y$$

The channel estimation MSE is defined as

$$MSE = E\left[ (b - \hat{b})^H (b - \hat{b}) \right] = 26^2 tr\left( S^H S \right)^{-1}$$

The MMSE is expressed as:

$$MMSE = \frac{26^2 L}{(N_t - L - H)}$$

which is achieved if and only if

$$S^H S = \begin{bmatrix} S_1^H S_1 & S_2^H S_1 \\ S_1^H S_2 & S_2^H S_2 \end{bmatrix} = (N_t - L + 1)I_{2L}$$

## Complexity Reduction

A straightforward method of designing two optimal training sequences $S_1$ and $S_2$ of length $N_t$ for estimating two channels each of $L$ taps is to design instead a single training sequence of length $N_t^1 = N_t + L + 1$ to estimate a single channel with $L^1 = 2L$ taps as following[3]

$$y = S(L^1, N_t^1)h(L^1) + n$$

Again, for optimality

$$S^H(L^1, N_t^1)S(L^1, N_t^1) = (N_t - L + 1)T = L$$

and the sequences $S_1$ and $S_2$ are constructed as

$$S_1 = \begin{bmatrix} S(0) & \dots & S(Nt) \end{bmatrix}, \; S_2 = \begin{bmatrix} S(t) & \dots & S(Nt + L) \end{bmatrix} \sqrt{a^2 + b^2}$$

Thus, the multiple-training-sequence design problem can now be reduced to designing a single, but longer, optimal sequence that achieves the MMSE when estimating the longer channel impulse response with $L^1$ taps. A similar approach can be followed for more than two transmit antennas.

## A Block Code for Training Symbols

The code can be described by a block matrix applied to the training matrix that corresponds to the input training sequence $S$. The received output can be expressed as

$$\begin{bmatrix} y_1 \\ y_2 \end{bmatrix} = \begin{bmatrix} S & 0 \\ 0 & S \end{bmatrix} \begin{bmatrix} -I_L & I_L \\ I_L & I_L \end{bmatrix} \begin{bmatrix} h_1 \\ h_2 \end{bmatrix} + \begin{bmatrix} n_1 \\ n_2 \end{bmatrix}$$

Multiplying the received output with the transpose-conjugate matrix

$$\begin{bmatrix} y_1 \\ y_2 \end{bmatrix} = \begin{bmatrix} 2S^H S & 0 \\ 0 & 2S_2^H S_2 \end{bmatrix} \begin{bmatrix} h_1 \\ h_2 \end{bmatrix} + \begin{bmatrix} \bar{n}_1 \\ \bar{n}_2 \end{bmatrix}$$

$$\begin{bmatrix} \bar{n}_1 \\ \bar{n}_2 \end{bmatrix} = \begin{bmatrix} -S^H & S^H \\ S^H & S^H \end{bmatrix} \begin{bmatrix} n_1 \\ n_2 \end{bmatrix}$$

Also,

$$S^H S = (N_t - L + 1) I_L, and\, then$$

$$\begin{bmatrix} y_1 \\ y_2 \end{bmatrix} = 2(N_t - L + 1) \begin{bmatrix} h_1 \\ h_2 \end{bmatrix} + \begin{bmatrix} \bar{n}_1 \\ \bar{n}_2 \end{bmatrix}$$

## BLAST

Bell-Laboratories Layered Space–Time Systems (BLAST) is designed for high-data-rate transmissions and is particularly effective when a large number of antennas are deployed at both the transmitter and the receiver.[6]

### Introduction

Current wireless systems are facing capacity limitations, and it is imperative to overcome this bottleneck so that the emerging wireless services could be accommodated. Recent research has revealed that deploying multiple antennas at both transmitter and receiver sides increases channel capacity significantly without sacrificing bandwidth or power. A number of MIMO (Multiple Input/Multiple Output) antenna transmission/reception schemes have been proposed and, particularly, space–time codes are getting special attention.

In BLAST, the data stream is demultiplexed into independent substreams that are referred to as layers. These layers are simultaneously transmitted through multiple transmit-antennas and, at the receiver, they are successively detected using the interference cancellation and nulling algorithms. The central paradigm behind BLAST is the exploitation, rather than the mitigation, from multipath effects to achieve very high spectral efficiencies (bits/s/Hz), significantly higher than are possible when multipath is viewed as an adversary rather than an ally.

The work presented in will concentrate on a simplified BLAST architecture that is capable of transmitting/receiving at much higher data rates as compared to normally used transmission/reception schemes for mobile communication, at very low SNR without degrading the BER for a dispersive fading channel. The very first BLAST structure was proposed by Foschini and is known as Diagonal BLAST (D BLAST). It utilizes a number of transmitting and receiving antennas over which data is transmitted or received by the help of a layered code in which code blocks are dispersed across diagonals in space–time. In this structure, data rates increase linearly by increasing the number of transmitting (Tx) and receiving (Rx) antennas (assuming equal number of Tx, Rx antennas), with these rates approaching up to 90 percent of Shannon Capacity.[2] However, this scheme suffers from many implementation problems — specifically, its complexity. Therefore, another scheme known as Vertical BLAST (V BLAST) has been preferred for our architecture, which is simple to implement and is less complex.[6]

## System Overview

The input data stream is demultiplexed into $n$ data substreams; each stream is then multiplied with a specific code and transmitted through one of the $n$ transmitters with symbol rate $1/T$. All transmitters are ordinary vector-valued transmitters. Power transmitted by each transmitter is proportional to $1/n$ so that the total power transmitted is always constant irrespective of the number of transmitters.

All 1 to $m$ receivers are ordinary receivers. Each 1 to $m$ receiver receives all the signals transmitted by 1 to $n$ transmitters. Each transmitter uses all of the system bandwidth, and the overall system bandwidth is much less than that in the case of CDMA, i.e., the same as that used by any conventional transmitter (such as QAM).

## System Architecture

We assume that $n = m$ to facilitate simple iterative nulling and cancellation at the receiver. At the transmitter, the incoming information stream is serial-to-parallel converted to $n$ substreams. Each substream is associated with a transmit antenna. At each time instant, one symbol from each substream is transmitted from its corresponding transmit antenna, resulting in $n$ symbols transmitted simultaneously.[5]

The wireless channel is assumed to be rich-scattering. The fading between each transmit and receive antenna pair are assumed to be independent. The channel is also assumed quasi static, and the channel parameters are assumed to have been estimated at the receiver before the detection procedure. The received signal at the $m$ receive antennas can be organized into a vector after matched filtering and symbol rate sampling:

$$\mathbf{y} = [y_1 \cdots y_m]^T$$

The transmitted signal from the $n$ transmit antennas can also be organized in vector form:

$$\mathbf{S} = [s_1 \ldots s_n]^T$$

The received signal $\mathbf{y}$ can be expressed as a linear combination of the transmitted signal $\mathbf{S}$:

$$\mathbf{y} = \mathbf{HS} + \mathbf{z}$$

where $\mathbf{H} \in C^{m \infty n}$ is the complex channel matrix, and $\mathbf{z} \in C^m$ is the spatially and temporarily white zero-mean Gaussian noise vector collected from the $m$ receive antennas, with autocorrelation $2\mathbf{I}$. The elements of $\mathbf{H}$ are independent of one another due to the rich-scattering environment. The channel matrix $\mathbf{H}$ can be partitioned into its columns corresponding to the $n$ transmitted signals, and it is denoted as $\mathbf{H}_s$:

$$\mathbf{H}_n = [\mathbf{h}_1 \ldots \mathbf{h}_n]$$

## Detection Algorithm

The algorithm consists of the following steps repeated $M$ times:[5]
For $K = n$ to $1$:
    Step 1: Calculate the inverse of the correlation matrix as

$$(\mathbf{R}^K)^1 = (\mathbf{H}^K * \mathbf{H}^K)^1$$

    Step 2: Because the user's SNRs are inversely proportional to their respective diagonal entries of $(\mathbf{R}^K)^1$, find the smallest diagonal entry. Let $\mathbf{I}$ be the index of the smallest diagonal entry. Reorder $\mathbf{H}^K$ such that the th column and the last ($K$th) column are interchanged:

$$\mathbf{H}^{K'} = [\mathbf{h}_1 \ldots \mathbf{h}_K \ldots \mathbf{h}_{K1} \ \mathbf{h}] \overset{def}{=} [\mathbf{H}^{(K1)})\mathbf{h}]$$

where the deflated channel matrix $\mathbf{H}^{(K1)}$ is the same as $\mathbf{H}^{K'}$ with the last column $\mathbf{h}$ deleted.
    Step 3: Calculate the pseudoinverse matrix $(\mathbf{H}^{K'})^t$. Let the nulling vector $w$ be the last row of $(\mathbf{H}^{K'})^t$. The transmitted signal is detected as the closest point in the signal constellation

$$\tilde{S} = dec(\mathbf{wx})$$

where dec(·) is the slice function, which depends on the modulation used.

Step 4: Perform interference cancellation by subtracting the detected signal from the received signal:

$$\mathbf{y} \leftarrow \mathbf{y} - \tilde{S}_n \, \mathbf{h}_n$$

Choosing the signal with the largest SNR at each step for nulling and cancellation achieves the global optimization that minimizes the probability of symbol errors. So the optimal ordering is the ordering of decreasing SNR.

The original decorrelating decision-feedback multiuser detector was used for detecting multiple user signals of a synchronous CDMA system.[9] By making the connection between a BLAST system and a synchronous CDMA system, decorrelating decision-feedback methods can be applied to BLAST systems as well.

The received signal vector $\mathbf{x}$ is correlated with the conjugate transpose of the channel matrix. This correlation is analogous to the matched filter bank front-end of a CDMA multiuser receiver.

The correlator output $\mathbf{a} \in C^n$ is:

$$\mathbf{a} = \mathbf{H}^*\mathbf{y} = \mathbf{RS} + \mathbf{z}_1$$

where $\mathbf{R} = \mathbf{H}^*\mathbf{H}$ is a $n \times n$ cross-correlation matrix, and $\mathbf{z}_1$ is a zero-mean Gaussian noise vector with autocorrelation $2\mathbf{R}$.

The cross-correlation matrix can be Cholesky decomposed as $\mathbf{R} = \mathbf{LL}^*$, where $\mathbf{L}$ is a lower triangular matrix and $\mathbf{L}^*$ is its conjugate transpose. A filter with impulse response $\mathbf{L}^{-1}$ is applied to the correlator outputs $\mathbf{a}$ to whiten the noise:

$$\mathbf{a}' = \mathbf{L}^{-1}\mathbf{a} = \mathbf{L}^*\mathbf{S} + \mathbf{z}$$

Because $\mathbf{L}^*$ is upper triangular, the $k$th component of $\mathbf{a}'$ can be expressed as:

$$a_k = L_{k,k}^* S_k + {}^n_{i=k+1} L_{k,i} S_i + z_k$$

which contains only interference from ($n$ $k$) signals.

The last component $\mathbf{a_n}'$ contains no interference, so a decision for this transmitted signal can be made first: $\tilde{S}_n = dec(\mathbf{a_n}')$. The next signal can be detected by subtracting the interference contribution from the $n$th signal using the previous decision, i.e.,

$$\tilde{S}_{n-1} = dec(\mathbf{a_{n-1}}' \cdot \mathbf{L}^*_{n-1,n} \, \tilde{S}_n)$$

This procedure is repeated until all signals are detected. The above decorrelating decision-feedback method first cancels the interference using the feedback of previous decisions and then makes a decision on the current signal. The detection and decision-feedback are performed in decreasing order of received signal energies in the original decorrelating decision-feedback CDMA multiuser detector.

## REFERENCES

1. S. Alamouti, A Simple Transmit Diversity Technique for Wireless Communications, *IEEE Journal in Selected Areas in Comm.*, October 1998.
2. V. Tarokh, H. Jafarkhani, and A.R. Calder Bank, Space-Time Block Codes from Orthogonal Designs, *IEEE Trans. Inf. Theory*, vol. 45, no. 5, July 1999.
3. C. Fragouli, N. Al Dhahir, and W. Turin, Training Based Channel Estimation for Multiple-Antenna Broad Band Transmissions, *IEEE Trans. Wireless Commun.*, vol. 2, no. 2, March 2003.
4. N. Al Dhahir et al., Increasing Data Rate over Wireless Channels, *IEEE Commun Mag.*, September 2002.
5. M. Asad et al., A Simplified Blast Architecture for Achieving High Data Rates Over Dispersive Fading Channels, *IEEE International Multitopic Conference (INMIC 2003)*, Islamabad, Pakistan, 2003.
6. G.J. Foschini, Layered Space-Time Architecture for Wireless Communication in a Fading Environment When Using Multiple Antennas, *Bell Lab. Tech. J.*, 1(2):41–59, Autumn 1996.
7. P.W. Wolnioansky, G.J. Foschini, G.D. Golden, and R.A. Valenzuala, VBLAST — an architecture for achieving high data rates over rich scattering wireless channels, *Proc. International Symposium of Signals, Systems, and Electronics, ISSSE'98*, Pisa Italy, 1998.
8. W. Zha and S.D. Blostein, Modified Decorrelating Decision-Feedback Detection of BLAST Space-Time System.
9. A. Duel-Hallen, Decorrelating decision-feedback multiuser detector for synchronous code-division multipleaccess channel, *IEEE Trans. Commun.*, vol. 41, no. 2, pp. 285–290, February 1993.
10. W.-J. Choi, R. Negi, and J.M. Cioffi, Combined ML and DFE decoding for the V-Blast System, *Proc. IEEE ICC'00*, New Orleans, June 18–22, 2000.
11. A.H. Sayed, *Fundamentals of Adaptive Filtering*, Wiley Interscience, New York, 2003.
12. B. Hassibi, *A fast square-root implementation for BLAST, 34th Asilomar Conference on Signal, Systems and Computers*, Pacific Grove, CA, October 2000.
13. B. Hassibi and B. Hochwald, High-rate linear space-time codes, *IEEE International Conference on Acoustics, Speech, and Signal Processing*, Salt Lake City, UT, May 2001.
14. Y. Xin, Z. Wang, and G.B. Giannakis, Space-time diversity systems based on unitary constellation-rotation precoders, *IEEE International Conference on Acoustics, Speech, and Signal Processing*, Salt Lake City, UT, May 2001.
15. B.K. Ng and E. Sousa, Space-time spreading multilayered CDMA systems, *IEEE Global Telecommunications Conference (GLOBECOM 2000)*, San Franscisco, CA, November 2000.

16. Damen, A. Chkei, and J.-C. Belfiore, Lattice code decoder for space-time codes, *IEEE Commun. Lett.*, vol. 4, no. 5, pp. 161–163, May 2000.

17. J.G. Proakis, *Digital Communications*, third edition, McGraw-Hill, New York, 1995.

18. T.S. Rappaport, *Wireless Communications: Principles and Practice*, second edition, Prentice Hall, Upper Saddle River, NJ, 2001.

# 9

## WLAN AND 3G CELLULAR CONVERGENCE

*Sotiris I. Maniatis, Lila V. Dimopoulou, and Iakovos S. Venieris*

### INTRODUCTION

During the past few years, mobile communications has significantly proliferated throughout the world. Second-generation (2G) mobile systems have achieved a great penetration into the mobile voice and low-bit-rate (around 9.6 kbps) data services market. However, the capabilities of 2G networks with regard to data services are very limited, and costs have been discouraging, especially when compared with wired data technology. Because mobile communications has turned into a necessity nowadays, the commercial and research communities have begun to focus on the evolution of third-generation (3G) cellular mobile systems that can offer high-speed and sophisticated wireless data services in mobile environments at a lower cost.

3G technology is a direct evolution of 2G, with an intermediate step widely known as 2.5G cellular technology. In particular, using the fundamental infrastructure of 2G, 2.5G systems offer wireless data services at approximately 100 kbps, which still does not meet the requirements of today's multimedia applications and the expectations of business users. 3G technology offers data services at speeds ranging from 300 kbps to 2 Mbps, which are sufficient for market needs, while retaining the wide area coverage and high mobility as well as the established user subscriber base of 2G systems. 3G mobile networks are under continuous standardization efforts of the relevant forums, in such a way that their architecture and basic operational procedures are well defined. In spite of their promising new capabilities and the well-adapted strengths inherited from previous generations, we have witnessed a relatively slower than expected pace in their deployment and market penetration. The main reasons for this are the very high costs of radio spectrum licensing and the extra investment needed to purchase or upgrade new equipment, as well as the high operational and maintenance costs. Still, mobile operators plan to fully deploy and exploit 3G mobile networks.

Parallel to cellular technology, wireless local area network (WLAN) technology has not only evolved but also succeeded in penetrating the market of high-speed wireless data services. WLAN technology offers a relatively low-cost, simple, and easy-to-deploy wireless infrastructure allowing for high-speed access to the Internet. WLANs free the operator or the enterprise from the necessity of installing significant quantities of wiring, while enabling users to wirelessly connect to the WLAN infrastructure. At the same time, users benefit from high data service speeds ranging from 2 Mbps to 54 Mbps, which are significantly greater than those of 3G, rendering it more appealing to end users. Unlike 3G, WLAN operates over the unlicensed radio spectrum, greatly reducing the costs. However, this feature decreases WLAN's availability and reliability, because the radio spectrum can be shared, resulting in potential interference from other users of the band. Moreover, WLAN has a much more limited area coverage than 3G, making it suitable for covering small areas — often called *hot spots* — with low mobility requirements. Apart from enterprises and homes, public places with a substantial number of visitors, such as airports, stations, hotels, etc., are fine examples of hot-spot areas. In Reference 37, it is stated that there will be over 70,000 hot spots worldwide by the end of 2003. Lastly, WLAN standardization is confined to the wireless access segment. There is actually no standardization activity either for an overall architecture or for supporting specific data services apart from basic Internet access.

When considering 3G cellular and WLAN technologies, one might think that these technologies compete with each other, because they both provide high-speed wireless data services to mobile users. However, on closer inspection, a great degree of complementarity can be identified. We can envision an environment that enables the convergence of these technologies in such a way that WLAN is used in a hot-spot area with low mobility to offer high-speed data services, whereas 3G is mostly used over a larger geographical area supporting high mobility but not such high speeds, although the speeds are still sufficient for most of the prevailing data services. From the 3G operators' point of view, WLAN is a complementary access medium that is selectively used in hot spots and strategic locations with high network usage, and is offered to their already established subscriber base. From the users' point of view, WLAN provides data services of enhanced overall performance in hot-spot locations, in hot-spot locations in a transparent way, meaning that switching between 3G and WLAN does not require any user intervention or additional subscriptions.

It must be pointed out that convergence is necessary so that transition from one access medium to the other becomes as transparent and seamless to the user as possible. An environment that allows either 3G or WLAN connectivity — a decision mainly driven by coverage availability — is already in place, but the average user is still unable to take advantage of it. On the other hand, a more convergent, or integrated, approach allows for single authentication and accounting facilities, user mobility, ongoing session continuity, possibly smooth quality of service (QoS), and an enhanced data services set for end users. An integrated environment certainly requires more advanced interworking functionality between WLAN and 3G, both at the network and mobile terminal side.

In this chapter, the topic of WLAN and cellular convergence is presented. The section titled "WLAN and 3G Cellular Convergence Architecture" focuses on two generic interworking architectures, namely *tight coupling* and *loose coupling*, concentrating on their advantages and disadvantages as well as their differences. The differences lie not only in topology but also in the way certain fundamental operations, such as mobility, security, and QoS, among others, are handled. The section titled "Standardization Activities" presents the standardization activities in the field, whereas the section titled "An Architecture Beyond 3G" covers a visionary convergence architecture going beyond 3G, and highlights some important issues and possible research roadmaps to the future.

## WLAN AND 3G CELLULAR CONVERGENCE ARCHITECTURE

In the literature, there are two generic approaches toward an interworking architecture between WLAN and 3G cellular technologies, which are referred to as tight coupling and loose coupling. Both approaches have commonly known limitations and advantages. They also exhibit significant operational interworking differences. Before we present the two alternatives, and to make the presentation of the architectures as clear as possible, this section will first describe the primary architectural modules and operations of WLAN and 3G cellular systems.

### UMTS Architecture Primer

Currently, two 3G cellular technologies are in use: UMTS and CDMA2000. The Universal Mobile Telecommunications System (UMTS) has been defined by the International Telecommunication Union (ITU) and has evolved from the corresponding 2G (Global System for Mobile Communications, GSM) and 2.5G (General Packet Radio Services, GPRS) technologies. UMTS is subject to a continuous and worldwide standardization effort under the umbrella of the 3G Partnership Project (3GPP). The CDMA2000 system is a natural evolution of the North American Standard IS-95 and is being standardized by the 3GPP2 forum. At the time of this writing, 3GPP is ahead of 3GPP2 in the specification of the 3G–WLAN interworking architecture, so we restrict our discussion to UMTS, keeping in mind that analogous principles can be applied to CDMA2000 and other members of the 3G family.

Figure 9.1 presents the basic architectural modules, interfaces, and segments of a UMTS system. A 3G mobile operator owns and manages one public land mobile network (PLMN) infrastructure, which is considered an autonomous administrative domain. The PLMN is logically divided into a core network (CN) and an access network (AN). The CN is further divided into a circuit-switched (CS) domain for voice-related user traffic, and a packet-switched (PS) domain for data-related user traffic. For the sake of this chapter, we restrict ourselves to the PS domain only. The CN is further enriched with an Internet multimedia subsystem (IMS) offering multimedia IP services to the UMTS mobile terminals. In UMTS parlance, the latter are known as *user equipment* (UE).

A UE communicates with base stations — access points — that are within its wireless range over the Uu interface. A UMTS base station is called *Node B* and serves one or more cells. In the uplink, user data is transformed into radio network layer frames and transferred over the air interface to the appropriate Node B. Subsequently, this Node B forwards the user frames (after performing specific layer 2 operations) to the serving radio network controller (sRNC) over the Iub interface. When the UE has more than one radio connection with its serving RNC for carrying user data, through Node Bs belonging to its active set and connected to RNCs other than the serving one — a functionality known as *macrodiversity* — the sRNC gathers the radio network layer frames for the host from all other RNCs over the Iur interface and reassembles the user packet. The serving RNC component is the connection point from the AN toward the CN.

The PS domain of the CN consists of two fundamental nodes: the serving GPRS support node (SGSN) and the gateway GPRS support node (GGSN), which, assisted by the home subscriber server (HSS), offer the necessary operational functions and data services to the mobile host. The SGSN is the connection point from the CN toward the AN, whereas the GGSN is the gateway to external IP networks such as the Internet. The HSS is a support module with the role of the master database that holds subscription information for each user. The operational functions mainly include authentication and authorization, accounting and billing, security,

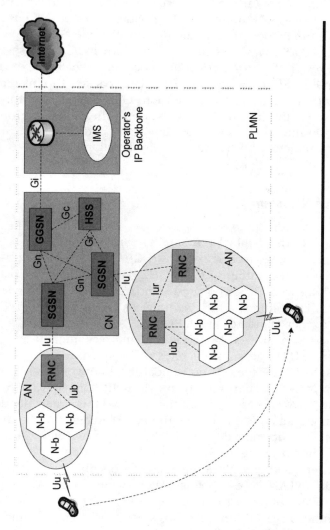

**Figure 9.1   3GPP UMTS architecture.**

mobility, and QoS, whereas data services primarily concern connection to the Internet and access to services of the 3G operator (such as multimedia services over the Internet multimedia subsystem) or other external providers.

Data transfer in the CN is based on the principles of the GPRS tunneling protocol (GTP). Actually, two GTP tunnels are established: one between the RNC and the SGSN (at the Iu interface) and a second one between the SGSN and the GGSN (Gn interface). Provided that the user has already registered with the PS domain for gaining access to the GPRS services (by means of the *GPRS attach* procedure), the formation of these tunnels comes as the result of a *PDP activation* procedure, which is executed by the host when it needs to transfer user data. User packets (uplink direction) from the serving RNC are encapsulated into GTP packets and forwarded toward the GGSN through the SGSN, whereas the GGSN performs the encapsulation in the reverse direction. Inter-RNC and inter-SGSN mobility are handled by the reestablishment — i.e., redirection — of the concerned GTP tunnels.

## A Typical WLAN System

Two dominant WLAN standards exist today: the family of IEEE 802.11 and HIPERLAN/2. The original IEEE 802.11 standard operates in the 2.4-GHz band and achieves speeds of up to 2 Mbps, whereas subsequent enhanced versions, operating in the 5-GHz band, will support data rates of 54 Mbps along with QoS. The latest standard (802.11i) further supports enhanced security features, which had always been the supposed drawback of the former standards. HIPERLAN/2 — also operating in the 5-GHz band — is standardized by the European Telecommunications Standards Institute (ETSI) and supports rates of 54 Mbps.

In contrast to UMTS, there is no standardized network architecture for WLANs. Each WLAN system merely provides IP connectivity to mobile hosts over the WLAN access network through two fundamentally different operational modes: a typical infrastructure-based architecture or an ad hoc one. Most of the currently deployed WLAN systems follow the former paradigm, as shown in Figure 9.2.

The Ethernet backbone plays the role of a distribution network interconnecting the WLAN access points (APs) to the WLAN Access Router (WAR). The WAR connects to an IP backbone that offers the basic IP operational functions to the mobile hosts, such as domain name system (DNS) and Dynamic Host Configuration Protocol (DHCP) functionality, as well as authentication, authorization, accounting, and billing. The main data service provided to users is connectivity to the Internet, although some basic application-level services can also be provided, e.g., by means of a Hypertext Transfer Protocol (HTTP) server.

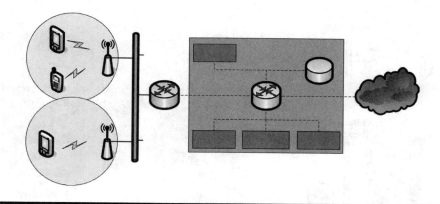

**Figure 9.2  A typical WLAN architecture.**

## Interworking Architectures: Tight and Loose Coupling

Currently, in the literature, there are two primary types of WLAN–3G interworking architectures, which were originally proposed by ETSI in Reference 8: tight coupling and loose coupling. The main difference between these two types stems from the "point" within the architecture where the two systems converge. To be more precise, in tight coupling, the WLAN access segment is considered to provide a direct alternative wireless access to the UMTS AN, in such a way that traffic coming from the WLAN passes through the UMTS CN. On the other hand, in loose coupling, there is no user plane overlapping but only utilization of specific operational functionality of the UMTS system, such as subscriber management. Figure 9.3 and Figure 9.5 depict these two interworking alternatives.

Starting with tight coupling, the interworking point is usually placed between the WLAN access segment and the SGSN, although some researchers have also proposed the point between the WLAN access segment and the GGSN. In both cases, some kind of interworking unit is needed to hide the WLAN access segment from the UMTS core. In the former case, the interworking unit resembles a pseudo-RNC that is connected to the SGSN through the standard Iu interface, whereas in the latter case, it is similar to a pseudo-SGSN connected to the GGSN through a standard Gn interface. Because the tight-coupling mode assumes that WLAN access is an integral part of the UMTS system, the protocol stack of the dual-mode mobile device must be UMTS specific at least up to the point of the radio link control and medium access control layer (RLC/MAC). At that point, a UMTS Adaptation Function (UAF) is needed to make the underlying WLAN access transparent to the upper layers. The user plane protocol stack for the mobile device and the UMTS interworking unit (UIU) are depicted in Figure 9.4.

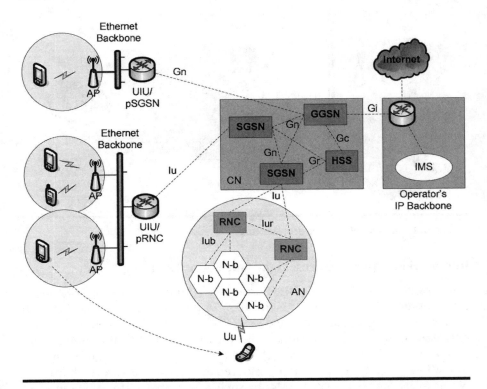

**Figure 9.3   Tight-coupling convergence architecture.**

The apparent advantage of tight coupling is that all UMTS operational mechanisms can be directly reused. Issues such as security, mobility, and QoS management are addressed by default. Because these mechanisms are used without modifications, their performance remains unaffected; however, the implementation details of the UIU and UAF may result in performance degradation. Furthermore, the UMTS core network resources and services can also be reused, such as the subscriber base and management mechanisms of the HSS or the multimedia services of the IMS. On the downside, the analysis, design, and implementation of the UIU and UAF are rather complex tasks. Moreover, capacity-related matters may emerge. An SGSN is designed to support a specific number of hosts generating specified traffic rates. In the case of a pseudo-RNC that hides a WLAN access segment, hosts will generate traffic at rates much higher than those specified. As a consequence, an advanced network planning tool must be used prior to system deployment. Lastly, a tight interworking paradigm requires that the WLAN be actually owned and operated by the 3G operator, because integration is needed at the system level — the 3G core has to directly expose its interfaces to the WLAN. Of course, this is considered a business matter rather than a disadvantage.

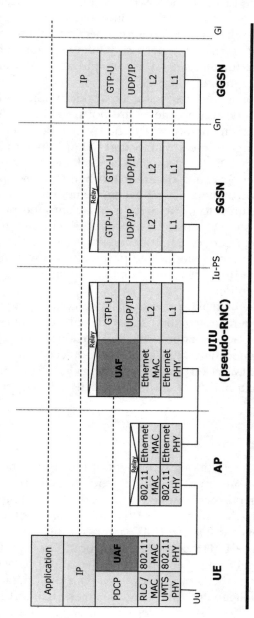

**Figure 9.4   User Plane Protocol Stack for Tight Coupling**

In contrast to tight coupling, the loose-coupling paradigm (Figure 9.5) does not need any kind of user plane integration. In terms of the reference architectures presented earlier, the interworking point is located at the Gi interface of the UMTS architecture. It is imperative that a separate backbone IP-based network serve the interconnection of the two wireless access networks by providing the necessary interworking mechanisms for the main operational functions such as authentication and authorization, billing, and mobility, among others.

With loose coupling, an explicit interworking unit and adaptation function at the network and host level are not needed. The deployment, operation, and management of UMTS and WLAN networks are mostly independent, allowing different UMTS and WLAN operators to achieve business-level agreements for providing a total interworking service to their users. However, because different mechanisms and protocols are used within each environment, the IP backbone network needs to provide certain interworking units to make these mechanisms and protocols operate together seamlessly. The operator may choose pure IP-based protocols for implementing the interworking functionality to alleviate this distressful situation. However, interworking at the protocol level will most probably result in degraded performance, in comparison with the total integrated solution of tight coupling.

Because the implementation needs of loose coupling seem to be less complex and time demanding than those of tight coupling, the current trend is to follow the "loose" approach. The provided interworking service will utilize the authentication, authorization, and accounting (AAA) mechanisms of UMTS (USIM-based) through the introduction of an AAA proxy server at the IP backbone, and adopt Mobile IP (MIP) protocol to cater to mobility.

## Interworking Issues

Independent of the interworking type, tight or loose, specific fundamental operational mechanisms must work together to support the seamless provision of data services. As noted earlier, tight coupling achieves an integrated mode of operation, where the mechanisms of UMTS can be directly reused, provided that the UMTS interworking unit and the mobile device implement the UMTS adaptation function. On the other hand, in loose coupling, interworking of mechanisms and protocols must be achieved externally at the protocol level. Among the operational mechanisms, mobility management (MM) and seamless handoff are of prime importance, because they support terminal mobility and allow the continuation of active user sessions. Other operational mechanisms include security issues, QoS, and support for enhanced applications, among others.

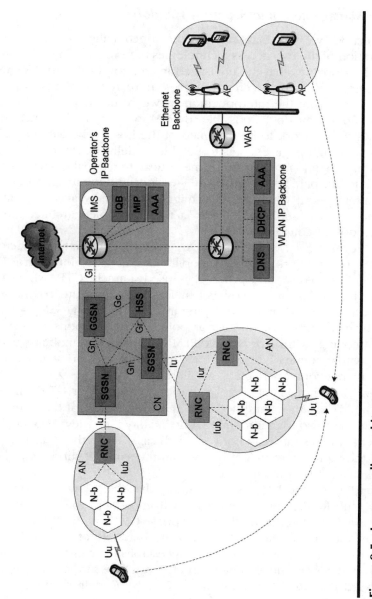

**Figure 9.5  Loose-coupling architecture.**

This section focuses on mobility, but it also provides brief insights into the other mechanisms.

### Mobility Management and Seamless Handoff

In this section, we will elaborate on the functionality required for the interoperation of the two access technologies in terms of MM and seamless handoff. It is obvious that the envisaged integration of 3G and WLANs is a challenging task, which certainly poses more requirements when the objective is to make the interoperation between the two technologies as seamless and as efficient as possible. By MM, we denote the functionality that allows the network to be informed of the host's whereabouts (either directly or through paging) so as to be able to deliver data to it. Handoff, on the other hand, involves the time-critical functionality that aims at maintaining transparency to active user sessions and minimizing the experienced interruption, when the host migrates to a new access point. Let us see how this is achieved, first in a tightly coupled architecture and then in a loosely coupled one.

In the former case, where the WLAN network is directly connected to the UMTS core network, seamless mobility is quite straightforward. Here, the well-defined GPRS MM protocol is used, namely GTP, which deals with data forwarding in a two-level hierarchy architecture composed of GGSN, SGSN, and RNC nodes. Keeping in mind that the GGSN — serving a fixed IP gateway — is the anchor point for the sessions of a mobile terminal moving within or across PLMNs, the host's mobility will be handled within the mobile operator's infrastructure and will leave the external IP network unaffected. As illustrated in Figure 9.6, intra-SGSN and inter-SGSN movements involve the establishment of GTP tunnels to update the MT's path within the UMTS network. Note that the inter-SGSN movement may also trigger roaming functionality if the new SGSN belongs to different administrative authorities. In all cases, from the IP layer's point of view, the mobile terminal moves within the same IP subnet and hence performs layer 2 (L2) movement.

Depending on the UMTS-specific entity that the WLAN network emulates, i.e., the RNC or SGSN, the UIU will implement the corresponding protocol stack for interfacing the appropriate UMTS node. When the UIU plays the role of an RNC (UIU/pRNC), the handoff of an MT from W-CDMA to 802.11 access will trigger the establishment of a GTP tunnel toward the SGSN (in movement 1) and the possible further update of the path toward the GGSN (in movements 2 or 3). Certainly, the update of the path is time consuming, which means that until it is performed, data following the old path are lost. This is handled within the UMTS network by means of a temporary forwarding tunnel (part of the serving RNC

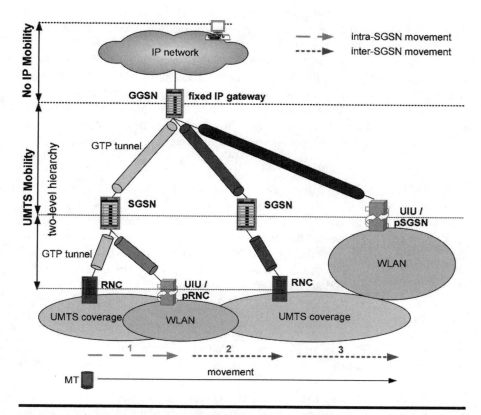

**Figure 9.6   MM in a tight-coupling approach.**

relocation procedure) established between RNCs for the redirection of data packets toward the new RNC.\* Buffering techniques are also employed here until the MT establishes the radio connection with the new RNC and starts receiving data. In the same manner, a forwarding tunnel can be set up between the RNC and the UIU for the temporary forwarding of packets, utilizing the standardized UMTS procedures.

When the UIU assumes the role of an SGSN (UIU/pSGSN), the handoff to WLAN access will trigger the establishment of a tunnel with the GGSN. However, the temporary forwarding of data between the RNC and UIU during the handover cannot take place as before, because the UIU

---

\* In UMTS networks, it is *soft handoff* that enables a seamless type of handover at layer 2. The forwarding tunnel only plays a supporting role for the time period that the involved RNCs switch roles. However, soft handoff is not applicable to inter-technology handoffs, where the forwarding of data is a more critical task.

emulates an SGSN. The standardized handoff procedure needs to be modified to allow the seamless integration of the two access technologies.

In the case where the WLAN is not directly connected to the UMTS core network — that is, it maintains no data interfaces with the UMTS CN — intersystem mobility management is coordinated at the IP layer using MIP.* Whereas different mechanisms will handle MM in the UMTS and the WLAN portions of the network, MIP will allow their seamless interoperation. The aim here is also to minimize, if not eliminate, the interruption in transport-level services running on the user's terminal while the latter moves across the heterogeneous parts of the interworking architecture.

In the WLAN–UMTS interworking architecture, let us imagine the WLAN network connected to the operator's IP network directly — the PLMN operator installs its own WLAN — or over public infrastructure such as the Internet. In all cases, the WAR is equipped with FA functionality to enable IP mobility for visiting MHs that wish to gain 802.11 access for their services. Similarly, GGSN acts as an FA as well. Let us assume that the dual-mode terminal supporting W-CDMA and 802.11 access (illustrated in Figure 9.7) has operator B as its home operator. When it powers up and attaches to a network to initiate a session with another host (a CH), mobile or stationary, it is located within the boundaries of operator A's network, and more specifically, under UMTS coverage. Note that the UMTS network, as well as the WLAN network, is an IP subnet from the IP layer's point of view.

The MH attaches to the visiting IP subnet and registers with its HA,** $HA_B$, in the home operator's domain to report its actual location, i.e., the

---

* *Brief introduction to MIP*: its fundamental principle is that a mobile host should use two IP addresses: a permanent address — the home address, assigned to the host and acting as its endpoint identifier — and a temporary address — the care-of address (CoA) — providing the host's actual location. Correspondent hosts (CHs) address the MH at its home address, whereas MIP-aware entities enable the delivery of packets to the host's actual location. These entities are the home agent (HA) and the foreign agent (FA). The HA maintains location information for the host. It resides in the host's home subnet and is responsible (when the MH is away from home) for intercepting packets addressed to the host's home address and tunneling them to its CoA, which is its current point of attachment. The FA, on the other hand, resides in the visited network of the MH and provides routing services to the latter, if needed — in other words, it detunnels the packets sent by the HA and delivers them to the host.

** More than one HA may exist in the operator's IP domain. As an example, HA functionality can be located at the WAR and GGSN nodes, in addition to other IP subnets within the domain. This will also enable load balancing and improve robustness.

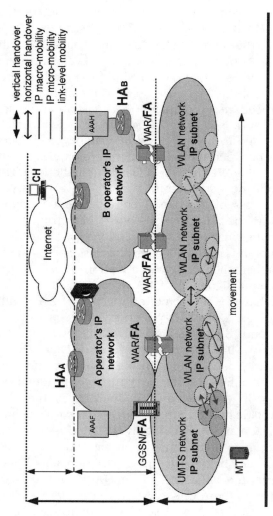

**Figure 9.7    MM in a loose-coupling approach.**

serving FA's address. In our example, the MH is configured with a static home address assigned by its home operator and an HA, although this is not always the case, as is described in the following text. Packets addressed to the host's home address will reach the host after being intercepted by its HA and redirected to its serving FA. When the host decides to switch from UMTS connectivity to WLAN connectivity while maintaining its active sessions, it registers again with $HA_B$ as a result of migrating to a new IP subnet. When the HA becomes aware of the host's new location, it starts tunneling packets to its new FA, i.e., the one collocated with the WAR. In this way, session continuity is enabled; however, the latency involved in these handoffs can be above the threshold required for the support of delay-sensitive or real-time services on account of the time needed for the registration procedure to complete (the registration message traverses the wide-area Internet). Note that this statement holds true if we assume that some time — due to layer 2 handover latency — is needed for the terminal to switch between the two access technologies. If this is not the case and the user gains WLAN connectivity in parallel to UTRAN connectivity, then the time needed for MIP registration to complete does not affect the seamlessness of the handover (the user continues to receive packets from UTRAN because he does not depart from the link).

Before elaborating on the possible solutions for assuring the seamless operation of IP mobility, we shall briefly discuss some extensions to the base MIP protocol that will allow (1) its adoption by cellular infrastructures and (2) the dynamic assignment of HA as a form of route optimization. 3GPP and 3GPP2 have been promoting the use of AAA with Mobile IP, and as a consequence, considerable work has been carried out in the IETF toward this end. Such integration will allow cellular operators to verify the identity of MHs and authorize connectivity in the absence of preconfigured security associations. In the envisaged functional architecture, the MH is identified by its Network Access Identifier (NAI), an identifier of the form user@realm, and there is no need for it to be configured with a static home address. In the context of such an integrated architecture, the HA may be dynamically assigned — either at the host's home or foreign domain — in conjunction with the home address. Recall that the home address belongs to the HA's subnet address space. The entity that performs the allocation of an HA is typically considered to coordinate closely with the AAA server, which triggers the allocation when a new registration is received. AAA servers in the foreign network (AAAF) and the home network (AAAH) interface with the foreign agents and home agents, respectively. The whole procedure is initiated when the MH registers with the FA; the latter then invokes the AAA infrastructure for authenticating and authorizing the host.

As pointed out in the preceding text, MIP may cause delays in regaining IP connectivity — that is, in being able to receive and send data — when the user changes his or her point of attachment. This is mainly attributed to the fact that the protocol does not differentiate between local and global mobility, also referred to as *micro* and *macro mobility*. Although only a part of the IP path might be affected due to the host's movement, e.g., movement within a small region, the HA is always notified, which implies that the whole IP path is being restored. In our example, although the MH moves within the region of operator A's domain, it exposes its movement to the distantly-located HA, introducing (1) an unnecessary load into the network and (2) extra delay in updating its IP path. It would be much more efficient if the local movement-related changes and signaling were confined to the local operator's IP domain. To this end, numerous IP micro mobility protocols have been proposed, with *MIPv4 regional registrations* — the IETF's proposal — being dominant in IPv4 networks. These protocols introduce another level of hierarchy to IP MM, usually by placing a mobility gateway at the borders of their domain, and handle the intradomain movement of hosts, i.e., the local reestablishment of IP paths. In this way, the handoff delay is reduced because the *path update* signaling does not reach the HA but the domain's mobility gateway. MIP signaling is only triggered when global mobility is involved and the whole IP path (from the HA to the MH) needs to be updated.

Although the local MM protocols reduce the handoff latency for intra-domain mobility, their main purpose has been to minimize the volume of path update signaling traversing the wide-area Internet and not to perform *handoff control*. As a matter of fact, these two functions are considered orthogonal and the current trend is for them to be implemented by different protocols. Handover involves the time-critical operation that "locally" redirects packets to the host's new point of attachment, whereas path update reestablishes the path after the handoff has been performed and IP connectivity regained. The most common methods for handoff control have been the establishment of temporary tunnels among access routers (ARs) for traffic forwarding, and the bicasting of data packets; both techniques may use buffering for handling the possible loss of synchronization between layer 2 and IP layer handoff. This loss of synchronization might occur, for example, when the IP handoff has taken place and packets are being forwarded to the new access router by means of a tunnel, whereas the MH has not yet established L2 connectivity with the new AR and is not able to receive any data. As a consequence, the forwarded packets need to be buffered for some time; otherwise, they are dropped.

The IETF MIP4 working group has consolidated the Internet draft *Low Latency Handoffs in Mobile IPv4* for handling IP layer handoffs in IPv4 networks. This scheme depends on obtaining timely information from layer 2 (L2 triggers) regarding the progress of an L2 handover, either on the network's or on the host's side. When the MH is still attached to the old AR and is informed of an impending L2 handover — which will also cause an IP handover — it may prebuild its registration state on the new AR/FA prior to the underlying L2 handoff. As soon as the host attaches to the new link, data already follows the new path. Alternatively, a tunnel may be established among the concerned ARs/FAs for the service to continue uninterrupted. Here also, L2 information, either at the old or new AR, triggers the tunnel setup. The MH does not participate at all in the IP layer handoff procedure, which is typically followed by the registration of the host. Such a handoff scheme can certainly be applied to the interworking architecture for minimizing the handoff latency and packet loss experienced by user sessions. What is needed, however, is the identification of UMTS- and WLAN-specific messages that will form the L2 triggers.

### Other Interworking Issues

#### Security

We restrict our discussion to the following major topics related to security: AAA, confidentiality, and integrity. Authentication and authorization are always the first steps before a user can have IP connectivity and access data services over a UMTS or WLAN system. Accounting and billing are also fundamental parts of the overall service procedure. Confidentiality ensures that malicious users cannot intercept, copy, or replicate information, whereas integrity secures the accuracy of information and protects it from illegitimate modifications.

In UMTS there are strict and well-defined procedures related to the aforementioned issues, mostly relying on the capabilities of the USIM integrated circuit card (UICC) found in mobile devices, as well as on the capabilities of the HSS. The UICC contains the USIM application that is responsible for running the UMTS authentication and key-agreement (AKA) procedure. AKA is based on challenge–response mechanisms and symmetric cryptography.

A central requirement for a WLAN–3G interworking architecture is that the security levels of UMTS not be compromised. On top of that, cellular operators have an already established subscriber base and subscriber management system, which it is highly desirable to reuse. Therefore, it is of prime importance that security procedures in the new convergent environment make use of the UMTS capabilities, and in particular, of the UMTS AKA procedure. The latter is not restricted only to UMTS, but can run over other transport mechanisms. For this purpose, the Extensible Authentication Protocol (EAP) can be used. EAP is specified by the Internet

Engineering Task Force (IETF) in RFC2284.* It is a framework for authentication, and most important, it supports more than one authentication method. It does not need IP to run because it is a protocol running on top of the data-link layer. Currently, there are implementations of EAP over PPP as well as over IEEE 802-based wired and wireless networks.

For the UMTS AKA procedure to be reused over a WLAN, the fundamental requirement is that mobile devices execute this procedure over the WLAN. This is accomplished with the aid of the EAP framework, as specified in the EAP AKA Internet draft. The framework only requires the mobile devices to have access to the UICC/USIM smart card.

Apart from authentication, the UMTS AKA procedure ensures confidentiality and integrity by generating the necessary session keys (128 bits in length). Having derived the common keys, symmetric cryptographic procedures can be applied to achieve the required levels of confidentiality and integrity.

## QoS

One of the major advancements of UMTS is the support of end-to-end IP services with guaranteed QoS. 3GPP standards propose a layered architecture for the support of end-to-end QoS, through the interaction of *bearer services* established between UMTS modules at different layers in both the AN and the CN. Each bearer service specifies the control signaling, user plane transport, and QoS management functionality, among others. Differentiation of user traffic is provided through the support of four QoS classes (conversational, streaming, interactive, and background), with delay sensitivity being the main distinguishing factor among them.

In a typical WLAN architecture, QoS is provided in both the IP backbone through standard IP mechanisms and the WLAN access segment, according to the capabilities defined in the standards, such as the WLAN QoS mechanisms specified in 802.11e. A lot of research showing traffic differentiation within the WLAN access is also found in the literature.

In a WLAN–3G convergent environment, interworking at the QoS level can be achieved in two ways, depending on the interworking type. In a tight coupling–based architecture, the overall QoS is controlled according to the standard UMTS paradigm. However, the adaptation functions in the mobile device and the UIU must take the necessary steps to set up and maintain specific levels of QoS within the WLAN access segment, in such a way that the QoS offered at that level is equivalent to that found in UTRAN. In loose coupling, an external interworking QoS broker (IQB)

---

* The Internet draft Extensible Authentication Protocol (EAP) <draft-ietf-eap-rfc2284bis-09.txt> will obsolete RFC2284, when approved.

mechanism (see Figure 9.5) must exist in the operator's IP backbone to take care of the interworking procedures between the QoS architectures implemented in the UMTS and WLAN networks. Because there is no such standardized mechanism in IETF, such as EAP AKA for authentication or MIP for mobility, the solution is left to the individual operator.

### Support of Enhanced Applications

The convergent architecture can offer new capabilities to applications and services running in this environment. To start with, applications would be able to adapt themselves to the underlying network, offering the best available data service at each time. For example, a video streaming application could take advantage of the occasional connection through the high-speed WLAN access and offer better video quality to the end user. Moreover, we anticipate that the advanced data service set found currently in 3G networks can still be used over the WLAN access. Although this is somehow straightforward in the tight-coupling scenario, in loose coupling, some interworking modules in the operator's IP backbone would be needed to offer the required interworking functionality. For example, some kind of a Session Initiation Protocol (SIP) proxy server could be used to interconnect the WLAN system to an IMS, so that SIP-based multimedia services can be extended to users connecting through the WLAN.

## STANDARDIZATION ACTIVITIES

3G cellular networks undergo an intense standardization process in such a way that only specific implementation details are left open for vendors to decide. Following this concept, and having recognized the necessity of providing a convergent WLAN–3G environment, the standardization bodies of primarily cellular networks are in the process of standardizing the interworking between 3G and WLAN networks. The foremost goal of standardization is to define the interworking interfaces, so that equipment from different vendors can operate together independent of the type of WLAN and cellular network. This section presents in brief the results of work done by European Telecommunications Standards Institute (ETSI) through the project broadband radio access networks (BRAN) working group and subsequently focuses on the outcomes of 3GPP efforts.

### ETSI BRAN

ETSI BRAN issued the technical report TR 101 957 in August 2001, titled "Requirements and Architectures for Interworking between HIPERLAN/2 and 3rd Generation Cellular Systems." The purpose of the technical report was to identify the requirements and architectures for proper interworking between the High-Performance Radio Local Area Network (HIPERLAN/2) and 3rd generation cellular systems (in particular, UMTS release 3).

The report first describes some general interworking requirements and then focuses on the areas of subscriber data, mobility and handover, and end-user devices. Subsequently, it discriminates between two interworking levels: tight and loose coupling. This is one of its main contributions, as this terminology has been extensively used since then. The report continues with the examination of these two levels in terms of the proposed system architecture along with the definition of basic interfaces, and then treats in detail issues such as security, mobility and handover, QoS, and user-traffic management.

## 3GPP

3GPP has been analyzing the issue of WLAN–UMTS interworking since September 2001. The primary goal is to extend the operational and service-provisioning environment of 3GPP to WLAN networks, in such a way that WLAN takes on the role of a complementary radio access technology to UTRAN. The desired result is to enable 3G operators to broaden their working environment and seamlessly provide WLAN access to their subscribers. Interworking should be achieved with no modifications to WLAN standards and with only minimal changes to 3GPP specifications. 3GPP is currently in the process of elaborating a set of specifications to be published with 3GPP release 6. These specifications deal with a feasibility study, the identified requirements for interworking to take place, and the resulting interworking architecture along with its interfaces and main procedures, while emphasizing security issues.

### 3GPP Interworking Scenarios

One of the major outcomes of the standardization process was the identification of six interworking scenarios. The first scenario is the simplest and easiest to be implemented, and each scenario that follows builds on top of the previous one. The intention is not only to follow a stepwise approach in the integration of WLAN with the 3GPP system but also to offer different levels of interworking to 3G operators. The six scenarios are briefly described in the following text.

The first scenario, *common billing and customer care*, assumes a single-customer relationship between the WLAN and 3GPP systems. In essence, customers receive a single invoice for the 3G and WLAN services they consume. It is apparent that no modifications or additions to the 3GPP specifications are needed here.

In scenario 2, *3GPP system-based access control and charging*, the 3GPP system provides AAA functionality to WLAN users. The main advantage for the 3G operator is that this allows subscribers to have WLAN access, while exploiting the subscriber management and maintenance procedures of the cellular network. This scenario adopts a new AAA proxy function to serve as a mediator to the WLAN system.

Scenario 3, *access to 3GPP system PS-based services*, goes a step further and allows a user connecting through a WLAN access segment to use the services of the 3GPP PS domain, such as instant messaging, location-based services, and multimedia services offered by the IMS, among others. This scenario further necessitates the introduction of new modules within the 3GPP system handling the interworking at the service level.

Scenario 4, *service continuity*, extends the previous scenario by allowing an ongoing service to be maintained in case of user migration between the WLAN and UMTS portions of the network, without the need for user intervention. The main implication of this is that it necessitates the existence of an intelligent decision-making mechanism for selecting the most suitable wireless access segment every time. However, the switch between WLAN and UMTS may result in extremely variable service quality due to the inherently different characteristics of the two systems, although it is possible that some services may be forced to terminate.

Scenario 5, *seamless service*, improves on scenario 4 by allowing a smooth transition from one wireless access segment to another. To accomplish this, it provides enhancements to the basic operational functions to maximize performance and minimize data loss or break of a session during a transition.

In scenario 6, access to 3GPP CS services, users can access services of the 3GPP circuit-switched domain over a WLAN.

### 3GPP WLAN Subsystem

The 3GPP WLAN subsystem comprises the essential architectural modules that provide 3GPP subscribers with access to 3GPP services via a WLAN access segment. In other words, it specifies the interworking architecture — mainly in terms of new modules, interfaces, and procedures — designated to meet the requirements posed by the interworking scenarios briefly described in the preceding text. The first scenario does not require any modification to the 3GPP standards, whereas scenarios 2 and 3 necessitate the addition of new modules to the 3GPP architecture. Until the time of this writing, the focus has been on the interworking architecture based on scenarios 2 and 3; the remaining scenarios will be developed in the future.

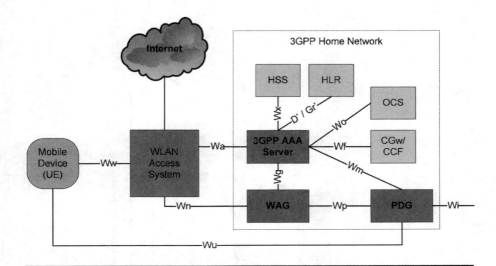

**Figure 9.8   Nonroaming 3GPP WLAN interworking architecture.**

Figure 9.8 and Figure 9.9 depict the two reference models of the 3GPP WLAN subsystem architecture, one for the nonroaming case and the other for the roaming case. The former refers to the situation in which the WLAN access system is owned by the 3GPP home network provider offering subscription services to the user. The roaming scenario supposes that the WLAN access system is attached to a 3GPP-visited network, which must have roaming agreement with the user's home network. The fundamental modules and interfaces of the architecture are explained briefly in the following text.

The pair consisting of *3GPP AAA Server* and *3GPP AAA Proxy* is used to provide the AAA functionality needed for scenarios 2 and higher. The former resides in the home network and is responsible for communicating with the HSS (or HLR) so as to retrieve subscriber information and perform the required USIM-based authentication and key-agreement operations, according to the EAP-AKA procedure. In other words, the 3GPP AAA server implements the network-resident functionality of the EAP-AKA procedure, while the client side is executed by the mobile device. The 3GPP AAA server also maintains the status of the mobile device, that is, it indicates whether the device is attached to the 3GPP WLAN subsystem and consequently is being served by the 3GPP WLAN interworking network. Besides, it produces reports relevant to per-user accounting and charging information and sends them to the charging collection function (CCF) of the charging gateway (CGw). The 3GPP AAA proxy resides in the visited network and mediates in the communication between the

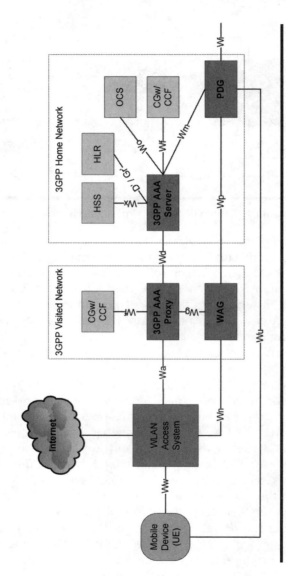

**Figure 9.9 Roaming 3GPP WLAN interworking architecture (PDG in home network).**

WLAN — equipped with AAA client functionality — and the 3GPP AAA server concerning AAA matters. In other words, it relays AAA information between the WLAN and the 3GPP AAA server and may, in addition, perform protocol conversion in case different protocols are used at the Wa and Wd reference points. Further, it generates charging reports toward the CCF/CGw of the visited network for roaming users.

The packet data gateway (PDG) and WLAN access gateway (WAG) modules are used for scenarios 3 and higher, to enable access to the 3GPP services of the PS domain. Before we detail the functionality of these two modules, some clarifications about IP connectivity and packet-based service access for scenario 3 are needed. In scenario 2, the mobile device makes use of one local IP address — belonging to the WLAN's address space — whereas for scenario 3, the mobile device is configured with two IP addresses: a local and a remote one. The local IP address is used to access external IP networks (e.g., the Internet) without the intervention of the 3GPP WLAN subsystem infrastructure. However, to access 3GPP PS-based services, the remote IP address — belonging to the address space of the 3G operator's IP domain — is used. Packets destined for this address finally reach the user by means of a *tunnel** established between the mobile device — local address — and the PDG module. As such, packet encapsulation and decapsulation techniques are used to transfer user data traffic from the mobile device to the 3GPP network and vice versa. Several tunnels may be established to access PS-based services of several external IP networks simultaneously. Apart from the IP address allocation and usage paradigm, to access the cellular operator's services, a service authorization procedure must be executed.** Service authorization over a WLAN follows the same *Access Point Name* (APN) concept already defined in the 3GPP standards. The mobile device makes use of the so-called W-APN (WLAN APN) name to indicate to the network the service (or set of services) it desires to access.

According to the principles briefly explained in the preceding text, the primary role of the PDG module is to facilitate a successful activation of a selected PS-based service for the mobile device. The activation of a selected service mainly consists of the following steps:

---

* The reasons for using the tunneling approach are manifold: it allows the operator to provide 3GPP services (such as MMS and IMS), which can be accessed only over its private PLMN, to users residing in a distant WLAN. Moreover, with tunneling, the operator can collect independent charging information and apply the desired policies.

** The service authorization procedure is different from the WLAN access authentication and authorization procedure, which always takes precedence.

1. Determination of the PDG's IP address, which is to be used for tunneling by the mobile device, depending on the W-APN and user subscription information.
2. Allocation of a remote IP address to the mobile device, in case one is not already allocated, and creation of the required tunnels.
3. Registration of the terminal's local IP address with the PDG for the latter to bind this address with the remote one.

Regarding user data traffic, PDG is responsible for performing all the supporting operations, such as address translation and mapping, packet encapsulation/decapsulation and routing, packet filtering and QoS control, and generation of charging information. It is worth noting that the PDG may reside in either the home or the visited network, depending on whether the user can have access to the PS services of the visited operator's network. Figure 9.9 depicts only the case in which the PDG is located in the home network.

The WAG module is a gateway via which user data is routed between the WLAN access system and the PDG. It resides in either the home network for the nonroaming case, or the visited network for the roaming case. WAG enforces packet routing through the PDG and collects per-tunnel traffic information (volume, time, etc.) to enable the visited network to generate charging information. Apart from routing enforcement, the WAG may also perform policy enforcement to filter out packets that do not conform to the security or QoS levels of the established tunnels.

The 3GPP WLAN subsystem supports the two basic charging 3GPP methods, which are *prepaid* and *postpaid*. The former charging method is being used by the online charging system (OCS), whereas the latter is being used by the CGw. It is worth noting that charging for a 3GPP PS-based service is service specific and not access specific, meaning that a service will have the same fee independent of the utilized access technology. In the case of IP-based services, which go through the PDG, the latter is responsible for accurately collecting and calculating charging information. This is why PDG can communicate with OCS and CGw over the corresponding reference points.

### Reference Points

The architecture defines some new reference points (RP) over which specific protocols must be implemented to provide the required functionality. Table 9.1 summarizes and briefly describes these reference points. In short, the Wn, Wp, Wu, and Wi reference points are used for the user data plane, whereas the rest are used for control plane functionality.

**Table 9.1 The 3GPP WLAN Subsystem Reference Points: A Brief Description**

| Reference Points | Description |
|---|---|
| Wa | It connects the WLAN access system to the 3GPP AAA server or proxy. The protocol running over Wa is used to transfer AAA information. |
| Wd | It lies between a 3GPP AAA proxy and a 3GPP AAA server. It is equivalent in functionality to Wa. |
| Wx | It is located between the 3GPP AAA server and the HSS. The protocol used in Wx enables communication between these two modules. |
| D'/Gr' | This reference point is equivalent to Wx, but it is used in releases prior to release 6 to enable communication between the 3GPP AAA server and the HLR. |
| Wo | It is placed between the 3GPP AAA server and OCS. The protocol applied here transfers online charging information to OCS. |
| Wf | It is located between the 3GPP AAA server and CCF/CGw. The protocol crossing this RP is used to transfer charging information to the collection function of CGw. |
| Wg[a] | It is an AAA interface between the WAG and the 3GPP AAA server, to enable policy enforcement in the WAG. |
| Wn[a] | It is placed between the WLAN access system and the WAG. The protocol running over this RP ensures that user traffic belonging to a tunnel passes through the WAG. |
| Wp[a] | It is located between the WAG and the PDG. |
| Wu[a] | It is located between the mobile device (UE) and the PDG, and its functionality is to enable UE-initiated tunnel establishment and transmission of user packets through the established tunnel. Transport for the protocol running in Wu is provided by the Wn and Wp RPs. |
| Wi[a] | This reference point is similar to the Gi reference point mentioned earlier, and it permits connection to external packet data networks. |
| Wm[a] | It is a reference point between the PDG and the 3GPP AAA server, to enable tunnel establishment and configuration as well as service authorization in the PDG. |
| Ww | It connects the mobile device to the WLAN access system. |

[a] Used only for scenario 3.

# AN ARCHITECTURE BEYOND 3G

The research community is showing a great interest in the so-called beyond-3G (b3G) or 4G architectures. A lot of definitions exist that try to identify the concept of b3G communications, but the one that receives the most acceptance deals with the ability to offer data services over virtually any wireless access system, ubiquitously, transparently, and seamlessly. The convergence between WLAN and 3G cellular systems is definitely a significant step toward the realization of a b3G architecture. However, it poses some restrictions on the generality of the previous definition, because it depends on the currently defined architectures for 3G cellular and WLAN networks.

Trying to delineate a more visionary approach, one could claim that it is desired to combine the high degree of integration and the resultant performance of the tight-coupling mode with the easy deployment and the high degree of IP suite reusability of the loose-coupling one. It is the opinion of the authors that the first step in this direction is the substitution of the 3G core network with a pure IP-based network. The core network uses only native IP (IETF-based) protocols for every operation: network address assignment, network management, mobility, QoS, AAA operations, etc. The intention is to decouple the wireless access technologies from the core network serving them. The various access segments will connect to this unified core through a generic reference point, communicating with special routers that will have the ability to perform any adaptation functions needed for the underlying wireless technologies. Figure 9.10 depicts the envisioned architecture of a b3G environment.

The *unified access router* (UAR) is a special router located at the edge of the core network toward the wireless access segments and is capable of serving diverse access technologies. To help accomplish this task, the UAR offers a generic interface to the ANs, which is called the *Generic Wireless Interface to IP* (GWIIP). The access technologies have to implement this interface to provide an access-agnostic service to the core. However, this requirement may necessitate some modifications to the protocol stacks defined in the standards in both the wireless access system and the mobile device.

From Figure 9.10, it is obvious that the architecture actually follows the tight-coupling paradigm. The major difference is that the fundamental operations are provided by pure IP protocols and not 3GPP-based ones, mitigating in this way the need to implement the "heavy" interworking unit and the adaptation functions of tight coupling, while retaining the usage of the loose-coupling standard protocols. In contrast to loose coupling, the b3G system relies on only one core network with unified operations, so it further eliminates the need for interworking at the protocol level. To give an example, MM in loose coupling needs an overlay protocol

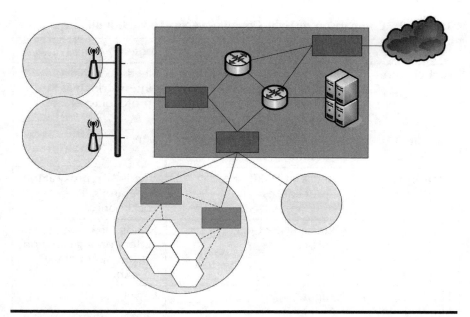

**Figure 9.10   Beyond-3G architecture.***

layer — offered by, for example, MIP — to make the mobility protocols in UMTS and WLAN work together, adding in this way another layer in the mobility hierarchy. In the b3G architecture, however, there is only one IP-based core network that caters to all the required MM procedures, resulting in a straightforward and more efficient interworking system.

## CONCLUSIONS

The evolution and proliferation of wireless communications favors the convergence of WLANs and 3G cellular networks in a unified environment capable of seamlessly offering data services to end users. The intention of this chapter was to give a presentation of the issues concerning WLAN and 3G interworking. The basic concept of such interworking is to enable common authorization, accounting, and billing, while allowing users to move freely in this convergent environment. Interworking can be achieved by two different architectural models: tight coupling and loose coupling, which are differentiated by the level of integration — or interdependence — between WLAN and 3G. A brief comparison of these two models, mainly in terms of their advantages and disadvantages, is presented in Table 9.2.

---

* dRNC stands for *differentiated RNC*, meaning that it is not a standard 3G RNC, because it has to connect to the core network through a new interface, the GWIIP.

**Table 9.2  Comparison of Tight Coupling versus Loose Coupling**

|  | Tight Coupling | Loose Coupling |
|---|---|---|
| Architecture | Needs system interworking units and adaptation functions in the mobile device and the network | Needs external interworking at the protocol level |
| Mobility | Reuse 3GPP GTP-based mobility | MIP |
| Security | Reuse of 3GPP AKA procedures | Needs AAA proxy at the operator's IP backbone |
| QoS | Reuse 3GPP QoS procedures | Needs QoS interworking Function at the operator's IP backbone |
| Level of integration | Higher | Lower |
| Complexity of implementation | Higher | Lower |
| Performance | Higher | Lower |
| Common radio resource management | Easier | Not possible |
| Business model | 3G operator owns WLAN | Independent WLAN ISPs |

Apart from the general discussions that take place regarding WLAN and cellular interworking, the standardization forums are also working on the issue. The most notable effort belongs to 3GPP, which leads the process of providing a complete interworking architecture, in terms of interfaces, procedures, and protocols. The architecture will be issued with 3GPP release 6. It will enable 3G operators to integrate WLAN systems into their cellular infrastructure and provide their vast number of customers with a full service portfolio enhanced with very high-speed data rates at popular or selected locations, such as airports, hotels, etc.

The future trend and the current focus of researchers are on the definition of an architecture beyond 3G. Such an architecture will most probably not rely on the existing 3G architecture, but it will feature the extended usage of the IP protocol and purely IP-based operations. In this environment, various wireless access technologies will be used in a complementary way, whereas interworking will be ensured through the IP-based core network interconnecting them. However, b3G architectures are still in their infancy, and much research has still to be pursued for them to mature. Until then, the 3GPP solution will provide the necessary basis for WLAN–3G interworking implementations so as to satisfy increased user demands.

# BIBLIOGRAPHY

1. 3GPP TS 23.060 v6.3.0, General Packet Radio Service (GPRS); Service Description; Stage 2, December 2003.
2. 3GPP TS 22.234 V6.0.0, Requirements on 3GPP system to Wireless Local Area Network (WLAN) Interworking (Release 6), March 2004.
3. 3GPP TR 22.934 v6.2.0, Feasibility study on 3GPP system to Wireless Local Area Network (WLAN) Interworking (Release 6), September 2003.
4. 3GPP TS 23.234 v2.4.0, 3GPP system to Wireless Local Area Network (WLAN) Interworking; System Description (Release 6), January 2004.
5. 3GPP TS 33.234 V1.0.0, Wireless Local Area Network (WLAN) Interworking Security (Release 6), December 2003.
6. 3GPP TS 23.107 V6.1.0, Quality of Service (QoS) Concept and Architecture, (Release 6), March 2004.
7. 3GPP TS 23.207 V6.2.0, End-to-end Quality of Service (QoS) Concept and Architecture, (Release 6), March 2004.
8. ETSI TR 101 957 V1.1.1, Requirements and Architectures for Interworking between HIPERLAN/2 and 3rd Generation Cellular systems, August 2001.
9. Ahmavaara K. et al., Interworking Architecture between 3GPP and WLAN Systems, *IEEE Communications Magazine*, no. 11, November 2003, pp. 74–81.
10. Ala-Laurila, J. et al., Wireless LAN Access Network Architecture for Mobile Operators, *IEEE Communications Magazine*, vol. 11, November 2001, pp. 82–89.
11. Arkko, J. and Haverinen, H., EAP AKA Authentication, Internet Draft, draft-arkko-pppext-eap-aka-11.txt, October 2003.
12. Buddhikot M.M. et al., Design and Implementation of a WLAN/CDMA2000 Interworking Architecture, *IEEE Communications Magazine*, no.11, November 2003, pp. 90–100.
13. Calhoun, P. et al., Diameter Mobile IPv4 Application, Internet draft, draft-ietf-aaa-diameter-mobileip-16.txt, February 2004.
14. Campbell, A.T. et al., Comparison of IP Micro-Mobility Protocols, *IEEE Wireless Communications Magazine*, 9, 1, February 2002, pp. 72–82.
15. Chiussi, F.M. et al., Mobility Management in Third-Generation All-IP Networks, *IEEE Communications Magazine*, no. 9, September 2002, pp. 124–135.
16. Doufexi, A. et al., Hotspot Wireless LANs to Enhance the Performance of 3G and beyond Cellular Networks, *IEEE Communications Magazine*, no. 7, July 2003, pp. 58–65.
17. Floroiu, J.W. et al., Seamless Handover in Terrestrial Radio Access Networks: A Case Study, *IEEE Communications Magazine*, no. 11, November 2003, pp. 110–114.

18. Glass, S. et al., Mobile IP Authentication, Authorization, and Accounting Requirements, RFC 2977, October 2000.
19. Gustafsson, E. et al., Mobile IPv4 Regional Registration, Internet draft, draft-ietf-mobileip-reg-tunnel-08.txt, November 2003.
20. Hiller, T. et al., CDMA2000 Wireless Data Requirements for AAA, RFC 3141, June 2001.
21. Honkasalo, H. et al., WCDMA and WLAN for 3G and Beyond, *IEEE Wireless Communications*, no. 2, April 2002, pp. 14–18.
22. Hui, S.Y. and Yeung K.H., Challenges in the Migration to 4G Mobile Systems, *IEEE Communications Magazine*, no. 12, December 2003, pp. 54–59.
23. Kim, Y. et al., Beyond 3G: Vision, Requirements, and Enabling Technologies, *IEEE Communications Magazine*, no. 3, March 2003, pp. 120–124.
24. Koien, G. et al., Security Aspects of 3G-WLAN Interworking, IEEE Communications Magazine, no. 11, November 2003, pp. 82–88.
25. Luo, H. et al., Integrating Wireless LAN and Cellular Data for the Enterprise, *IEEE Internet Computing*, March–April 2003, pp. 25–34.
26. Luo, J. et al., Investigation of Radio Resource Scheduling in WLANs Coupled with 3G Cellular Network, *IEEE Communications Magazine*, June 2003, pp. 108–115.
27. Malki, K., ed., Low Latency Handoffs in Mobile IPv4, Internet draft, draft-ietf-mobileip-lowlatency-handoffs-v4-08.txt, January 2004.
28. Manner, J. et al., Mobility Related Terminology, Internet draft, draft-ietf-seamoby-mobility-terminology-06.txt, February 2004.
29. Marques V., An IP-Based QoS Architecture for 4G Operator Scenarios, *IEEE Wireless Communications*, no. 3, June 2003, pp. 54–62.
30. Pahlavan, K. et al., Handoff in Hybrid Mobile Data Networks, *IEEE Personal Communications Magazine*, no. 2, April 2000, pp. 34–47.
31. Pattara-atikom, W. et al., Distributed Mechanisms for Quality of Service in Wireless LANs, *IEEE Wireless Communications*, vol. 10, no. 3, June 2003, pp. 26–34.
32. Perkins, C.E., Ed., IP Mobility Support for IPv4, RFC 3344, August 2002.
33. Perkins, C.E., Mobile IP at IETF, ACM SIGMOBILE *Mobile Computing and Communications Review*, 7, 4, October 2003, pp. 1–4.
34. Perkins, C. E., Mobile IP joins forces with AAA, *IEEE Personal Communications Magazine*, 7, 4, Aug. 2000, pp. 59–61.
35. Salkintzis, A.K. et al., WLAN-GPRS Integration for Next-Generation Mobile Data Networks, *IEEE Wireless Communications*, 9, 5, October 2002, pp. 112–124.
36. Tachikawa, K., A Perspective on the Evolution of Mobile Communications, *IEEE Communications Magazine*, no. 10, October 2003, pp. 66–73.
37. Varma, V.K. et al., Integration of 3G Wireless and Wireless LANs, Guest Editorial, *IEEE Communications Magazine*, no. 11, November 2003, pp.72–74.
38. Zhang, Q. et al., Efficient Mobility Management for Vertical Handoff between WWAN and WLAN, *IEEE Communications Magazine*, no. 11, November 2003, pp. 102–108.
39. Zhuang, W. et al., Policy-Based QoS Management Architecture in an Integrated UMTS and WLAN Environment, *IEEE Communications Magazine*, no. 11, November 2003, pp. 118–125.

# III

# APPLICATIONS

# 10

# MOBILE WLAN APPLICATION SERVICES

*Alfons H. Salden, Cristian Hesselman,*
*Ronald van Eijk, Andrew Tokmakoff, Mortaza Bargh,*
*Johan de Heer, and Hartmut Benz*

## INTRODUCTION

The costs of deploying wireless LANs (WLANs) are low and continue to drop; however, the costs of creating and managing value-added mobile WLAN application services remain considerably higher, compared with other communication networks. The creation and management of mobile application services for WLANs may require much more expertise than for other networks because they may be functionally much more complex. Such services may not only be owned and managed by many different types of organizations, but they may also have to be adapted to many types of users, terminals, communication networks, and application services.

Modeling is therefore a must before one may even think of deploying mobile WLAN application services. A model for provisioning mobile WLAN application services can help tackle increasingly complex application service issues. Such a model may combine models for businesses, public organizations, information and communication technologies, and human needs and behaviors. It can considerably speed up the development, deployment, and management of really useful mobile application services. These services can in turn support, automate, and sustain E-commerce, E-business, and collaboration more naturally. They can support individuals, enterprises, or institutions in carrying out their specific activities across various types of communication and computing networks. Given the

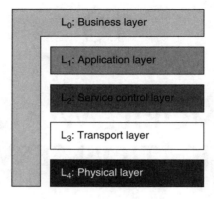

**Figure 10.1   A layered functional model for provisioning of mobile application services.**

increasing complexity of business, organizational, information, and communication networks, it is clear that the success of future mobile WLAN application services depends heavily on such a model.

In this chapter, a simple model for provisioning mobile WLAN application services is presented. This model will subsequently serve the design and implementation of mobile service applications deployable on WLANs and other communication networks. The model and application services take into account issues that could arise across business layer $L_0$, application layer $L_1$, service control layer $L_2$, transport layer $L_3$, and physical layer $L_4$ (see Figure 10.1).

Here, the business layer $L_0$ includes networks of businesses, institutions, and users. During modeling, all relevant business layer issues, application layer issues, service control layer issues, and transport layer issues are resolved. The services described demonstrate, in particular, the added features of WLAN networks.

## Setting the Stage

A model for provisioning mobile WLAN application services should abstract the application, service control, transport, and physical network dynamics away from the specific details of the business. It should enable enterprises or institutions to reach a consensus on how to seamlessly collaborate on those networks. It should support even automation of roles, functions, and relations that exist among enterprises and institutions, i.e., implementation of an interfacing service infrastructure. The reason for modeling is that an application service may not be suited to every mobile business or institutional context. For instance, at hot spots such as airports, a Universal Mobile Telecommunications System (UMTS) provides too little network capacity to stream multimedia to thousands of travelers. Consequently, there is no viable business model — unless the intention is to provide multimedia streaming services at exorbitant prices and to only a select few.

However, multimedia streaming across WLANs is a viable option, because WLANs may provide high enough communication network bandwidths even for multiple end users. Another advantage of WLAN is that it may provide, both indoors and outdoors, higher positioning accuracy than, for example, UMTS.[1] For instance, WLAN can give tourists access to location-based services such as exposition and map services of museums.[2] Furthermore, such tourists can, via their Bluetooth-enabled personal digital assistants, make connection with radio-frequency identification (RFID) systems that are attached to museum objects. This way, they can acquire other types of WLAN museum services that could provide detailed information about the sculpture they are standing in front of. Besides its relatively high network capacities, a third advantage of WLANs is their relatively low deployment and maintenance costs. A fourth advantage of WLAN is that a company or institution does not have to gain license rights for the frequency band used. Therefore, enterprises or institutions can always be in full control of their own communication and computing networks. They can restrict access of their WLANs to their own personnel or customers. Last but not least, they can strictly reserve communication network bandwidth and computing network resources of mobile application services to their own personnel or customers.

Enterprises, institutions, and consumers all around the world demand ever more complex mobile (WLAN) application services at hot spots;[3] the WLAN market is expected to double in the next two years (by 2006). The WLAN hot spots are located at airports and railway stations; business or consumer facilities, e.g., conference centers, hotels, stadiums, shopping malls, cafés, and restaurants; at schools, university campuses, and in residential areas. The mobile application services at WLAN hot spots can be categorized as person-to-person communication services, entertainment services, mobile transaction services, mobile information services, and business solutions. According to Axiotis and coworkers,[3] any service category can in turn be characterized under one or more quality-of-service (QoS) classes for synchronicity, interactivity, and end-device capability.

However, they[3] do not provide means for precisely and easily modeling mobile WLAN application services. Future mobile WLAN application services need to be readily adaptable, for example, to network load or user profiles. They also need to be dynamically reconfigurable, for instance, when business relations and conditions change during a multimedia streaming service session. Another requirement is that these services need to be commercially exploitable or publicly accessible. Consequently, sophisticated mobile application service modeling and modeling languages are needed.

Nowadays, a plethora of modeling languages exist for almost any product, service, and application domain. Each of these modeling languages has its own concepts, tool support, and visualization techniques. Each is tailored to the needs of specific stakeholders or users. Although there is no universal modeling language available yet, an initiative to arrive at an alignment across business, application, service control, transport, and physical layers is underway.[7]

Analogously, for business–IT alignment,[7] a model for provisioning mobile WLAN application services should be very flexible. It should allow for fast development, deployment, and management of such services across all layers of the layered functional model (see Figure 10.1). Those services should in turn be readily adaptable or reconfigurable to various contexts and changes across all layers. In addition, such a model should allow for tailorability to individual or groups of users. It should be possible to make the services depend on the goals, preferences, situational contexts, and cognitive capabilities and behaviors of end users or organizations.

Scenario analysis and use-case design[8] have proven themselves to be useful instruments in developing (mobile) human–computer interaction services. Such services may then be conceived and built with specific business or institutional purposes in mind.[12] However, participatory user-oriented design (UOD) will be indispensable in finding useful and commercially exploitable mobile WLAN application services. To this end, cognitive engineering environments can help determine ergonomic as well as cognitive capabilities of users or groups. Subsequently, these environments can also help adapt mobile WLAN application services to ergonomic and cognitive capabilities.[9–11] This way, UOD can significantly reduce cognitive loads or emotional stress. Moreover, the best practices for UOD can be inferred from statistical analysis of HCI observational data.[13]

## Outline

Because mobile WLAN application services may involve cross sections of the layered functional model (see Figure 10.1), a conceptual model for reasoning about those services is presented in the section titled "A Simple Model for Mobile Application Services." In the subsequent sections, this conceptual model is used to develop and deploy value-added mobile WLAN application services. These services appear as instantiations of our conceptual model at specific granularities of the layered functional model. In the section titled "Private and Secure B2B Scheduling Services," a mobile WLAN application service is derived for scheduling private and security-sensitive B2B meetings; the focus is on support and automation of B2B relations across WLANs within layers $L_0$ and $L_1$. Here, software agents enhance standard Microsoft Outlook scheduling services for employees who are on the road. As they may drive a car, they will not have their hands free or have time to interact with their ordinary desktop Outlook scheduling service. The added value of the service is that the agents autonomously figure out and notify the most convenient meeting place and time, given the schedules of the employees. Here, business security and privacy policies of the employees of the different companies are taken into account. In the section titled "A Medical Teleconsultation Service," a medical teleconsultation service (MTCS) across WLANs is described; the focus is on supporting (a)synchronous (multiparty) communication and collaboration across WLANs for healthcare personnel, within all layers $L_0$–$L_4$. In the section titled "Content Distribution in an Internet with Wireless LANs," a multimedia streaming service for entertainment purposes is modeled and realized; the focus is on supporting and automating, within all layers $L_0$–$L_4$, roaming of mobile multimedia streaming across heterogeneous communication and computing networks and business domains. Finally, in the section titled "Mobile Radio Broadcasting Services," the personalization and exploitation of a commercial mobile music service is designed and implemented; the focus is on building, within layers $L_0$ and $L_1$ on top of a service platform, sophisticated user profiling and exploitation services for mobile music services across WLANs and other communication networks.

## A SIMPLE MODEL FOR MOBILE APPLICATION SERVICES

A simple model for provisioning mobile WLAN application services can be described in terms of a so-called E-business or E-institutional model.[4] Such a model identifies a minimal number of potential business or institutional entities, among which there exist business or institutional relations such as AAA relations, trust, and roaming agreements. This model not only captures business aspects, e.g., roaming agreements, but also describes user and technical aspects. Our model for mobile WLAN application services is a cross section of such an E-business or E-institutional model. A mobile WLAN application service is a service for nomadic users having access to a WLAN. Note that it does not imply that such mobile application services are only accessible and available on WLANs. On the contrary, some of these services may very well be provided on other communication networks as well.

The business or (public) institutional entities in our mobile (WLAN) application service model may involve many types of actors or software agents, communication and computing networks with related technologies, and (mobile) application services. They may cover specific cross sections of the layered functional model (see Figure 10.1). These entities may play several roles and functions within an enterprise or institution. They have many relations that may be subject to various conditions or constraints. The complexity and dynamics of mobile WLAN application services manifest themselves as an optimal functional network hierarchy at an appropriate granularity for an enterprise or institution. This means that roles, functions, relations, and conditions across the functional layered model are implemented according to a particular network hierarchy and at critical granularities. In this way, business or institutional objectives may be achieved in an optimal manner.

## Roles and Functions

In the functional layered model (see Figure 10.1) actors in layer $L_0$ may be customers, institutions, or companies. They may play many roles and carry out several functions within an institution or enterprise. For example, they may be end users who enjoy multimedia streaming and pay only for their WLAN access. Software agents may support, automate, execute, and establish business or institutional roles, functions, or relations. They may do so following certain given business policies or conditions. Note that both actors and agents may have and share different communication languages and levels of intelligence.

The computing networks in layer $L_4$ may consist of several types of mobile devices, access points, access routers, and computer networks. They all may run on different operating systems. The information and communication technologies mainly residing in layer $L_2$ (AAA and session control) and layer $L_3$ (IP routing and mobility) may be developed for wireless or fixed networks. Among the wireless technologies, one may discern those for UMTS, GPRS, WLAN, GSM, and others. The wired technologies may cover those for dial-up, xDSL, and core backbone GIGA bandwidth networks. Note that these technologies may have layer $L_4$ components.

## Relations and Conditions

The mobile WLAN application services mainly residing in layer $L_1$ may cover several service categories and QoS classes.[3] These services may have to be adaptable, reconfigurable, and tailorable to specific aspects of one or more of the business or institutional entities mentioned in the preceding text. This tailorability immediately links to modeling business or institutional relations.

While roaming across heterogeneous communication and computing networks, mobile WLAN application services need to be rescaled according to specific business or institutional relations, depending on conditions such as trust among the involved parties and the available bandwidth. For example, intellectual property and usage rights among actors or software agents on communication networks or mobile application service usage can considerably influence the QoS. Service–revenue relationships and strategies among customers, employees, and providers can be equally important.

Business and institutional relations and conditions may be automated as security, privacy, and customer preference schemes; usage rights policies; and financial clearing policies. In this respect, E-contracting may help, for instance, in arranging and executing roaming agreements and service level agreements among many providers, nomadic customers, or employees.[6] E-contracting may involve user, network, and end-device profiling, mobility management, AAA, and QoS management.

A service platform provider may offer such E-contracting services.[5] For instance, to support seamless roaming, it may even be mandatory to anticipate, predict, and resolve E-business issues across all layers of the layered functional model (see Figure 10.1). A service platform provider could then, for example, proactively transfer security contexts to other domains to ensure seamless handover. However, a group of network owners, providers, Internet service providers, and application service providers could set up E-contracts and policies themselves.[4] In such cases, they have to negotiate and settle similar service-provisioning issues as any such service platform provider does. If E-contracting is not a core competence of the parties involved in an enterprise, then they may consider outsourcing such business functions.

## Functional Network Hierarchy and Granularity

In the picture of our conceptual model (see Figure 10.2) one distinguishes roles A, functions $R_i$, relations R, and conditions $C$. A role A may be played by any entity across all layers in an enterprise or institution. This entity may be an actor or agent with specific tangible and intangible assets. As stated in the preceding text, these assets may consist of various physical, information and communication, technological, and organizational infrastructures. Roles A may be decomposed in terms of business or institutional functions $R_i$.

A business or institutional relation R(A,B | $C$(A,B)) may represent value propositions, strategies, service provisioning, revenue exchange, or policies among roles A and B given conditions $C$. $C$ can be prepayments being done or trust being provided by a certificate authority.

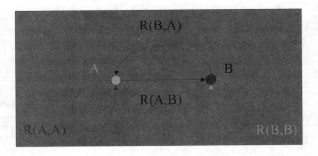

**Figure 10.2  A simple model for provisioning of mobile (WLAN) application service R(A,B) by role A to role B upon or before delivery of remuneration R(B,A) by role B to role A.**

A role A can create value for itself, which means that an entity in the role A executes a set of business functions R(A,A | $C$(A,A)) for its own purposes — such as human resource development. One could identify an internal relation R with a set of business functions $R_i$: R = ($R_i$). A role A can also create value for other roles, such as for B. In other words, an entity in the role A establishes a business or institutional relation R(A,B | $C$(A,B)) with another entity in the role B. Note that establishing a relation coincides with a service or revenue stream. The value perceived by one of the roles is a subjective quantity. Furthermore, symmetries of R or $C$ with respect to roles do not hold, e.g., R(A,B | ...) ≠ R(B,A | ...). This means, for instance, that an application service cannot be identical to the price a customer pays for it. Last but not least, R or $C$ may involve many entities. For example, financial clearance of digital rights may concern several entities within an enterprise. Moreover, establishing relation R(A,B | $C$(A,B)) may require some time, depending on the complexity of setting up relations R or realizing conditions $C$.

Besides the aforementioned functional network decomposition of mobile WLAN application services, our conceptual model also allows for their hierarchical decompositions. This means that the order of responsibilities among business or institutional entities may be spelled out.

In addition to the functional network hierarchy, our model allows the entities within the layered functional model (see Figure 10.1) to be conceived and implemented in every layer $L_i$ at different granularities. This means that roles A and functions $R_i$ can be created, split, merged, or extended whenever needed. Similar remarks hold for relations R and conditions $C$. The relevant mobile (WLAN) application service issues intimately relate to the hierarchy and the granularity of the layered functional model for an enterprise or institution. A hierarchy defines the order in the network of responsibilities spread around an E-business or E-institution. The granularity defines the level of service details needed to achieve business or institutional objectives. Adaptability, reconfigurability, and tailorability of these services concern the complexity and dynamics of these hierarchies and granularities of roles A, functions $R_i$, relations R, and conditions $C$.

In the sections titled "Private and Secure B2B Scheduling Services," "A MTCS," "Content Distribution in an Internet with WLANs," and "Mobile Radio Broadcasting Services," we give simple models of typical mobile WLAN application services in terms of our conceptual model. If necessary, investigation of the usability, utility, and commercial exploitability of the application services are substantiated by scenario analysis, use-case design studies, and participatory user-oriented design studies. In particular, in the section titled "Content Distribution in an Internet with WLANs," business relations among a group of network owners, providers, Internet service providers, and application service providers are modeled to ensure seamless multimedia streaming across various types of networks and domains.[4] In the section titled "Mobile Radio Broadcasting Services," service platform provisioning[5] is modeled to customize, personalize, and exploit mobile WLAN application services. In all these sections, the complexity in adapting, reconfiguring, and tailoring the mobile WLAN application services become evident.

## PRIVATE AND SECURE B2B SCHEDULING SERVICES

Currently, scheduling services are accessible on small wireless devices. However, the time it would take to upload and update graphical scheduling information of, for example, Microsoft Outlook over a slow UMTS network is not acceptable from a business perspective. Scheduling information is typically something that a traveling employee wants real-time on the navigation screen of his or her car. Otherwise he might miss an exit to a rescheduled meeting location. Therefore, a scheduling service can profit from a faster WLAN connection. Such a connection has the additional advantage that it can be offered and controlled by a company itself. Therefore, it is cheaper than UMTS and forms a more convenient communication network infrastructure for providing business solutions to its own personnel.

Next-generation WLAN services in business-to-business (B2B) settings will make use of context awareness, personalization, and adaptation technologies and, therefore, put very high demands on privacy protection as well as business security features of such services. Although WLANs still lack strong privacy, security, and QoS guarantee means, these flaws are just due to protocol and system legacies that call for revisions. To deal with privacy protection and business security, a middle-agent framework can be used that allows parties to securely exchange personal or business-sensitive contextual information across WLAN networks. An enhanced Microsoft Outlook scheduling service can be presented, in which the middle agents collectively arrange an update of a meeting between employees. The service does so by adapting the location and time on the basis of privacy policies of the traveling employees themselves and the business policies of the companies they work for.

## Scheduling Scenario

Assume that three persons, each from a different organization (A, B, and C), have scheduled a meeting at a business science park. WLAN has been rolled out in this park. They all have to travel to this park to attend the meeting, and when it is finished, each of them has to drive home or to a second meeting. However, because of a flight delay, one attendee is not able to arrive at the planned meeting in time. Then, using the WLAN network offered by the airport, the attendee activates his or her schedule agent on a remote agent platform to rearrange the meeting at another location and at a later time. Scheduler agents of the three attendees are activated on the agent platform and the location and time of the new meeting are negotiated, considering user-related information like current locations, time constraints, and user policy. Employees may not be willing to reveal their current location or schedule to others because they themselves or their companies do not allow direct access to schedules (depending on business and user policy). After some negotiations, the users receive a notification of the new appointment, including the organizer of the meeting, who is walking over to the WLAN-supported business and science park. The server of Microsoft Outlook will subsequently stream all the graphical schedule information to the mobile client, causing most of the network traffic. However, WLAN bandwidth is certainly enough for this purpose.

The scenario presented in the preceding text illustrates a very simple value network (see Figure 10.2). Organizations A, B, and C share the standard Microsoft Outlook scheduling service. They offer each other an enhanced scheduling service R(A,B,C | C(A,B,C)) on top of their own Microsoft Outlook services. These enhanced services can be provided only after privacy and security conditions C(A,B,C) are met. Those conditions are reached when a contract among organizations A, B, and C contain privacy and security policies of all the organizations and are endorsed and followed. In this scenario, the organizations do not allow each other to gain direct access to their business or private agendas. The organizations have outsourced contracting issues to an agent service platform provider, who provides the enhanced scheduling service R(A,B,C | $C$(A,B,C), $C$ (A,B,C,Q)). All organizations trust this third party Q together with its agent platform: condition $C$ (A,B,C,Q) is assumed to be effective.

## Scheduling Agent System Architecture

The JADE-LEAP agent platform[15] helps developing agent-based applications that can run on WLAN networks. On this Java-based and FIPA-compliant agent platform,[16] agents communicate by sending messages in the FIPA Agent Communication Language (ACL) across WLAN networks, which connect different runtime environments on local servers or running WLAN terminals. One or more local servers will host, besides databases and profile information, the local containers (runtimes) of the JADE-LEAP platform. On local servers, the platform will host one scheduler server agent for each user that acts and negotiates on its behalf and a database access agent (DAA) that handles requests for travel-time information, profiles, and schedules. Each portable client device runs a JADE-LEAP peripheral container that hosts one single agent; the meeting scheduler GUI agent enables the user to activate his or her own scheduler agent on the server.

## Privacy or Business Security Strategies

In its simplest form, a subservice is requested by a requester (R) from a provider (P) that has access to the resources to deliver this service. A requester agent (RA) and a provider agent (PA) represent R and P, respectively. Note that such agents may not always be software agents, but a piece of hardware, the user herself or himself, etc. RA and PA are in possession of their own preference and capability information, respectively. An agent that deals with preference or capability information that is neither a requester nor a provider is called a middle agent (denoted by MA). Any agent (RA, PA, or any other third party agent, e.g., MA) that is informed of both preferences and capabilities is in a position to make a decision. Using this approach, the privacy issues involved in subservice brokerage can now be resolved by ensuring that only entrusted parties can access the preference and capability information.

Upon close investigation of the nine classes of MAs mentioned[14] (each corresponding to one specific combination of RA, MA, and PA, being aware of the preferences and capabilities at the time of decision making), one can distinguish three main privacy enforcement strategies, each with one most probable decision-making authority:

■ Strategy 1 — None of the actors has both preferences and capabilities information at his or her disposal. Here, to reach an agreement, negotiation strategies that are widely studied in AI and multiagent systems have to support the application. As a result, the RA and the PA can withhold the sensitive information regarding the preferences and capabilities (for example, the current location of the employee).

■ Strategy 2 — An MA is aware of both preferences and capabilities. As an entrusted entity, such an MA is allowed to make a decision on behalf of the others.

■ Strategy 3 — An RA or PA, respectively, is solely in the position of making a decision based on full information of preferences and capabilities.

Note that it depends on the organizational role of an agent whether a decision-making authority will be in favor of the requester, the provider, or both.

## Enforcing Privacy and Business Security

The agent that triggers the rescheduling process, referred to as the initiator, can be considered an RA that requires resources (time) from the responding agents to set a meeting (see Figure 10.3). The responders free up time and thus can be considered PAs. It is assumed that RAs and PAs do not directly share their strategies or the main part of their user preferences. Also, they do not give this information to an MA. Generally, RA (the initiator) is not aware of the capabilities (available time–space slots) of PA (the responder) or even RA may not be authorized to reschedule the meeting by itself. Therefore, RA and PA will have to negotiate. Scheduling agents are negotiating agents (strategy 1).

The DAA provides an SA with (1) the schedule of the corresponding employee's meeting times and locations and (2) the travel times to the location of the next and consecutive meetings. The DAA logs into the Microsoft Exchange Server as one of the users and extracts information from the corresponding schedule. The DAA is a trusted third party in our implementation, and it will refuse any direct request for schedules or locations from outside agents. In other words, it will only access the schedule of the user who corresponds to the SA that made the schedule request. It will do so only if this agent provides the proper username and the corresponding password required to access the schedule.

Based on the FIPA Iterated Contract Net Interaction Protocol (ICNI protocol; see FIPA), a fully functional interaction algorithm was developed and implemented for the SAs, including the possibility to query, collect,

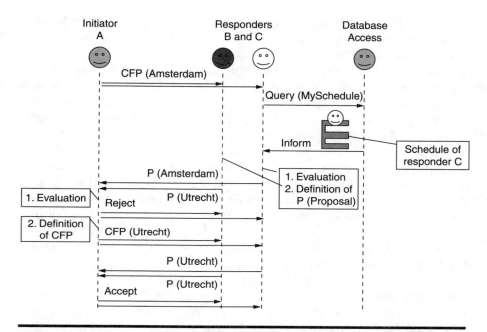

**Figure 10.3  Sequence diagram illustrating scheduling agents A, B, and C and the database access agent (DAA), which is a middle agent (MA).**

and process information from other agents.[17] An important feature in the ICNI protocol is the distinction between the initiator and the responder. The initiator (from A) starts and manages the interaction by defining and sending call for proposals (CFP) and by rejecting or accepting proposals from the other entities in the interaction, i.e., the responders (from B and C).

To summarize, B2B WLAN services require the implementation of business policies, customer preferences, and the embedding of privacy enforcement in the service provisioning process. To enable this, a third party may have to be contracted — in this case, an agent platform provider. The environment presented here ensures the privacy of the involved parties using agent-based brokerage mechanisms. Based on such privacy-protecting MAs, a scheduling service has been developed that reschedules a meeting for traveling employees from different companies.

## A MTCS

The district nurse is one of the most mobile persons in healthcare, visiting patients and providing them with medical and nonmedical care in the intimacy of their homes, and averting the need for expensive treatment in hospitals or nursing homes. As a mobile outpost of healthcare, a district nurse can neither quickly nor easily consult colleagues or specialists for advice or a second opinion. Also, direct access to electronic patient data is not easily available.

Live videoconferencing offers a solution when a district nurse needs guidance from a remote colleague or a specialist. The elderly people that a district nurse mostly takes care of frequently suffer from venous leg ulcers, open wounds, or fluid retention (edema). Treatment primarily consists of regularly inspecting the wounds and applying a compression bandage.

MTCS aims to demonstrate the use of a wireless broadband infrastructure for medical professionals. The service provides them with effective communication tools and hence increases the efficiency and quality of their work. The medical professionals whom MTCS addresses are district nurses, general practitioners (GPs), and dermatologists. Other involved groups are healthcare patients, district nursing organizations, regional hospitals, and health insurance companies.

With the support of the MTCS, the visit of a district nurse to a patient could look like this: after the old bandages are removed and the wound is cleaned, its state is documented with a photo from the digital camera integrated in the mobile wireless device that the nurse carries. After evaluating the development of the wound by comparing it with earlier images in the patient's wound-log, the nurse decides to go for a second opinion and immediately sees which colleagues are currently available for a videoconference (see Figure 10.4). During the videoconference that follows, the two nurses discuss the issue and decide that they require advice from the patient's dermatologist, who is not available for a videoconference at this moment. Therefore, the nurse captures a few minutes of video and some specific photos of the wound, annotates them with a few comments and digital ink marks,[18] and sends them to the dermatologist. Later that day, the dermatologist replies that this development is not very unusual and that there is no need to see the patient until the next regular visit in two weeks. The dermatologist's answer automatically becomes part of the patient's wound-log such that every nurse attending to the patient can act on it.

In this scenario, the MTCS increases the quality of care because it allows getting second opinions with minimal delay and supports the visual tracking of the wound's progress. Currently, the nurse will have to visit the patient personally, which frequently introduces a delay of one or two days. Furthermore, the patient is spared the straining and time-consuming transport to the dermatologist in the hospital.

A simpler, low-bandwidth, e-mail-based service, similar to MTCS, is presented in Reference 20. Images taken at the patient's site are sent to the GP by regular e-mails using a digital camera and GSM smart phone.

**Figure 10.4  A district nurse treating a patient using the MTCS.**

Videoconferencing is not supported, but nurses and doctors could use a regular phone call to discuss a specific case. A classic setting of a fixed (wired) teledermatology scenario is described in Reference 20 and some chapters of Reference 19. There are a large number of Web sites dedicated to telemedicine in general and teledermatology in particular, e.g., see Reference 21 to Reference 23.

## Wireless Access and Infrastructure

Wireless and ubiquitous access to the MTCS is a paramount requirement of district nurses. The wireless infrastructure that is already available in Enschede, the Netherlands, and rolled out in the city center and adjacent neighborhoods in 2004–2005 provides an ideal test bed consisting of several hundred 802.11b WLAN access points. A glass fiber forms the backbone network between the various hot-spot locations, the hospitals, and the district nursing centers. To provide ubiquitous access to the service, the WLAN infrastructure is complemented with GPRS or UMTS in suburban and rural areas. An adaptive client application scales the available features to the available bandwidth, which means that videoconferencing, which is only supported with WLAN and when UMTS has sufficient bandwidth available, is reduced to an audio connection with best-effort image sharing when only GPRS or low-bandwidth UMTS is available. Thus, the mobile application service is not WLAN specific, but can benefit from it enormously, especially when videoconferencing is invoked.

## User Requirements and System Requirements

Early in the project, we systematically interviewed district nurses, dermatologists, and GPs to determine their requirements and the roles, functions, and relations that the MTCS needs to provide. The main requirements for the MTCS, resulting from these interviews, are:

- Mobile devices with wireless access both in the city and in rural areas
- Teleconferencing with true-color images to get second opinions
- Multimedia messaging with true-color images, video, and audio to communicate with colleagues and doctors who are unavailable for a teleconference
- Presence and location information about potential communication partners to determine easily who are available and who should not be bothered
- Patient-centered wound-log documenting the patient's progress with images, videos, and all multimedia messages relating to the patient

- QoS management and priority communication in public WLAN hot spots, i.e., higher QoS than other users in a public WLAN
- Secure communication and storage to protect sensitive patient data
- Capability to extend the service, in the future, to cover logistic management (appointments, routes, and ordering of pharmaceutical products), accounting and billing (when, where, and what), and access to a patient dossier (GP, hospital)

The main end-user roles are: the advisor (i.e., specialist nurse or doctor), the advice seeker (i.e., district and specialist nurse), and the progress monitor (i.e., nurses reporting to the wound-log). Additionally, we have the roles of application service provider and infrastructure provider (i.e., the WLAN), both of which require financial compensation for their services. Financial beneficiaries are the health insurance companies (transportation cost, treatment time) and, to some extent, the home care organizations (efficiency). Both are potential sources to finance the service. Home care patients benefit from improved medical care with the potential of faster recovery and fewer costly journeys to hospitals or their GPs. Additional indirect benefits can result from using MTCS for public relations both by health insurance companies and home-care organizations.

The final commercial relationships and the business case for a large-scale introduction of the service will be developed in the second phase of the project, which will provide extended testing and evaluation with real patients in selected neighborhoods.

## CONTENT DISTRIBUTION IN AN INTERNET WITH WLANS

In the Internet of the near future, WLANs (typically based on 802.11) and other wireless networks (e.g., UMTS)[24] will enable mobile hosts to receive channels of live and scheduled multimedia content (e.g., radio or TV broadcasts). These channels can, for instance, be distributed through multiple proxy servers, with mobile hosts handing off from one server to another as a result of mobility (because different proxy servers serve different networks).[25] This idea can be extended to the distribution of channels through multiple aggregators. An aggregator is an intermediary service provider that operates a pool of proxy servers to aggregate channels from sources and to deliver them to mobile hosts.[26,27] As a result, mobile hosts can potentially receive a channel from multiple alternative aggregators, possibly through multiple alternative wireless networks (e.g., in a hot spot). A mobile host must therefore be able to select an aggregator from which it wants to receive a channel (e.g., based on the quality the user prefers). While receiving a channel, mobile hosts have to be able to select another aggregator and hand off to it, for instance, because the

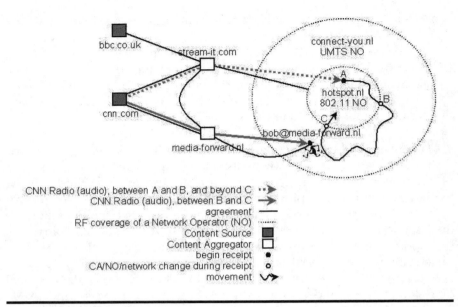

**Figure 10.5    Roaming scenario.**

current aggregator becomes unavailable as a result of being out of network range. The aggregator switching system for mobile receivers of live multimedia streams (ALIVE) realizes this behavior. To improve user-friendliness,[28] the ALIVE system selects aggregators automatically.[29]

## CORD Business Model

The CORD business model is a submodel of a more generic E-business model (see section titled "A Simple Model for Mobile Application Services") that revolves around three business roles:[26,27] content source, content aggregator, and network operator.

A content source is the origin of one or more channels. A content aggregator receives channels from sources and forwards them to mobile hosts in a way suitable for the limited capabilities of wireless links and mobile hosts. The proxy-style distribution scheme via aggregators increases scalability in the absence of IP multicast.[30] Sources and aggregators primarily process and forward application-level data units, typically in the form of RTP packets.[31] Figure 10.5 shows an example in which the source cnn.com* distributes audio channel CNN Radio via the aggregators stream-it.com and multimedia-forward.nl.

---

\* The domain names in this section are for illustrative purposes only.

Users receive channels from aggregators via network operators. The primary function of a network operator is to provide IP-level connectivity to mobile hosts. In Figure 10.5, the operators hotspot.nl and connect-you.nl operate an 802.11 LAN and an overlaying UMTS network, respectively. Bob's mobile host is equipped with an 802.11 interface and a UMTS interface. As a result, it can, for instance, receive CNN Radio from stream-it.com through its 802.11 interface in the hot spot (at point A), while it can receive CNN Radio from media-forward.nl through its UMTS interface. Figure 10.5 does not show the Internet backbone providers that may interconnect sources, aggregators, and network operators.

A user has to establish an agreement with an aggregator to gain access to the aggregator's channels. This aggregator is called the user's home aggregator (e.g., Bob's home aggregator in Figure 10.5 is media-forward.nl). A home aggregator is responsible for authenticating its users.

A home aggregator sets up application-level roaming agreements[26,27] with foreign aggregators so that its users can also receive channels from the foreign aggregators (e.g., to enable Bob to receive CNN Radio from the foreign aggregator stream-it.com).

Aggregators can be local or global. Local aggregators are available through a limited number of network operators (e.g., stream-it.com is only available in hotspot.nl), whereas global aggregators (e.g., media-forward.nl) are available on the entire Internet. A mobile host has to handoff to another aggregator when it receives a channel from a local aggregator and leaves its service area (e.g., Bob's mobile host has to handoff to media-forward.nl when it leaves the service area of stream-it.com at point B). An aggregator is a local one when it has an agreement with one or more network operators (see the agreement between stream-it.com and hotspot.nl).

Each aggregator offers different versions of the same channel (e.g., using different bandwidth levels or compression formats). This enables them to serve different types of mobile hosts that connect to the Internet through different types of (wireless) networks and to be able to deal with different user requirements (e.g., pertaining to cost or perceived quality). The description of a channel version is called a configuration. For instance, it describes the number of streams, their required network bandwidth, and their encoding format.[32]

The agreement between a user and his or her home aggregator indicates the configurations at which the user can receive channels from the aggregator. An application-level roaming agreement specifies the configurations at which a user from the home aggregator can receive channels from the foreign aggregator. For example, the roaming agreement between media-forward.nl and stream-it.com could specify that users of media-forward.nl can only use the lowest-quality channel configuration

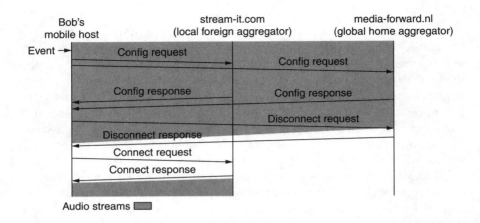

**Figure 10.6 Typical protocol behavior at point C.**

of stream-it.com to receive channels. In summary, users can receive the same channel from different aggregators at different configurations, possibly through different interfaces.

## The ALIVE System

The ALIVE system enables mobile hosts to automatically hand off to another aggregator while receiving a channel. It consists of an application-level protocol[26,27] and a decision-making component[33] that determine which aggregator provides the best version of a channel.

The ALIVE protocol enables mobile hosts to determine which aggregators support which configurations. A mobile host invokes the protocol when it is looking for a better configuration of the channel it is receiving, for instance, when it moves into a subnet (an aggregator with a better configuration may appear) or moves out of one (aggregators may disappear, resulting in a new and better aggregator). The assignment of a (new) IP address to one of the host's network interfaces (e.g., to Bob's 802.11 interface at point C) and the loss of network connectivity (e.g., of the 802.11 network at point B) could signal these two events, respectively.

Figure 10.6 shows Bob's mobile host sending a so-called configurations request to stream-it.com (via its 802.11 interface) and to media-forward.nl (through its UMTS interface) at point C to determine which configurations of CNN Radio they can offer. Using the responses of the aggregators, the decision component determines that stream-it.com provides a better version of CNN Radio than media-forward.nl. It therefore sends a disconnect request to media-forward.nl and a connect request to stream-it.com. As a result, Bob's mobile host now receives the 'better' version of CNN Radio from stream-it.com via hotspot.nl's 802.11 network. The protocol's behavior is similar at points A and B, except that stream-it.com is unavailable at point B.

When an aggregator receives a configuration request from a foreign user, it first authenticates the user with his or her home aggregator (e.g., stream-it.com authenticates Bob with media-forward.nl). Using the roaming agreement with the home aggregator, the foreign aggregator determines which of its local configurations the user is allowed to use. Aggregators cache this information so that users do not have to be authenticated or authorized on a per-request basis, which improves scalability and reduces response time. Home–foreign aggregator interactions are typically based on an AAA protocol (e.g., DIAMETER[34]). For simplicity, they are not shown in Figure 10.6.

## Implementation

Figure 10.7 shows a test bed that realizes the ALIVE protocol on top of the Session Initiation Protocol (SIP).[35] The test bed uses the Session Description Protocol (SDP)[36] to describe configurations. The SIP/SDP code (in C) is from the open SIP project.[37]

The server hosts a process that represents a global foreign aggregator ($P_{GFA}$) and another that represents a local foreign aggregator ($P_{LFA}$). The Free Radius server[38] represents a home aggregator (HA) and authenticates users. The laptop represents a mobile host such as that of Bob. It is equipped with an 802.11 interface, a fixed Ethernet interface, and a GPRS interface (via Bluetooth over USB). The server, the Radius server, and the laptop are Linux machines. $P_{GFA}$ and $P_{LFA}$ execute the server-side software of the CORD protocol (see Figure 10.6) on top of an SIP user agent server. The laptop communicates with $P_{LFA}$ via its 802.11 interface and with $P_{GFA}$ through all three of its interfaces.

The server also runs VIC, the video conferencing application,[39] as a multimedia server. Mobile hosts send SIP INVITEs (the connect requests of Figure 10.6) to $P_{GFA}$ or $P_{LFA}$ to have them change the IP address to which the VIC server sends its packets,[40] or to request them to stream at another configuration. $P_{GFA}$ and $P_{LFA}$ communicate with the Radius server to authenticate the user. The Radius server returns an SDP description of the configurations the user can use at his or her home aggregator.

The laptop runs three processes: one that executes the client-side software of the system on top of an SIP user agent client ($P_C$), another that executes VIC as a client, and a third that runs a mobility manager[41] that keeps track of the state of the laptop's network interfaces (e.g., which interfaces currently have link-layer connectivity). The mobility manager sends events to protocol software (e.g., network disconnects), which then

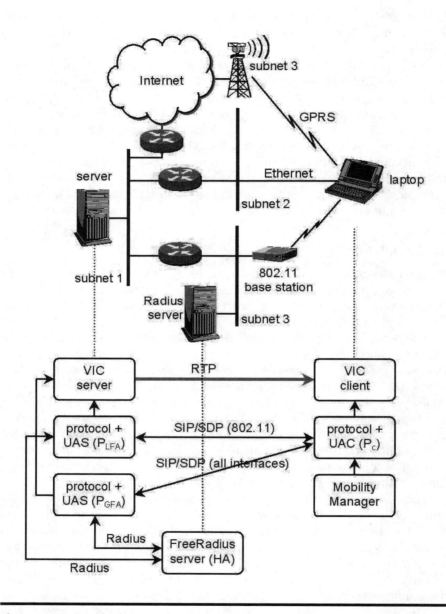

**Figure 10.7   Test bed.**

results in the protocol being executed (see Figure 10.6). For example, unplugging the Ethernet cable will result in the mobility manager sending an event to the protocol software, which then queries the aggregators that the mobile host can still reach ($P_{LFA}$ via the wireless LAN interface and $P_{GFA}$ via the 802.11 and the GPRS interfaces). Next, the software on the laptop, for instance, sends a connect request (an SIP INVITE message) to $P_{LFA}$, which will configure the VIC server with the address of the laptop's wireless LAN interface instead of with the laptop's Ethernet interface.

## MOBILE RADIO BROADCASTING SERVICES

From the perspectives of 2.5 and (beyond) 3G telecommunication network providers, application service providers, and content service providers, the commercial success of broadband mobile application services is highly dependent on network accessibility, mobile application service availability across networks, and its acceptance by the various business entities (end users in particular) involved in content provisioning. However, the above providers cannot simply base their exploitation model for mobile application services on the existing Internet model. The investments for communication network infrastructures, costs of manual rollout of mobile application services, given the large variety of device and network characteristics, and the stakes of content providers are exorbitantly high. Therefore, free content distribution is not really an option in a viable exploitation model for the mobile domain. With the current growth of the WLAN market, the Internet model will become even more vulnerable; the distribution of illegal content across independent peer WLANs will require digital rights management (DRM) solutions.[42]

A service platform provider has been proposed as a mediator between heterogeneous business entities.[5] A service platform provider can do it himself or outsource the handling of E-contracting in terms of SLAs, network-roaming agreements, DRM, personalization, or customization. He can arrange E-contracting and alike also for mobile WLAN application services and content provisioning services across heterogeneous communication networks. CallSong is an example of such a mobile application service for music distribution. It is built on top of the mobile application service platform called "Uluru."[43]

### A Mobile Application Service Platform

A mobile application-service platform provider is an intermediary between content providers and access network providers who can provide multimedia personalization, DRM, accounting and payment, content management, and content distribution services to content providers and mobile networks providers. Content providers and communication network providers can create innovative mobile multimedia applications on top of

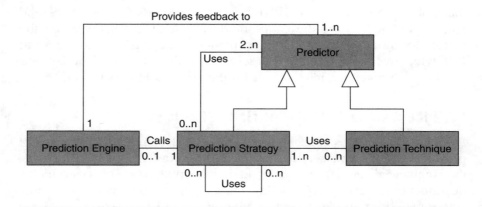

**Figure 10.8  Relations between a prediction engine, strategies, and techniques.**

such a service platform. Personalization and financial exploitation services are essential functionalities that are crucial for acceptance of mobile application services by the various business entities. End users require tailored services, and the providers demand to be remunerated through financial clearance of their mobile application service offerings.

For personalization purposes, mobile application platforms may provide tools that application service developers can use to integrate common mobile application services and prediction engines about user preferences and behaviors. Prediction engines can indicate how usable or "interesting" a mobile application service or multimedia content item will be for individual users or groups of users. The techniques used in such engines can be based on social filtering, case-based reasoning (CBR), information filtering, item–item filtering, genre least mean square filter, stereotype comparisons, or strategies thereof.[44] In Figure 10.8 the prediction engine provides feedback to one or more predictors, and every prediction technique is used by one or more strategies. Developers are free to choose from a set of prediction techniques and strategies.

For exploitation purposes, a developer may use Parlay's Charging API specification to integrate the mobile application service platform and mobile operator payment systems.[45] Using a Parlay bridge, the platform can make payment reservations and also act on such reservations. A charging engine on the platform can take various usage metrics and exploitation policies for determining different charging schemes associated with a certain usage. For example, the user's prepaid mobile phone account can be immediately charged whenever a service has been successfully delivered. Alternatively, usage metrics can be "best-guess expected-usage" figures, and the calculated charge can be used to make a "reservation" on a prepaid account.

## A CallSong on Uluru

CallSong (see Figure 10.9), on the service platform Uluru,[43] is a personalized and commercialized mobile audio broadcasting service that is based on a service platform model that makes use of numerous third-party components-off-the-shelf (COTS), as described in the preceding text.

The NCRV (a major Dutch public broadcasting organization) showed an early interest to participate in Uluru user trials. Their team examined the possibilities of Uluru and compared it with their needs, resulting in a mobile broadcast application. Nowadays, CallSong customers can listen to, rewind, and fast-forward selected songs and editorial items on human interest, politics, and music in personalized playlists. Furthermore, they can choose their own preferred charging policies. Last but not least, they can share their favorite songs with fellow CallSong community members (see Figure 10.10).

The trials focused on creating mobile application services for future digital audio broadcasters having access to 3G networks and WLANs. It showed that personalized commercial multimedia broadcasting services require a different way of content creation and management, particularly when there is a lack of available bandwidth on networks other than WLANs. By utilizing a general mobile application service platform such as Uluru, application developers can leverage preexisting, stable, and well-tested platform functionalities such as personalization, content management, charging, and billing to reduce their time to market and, thus, improve profitability.

## CONCLUSIONS

WLANs may provide users and organizations access to similar, but far more attractive and intricate, mobile application services than other types of wireless networks such as UMTS. For example, WLAN hot-spot owners and providers can locally support, sustain, and reserve very high network bandwidths against relatively low costs for their own employees or customers. Furthermore, by rolling out and managing their own WLAN hot spots, both enterprises and public institutions can exploit their own mobile application services. From a functional perspective, these application services can be very sophisticated and advanced. In general, these services are location-based, and enterprise- or institution-specific. Therefore, mobile WLAN application services require not only rigorous modeling of the hierarchy and granularity of business or institutional roles, relations, and conditions, but also call for modeling of the hierarchy and granularity of the information and communication technological architectures and infrastructures. Moreover, their modeling not only concerns the current

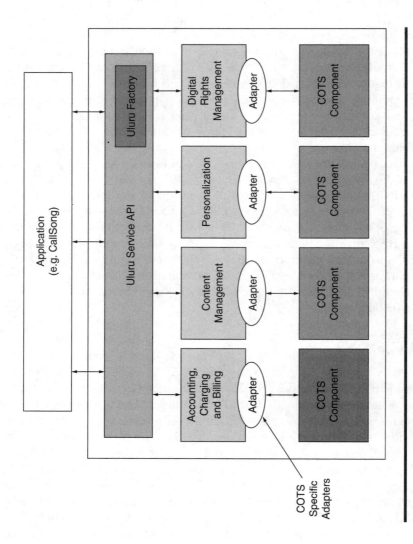

**Figure 10.9   Relation between CallSong and Uluru.**

**Figure 10.10    Screen shot of the CallSong trial application.**

but also the future hierarchies and granularities that are needed to sustain parts of an enterprise or institution.

All the projects presented in this chapter started off by identifying the essential network architecture and the (wireless) infrastructural, business, institutional, and user requirements before developing and deploying mobile WLAN application services. Scenario analysis, use-case design, and participatory user-oriented design studies laid bare the related mobile WLAN application service requirements. In particular, the usefulness and commercial exploitability of the CallSong service on Uluru were investigated. For the MTCS, the studies are underway.

Mobile WLAN application services may be functionally very complex. The complexity manifests itself as service adaptability with respect to new types of requirements. This feature is characteristic of all reported application services. For example, both institutions and enterprises can re-tailor privacy and business security rules for their own Microsoft Outlook scheduling services (see the sections titled "Private and Secure B2B Scheduling Services" and "An MTCS"). They can do so for their employees or representative software agents. This way, they can actually reduce cognitive loads and emotional stress of businessmen on the road or district nurses who are too busy to rearrange meetings. Similarly, mobile multimedia streaming services can be commercially exploited in a different manner on the basis of other consumer marketing and pricing strategies (see sections titled "Content Distribution in an Internet with WLANs" and "Mobile Radio Broadcasting Services").

An issue of service adaptability that requires further research is that service changes have to be manually handcrafted even today. Business security and exploitation schemes are implemented as E-contracts on, for example, Uluru by a service creator or provider or by the parties themselves.

Besides complexity, another issue that has to be resolved is the dynamics and heterogeneity of (mobile) application service offerings across business, communication networks, Internet domain, and application-service layers. The sections titled "Private and Secure B2B Scheduling Services" and "Content Distribution in an Internet with WLANs" provide good examples of other service configurations manifesting themselves in the usage of agent communication languages and in provisioning different quality of service classes, respectively.

WLAN applications can make use of the generic functionality of the so-called service platforms (see the section titled "Mobile Radio Broadcasting Services"). A service platform can handle E-contracting of SLAs, network-roaming agreements, DRM, personalization, or customization. The existence of such a platform ensures that the developer of a WLAN application can focus on the application functionality. Many service platforms are being developed at this moment, for example, by (mobile) telecom operators, Internet service providers, or by financial or other commercial parties, and will be offered to application developers in the near future. Besides this, software agent platforms will be available for the developer and will offer generic functionalities such as yellow pages, service discovery, and communication and negotiation protocols.

All applications presented in this chapter show the complexity of designing and deploying mobile WLAN application services. These applications also show that modeling is important to design and build such complex and dynamic services. Such a model should deal with all functional layers that are relevant, from the physical layer up to the business layer. The model should especially focus on business value propositions to roll out a mobile WLAN application service that can be commercially successful.

However, both the problems, of handling complexity and of dynamics of mobile (WLAN) application services, require an extensive modeling of enterprises or institutions over time. Keeping up with the ever-increasing application service demands may call for embedding and embodying collective knowledge and intelligence into autonomous multiagent systems and platforms within E-business and E-institutional networks.

## REFERENCES

1. Bahl, P. and Padmanabhan, V., RADAR: An In-Building RF-Based User Location and Tracking System, In *Proceedings of IEEE Infocom 2000*, Tel Aviv, Israel, 2000.
2. van Eijk, R. et al., Handling Heterogenity in Location Information Services, In *Proceedings of Communication Networks and Distributed Systems Modeling and Simulation conference (CNDS 2004)*, San Diego, CA, 2004.
3. Axiotis, D. et al., Integrated Services, Hot spot and User Equipment Forecast for Interworking 3G and WLAN Networks, in *B3G Report*, Systems Beyond 3G Cluster, 2003.
4. Verhoosel, J., Stap, R., and Salden, A., A Generic Business Model for WLAN Hotspots — A Roaming Business Case in The Netherlands, In *Proceedings of the First ACM International Workshop on Wireless Mobile Applications and Services on WLAN Hotspots (WMASH 2003)*, San Diego, CA, 2003.
5. Laarhuis, J., Towards An Enterprise Model for 4G-Environments, submitted to *IEEE Transactions on Networking*, 2004.
6. Angelov, S. and Grefen, P., A Framework for the Analysis of B2B Electronic Contracting Support, In *Proceedings of the 4th Edispuut Conference — Multidisciplinary Perspectives on Electronic Commerce*, Amsterdam, The Netherlands, 2001.
7. Jonkers, H. et al., Towards a Language for Coherent Enterprise Architecture Descriptions, In *Proceedings of the 7th IEEE International Enterprise Distributed Object Computing Conference (EDOC 2003)*, Brisbane, Australia, 2003.
8. Schank, R., *Scripts, Plans, Goals, and Understanding*, Lawrence Erlbaum Associates, Hillsdale, NJ, 1977.
9. Falzon, P. (Ed.), *Cognitive Ergonomics: Understanding, Learning and Designing Human-Computer Interaction*, Academic Press, London, 1997.
10. Vicente, K., *Cognitive Work Analysis: Towards Safe, Productive, and Healthy Computer-Based Work*, Lawrence Erlbaum Associates, Mahwah, NJ, 1999.
11. Vicente, K., *The Human Factor: Revolutionizing the Way People Live with Technology*, Alfred Knopf, Toronto, Canada, 2003.
12. Barwise, J. and Perry, J., *Situations and Attitudes*, MIT Press, Cambridge, MA, 1983.
13. Pirolli, P. and Card, S., Information Foraging, *Psychological Review*, 106, 643, 1999.
14. Decker, K., Sycara, K., and Williamson, M., Middle Agents for the Internet, In *Proceedings of the 15th International Joint Conference on Artificial Intelligence*, Nagoya, Japan, 1997.
15. Bergenti, F. and Poggi, A., LEAP: a FIPA Platform for Handheld and Mobile Devices, presented at *ATAL 2001*, http://leap.crm-paris.com/.

16. FIPA, Foundation for Intelligent Physical Agents, http://www.fipa.org/.
17. Bargh, M. et al., Agent-Based Privacy Enforcement of Mobile Services, In *Proceedings of the International Conference on Advances in Infrastructure for Electronic Business, Education, Science and Medicine and Mobile Technologies on the Internet (SSGRR2003w)*, L'Aquila, Italy, 2003.
18. Benz, H., Casual Multimedia Process Annotations — CoMPAs, CTIT, Ph.D. thesis, University of Twente, Enschede, The Netherlands, 2003.
19. Wootton, R. and Oakley, A. (Eds.), *Teledermatology*, The Royal Society of Medicine Press, London, 2002.
20. Lamminen, H. and Voipio, V., Mobile Teledermatology, in Wootton, R. and Oakley, A. (Eds.), *Teledermatology*, Chapter 21, The Royal Society of Medicine Press, London, 2002.
21. E-medicine, http://www.emedicine.com/.
22. Telemedicine Information Exchange, http://tie.telemed.org/.
23. Nordunet2 Telemedicine, http://nordunet2.nhn.no/.
24. Haardt, M. and Mohr, W., The Complete Solution for Third-Generation Wireless Communications: Two Modes on Air, One Winning Strategy, *IEEE Personal Communications*, 2000.
25. Dutta, A. et al., MarconiNet supporting Streaming Media over Localized Wireless Multicast, In *Workshop on M-Commerce 2002*, Atlanta, 2002.
26. Hesselman, C. et al., A Mobility-aware Broadcasting Infrastructure for a Wireless Internet with Hotspots, In *Proceedings of the First ACM International Workshop on Wireless Mobile Applications and Services on WLAN Hotspots (WMASH'03)*, San Diego, CA, 2003.
27. Hesselman, C. et al., Delivering Live Multimedia Streams to Mobile Hosts in a Wireless Internet with Multiple Content Aggregators, *Mobile Networks and Applications Journal (MONET)*, special issue on *Wireless Mobile Applications and Services on WLAN Hotspots*, 2005.
28. Kleinrock, L., An Internet Vision: the Invisible Global Instrastructure, *AdHoc Networks Journal*, 1, 3, 2003.
29. Clark, D. and Wroclawski, J., The Personal Router Whitepaper, Version 2.0, 2000, http://www.ana.lcs.mit.edu/papers/PDF/PR_whitepaper_v2.pdf.
30. Chennikara, J. et al., Application-Layer Multicast for Mobile Users in Diverse Networks, In *Proceedings of IEEE Globecom 2002*, Taipei, Taiwan, 2002.
31. Schulzrinne, H. et al., RTP: A Transport Protocol for Real-Time Applications, *RFC 1889*, 1996.
32. Schulzrinne, H., RTP Profile for Audio and Video Conferences with Minimal Control, *RFC 1890*, 1996.
33. Kamilova, M. et al., Using Policies for the Automatic Selection of Service Providers in a Wireless Internet, submitted to *1st International Workshop on Streaming Media Distribution over the Internet (SMDI04)*, Athens, Greece, 2004.
34. Calhoun, P. et al., Diameter Base Protocol, Internet draft, 2003, draft-ietf-aaa-diameter-17.txt.
35. Rosenberg, J. et al., SIP: Session Initiation Protocol, *RFC 3261*, 2002.
36. Handley, M and Jacobson, V., SDP: Session Description Protocol, *RFC 2327*, 1998.
37. SIP webpage, http://www.gnu.org/software/osip/.
38. Free Radius webpage, http://www.freeradius.org/.
39. VIC at UCL, http://www-mice.cs.ucl.ac.uk/multimedia/software/vic/.

40. Wedlund, E. and Schulzrinne, H., Mobility Support Using SIP, In *Proceedings of the 2nd ACM/IEEE International Conference on Wireless and Mobile Multimedia (WoWMoM'99)*, Seattle, WA, 1999.
41. Peddemors, A., Zandbelt, H., and Bargh, M., A Mechanism for Host Mobility Management Supporting Application Awareness, In *Proceedings of the 2nd International Conference on Mobile System, Applications, and Services (MobiSYS'04)*, Boston, MA, 2004.
42. Burnett, I. et al., MPEG-21: Goals and Achievements, *IEEE Multimedia*, Vol 10, No 4.
43. Tokmakoff, A. et al., Uluru: A platform for adaptive mobile multimedia applications, In *Proceedings of the IEEE International Conference on Multimedia and Expo (ICME'2004)*, Taipei, Taiwan, 2004.
44. van Setten, M. Veenstra, M. and Nijholt, A., Prediction Strategies: Combining Prediction Techniques to Optimize Personalization, In *Proceedings of TV'02: The second workshop on Personalization in Future TV*, Malaga, Spain, 2002.
45. Alur, D., Crupi, J., and Malks, D., *Core J2EE Patterns: Best Practices and Design Strategies*, Prentice Hall, Upper Saddle River, NJ, 2001.

# 11

## MOBILE COMMERCE AND ITS APPLICATIONS

*A.F.M. Ishaq and Mohammad Mohsin*

### INTRODUCTION

Over the last few years, the wireless communications industry has seen phenomenal growth in all spheres, including in terms of capabilities of mobile devices, middleware development, standards and network implementation, and user acceptance [Va02]. Mobile devices have been the fastest-adopted consumer products of all time [Mo03]. The increase in adoption and use of these devices for data services has created an opportunity for E-commerce to leverage the benefits of mobility. In Great Britain, almost half of the children aged between seven and sixteen now have mobile phones. Text messaging is rapidly becoming the favorite method of communication, with an average of 2.5 messages per phone sent every day [Ka02]. Japanese figures for mobile technology use and adoption are more impressive as more than one out of every four Japanese are actively using NTT DoComo's i-mode alone [Da03]. The adoption of wireless technology in organizations is also on the rise. Among large U.S. companies, 20 percent of the workforce is already mobile, and over 80 percent of European corporations consider mobile devices and applications to be very important for their business [Ju03]. As organizations experience more volatile marketplaces, global competition, shortened product life cycles, customer pressures for tailored offerings, and tighter performance standards, they increasingly depend upon new technology solutions [Fe98]. Mobile commerce (M-commerce), which exploits the features of wireless technology, is yet another milestone on the technology curve that has the potential to leverage or to transform existing business processes.

This chapter discusses the many facets of M-commerce, encompassing a number of dynamic and evolving services and applications, in the context of the various individual and organizational needs that can be better fulfilled by this technology. These needs vary from the basic level such as provision of rapid access to current business applications, irrespective of time and place, to more sophisticated mobile functionality of location-specific information and real-time push notifications. Our main contribution lies in mapping these applications on a strategic grid framework, which has resulted in these applications being categorized on the basis of their impact on individuals and organizations. This will be useful for placing and seeing the M-commerce portfolio of applications in their proper context. We also envisage that this chapter will assist strategic managers, IS planners, architects, and strategists in adoption of the right application for their individual and organizational needs.

M-commerce is electronic commerce over wireless devices [Br03]. It encompasses all activities related to a (potential) commercial transaction conducted through communications networks that interface with wireless devices [Ta02]. The definition we have adopted describes the role of M-commerce in a broader context as exchanging products, ideas, and services between mobile users and providers [Ne03]. This broader definition extends M-commerce to mobile business, and it also involves business-related communication in which financial transactions do not necessarily occur.

M-commerce promises to and is poised to change the way we all live, play, and do business [Wi01]. Researchers believe that M-commerce is another structural shift in technology that can lead to enhancement in value for the customer [Ka02]. M-commerce provides business workers and residential users with greater opportunities to connect to the Internet, outside the limited area dictated by fixed networks. The surge in the use of portable communication devices and the wider reach and range of the Internet are expanding the E-commerce market in which mobile commerce is expected to grow [UN02].

M-commerce differs partially from E-commerce due to the special characteristics and constraints of the mobile devices and wireless networks. M-commerce services and applications can be adopted through different wireless and mobile networks, with the aid of several mobile devices. These can be characterized broadly as wireless LANs (WLANs), personal area networks (PANs), and ad hoc devices. Figure 11.1 shows some scenarios for mobile financial services, in which the use of both cellular and LAN networks is used to carry out transactions. The trend in the wireless industry is to provide integrated services, including voice,

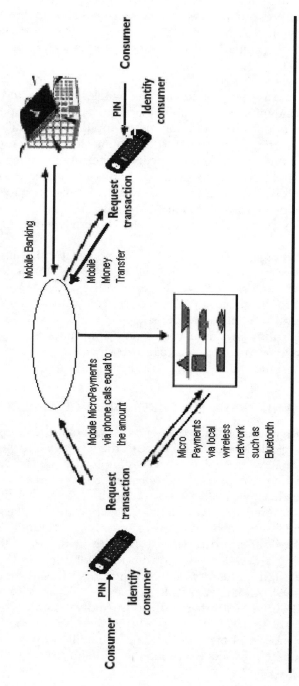

**Figure 11.1  Micro and macro payment services using LAN and WAN networks.**

messaging, entertainment, shopping, and personal organization, over wireless handsets. At present, the cellular systems have relatively low data rates compared with WLANs and PANs; therefore, it is not suitable for LAN applications. A comparison of existing wireless technologies is presented in Table 11.1.

Wireless devices are expected to penetrate the consumer market and may make the home a major test bed for new wireless devices and services. The access to WLAN devices in public places through access points (hot spots) holds a lot of promise for M-commerce applications. Similarly, PANs enable all devices, static or moving, to communicate with each other. Ad hoc devices serve one particular purpose, typical examples being a garage-door opener, remote control to operate various consumer devices, etc. The standards that have evolved for these wireless devices enable anyone to set up a network easily and share a broadband Internet connection among several devices equipped with wireless Ethernet cards or Bluetooth devices.

This chapter is organized in the following way. After the introductory section, we discuss the evolution and distinguishing features of M-commerce. Within the same section, we also discuss the different M-commerce players that lead to the creation of an M-commerce value chain. The third section introduces and explains the strategic grid for categorizing M-commerce applications. A range of existing applications and probable future applications of M-commerce that are expected to deliver tangible business and consumer benefits are included in this section.

## EVOLUTION AND MAIN FEATURES OF M-COMMERCE

Individual and organizational needs have always been the main impetus for any technological development. Today, the need of individuals and organizations is to have the ability to access and process information from anywhere, at any time, conveniently. M-commerce is a giant step toward fulfilling this need. However, to take full advantage of this technology requires us to look into the future and come up with some applications that will fulfill the future needs of individuals and organizations. Predicting the future is risky. Rutherford B. Hayes, the 19th president of the United States, after seeing a demonstration of the telephone in the 1880s, commented that although it was a wonderful invention, businessmen would never use it [Ap99]. Another example is that of Bill Gates of Microsoft, who proclaimed in the 1980s that 640 K of memory "ought to be enough for anyone" [Wi01]. In 1996, it was predicted that the number of global users of cellular mobile phones was around 170 million by the year 2000. In reality, the number of users exceeded the 1-billion mark in 2000 [Va00].

**Table 11.1  Comparison of Several Wireless and Mobile Networks**

| Issue | LANs | Loops | PCs | Mobile IP | ATM | Satellites |
|---|---|---|---|---|---|---|
| Coverage | Local area | Local[a] or metropolitan | Metropolitan | Wide area | Wide area | Wide area |
| User bandwidth | 1 to 20 Mbps | 1 to 20 Mbps | 19.2 kbps | Network dependent[b] | 1 to 20 Mbps | 19.2 kbps to several Mbps[c] |
| Application | Data and voice | Data and voice | Data and voice | Data and voice | All | Data and voice |
| Limitations | Limited area | Interference | Bandwidth | Limited applications | Cost | Initial cost |
| Status | In use | Emerging | In use | Emerging | Emerging | Emerging |

[a] Depending on the underlying technology (such as 3–10 mi for LMDS).
[b] Depends on the underlying wireless network.
[c] Higher limit for satellites such as Teledesic.

*Source:* From Varshney, U. and Vetter, R., Mobile commerce: A new frontier, *IEEE Computer*, October 2000. With permission.

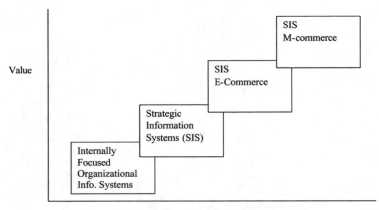

**Figure 11.2  Stages of evolution of M-commerce.**

To predict the future, it is useful to trace the stages of evolution that have led to the current situation. Most studies have identified three eras of IS evolution. The initial two eras of data processing and management information systems (MIS) focused on operational efficiency and management effectiveness within the organization, i.e., the focus was internal. The third era of strategic information systems (SIS) focused on transformation and integration of processes and information systems, not just from an internal organizational perspective but also in the context of the industry. The vision of interorganizational systems was explored as early as 1966 by Kaufman [Ka66]. Today, many of the most dramatic and potentially powerful applications of IT are based on the concept of interorganizational systems. E-commerce technologies enabled companies to incorporate buyers, suppliers, and partners in the redesign of their key business processes, thereby enhancing productivity, quality, and speed. Now, with the emergence of M-commerce, the location-dependence limitation of E-commerce has been eliminated, resulting in more flexibility besides retaining most of the other advantages of E-commerce. Figure 11.2 illustrates the stages of evolution of M-commerce in the era of SIS. The SIS era now shows a pattern of structural migration from PC-centered models to mobile person-centric models. As mobile computing devices supported by broadband access and new wireless networks are becoming commonplace, companies need to adopt M-commerce to derive competitive advantage.

To develop and discuss where and how M-commerce should be used, it will be useful to refer to the model of Choi and colleagues [Ef00], which illustrates the relationship between the products, delivery agents, and processes that exist in both electronic and physical markets (Figure 11.3).

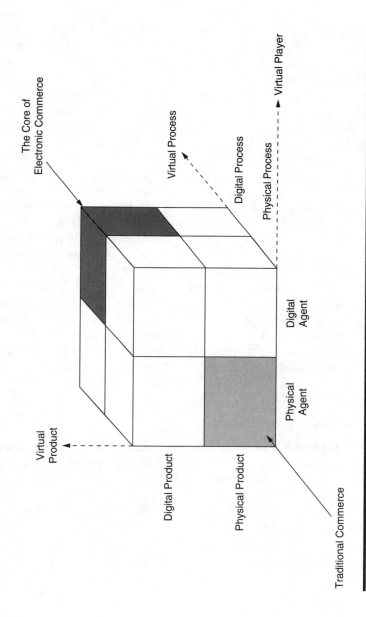

**Figure 11.3  A model of E-commerce market areas. (From Efraim Turban et al., Electronic Commerce: A Managerial Perspective, Chap. 1, p. 5, Prentice Hall International Edition, 2000. With permission.)**

In this model, products are differentiated based on whether they are physical or virtual (i.e., electronic). The process dimension in the model differentiates digital from physical processes. A digital process is one associated with using the Internet to access the Web, whereas a physical process involves a physical action associated with carrying out a commercial transaction. The third dimension is the nature of the delivery agents involved in the transaction. A Web-based auction store is digital, but the corner supermarket is physical.

The addition of location independence to the quadrant where the products, delivery agents, and processes are all digital, provides the potential for commerce to be engaged in anytime, anywhere, and for almost any conceivable product or service. This additional variable changes the processes. In wired commerce, it is the user who moves to the computer to do business, whereas in M-commerce, it is the service that moves to the user as per user requirements. Thus, M-commerce is not the replacement of E-commerce, but it is the extension of services to places where traditional E-commerce services are not available.

The key factor in all definitions of M-commerce is mobility, which is dependent upon the range of the device and the technology being used. For example, a home cordless phone user has a much more limited allowable range of movement away from the base unit compared with the cellular mobile user. Applications are being developed in which seamless integration of wide area network (WAN), LAN, and PAN has become possible, allowing high-speed data transmission, which is desirable in the case of M-commerce applications. For example, mobile phones having an embedded Bluetooth device now can use PAN communication to an in-house access point and transfer to cellular when outside the house.

## Features of M-Commerce

The essence of M-commerce revolves around the idea of reaching customers, suppliers, and employees, regardless of where they are located. It is directed toward delivering the right information to the right place at the right time. Many of the features of M-commerce are nonexistent in E-commerce, such as the ability to access the Internet from any location at any time, to pinpoint an individual mobile terminal user's location, to access information at the point of need, and the availability of information as per user needs. A few of the unique features are discussed in the following text.

### Ubiquity

Ubiquity is the primary advantage of M-commerce. As the service is always available, it gives the ability to get any information, from any place, and at any time, through Internet-enabled mobile devices.

### Reach Ability

Knowledge workers are moving, commuting, and working constantly. M-commerce lets them transact and process information as and when required with no time delays.

### Personalization

In an era of information overload, and with the availability of an enormous number of information services and applications, the relevance of information that users receive is of great importance. M-commerce applications can be tailored to the personal information needs of the users.

### Data Portability

Users do not have to memorize information but can instead store profiles of products, company addresses, information about restaurants and hotels, banking details, payment and credit card details, and security information, and access them when needed for purchases — all from their mobile handsets.

### Localization

Organizations and companies can be aware of the whereabouts of users and convey instructions and offers specific to their location. Some wireless infrastructures support simultaneous delivery of data to all mobile users within a specific geographical region. This functionality offers an efficient means to disseminate information to a large consumer population.

## M-Commerce Value Chain

The conduct of commercial activity and the behavior of any user undertaking it have been described in a very appealing model by Webraska [Ki03], which is shown in Figure 11.4. According to it, user behavior can be divided into four areas:

Find it: Finding the object of interest at an interest point
Route it: Determining the route to get to that interest point
Share it: Sharing information about this point of interest with other friends or associates
Buy it: Completing a transaction related to the selected location

To cater to this commercial activity of identifying, locating, and routing buying needs, various companies having their own specialty domains are

**Figure 11.4 User interaction model. (From Mitchell, K. and Whitmore, M., Location-Based Services: Locating the Money, Webraska Mobile Technologies, Australia, Idea Group, 2003. With permission.)**

preparing themselves to follow that trail. Huge investments in infrastructure development are being made to enable these functions. Durlacher in his M-commerce report [Du02] classified wireless technologies into mobile client devices providing presentation services and communications infrastructure. The report's value chain model, as shown in Figure 11.5, identified eleven participants, namely, technology platform vendors, infrastructure and equipment vendors, application platform vendors, application developers, content providers, content aggregators, mobile portal providers, mobile network operators, mobile service providers, handset vendors, and customers in the wireless industry constituting the M-commerce value chain. The above value chain participants can also be consolidated into three groups, such as technology providers, network operators, and service providers responsible for content and transaction services.

## Technology Providers

Technology providers are companies that manufacture and provide the wireless infrastructure equipment, including the devices for enabling the infrastructure, middleware, and business applications. We have included the technology platform vendors, infrastructure and equipment vendors, application platform vendors and handset vendors in this category.

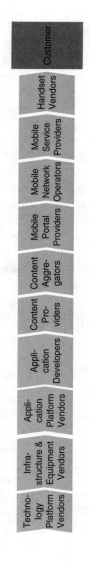

**Figure 11.5 M-commerce value chain. (From Durlacher, Mobile commerce report, www.durlacher.com, 2002. With permission.)**

Technology platform vendors deliver the operating systems (OSs) and microbrowsers for mobile devices such as smart phones and communicators. Major OS market leaders are Microsoft with Windows CE and Symbian with Palm OS.

At present, the leading suppliers of mobile network infrastructure equipment are Alcatel, Motorola, Ericsson, Siemens, Nokia, and Lucent Technologies. The dominant technology to date has been wide-area cellular; however, alternative technologies such as WLAN and PAN are being actively developed to deliver broadband services at higher data rates than is possible through cellular technology.

Application platform vendors are responsible for providing middleware infrastructure, i.e., Wireless Access Protocol (WAP) gateways at either the mobile operator's site or at the business customer's site, to provide seamless transfer of data from one technology to the other. In addition to the traditional standards bodies, a number of interest groups are setting *de facto* standards by assembling key players and agreeing on workable development conditions much faster than the traditional standards bodies.

Application developers are technology providers who build applications on existing available platforms. Handset vendors are critical in the value chain. The usability issue has become important for mobile devices because of their small user interfaces and personal preferences of the users. Handsets that can support a wide variety of services are also becoming a critical bottleneck for the M-commerce value chain. Handset vendors are being pushed to provide support for many value-added services and develop a wider variety of products that can support future applications.

### Network Operators

Prior to the emergence of WLAN technologies, the situation for network operators, i.e., the owners of the networks, was quite favorable. They were best positioned to benefit from the emerging M-commerce services because they already owned a billing relationship with the customer, and they controlled the portal that is preset on the SIM card when it is distributed. However, today it is the mobile network operators who benefit most from the value chain as any increase in mobile services will lead to increase in revenues because of the increase in duration of call time or the volume of traffic. NTT DoCoMo iphone is the best practical example of this. However, with the emergence of mobile LANs (MLANs), this advantage is at risk as new players are expected to share the revenue because of WLANs' capability of high-speed data access, particularly in hot-spot areas.

### *Service Providers Responsible for Content and Transaction Services*

Service providers work in close collaboration with network operators. They include content providers, content aggregators, mobile portal providers, and mobile service providers. The traditional content providers in the case of E-commerce, such as Yahoo!, Google, and excite, etc., are better placed because it is convenient for them to shift to the mobile domain. Similarly, another category of content aggregators who repackage available data for distribution to wireless devices is evolving. The added value is in delivering content in the most appropriate package. Content providers and content aggregators serve the consumer market because the data for business is kept internally by corporations and usually is not available for external consumption. Banks and other financial institutions are initiating services to corporations and businesses through the mobile medium. They need to customize their service offerings, such as intelligent information alerts for well-defined business and individual needs. Mobile portals are another channel for pulling information. They provide a window to a host of other useful applications and content from various providers to become the user's prime supplier of Web-based information that is delivered to the mobile terminal.

## M-COMMERCE STRATEGIC GRID FRAMEWORK

Applegate and coworkers [Ap99] stated that "in different settings, IT can profoundly affect one or more of a firm's value-chain activities, sometimes simply by improving effectiveness, sometimes by fundamentally changing the activity, and sometimes by altering the relationship between activities." M-commerce technologies are also having a deep impact on the working and personal lives of individuals. In fact, it is the personal market rather than the business market that has driven many of the mobile applications in organizations [Me03]. To assist the management in understanding the contribution of existing and potential applications to the individual and to business, a number of approaches to categorizing systems have been published in the literature. The framework we have adopted is a hybrid of the work of Ward and colleagues, and Kalkota and Robinson [Wa95, Ka02].

Figure 11.6 is useful in analyzing the mix of M-commerce applications and their intended impact on organizations and individual consumers. The applications falling in the key operational areas having a supply chain focus are specific to organizations only. The other three quadrants have a host of applications servicing individual consumers and organizations. M-commerce applications can be broadly divided as follows:

| STRATEGIC | HIGH POTENTIAL |
|---|---|
| CUSTOMER FOCUS<br><br>Applications that are critical for achieving future success of an individual or business | NEW INNOVATION OPPORTUNITIES<br><br>Applications that may be important in achieving future success of an individual or business |
| Applications on which the business depends for success<br><br>SUPPLY CHAIN FOCUS | Applications that are valuable but not critical to the to the individual's or business's success<br><br>CONSUMER/OPERATIONAL FOCUS |
| KEY OPERATIONAL | SUPPORT |

**Figure 11.6   Categorization of M-commerce applications.**

*Support Applications Having an Operational Focus:* Support applications improve the performance of individuals and business activities, helping achieve productivity or efficiency improvements that deliver mainly economic benefits. However, such types of applications are not vital to achieving current or future objectives. The companies that invest heavily in IT applications are moving toward wireless access for existing applications to better justify their investments. The motive behind this transition to wireless solutions is the most cost-effective use of existing IS resources. Profitability and productivity increase as employees are not time and location bound to access and process information. Individuals get increased convenience, performance, productivity, and, sometimes, reduction in cost as well. Some of the applications falling under this category provide simple mobile access (e.g., short message service [SMS], e-mail). More sophisticated business applications differ in their level of IT integration with existing organizational IT systems. Examples of applications falling in this category are personal information management, enterprise information portals, and extensions of existing applications available at the enterprise.

*Key Operational Applications Having a Supply Chain Focus:* Applications falling under this category are specific to organizations and businesses. The smooth operation of these applications is extremely important, and a failure in the working of these applications results in immediate and significant problems for the organizations. These kind of applications are embedded in the core activities of the business. Any weakness or problem in these systems can jeopardize operations because the business process cannot be completed satisfactorily if they malfunction. Many of the applications falling in this category not only impact their own business, but their influence also extends beyond the enterprise boundaries to the entire supply chain. Inefficiencies or problems of any participating member in the supply chain directly impact the business process. M-commerce leads to efficient flow of information, orders, products, and payments among the various players. Its adoption and use provides management with the ability to respond and react quickly to the disruptions in any stage by proactively adjusting strategy or plans regarding critical supply chain events. M-commerce applications falling in this category are inventory management, personnel management, asset tracking, and distribution and ordering applications.

*Strategic and Customer-Focused Applications:* Customer-focused applications (whether for an individual or an organization) are categorized as strategic applications. Strategic systems are intended to provide competitive advantage in meeting the future objectives of the organization. Strategic systems provide organizations with strategic advantages, thus enabling them to increase their market share, better negotiate with their suppliers, or prevent competitors from entering their territory. Competitive advantage can be gained if competitors do not have a system with equivalent or comparable qualities. These systems target the existing and future needs of the customer and should create an exceptional customer experience. Companies that have adopted M-commerce are utilizing the systems in a manner that is providing them an advantage over their rivals. Some of the applications falling into this category are mobile ticketing, mobile shopping, mobile banking, mobile trading, etc.

*High-Potential Applications Having New Innovation Opportunities:* This quadrant includes applications that are in the research and experimentation phases. Based on new and innovative ideas, a number of M-commerce applications that are predicted to have successful future use in enterprises and homes are evaluated. Evaluation may be by means of prototypes, or by assessing market acceptance of an idea, or by paper evaluation of the business and technological environment. M-commerce applications are the primary driver for the future of wireless technology. Only those applications will survive that fulfill the needs of the user. Developments in wireless technology have opened vistas of a host of possible M-commerce applications. In his book, *Future of Wireless Communications,*

William Webb has predicted that in the future, end users will demand the creation of unified, ubiquitous wireless communication systems that deliver data-rich content on demand, with low latency and high throughput. Moreover, applications will also become more intelligent and aware of the user's personal preferences. Applications will make use of intelligent agents to help fulfill user needs.

How wireless technology is being used and how it can be used in the form of M-commerce applications that can realize user needs is elaborated in the following text by mapping them to our adopted hybrid framework.

## Support Applications Having an Operational Focus

M-commerce was initially adopted by individuals and organizations that could see improvements in efficiencies and a cost benefit out of the use of wireless technology. Mobile consumer applications have been successfully promoted by NTT DoCoMo's iMode service, resulting in productivity gains as a result of greater convenience and satisfaction in many ways. For example, field staff can enjoy productivity gains by reducing their need for travel, reducing idle time, and increasing the availability of problem-specific information at the service site, thereby resulting in more efficient service execution. Improved time management also results because of better communication between the field staff and the dispatching office. It may also result in process improvement in the main office because the basic infrastructural requirements for M-commerce require the internal business processes to be automated to achieve real benefits. For example, costs can be reduced because of operational improvements in business processes such as accounting, invoicing, and administration. As shown in Table 11.2, we have included the following applications in this category.

### *Personal Life Management*

These services are for individual mobile consumers and include such applications as accessing e-mail accounts, chatting and instant messaging, contact information, entertainment, and information services. In fact, e-mail access can be the most revenue-generating application because hotmail alone has over 70 million users. Information services is another powerful attraction for individuals. People are using their mobile devices to obtain information on a wide range of subjects such as stock prices, flight schedules, news headlines and general news, financial news, sports news, entertainment news, environment news, important events, travel information, and much more. Information can be pushed to the mobile device, or the mobile device can pull information; for example, a consumer interested in cricket match scores types in a keyword and sends it to the SMS center and pulls back the information in the form of an SMS reply. Consumers can also connect to the Web site and subscribe to services that are later pushed back to the subscriber.

## Mobile Office

Applications falling in this category provide the employee with control over the various forms of communications they deal with in office or field work every day. The concept of these applications is that a mobile employee should no more be limited by the absence of information or data to perform the job even when he or she is outside the office. Examples of such types of applications are access to organizational e-mail communication, SMS, some form of scheduling functionality such as calendar, address book, tasks and journals, and access to corporate customer services. Field workers such as engineers, service professionals and maintenance staff, sales and marketing staff, etc., often need information on the spot. They can now make business decisions more quickly and effectively and provide better customer service as a result. Mobile information access during a meeting between a field worker and a prospective customer can reduce the back and forth communication that is so time consuming for both parties. A number of applications are being developed that are seamlessly connected to the enterprise or corporate databases and applications. A few companies are already using most of these features productively; for example, Kodak reduced the time to report sales figures and update them in the financial system of the company from 30 days to 1 day [Ka02].

## Mobile Access to Web Portals

Today, many large organizations have linked their enterprise resource planning (ERP) system to provide a single point of access to their stakeholders through Web portals. Companies normally have portals providing corporate, human resource management, and financial information. Many companies are now extending the functionality of these portals to mobile devices. The important decision in this regard is to come up with applications that extend the functionality of these portals to mobile scenarios. Mobile access to these portals can result in saving either time or cost, and consequently can improve productivity. People working in different specialties require information related to their domain. However, due to the specific features and constraints of mobile devices, the mobile application services offered through these portals are specifically designed and tailored

**Table 11.2 Categorization of M-Commerce Applications**

| Category of Application | Application Name | Details | Examples |
|---|---|---|---|
| **Support Applications Having an Operational Focus** | Personal life management | Applications fulfilling typical routine activities | E-mail, chatting, instant messaging, information services |
| | Mobile office | Applications extending office-like facilities | Accessing corporate databases, applications, and legacy systems from anywhere at any time |
| | Mobile access to Web portal services | Applications in which a portable mobile device becomes the means of access to the services | Posting and receiving information from corporate and individual Web portals |
| **Key operational Applications Having a Supply Chain Focus** | Inventory management or product locating and shopping | Applications attempting to reduce the amount of inventory needed by managing inventory in-house and on the move | Location tracking of goods, materials, parcels, and people |
| **Strategic Applications Having a Customer Focus** | Financial applications | Applications in which mobile devices replace or substitute the traditional financial transaction mechanisms | Banking, stock trading, and micropayments |

| | | | |
|---|---|---|---|
| Strategic Applications Having a Customer Focus | Advertisements | Applications in which mobile devices provide an additional, powerful marketing medium and channel | User preferences, context, and location-aware advertisements |
| | Mobile auctions | Applications allowing users to buy or sell certain items | Bidding in one or more auctions simultaneously in near real-time |
| | Reengineered applications | Applications that transform the existing business process, leading to radical improvements | Wireless telemedicine healthcare facilities |
| High Potential Applications Having New Innovation Opportunities | Proactive service management | Applications that anticipate services needed in the near future | Transmission of utility bills and information related to aging machinery components to vendors |
| | Mobile distance education | Applications extending distance or virtual education support for mobile users everywhere | Taking a class using streaming audio and video |
| | Entertainment services and games | Applications in which mobile technology is used for entertainment | Audio and video on-demand services and interactive games |

to individual information needs. For example, consider the scenario in which the services of a consultant (be it legal, medical, or project) are required at any time. Now, that consultant, when his or her mobile device is paged, can respond to the situation by accessing pertinent information through the legal, health, or projects application database available at the respective portals. The information sought by the consultant is consolidated through several internally maintained databases containing the related records regarding the case or situation. Another useful mobile application is for financial and accounting managers who, while on the move, need to view and monitor financial health indicators such as sales volumes, gross profits against a product, cash flows, and balance sheets. Employees of a company would be more interested in having access to applications related to the HR department that would simplify their administrative work. Major applications in this category include travel and expense management. Through an M-commerce solution, employees are able to plan their traveling itinerary and later submit details of their expenses for automatic updation in the HR department's records.

## Key Operational Applications Having Supply Chain Focus

In this strategic information era, companies have started to adopt interorganizational information systems to create more business value in the form of increased efficiency, low prices, and wider product selection. As businesses have become more competitive, and with fewer possibilities of increasing efficiencies further within the organization, they have started to develop a collaborative environment within the industry in which each specialist partner contributes to the entire business, thus helping in raising the efficiency of the entire industry. Thus, individual firms benefit from the overall increase in the revenues of the entire industry. This has resulted in the concept of the IT-enabled supply chain in which all the partners — suppliers, manufacturers, distributors, wholesalers, retailers, and customers — have a common stake in the business process. This simply means that a delay or problem at any of the partner organizations impacts all the common stakeholders. Improvements in the efficiency of supply chains are leading to stricter demands by customers. For example, now customers not only expect a product that conforms to their requirements but also expect to have real-time information about it, i.e., the location of the product as it moves through the logistics network, and the exact time it is going to be delivered to them. As customers have become more informed and knowledgeable about market conditions and the prevailing economic trends, they now also expect businesses to react to their decisions (say purchasing) in near real-time. These escalating customer demands are forcing corporations to adopt mobile technology to optimize and streamline the information flows in the supply chain. To fulfill these demands, organizations need to have flexible or adaptive supply chains that have the capability to accommodate these expectations. The location-based services feature of wireless devices provides unlimited opportunities to improve the business-to-business supply chain process. Kalakota and Robinson [Ka02] have placed all applications in this category under four groups:

1. E-procurement applications such as purchasing, approvals and payments
2. Supply chain execution applications comprising inventory management, logistics, and fulfillment
3. Supply chain visibility applications containing asset tracking and receiving
4. Data collection and service management applications comprising reverse logistics, field force automation, and dispatch management

By integrating the mobile terminal into the supply chain, it has become possible to improve all basic supply chain-related processes concerning purchase orders, invoices, bills of materials, order delivery, inventory management and utilization, customer services, etc. As an illustration, we take the example of an inventory management application.

### Inventory Management

Inventory management applications exploit the location-based services feature to add more value to the inventory management process. In fact, inventory here can signify men, products, materials, and other physical resources. A number of business models can be used that improve upon the inventory management process by utilizing wireless solutions. An example is the case of trucks carrying goods. The inventory information is broadcast to potential customers in real-time. Upon receiving the broadcast message, businesses needing inventories send orders for just-in-time deliveries. The order processing and inventory statistics are maintained inside the trucks, using a WLAN network because cellular or satellite signals do not work well inside the trucks. This model reduces the inventory space requirements and cost for both suppliers and businesses. Assembly line production units are another perfect candidate for the use of wireless technologies to best manage inventories, thus resulting in cost and production efficiencies. Here, wireless sensing and transmitting devices placed at different stages in the assembly line can dynamically adjust its speed depending upon the rate of consumption of components. As soon as a component is required, the device signals back the quantity required to the earlier stage in the assembly line.

## Strategic and Customer-Focused Applications

Strategic applications are those that align well with individual and organizational information needs. These are generally customer driven and intertwined with the paradigm of customer satisfaction and fulfillment. In other words, customers and their information needs are the driving force for developing strategic applications. A number of applications can be placed under the strategic and customer-focused category. A few of these are discussed in the following subsections.

### *Mobile Financial Applications*

Mobile devices are now increasingly being used to perform banking, payment, and other transactions. People find these devices to be a convenient mode for authorizing transactions in mobile banking, brokerage services, micro credit payments and money transfers, and confirmation of direct payments via the phone's microbrowser.

As shown in Figure 11.1, the user identifies himself or herself to the mobile device through secure identification mechanisms such as voice, personal identification number, or fingerprints. The device then authorizes a transaction to the service provider. Macropayments requiring more security use a secure public-key signature process, allowing for precise allocation of responsibility for fraud. Micropayments in the close proximity use infrared, Bluetooth, and radio frequency identification (RFID) technologies, and less secure form of authorization such as mobile subscriber identification or encrypted passwords. Vending machines using Bluetooth technology have been developed that allow the purchase of soft drinks, VCD titles, and chocolate bars by using a Bluetooth-equipped mobile phone. The payment mechanism for these services is so devised that the call charges include the cost of the service provided and are either subtracted from the prepaid cards or are included in the postpaid bill.

Financial institutions have found mobile devices to be an additional delivery channel for pushing information to customers on an event basis. Many of the brokerage houses have started using mobile solutions to gain an edge over their competitors. For example, in one application, shares exceeding certain price points could trigger messages asking whether to buy or sell. These types of services are expected to increase the revenue and profits of the firms because customers find it convenient to respond to information sent to their mobile devices.

### *Mobile Advertisements*

Wireless technology has placed a very strong tool in the hands of advertisers for proactive and direct marketing services. The demographic data of subscribers, their historic billing patterns, and mobile usage interests, coupled with the location-based feature of M-commerce, have made it possible for service providers to target customers with proactive services that may be of direct interest and value to them. The information can be pushed to or pulled by customers.

For example, as soon as a passenger holding a Bluetooth-enabled mobile device arrives at a destination, his or her device is sensed by the nearest hot spot and information regarding accommodation, weather, tourist spots, etc., are pushed to his or her device. Information can be made available on an event basis also; for example, when the car's scheduled service time approaches, the service station can transmit an alert. To reduce undesired messages, a subscriber can also opt for the pull method of information retrieval in which the subscriber sends an enquiry regarding the availability of a certain product or service. The service provider matches the query with the existing database of advertisers and transmits the required information back to the subscriber.

## Mobile Auctions

This class of application allows users to buy or sell certain items. Now bidders need not be present at the site to participate in auctions. This location independence of the auction has resulted in a new auction model in which people can bid even while on the move. Bidders can now participate in more than one auction simultaneously in near real-time.

## Reengineered Applications

The applications falling in this category focus on improving the quality of business services using mobile devices and wireless infrastructure. There are many business processes, such as insurance claims refund and medical treatment in remote and rural areas, that require a long time to complete just because of the absence of the location independence feature. Consider a scenario in which a patient at a remote place needs emergency medical services, and the consultant is also not available in the hospital. Using the wireless telemedicine facility, the paramedical staff can connect to the hospital's main database to view the patient's previous history and insurance coverage. The patient's history, with details of the current condition, can be transmitted on a near real-time basis to the consultant, who can then advise proper treatment. As a further improvement, the charges could be simultaneously billed to the health insurer, who could refund the amount at the same time. The normal procedure in such an emergency is to first take the patient to the hospital, which may require a long time; then locate the consultant; and finally, wait for the health insurer to refund the expenses.

## High-Potential Applications Having New Innovation Opportunities

### Current and Proactive Service Management

This class of application is based on gathering the current and future information needs of users and then offering specific services tailored to their needs. For example, in less developed countries, the meter readings for units of gas, electricity, and water consumed are collected from the customer premises by field workers specifically assigned by utilities for this task. The involvement of the human element increases the chances of errors in reported figures and is also cost and labor intensive. The process can be modified by fitting meter reading–sensing and transmitting devices to the meters to collect and transmit the utility consumption figures to a centralized database at the main office. The main system from the central office will then transmit the utility bill charges to the mobile device of the consumer, who can pay the billed amount to the utility company.

Varshney [Va02] has presented another interesting example, in which a car is fitted with a sensor that monitors the level of wear and tear in components. The information is then transmitted to all concerned businesses such as the vendor, repair workshop, distributor, and manufacturer. The repair workshops can then transmit proactive messages offering repair facilities at bargain prices. Vendors may be able to arrange spare parts in anticipation, and distributors and manufacturers could use the information for product analysis and improvement.

### Mobile Games and Entertainment

The mobile games and entertainment industry is waiting for the availability of more bandwidth capacity. With cellular mobile technology, only very simple games are possible. The arrival of third-generation cellular technology or the more aggressive deployment of wireless PAN and LAN networks may lead to enormous growth in the development of interactive and complex games. Subscribers already can download music and melody tones through their service providers. Mobile handsets integrated with MP3 portable players have been introduced, in which music titles are stored in the mobile device. People are still skeptical about the success of real-time streaming video applications mainly because of the limited current capacity of wireless networks and doubts about the appropriateness of mobile devices for viewing streaming video.

### Mobile Distance Education

Mobile devices are emerging as one of the most promising technologies for supporting learning scenarios — particularly collaborative learning scenarios. Experiments with content delivery using mobile devices have had some successes — for example, the BBC's Bitesize revision materials delivered via SMS to mobile phones [Ho03]. Mobile technology can be used to provide just-in-time training in specific skills. Field staff can be taught how to operate a machine or how to carry out machine trouble-shooting. M-commerce technology can supplement existing learning facilities. For example, students while on field trips will be able to download required classroom material and even submit their data and findings for near real-time analysis back to their laboratories.

## CONCLUSION

The general increase in the use and adoption of mobile and wireless technology has provided an impetus for businesses to conduct their activities using wireless infrastructure. M-commerce revolves around the idea of reaching customers, suppliers, and employees, regardless of where they are located. In E-commerce, it is the user who moves to the computer to do business, whereas in M-commerce, it is the service that moves to the user as per user requirements. Thus, M-commerce is not a replacement for E-commerce, but it is the extension of services to places where traditional E-commerce services are not available. We have grouped individual M-commerce applications into four categories and have mapped these on the basis of their impact on organizations and individuals. We also note that many of the future M-commerce applications will be using wireless PANs and LANs besides cellular technologies to provide digital connectivity among mobile computing devices, including mobile phones, laptops, PDAs, and home appliances.

## REFERENCES

Ap99    Applegate, L.M. et al., *Corporate Information Systems Management — Text and Cases*, 5th ed., Irwin/McGraw Hill, 1999.
Br03    Mennecke, B.E. and Strader, T.J., A framework for the study of mobile commerce, *Mobile Commerce: Technology, Theory and Applications*, Idea Group, 2003.
Da03    Macdonald, D.J., *NTT DoComo's i-mode Developing Win Win Relationships for Mobile Commerce*, Idea Group, 2003.
Du02    Durlacher, Mobile commerce report, www.durlacher.com, 2002.
Ef00    Turban, E. et al., *Electronic Commerce: A Managerial Perspective*, Chap. 1, p. 5, Prentice Hall International Edition, 2000.
Fe98    Feeny, D.F. and Wicocks, L.F., *Core IS Capabilities for Exploiting Information Technology*, Sloan Management Review, 1998.
Ho03    Hoppe, H.U., Joiner, R., Milrad, M., and Sharples, M., Wireless and mobile technologies in education, *Journal of Computer Assisted Learning*, 2003.

Ju03    Alanen, J. and Autio, E., *Mobile Business Services: A Strategic Perspective*, Idea Group, 2003.

Ka02   Kalakota, R. and Robinson, M., *M-business: The Race to Mobility*, McGraw-Hill, New York, 2002.

Ka66   Kaufman F., Data systems that cross company boundaries, *Harvard Business Review* (January–February, 1966).

Ki03    Mitchell, K. and Whitmore, M., *Location-Based Services: Locating the Money*, Webraska Mobile Technologies, Australia, Idea Group, 2003.

Me03  Mennecke, B.E. and Strader, T.J., *Mobile Commerce: Technology, Theory and Applications*, Idea Group, 2003.

Mo03  Mohsin, M. and Ishaq, A.F.M., Mobile Commerce — The Emerging Frontier: Exploring the Prospects, Applications, and Barriers to Adoption in Pakistan, International workshop on Frontiers of IT, Islamabad, Pakistan, 2003.

Ne03  Nenad J. et al., *M-Commerce a Location-Based Value Proposition*, Idea Group, 2003.

Th03  Zimmerman, T.G., *Wireless Personal and Local Area Networks*, IBM Almaden Research Center, San Jose, CA, Idea Group, 2003.

UN02  E-commerce and Development, Report, United Nations Conference on Trade and Development, 2002.

Va00  Varshney, U. and Vetter, R., Mobile commerce: A new frontier, *IEEE Computer,* October 2000.

Va02  Varshney and Vetter, R., Mobile commerce: framework, applications and networking support, *Mobile Networks and Applications*, Kluwer Academic, July 2002.

Wa95  Ward, J., *Principles of Information Systems Management,* Ch 3, p. 50, Routledge Series, 1995.

Wi01  Webb, W., *The Future of Wireless Communications*, Artech House, 2001.

# 12

## APPLICATIONS OF WLANS IN TELEMEDICINE

*Mohammad Ilyas and Salahuddin Qazi*

### INTRODUCTION

The role of wireless LANs (WLANs) in our daily activities is rapidly expanding. The world is increasingly becoming an information age society. Computers, digital handheld devices, digital information, communications, and software are not only being used in routine and mundane tasks such as those performed by traffic lights, rice cookers, home security systems, etc., but have also enhanced our capability to bring distant points closer to each other, prompting the term "global village." The use of the Internet essentially gives all of us the capability to connect with anyone, anywhere, and at anytime. This telecommunications marvel has made it possible to access distributed resources for collecting information, processing information, and dissemination of information in an efficient and cost-effective manner. Those familiar with the impact that information technology has had on the way we live our lives are also aware that there is much more to come. Recent advances in microelectronics, demand for and development of new applications, an amazing reduction in the size of digital devices and an equally amazing increase in the processing power of processors, and newer and efficient algorithms for information processing are just a few factors that are shaping the future of WLANs. Whatever advancements we have witnessed so far seem to be just a drop in the bucket, and there is much more to come.

One of the fields that is in a position to readily accept the conveniences of WLANs to serve our society better is medicine. This profession has always accepted newer technologies faster than any other field for two main reasons. First, medicine deals with a critical need of our society, and any emerging technology that is helpful in making the medical care more effective and responsive is readily deployed. The second reason is the availability of resources. There are many areas of the medical profession that have adopted the use of emerging communication technologies. Telemedicine is one. However, delivery of healthcare through telemedicine has some challenges to overcome and that will happen with time.[1] This chapter focuses on this important area.

Telemedicine is the practice of medicine over a distance. The distance could be as small as a few meters (adjacent rooms) or as large as thousands of miles or more. One of the obvious needs in telemedicine is the sharing of medical information (voice, data, and video). The evolution of sophisticated communication technologies (including WLANs) makes it easier to share digitized medical information in one-to-one, one-to-many, many-to-many, and many-to-one scenarios. Once information has been digitized, it can be stored, processed, and manipulated to make it reliable and more useful. Such processing may include compression, merging (with other pieces of information), extracting relevant information, and presenting it in comparison with other similar cases. All these aspects together are referred to as *information technology* and have numerous applications in healthcare and telemedicine. There are many challenges as well. With rapid evolution and emergence of information-handling technologies and communication technologies, many applications of these in telemedicine are becoming increasingly common.

Let us consider a typical scenario. An emergency call is made from a shopping mall indicating that an elderly patient needs medical attention. An ambulance is dispatched to the scene. As the ambulance arrives and the patient is transported to the ambulance, the medical treatment begins. Although awake, the patient is not able to provide his or her medical history, including the details of medications currently being taken. Usually, time is very critical in such situations and cannot be wasted. As the ambulance is equipped, in additional to medical equipment, with all the latest communication technologies, the work on gathering all the medical information about the patient begins immediately after the patient has been identified. In parallel, a nearby hospital is alerted about the arrival of the patient and the possible medical treatment that he may need. The ambulance also begins the process of communicating with healthcare professionals about the medical condition of the patient. The ambulance may even establish a video connection with the medical professionals at the hospital, or wherever they may be, to observe and assess the patient's condition. By the time the patient arrives at the hospital, all the medical history has been analyzed, a necessary team of medical professionals has been assembled, and a preliminary decision about the treatment plan has been made. This process can reduce the time used for diagnosis and the treatment can begin much sooner and may help tremendously in saving lives.

With the proliferation of the Internet and wireless communications, it is imperative that these technologies be integrated with medical and other such technologies. In this chapter, we discuss some possible applications of WLANs and the challenges faced in telemedicine.

## WLANS IN TELEMEDICINE

In delivering effective healthcare, one of the most important components is knowing the prior medical history of the patient. This includes biological data, information about prior treatments, surgeries, hospitalizations, current use of medications, and allergies to medications (if any), etc. The patient being treated may not be able to provide this information at the time it is needed. In addition, a medical professional for treating a particular medical situation may not be readily available. Wireless networks provide many interesting possibilities that may prove to be very useful in delivering medical care in such situations.[2]

Following, we present various applications of WLANs in telemedicine. There are many applications that are not included in this section because it is not possible to address all applications in one chapter.

### Gathering Medical Information

Based on their medical needs, individuals may visit several medical professionals (of different specialties) for care. Each of these practitioners files information about their patients. Similarly, hospitals keep all the records of patients who require hospitalization. In addition, patients fill their prescriptions at different pharmacies, which also keep records. All the bits and pieces of information that are scattered at various places may be necessary for providing effective healthcare to an individual. There are several ways of keeping this information handy and ready for use when needed. Obviously, it is not practical for individuals to carry their medical records with them in paper form all the time. Also, legalities associated with medical records require that this information should not be altered. Therefore, all mechanisms used for gathering, disseminating, or transporting medical information must adhere to all legal requirements. Advances in communication technology such as WLANs have provided many options for individuals to have their medical history available whenever it is needed.[3]

One possible option that has been explored in recent years is the use of *smart cards*, which are credit-card-sized plastic cards with some memory onboard and some intelligence. Such a card can be utilized for storing the latest information and individuals can keep it with them all the time. The card can contain possibly all the past medical history of an individual in addition to other relevant information such as health insurance. When visiting a medical facility or pharmacy, this card is presented to the medical staff at the time of check-in. The history is retrieved from the card and updated, based on the new information available during the current visit to the medical facility. At the time of checkout, the medical records are updated and stored on the smart card for individuals to carry with them. The same process is repeated during visits to pharmacies for prescriptions.

Graphical information such as x-rays and electrocardiograms (ECG and EKG) can also be stored on the smart cards and used when needed. Even video information generated from ultrasound scans can be stored on smart cards. This strategy is very useful but requires all health facilities to maintain equipment for reading from and writing on these smart cards. This technique works only if the individual in need of medical care has his or her card with him at that time.

Another possibility is to have the information stored in a database that is accessible through the Internet. When an individual checks in at a medical facility, his or her medical history and other relevant information can be retrieved from the database and used as needed. The information can be updated on a regular basis. In this option, individuals do not need to carry anything with them except an identity with a unique identification number. The information can also be accessed from remote places via wireless communication. With wireless communication available in almost every part of the world, Internet access will certainly be universal and greatly facilitate information access.[4,5]

For making an effective use of WLANs, the government agencies, health organizations, healthcare professionals, and healthcare facilities need to work together for full implementation of such an infrastructure.

## Mobile Multimedia

Exchanging multimedia (audio, video, and data) information between the patient and the healthcare facilities is very helpful in critical and emergency situations. An individual being transported to a hospital by an ambulance may exchange information using mobile multimedia communication facilities. A healthcare professional, in many situations, is in a much better position to diagnose and prepare a treatment plan for an individual if he has video information rather than just audio or data information. For instance, video information may be helpful in assessing the reflexes and viewing the coordination capability of a patient. Similarly, the level of injuries of a patient can be established better by visual information than just by audio or other descriptive information.

Real-time ultrasound scan of a patient's kidneys, heart, or other organs may be very helpful in preparing a treatment plan prior to the arrival of the patient being transported to a hospital. Such information can be transmitted through wireless communication networks from an ambulance to a hospital or to other healthcare professionals who are currently scattered at different places but are converging toward the hospital for treating the patient being transported. Due to bandwidth limitations, wireless communication networks may not be able to support exchange of high-quality video information. However, in many cases, such as with heart patients, quality of video information may not be as important as showing the movement of the heart. If the video information can be processed and the information about the movement of the heart can be highlighted before transmission, that will certainly provide tremendous help to medical professionals in making a timely decision about how to treat the incoming patient.[6]

## Imaging Technologies

Electronic imaging technologies play a significant role in medical diagnosis, treatment, and recovery phases of medical care. Ultrasound scanners, tiny cameras used in arthroscopic surgeries and in diagnostic tools, pathology, radiology, intestinal examinations, and brain scans are just a few examples of the use of imaging technologies. Having these electronic images in digital form makes it a lot easier to store, process, compress, and efficiently transport such information through wired or wireless environments without loss of important information. Information can also be shared among many healthcare professionals around the world using a virtual meeting environment, and their opinions can be sought for making the best possible decision about treatment of patients.

High-resolution, low-cost, and small cameras are very useful in creating images that are realistic and in delivering healthcare with no room for errors. Digital information from these cameras can be processed and compressed so that the images can be transmitted to or from remote places where sophisticated communication facilities may not be available. Providing healthcare in rural areas is a good candidate for using such imaging technologies. Storage and transmission of high-resolution images require tremendous resources in terms of memory and bandwidth. However, digital information produced by these cameras can be compressed significantly without loss of its contents, by using efficient compression algorithms and digital signal processing techniques that are available these days.[7]

## Wireless Medical Monitors

In a hospital's intensive care unit, in surgical facilities or delivery rooms, there is a large quantity of medical equipment connected by countless wires. The equipment in each room is also connected to a central facility in each ward where healthcare professionals and staff members monitor each patient's condition. If these wires were to disappear, each room would look much more open and have room for unhindered movement for the patient and caregivers. This can certainly be done by using WLAN technologies. Wireless transmitters with a small range of communication can be used for less power consumption, smaller size, and reduced interference with other similar devices. Bluetooth technologies and IEEE 802.11 environments are becoming very popular and can be easily deployed.[8]

For implementing such a communication environment, the devices that monitor a patient's condition need to be able to transmit the collected information wirelessly. The size of such transmitters today has become so small that soon the size of monitoring equipment will also become smaller. This may include portable ultrasound scanners (with wireless transmission capability), ECG and EKG, and monitors for brain activity, blood pressure, pulse, temperature, etc. Even stethoscopes could use wireless communication. The possibilities are endless. Use of these technologies makes it easier for storing, processing, transmitting, and sharing collected information, which can be easily retrieved for ready reference whenever needed. Information can also be used for training students and staff.

## Data Mining

The use of information technology in healthcare (including telehealth) will generate a huge amount of digital information that needs to be stored. The size of such storage (databases), which may potentially contain information for all patients worldwide, will certainly be huge. Although this seems to be a mammoth challenge, it opens up many potentially useful possibilities.

In treating patients, healthcare professionals look for similar cases and the treatments that were effective for those conditions. This information may be very useful for healthcare professionals in making a decision about treating the situation at hand. Extracting selective information that meets a specified set of parameters and deriving patterns and relationships between different data sets can be very useful. This process is referred to as *data mining.*

Searching for specific information from a huge database could be very time consuming, and time is limited in medical situations. Therefore, efficient data-mining techniques and algorithms are being developed, which will certainly help in reducing search time and in extracting meaningful information promptly from these databases. Mature statistical and data analysis techniques can be applied to make projections and plans for handling medical situations that we may come across in the future.[9]

## Data Acquisition — Homebound Patients

With the average life expectancy on the rise, there are increasingly people who are homebound or in assisted living facilities. Monitoring the health conditions of this segment of the population is a challenge. For every small change in their health, these individuals need to travel to a healthcare facility so that their conditions may be evaluated and, if needed, treatment may be administered. In many such cases, this visit may not be necessary. If a healthcare professional is able to regularly monitor these patients remotely, an educated decision can be made regarding necessity of travel to a healthcare facility.

WLANs can be used in monitoring health conditions of homebound patients. Easy-to-use computerized equipment can measure and transmit all the necessary information to a healthcare facility on a regular basis (as frequently as needed) and may include blood pressure, pulse rate, temperature, sugar level, and many other routine test results. Such information can be collected from various locations inside the residence and transmitted wirelessly to a central data acquisition system that is also inside the home. The data acquisition system can put the information in a desired format and forward it to the healthcare facility.

If the collected information is within normal parameters, the healthcare professional can review it as a routine task and store it for their record. However, if the information has an unusual pattern, a healthcare professional can be alerted to the fact that the information may need to be reviewed immediately. In that case, healthcare professionals can review and make a decision about the next step in dealing with the situation. The data acquisition system for collecting such information from homebound patients can also be Internet-enabled. That will give an added benefit to patients and their healthcare providers because that information will be accessible from anywhere, and hence there is a possibility of timely action based on that information.

## Robotics

Use of robotics in manufacturing facilities is very common. Widespread use of robotics in other fields including healthcare and particularly in telemedicine is not too far away. Major concerns surrounding the use of robotics has been reliability, quality, and, therefore, safety. As we witness further advances in technology, including WLANs, the use of robotics in healthcare will certainly increase. Unpredictable and random time delays in exchanging information through the Internet is one of the major factors in deploying robots for remote surgeries. A limited use of robotics is already in place for educating healthcare professionals. Feasibility of robotics in remote surgeries is being evaluated, and a number of minor surgeries have already been performed by robots controlled through a communication systems. The use of robotics in healthcare will open up tremendous opportunities for humanity. Other uses of robotics in healthcare include diagnostics and rehabilitation services. For example, robots can be used for physical therapy of patients during their rehabilitation.

## Smart Homes

Advances in wireless communication technologies also introduce some intelligence possibilities in the design of our residences, hence the coined term "smart homes." Such homes will be able to make some basic "decisions" that will be particularly beneficial to the elderly and homebound population. Some of the actions that smart homes can take include monitoring the movement patterns inside a home, recognizing someone's fall on a floor, recognizing other unusual situations and informing a relevant agency so that appropriate help can be provided, if needed. Locking the doors, turning the lights on or off, switching sprinklers on or off based on the weather conditions, controlling household appliances, tracking household objects, voice-activated commands for household functions, etc., are just a few applications. There are many more possibilities that smart homes can provide by making effective use of information technology.

## Smart Dresses

With rapid advancements in semiconductor technologies, the size of electronic devices is becoming smaller. At the same time, these devices are able to muster higher processing powers on tiny chips. This combination of advantages has led to the development of many applications that were not possible earlier, such as the wearable computer. A recent article in *IEEE Spectrum*[10] has highlighted this device; it uses tiny computers, and can be worn as a wristwatch that will have a monitor reflecting and displaying information on eyeglasses.[11] So, a person could be stopped in traffic and browsing the Internet at the same time. In fact, the idea of wearable computers is not really new, but the idea of a "smart dress," which consists of many tiny computers or sensors, is relatively recent. A smart dress is essentially a network of tiny computers that goes with you wherever you go. Tiny computers that are connected by tiny wires and, in some case, optical fibers, can exchange information with each other, process information, and take an action that they are programmed to do if all the prerequisites are satisfied.

A smart dress (or shirt) has many applications in healthcare as well as in other fields, including defense. In healthcare, a smart dress may be programmed to monitor certain health conditions and vital signs on a regular basis. The monitored information can be processed and appropriate action can be taken by your own dress, if needed. Your dress may even be able to indicate the exact location of the problem, e.g., your heart or your kidney. It may also be able to call for help if the seriousness of the situation warrants it. Such smart dresses will be very useful for our elderly population. Some business-minded individuals are already looking at fashionable designs for such dresses and possible marketing strategies!

## Wireless Ad Hoc and Sensor Networks

More useful applications of information technology in healthcare are possible in the area of wireless ad hoc networks and wireless sensor networks. The applications discussed in the earlier part of this section used wireless communication. The smart dress application is essentially a sensor network (but wired) with wireless communication potential. Wireless ad hoc networks are essentially communication networks in which communication devices are mobile and are not tied in any fixed topological infrastructure. As the devices move, the network topology changes. The devices in such networks are not only the source and destination of information being exchanged, but they also act as intermediate devices to relay information from one device to another that is not within communication range.[8] Wireless sensor networks are essentially ad hoc networks, but the devices are extremely small in size. These devices could be as small as a grain of rice and are self-sufficient in all respects — transmission, reception, processing, and power. These sensors can be programmed to suit any given application.

One of the many applications of wireless sensor networks is information gathering. In such applications, the sensors are essentially disposable and are used only once for an application. They can be very effective in disaster situations for search and rescue operations. They can also be used for gathering information about a particular disease or health-related situation. These sensors can be deployed by the hundreds or thousands over a selected area chosen for information collection and by air. Because of their tiny size, these sensors will remain suspended in the air for some time. During that time, they can collect information that they have been programmed for, process the information, share it among nearby sensors, reach a consensus, and transmit information to a central location. The information can then be analyzed at the central processing facility and a decision about the next step can be made.

## CHALLENGES

Despite the proliferation of communication technologies including WLANs, many challenges remain in their use in telemedicine. It is known that emerging technologies have entered into our lives in a relatively brief duration of time and have affected all aspects of daily activities. However, this rapid process has not given a large portion of our population the opportunity to educate itself about these technologies, and they have been caught by surprise. The full potential of these technologies will not be realized until people are educated about them, and their fears are put to rest. The use of WLANs in healthcare is growing at a healthy pace. However, delivery of healthcare (including telemedicine, remote surgeries, monitoring health conditions of homebound patients, etc.) is growing at a relatively cautious pace. Other aspects of healthcare services such as record keeping and computerized medical devices heavily use wireless communication technologies, but the physical touch and presence of a healthcare professional will not be so easily replaced by telemedicine.

One of the major challenges that telemedicine faces in expanding worldwide is related to regional economic conditions, cultural values, and traditions, all of which could affect the commitment to adopting telemedicine in certain parts of the world. The resources needed to implement telemedicine and ensure its acceptance (by educating the masses) may be prohibitive in many countries.

Another major challenge in many countries is the availability of resources for ongoing research on not only developing new information technology applications, but also on their social implications. Resistance to change and to the adoption of new applications of wireless communications will certainly be detrimental to utilizing its full potential. Such resistance can only be overcome through education.[12]

## SUMMARY AND CONCLUSIONS

This chapter has discussed the use of WLANs in healthcare and particularly in telemedicine. The use of WLANs in healthcare services started early, but its use in telemedicine (delivery of medical care remotely) has been rather slow. There are many reasons for that, including the lack of education of those who use healthcare the most, i.e., the elderly population. However, the impact of wireless communication technologies is such that no field can escape from it, and its use will bring many benefits to all involved.

# REFERENCES

1. Darkins, A.W. and Cary, M.A., *Telemedicine and Telehealth: Principles, Policies, Performance, and Pitfalls*, Springer: New York, 2000.
2. Laxminarayan, S. and Stamm, B.H., *Technology,* Telemedicine and Telehealth, Technology and Applications, *Telemedicine,* Issue 3, 2002, pp. 93–96.
3. Trafford, M., Consales, J., and Hamasu, C., The Role of Information Science and Knowledge-Based Resources in Delivering Telehealth Services, *Proceedings of the Medical Technology Symposium,* August 1998, pp. 394–400.
4. McKinney, W.P. and Bunton, G., Exploring the medical applications of the Internet: a guide for beginning users, *American Journal of Medical Science,* 306 (3), 141, 1993.
5. Furht, B. and Ilyas, M. Eds., *Wireless Internet Handbook: Technologies, Standards, and Applications,* CRC Press, 2003.
6. Klecun-Dabrowska, E. and Cornford, T., Telehealth acquires meanings: information and communication technologies within health policy, *Informations Systems Journal,* Vol. 10, 2000, pp. 41–63.
7. Solaiman, B., Cauvin, J.M., Guillou, C.L., Debon, R., and Roux, C., Enabling Technology for Telemedicine and Telehealth, *Proceedings of the 23rd Annual EMBS International Conference,* Istanbul, Turkey, October 2001, pp. 4113–4116.
8. Ilyas, M. Ed., *Ad Hoc Wireless Networks,* CRC Press, 2003.
9. Heathfield, H., Pitty, D., and Hanka, R., Evaluating information technology in healthcare, *British Medical Journal,* Vol. 316, 1998, pp. 1959–1961.
10. Marculescu, D., Marculescu, R., Park, S., and Jayaraman, S., Ready to Ware, *IEEE Spectrum,* pp. 28–32.
11. Winters, J.M., Wang, Y., and Winters, J.M., Wearable Sensors and Telerehabilitation, *IEEE Engineering in Medicine and Biology Magazine,* Volume 22, Issue 3, May–June 2003, pp. 56–65.
12. Lobley, D., The economics of telemedicine, *Journal of Telemedicine and Telecare,* Vol. 3, 1997, pp. 117–125.

# 13

---

# INTEGRATED WLAN DEPLOYMENT: IMPLEMENTATION OF A MOBILE WIRELESS DIABETES MANAGEMENT SYSTEM

*Akira Kawaguchi, Stewart Russell, Guoliang Qian, Cristina Miyata, and Jorge Becerra*

## INTRODUCTION

As mobile computing is becoming the dominant computing paradigm, the increasing ease with which PDAs can gain wireless access to the Web offers new opportunities and challenges to Web application designers. The application of this fast-developing technology in the healthcare industry will bring promising economic and technology contributions.[1] However, the use of handhelds combined with wireless LANs (WLANs) to gain access to information services, especially determining ways to implement systems that span multiple heterogeneous devices, both online and offline, requires further research.[2] Consider, for instance, *telemedicine*; that is, the use of telecommunications to provide medical information and services, ranging from basic medical information management to more sophisticated biometric information processing.[3,4] Wireless telemedicine is a new and evolving area in telemedical and telecare systems. It involves the exploitation of mobile telecommunication and multimedia technologies and their integration for developing new mobile healthcare delivery systems.[5] Smart mobile phones and WLANs now offer hospitals new options for increased flexibility, bedside documentation, and immediate access to data throughout the organization.

The need for better self-care practices has been recognized by many device manufacturers and software developers, and a growing body of researchers and manufacturers are working to develop new-generation wireless technology applications for the medical field. National health expenditures in the year of 2003 were estimated at about $1.5 trillion, or almost 15 percent of GDP, by the Centers for Medicare and Medicaid Services (www.cms.gov). This comprises the largest service sector in the U.S. economy; nearly 15 to 20 percent of medical practitioners now use palmtops, and a vibrant specialty software sector has emerged, making hundreds of palmtop medical applications available.[6–8]

The possibilities of mobile telemedicine applications increase as current limitations related to mobile computing environments are properly addressed and studied.[9] Areas of concern include highly variable communication quality due to environmental variations and handoff, management of data location for efficient access, restrictions of battery life and screen size, cost of connection, and increased security risks. The next generation of mobile communication environments will be able to effectively support high-speed wireless applications with proper mechanisms[10] and Mobile QoS (quality of service), which is a set of performances associated with links such as channel error rate and with mobile environments such as handoff-call dropping probability (HDP) and new-call blocking probability (NBP).[11] This chapter will discuss in detail the methodology and technical issues in realizing wireless-capable telemedicine applications. A case model considered for the study is *a wireless diabetes information management system*. The establishment of architecture suitable for the evaluation and measurement of its service quality is also discussed.

## Facts about Diabetes

Complications from diabetes is the seventh leading cause of death by disease in the United States.[12] Unregulated blood-glucose levels (BGL) in patients with type 1 and type 2 diabetes mellitus increase the likelihood of morbidity and mortality due to diabetic complications. The situation with regard to diabetes may become worse in the near future. For instance, New York City Health and Mental Hygiene Commissioner announced that the percentage of adult New Yorkers with self-reported diabetes has doubled since 1994, from 3.7 percent to 7.9 percent. Although more than 450,000 New Yorkers know that they have diabetes, it is estimated that approximately one third of all diabetes cases remain undiagnosed, suggesting that, in total, more than

675,000 City residents have diabetes (data was collected from the NYC Community Health Survey and presented in a new report, "Diabetes is Epidemic," available online at http://www.nyc.gov/health/survey).[13]

Research has shown that it is possible to reduce the complications associated with diabetes by careful management of BGL.[14] Clinical research has shown that maintenance of BGL at or near normal can greatly reduce this likelihood. Appropriate timely intervention by healthcare providers in response to records of unregulated BGL can maintain near-normal levels. However, the task of such management is tremendously complicated and frustrating for both patients and care providers, requiring tedious record keeping, timely review, and prompt action. The magnitude of the task of self-care and record keeping may be beyond the reach of many patients. Research also shows that the team-management model of tight blood-glucose control is far better than others.[15] It is proposed that the introduction of a wireless blood-glucose monitoring system into a team-management approach to diabetes care will lead to an improvement in public health.

The challenge is to bring this technique to diabetics at a reasonable cost, with reasonable incentive to the healthcare industry. The collecting of individuals' medical information in a database has been proposed earlier as a tool for care providers. However, it has not been conclusively shown that simply making this information available to providers has direct benefit to public health. What research has shown is that in the case of diabetes, when a health-management team is able to respond in a timely fashion to carefully kept blood-glucose records, an overall improvement of the patients' health is realized.[16] Recently, several wireless systems for automating BGL data reporting have been developed.[17] Each of these systems has the potential to significantly reduce the administrative burden to the patient. As promising as these systems are, without appropriate design consideration, interoperability with each other and with existing computer systems will be severely limited.

## A Case Study

The New York Center for Biomedical Engineering and Department of Computer Science at City College of New York, in conjunction with the Bayer Corporation, have developed a system called the Wireless Blood-glucose Monitoring System (WBgM) that will automatically transfer blood-glucose readings from a handheld glucose meter to a wireless personal digital assistant (PDA), and then to an Internet database.[17–22] This research effort is aimed at simplifying diabetes management through application of current technology. The first aim of this research is to establish a prototype of a WBgM. The device will automatically transmit blood-glucose levels to a centralized database, where it will be available to the diabetes management team for proactive response. The second goal is to show that this system will facilitate the timely delivery of this type of data and response. Iterative design procedures will result in a product that can lead to more effective management, record keeping, and team development.

The successful demonstration of a prototype WBgM has opened a new area of research in the development of improved communication between physicians and patients, and more accurate measure of patient compliance with doctor-prescribed protocols. With this system, the patient is able to transfer exact readings of BGL to a secure Internet database where the data is then available to the diabetes management team. Because each individual diabetes management team is a fluid organization, access to the data may be required for a shifting set of doctors and researchers. The secure Internet database with wireless access is ideally suited for this function. As a consequence of the diversity of the management team, multiple computing environments must be traversed in accomplishing data transfer and storage, including proprietary testing devices, operating systems, wireless systems, and storage facilities. Without interoperability of these elements, none of the inherent advantages of wireless computing can be fully realized. The Reference Model of Open Distributed Processing (RM-ODP),[23] a joint effort by the international standards bodies International Organization for Standardization (ISO) and International Telecommunications Union (ITU), has been developed to address a similar situation in wired computing and is an appropriate development model for our case. Our recent collaborative effort with researchers in the Department of Electrical Engineering at the Polytechnic School of São Paulo has guided us in utilizing their service level management (SLM) approach to facilitate the identification of the ODP development parameters for the WBgM system. The resulting ODP definition should serve as a basic model for wireless biometric management systems of this type.

In fact, a major impediment to the progress towards evidence-based medical practice, shared patient care, and resource management in healthcare is the inability to effectively share information across systems and between caregivers.[24,25] Electronic and paper healthcare records are held as islands of information in various independent information systems, each with its own technical culture and view of the healthcare domain. Our study presented in this chapter has been motivated by these issues. For the need of public use, system development practices must be based on an enterprise model that encompasses the capabilities required to process medical information by hospitals, governments, insurance parties, etc. The objectives of the WBgM system development are (1) to specify a generic and open means of combining healthcare records or dossiers consistently, comprehensibly, securely, and in a simple fashion, to enable the sharing of data between different information systems in different places, and (2) to produce tools and guidelines that can be used in the migration from legacy healthcare systems, as an evolution strategy for a region or member state, or as an exploitation plan for a healthcare site or commercial company.

To guarantee the agreed service level, to charge the customers correctly, and to improve the service providers' products, it is necessary to have an SLM process to articulate service level agreements (SLA), create products, monitor the services, measure their service level, and produce service level reports (SLR). An SLM system assists the service provider's SLM process in measuring the monitored services' quality, acting before an SLA is violated, and producing SLRs.

## Chapter Contents

This chapter provides a detailed account of the methodology and technical issues used to realize a wireless-capable diabetes information management system for practical use. The system envisioned is distributed across multiple platforms and functions with various wireless communication methods. Implementation techniques and key issues for data management discussed in this chapter are based on the experience we gained from this development. We also describe our approach to the system of wireless biometric data management to meet today's security challenges and comply with HIPAA (The Health Insurance Portability and Accountability Act) regulations activated on April 14, 2003. Specifically, we present our database design, which facilitates sharing of information, enhanced with strict privacy protection. We also discuss several technical issues that these goals raise. Our motivation in this research is to stimulate open discussions for the methodology to realize better healthcare, integration of currently available information technology, the need for standardization, and practical means to achieve timely deployment of the best diabetes information management system that is deemed increasingly urgent for public use. Thus, we also present in this chapter an ODP reference model-based method of SLM system architecture specification for Wireless Biometric Systems — in particular, the WBgM system. This is to explain how the RM-ODP viewpoints are used to identify QoS characteristics relevant to monitored services.

The rest of this chapter is organized as follows: The next section, "WBgM System Architecture," provides an overview of the systems requirements and the approach we have taken to design the system architecture. The section titled "System Development Practice" presents key techniques and their implementation details. The section titled "Technical Elements in Wireless Security" describes considerations and technical issues to comply with HIPAA regulations and security challenges. In the section titled "An Approach for Commercial Development," we present an ODP reference model-based method of SLM system architecture specification for the WBgM system. The RM-ODP viewpoints are also used to identify QoS characteristics relevant to monitored services. The final section, "Summary and Conclusions," will summarize the work and conclude this chapter.

## WBGM SYSTEM ARCHITECTURE

### General Requirement and System Design Approach

WBgM is essentially a data management system designed to improve communication between patient and doctor and encourage participation in self-care. More specifically, it is a tool that offers communication between three distinct operational environments in the clinical setting — the diabetes management team, the healthcare administration team, and the computing support team [see WBgM's general architecture illustrated in Figure 13.1(a)]. The patient takes a blood-glucose reading at prescribed intervals. The blood-glucose monitor or a handheld meter (e.g., the Bayer Glucometer Elite or Glucometer DEX) is connected to a PDA, such as a Palm OS™ or Microsoft Pocket PC™ device, by a serial cable. The patient may input data manually or automatically from the meter to a mobile wireless device for uploading to a secure cross-platform Internet database. When the diabetes program on the PDA is activated, the data (including the time and date stamp, ID, blood-glucose reading, and possibly a short note), can be transmitted to the Internet database, which stores the information for later retrieval by the diabetes management team. Uploading is done through a mobile or WLAN connection available in the patient environment. In addition, the user may (1) input service request parameters to view current and historic data on the local device, (2) activate the master expert system program (e.g., Bayer's WinGlucofacts™ Professional Edition expert system) that runs on the laboratory computer to analyze data, and (3) e-mail or fax a report from the expert system to a recipient doctor's office.

Periodically, the information for each patient is collated and processed in the form of graphs, charts, and other learning tools and delivered to each patient and each member of the diabetes management team. Conferencing about this information can then be handled in any number of ways specified by the team. A set of tools facilitate online analytic processing, and data-mining methods can produce a series of flags to call special attention to known critical states. WBgM electronic transactions consist of a simple set of formatted data that represents (1) patient profile

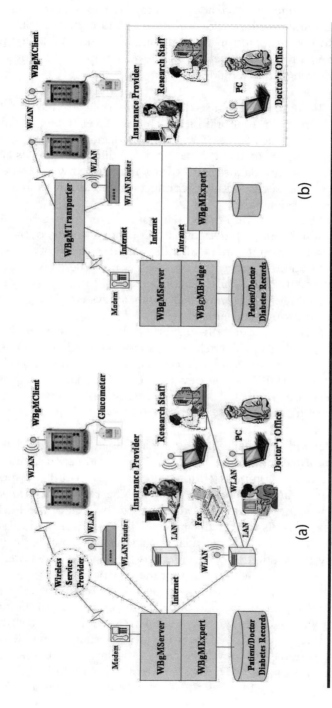

**Figure 13.1 WBgM system architecture design overview.**

(patient identity, name, contact address, etc.), (2) biometric data represented by time-stamped blood-glucose reading and coded nutrition characteristics uniformly accepted in medical practice, and (3) doctor profile similar to patient profile. The set of high-level components required for building a comprehensive WBgM system are:

*Wireless-BgMClient:* A PDA-based interactive application system that is capable of reading temporal blood glucose levels using a handheld meter as the input source. The gathered data, presented in a time series, category chart using a handheld's built-in feature, is sent by HTTPS (i.e., secure data stream) over a high-speed wireless network or PPP-capable secure modem connection and archived in an Internet-accessible database.

*Wireless-BgMServer:* A Web-based database server that stores all categorized items into a transactional repository, indexed by a user identification account and a URL indicating the source of those items. Raw data is tailored for storage by wireless BgMClient (WBgMClient). The Web site maintains histories of personal blood-glucose readings using the Oracle™9*i* commercial database system featured with a temporal indexing mechanism.

*Wireless-BgMExpert:* A decision-support application system built on the database running on a wireless BgMServer (WBgMServer) such as WinGlucofacts Professional Edition, which assists the diabetes management team (DBT) in the development of management protocols and which determines if there are measurable improvements to public health from the use of the WBgM system.

WBgMServer will be installed at a regional diabetes center to determine an effective information-sharing protocol. The healthcare administration team (HCAT) and the computing support team (CST) will be responsible for the installation and maintenance of WBgMServer. There are several points at which security could be compromised: by loss of the PDA unit, during wireless transmission, at the receiver base station, and in storage if the database is subject to an intrusion. Patient data confidentiality and system security are assumed to be enforced by product functionality for now, and more on the security configuration for each of the WBgM components will be detailed in the section titled "Technical Elements in Wireless Security." It should be noted that each component of the product must be developed to be fully evolvable. That is, not only must they be able to upscale to the advent of new technology such as wireless communication extensions, but they must also be designed in such a way that they can take advantage of new ideas from a multitude of sources, including added security features for wireless communication.

## Supporting Hardware, Software, and Communication Heterogeneity

Establishing reliable wireless data transmission is one of the most important tasks in WBgM development because the system needs to support three types of data transfer across a network: WLAN connection through a local access point, direct network connection through a wireless service carrier, and dial-in connection to a WBgMServer. Today's wireless communication is inherently noisy,[6,26,29] and a good communication protocol is needed at the application level, allowing the retention of only the most recent data items in the user's PDA to reduce its memory usage and to guarantee the durability of BGL measurements by storing the volatile data into the server database. Any data items updated and inserted by the user may thus need to be transferred (or retransferred upon communication failure) to the server whenever a session of wireless communication is established (or reestablished).

In addition, as consumers and healthcare providers have become more experienced with using the kind of diagnostic information that WBgMExpert makes available (such as temporal trend analysis, summary information, extraordinary change of measured data, etc.), there has been an increased demand for it. Currently, for all device manufacturers, data management for diabetes self-testing is handled at the level of the user. There is no provider of glucose analysis software that allows a wide-area-networked data-management system. Our recent conversation with a medical expert of the Diabetes Center at Mt. Sinai/NYU Health confirmed several of our motivating hypotheses about this point: there is a clear and present need for easily accessible data by multiple parties that does not compromise security. A collaborative effort with Bayer Corporation has identified that this system could be realized by developing a robust data-management system on the UNIX platform that would run as a data server and communicate with PC-based decision support systems produced by numerous companies. With this integration, the user will be able to (1) input service request parameters to view both current and historic data on the local PDA, (2) activate, from the PDA, a decision support program that runs as WBgMExpert to analyze data, and (3) e-mail data analysis reports by WBgMExpert to a recipient (doctor's office) specified on the PDA as an e-mail attachment, or with specific priority to print the report to the doctor's office fax or printer.

The central task of this integration is to realize multiplatform communications among the PDA client software, the UNIX-based enterprise database server, and the PC-based diagnostic application system. However, when using this system with an existing expert system such as the Bayer's proprietary PC application (as exemplified with Bayer's WinGlucoFacts), WBgMClient should not require a redesign. Rather, a separate module needs to be developed as an interface with the expert system. These considerations identify that the WBgM architecture needs to have the following two additional components [see Figure 13.1(b)]:

*Wireless-BgMTransporter:* Robust communication protocol between wireless client and database server. The Web access interface will be upgraded to provide industrial level security and privacy protection featuring password-protected, certificate-based authentication, and allowing differentiation by user role, such as doctor, server administrators, and patient.

*Wireless-BgMBridge:* An additional multiplatform data communication layer required to facilitate bidirectional communication of the UNIX-based Oracle-stored data and PC format (e.g., Microsoft Access) for use in a PC-based expert diagnostic system.

The fully integrated system will allow the uploading of personal medical data from a wireless PDA monitor to an Internet server, and the retrieval and distribution of the results produced by a decision support and diagnostic system into wireless PDA monitors, which can also be forwarded to a doctor.

## SYSTEM DEVELOPMENT PRACTICE

A comprehensive WBgM system that consists of the five components described in the previous section has been implemented at the City College of New York (CCNY). See Figure 13.2(a) for the overall configuration. This development is chiefly done in Sun Corporation's Solaris 8 UNIX™ and Microsoft's Windows 2000™ environment, and presently one Sun Microsystem's Ultra 10 workstation is serving as an access point for the WBgM online services. Oracle 9*i*, JDBC (Java Database Connectivity), and Java Servlets are used to implement WBgMServer. This section will summarize our implementation practice. More technical details can be found in Reference 18 and Reference 20.

### Development of WBgMServer and WBgMClient

WBgMServer runs with the Oracle 9*i* database management system, which provides features to integrate online transaction processing and data warehouse applications.[27,28] It also serves as a Web site for WLAN and "wired" Internet accesses, as well as PPP dial-in connections. Groups of users, classified as patients, doctors, researchers, and server administrators, each having a distinctive role in the WBgM services and data analysis capabilities, are granted access through password-protected, certificate-based authentication. Database design is relatively straightforward. The

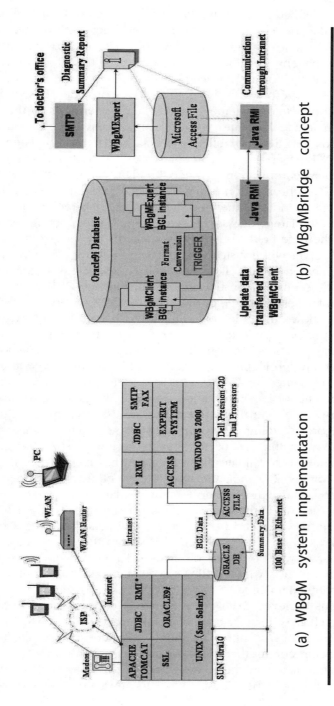

Figure 13.2  WBgM system implementation diagram.

primary tables defined in the WBgMServer database are UserProfile, DoctorInfo, and MainTable. The UserProfile table keeps information about users. The MainTable contains the biometric data records of all users, including the patient identification number, and the DoctorInfo table contains information for all physicians in the clinical group.

A nonwireless prototype of WBgMClient was provided by Bayer Corporation by contractual agreement, and several extensions to this application were made to facilitate bidirectional wireless data transmissions. The user needs to wirelessly reflect the volatile data accumulated in the PDA to the WBgMServer's permanent store. The transmission process mandates data synchronization between the PDA (client) and WBgMServer (server). Such synchronization is not done automatically. A note field is added as a flag to indicate the current status of each record. The server is able to differentiate processes, either inserting or updating, based on the value of the note fields found in the data stream sent from the client (this way the server will not store duplicate records). Specifically, a menu of the implemented WBgMClient has a wireless access button, which activates wireless push of unsent data on the PDA to WBgMServer and wireless pull of a desired range of the data from WBgMServer to the PDA. Figure 13.3(a) shows an implemented interface that allows input to a Palm monitor (the user is typing in his or her doctor's information).

Pushing the "wireless" button then activates a data transmission [see Figure 13.3(a) through Figure 13.3(d)]. A Send Data button and a Retrieve Data button, illustrated in Figure 13.3(b) and Figure 13.3(c), allow the user to retrieve his or her information from the WBgMServer database. A Show button enables the user to remotely access a valid time range of the metric records residing in the server database. Data transmissions are realized in the following way: (1) along with these requests, a `PatientID` value is transmitted to validate if the database has a record of the same `PatientID` value; (2) if not, the user is prompted to fill in his or her profile and medical data first; otherwise, the available time period is shown on the PDA's screen; (3) the user can subsequently specify the date range to retrieve data records; and (4) the server will send back the sequence of those records if the requested period is within the correct time range; an error message will be issued otherwise. These processes are implemented in combination with the PDA's LocalMainTable, a subset structure of WBgMServer's MainTable.

The records retrieved from the server may consume a large amount of space, which may in turn compete with space reserved for the data that is recently read from the meter and only kept in the PDA's LocalMainTable. It is important that the archived data, when retrieved from the server, be manipulated in the same way that the data newly uploaded from the meter should be. These two parts must be rendered together

(a) Doctor information   (b) Sending BGL data   (c) Sending in process   (d) Receiving in process

**Figure 13.3**  WBgMClient user interface.

seamlessly when responding to a user query that spans a history in the archived part and the one in the most recent reading. This is realized by creating a backup table to temporarily save the LocalMainTable's instance and by repopulating the table with server-retrieved data and current readings. This, however, means that for some non-critical fields, the user may be able to change the information. User-changed archived data on the PDA is automatically flagged for backup so that if the user retrieves and changes multiple archived datasets, changes will not be lost. The system will automatically restore all the data in the backup table to the LocalMainTable upon restart, or if the user activates the Read Meter process.

## WBgMTransportor Development

A handshake protocol has been developed to cope with the instability in wireless communications. This protocol guarantees lossless and consistent data transmissions between WBgMClient and WBgMServer, and enhances robustness against repeated communication failures. The mechanism works as follows: (1) A PDA device only accumulates the most recent BGL records, each of which has a flag, "sent" or "unsent," to indicate success or failure of the previous transmission to the server, (2) New and updated records are found by scanning an unsent flag, and these are sent to the server whenever a new wireless session begins, and (3) Unsent flags in the records are turned off upon receiving an acknowledgment from the server, which indicates a successful transmission and a successful commit of the transaction updating the server database.

The flag mentioned here is the note field added to the WBgMClient's PDA data files, Local-MainTable, described earlier. The note field is designed to ensure wireless data transmission and to maximize communication performance over the limited bandwidth by minimizing the total amount of data to send. Notice that there are two cases in which the WBgMServer could fail to turn back the acknowledgment: (1) data transmission failed before starting database update, or (2) acknowledgment transmission failed after committing a database update. Both cases involve retransmission of the unflagged data whenever the next wireless session starts. In the latter case, the protocol prevents duplicating database operations (insertions and updates) at the WBgMServer by checking if the sent record is already saved in the database.

## WBgMBridge Development

The primary task of the WBgMBridge development is to realize a multi-platform communication between a UNIX-based database server (Oracle) and a PC-based database (Access) to transfer the BGL data transparently for activating the WBgMExpert component. The approach taken is to utilize Java Remote Method Invocation (RMI), a facility that enables a Java program running on one machine to invoke a method of an object on a different machine through direct-stream communications. RMI bypasses HTTP server processes and has lower overhead.

Specifically, two RMI modules are developed as illustrated in Figure 13.2(b): one running on Windows 2000 and another on UNIX. Both modules invoke each other (acting like a bridge) to pass the data between machines. The UNIX module is activated by the call from the Windows side with parameter indicating a patient and a portion of the data to be extracted from the Oracle database. Notice that the conversion from the BGL data sent from WBgMClient to the format suitable for WBgMExpert is automatically done by the trigger being implemented in the Oracle database. The extracted data is then passed to the Windows WBgMExpert. On the Windows side, the received data is immediately imported into a Microsoft Access format, and the WBgMExpert application gathers the contents for analysis and produces a diagnostic report, which is subsequently transferred to a doctor's office as an e-mail attachment or fax using a Windows SMTP mail transmission layer. The entire process is automatic. The summary document can be transferred back to the UNIX machine and stored in the Oracle database as a binary large object so that it can be viewed later by Web browsers.

## TECHNICAL ELEMENTS IN WIRELESS SECURITY

The Health Insurance Portability and Accountability Act (HIPAA),[30–32] which has been in effect since April 14, 2003, has the aim of (1) assuring health insurance portability regardless of preexisting conditions, (2) reducing healthcare fraud and abuse, (3) enforcing standards for health information, and (4) guaranteeing the security and privacy of health information. A deployment and academic trial of the WBgM system must meet increasingly stringent electronic data privacy and security requirements. Ease of access is important for the system to provide patients and caregivers with greater flexibility, greater accuracy, and stricter accountability in the generation and management of patient data. Mobile data transmission is especially problematic when it comes to issues of security. Data transfer is by radio waves, and therefore impossible to shield from interception.[33] This section describes our approach and several technical issues for our system of wireless biometric data management to meet security challenges and comply with HIPAA regulations, with strict privacy protection.

## Accessibility of Electronic Medical Data

Electronic health transactions include health claims, health plan eligibility, enrollment and disenrollment, payments for care and health plan premiums, claim status, first injury reports, coordination of benefits, and related transactions.[30] The proposed HIPAA rule requires use of specific electronic formats developed by the American National Standards Institute (ANSI) for most transactions except claims attachments and first reports of injury (cited from searchable HIPAA regulations in www.hipaaadvisory.com). All the data entries in the WBgM database and access to them are coded based on the ANSI SQL standard relational database format. Our electronic transactions consist of a simple set of formatted data that represents (1) patient profile, (2) biometric data represented with time-stamped blood-glucose readings and coded nutrition characteristics uniformly accepted in endocrine medical practice, and (3) a doctor profile similar to the patient profile. We believe that this conforms well to both the letter and intent of HIPAA rules. Unfortunately, rapid advances in computer systems and implementation of electronic data repositories at medical care facilities have sometimes left medical ethics issues in the hands of computer scientists, and computer security issues in the hands of healthcare professionals.[34] In fact, this is one of the prime motivations behind the HIPAA legislation.[35]

The philosophy behind the ownership definition in our system has been carefully studied during development. Specifically, the patient has sole ownership of the patient profile and blood-glucose readings. This means that, similar to the use of E-commerce applications, the patient can utilize the system to keep track of his or her readings whenever the registration is completed and appropriate software is installed into the wireless-capable handheld. As illustrated in Figure 13.4, WBgM's database architecture allows the registered patient to grant a specific group of doctors access to profile data and blood-glucose readings. The intent of the system is to provide more than an automated monitoring system whereby remote doctors make decisions based on numerical data alone. It has been developed as an enhancement of care, as a method to involve the patient more actively in self-care. With any chronic condition, it has been shown that involving the patient as an active participant through increased self-care leads to better long-term health results.[12]

Accessibility is not just the concern of the patient. The diabetes management team must be involved in the review of records. Doctors, hospitals, and other members of the healthcare industry will also need to have access to certain portions of a patient's data history. The doctor's registration process to the system is verified by the system administrator. In certain circumstances, access privileges may be extended by the patient to researchers in clinical institutions for clinical studies where institutional

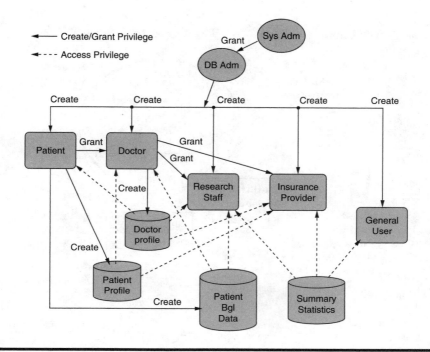

**Figure 13.4  Schematic of access and privilege granting hierarchy.**

privacy practices are followed. This allows the researchers to conduct studies on diabetes across patients, not bound by a particular doctor or hospital. The certificate method of authentication can be established electronically to ensure only authorized access to data for this type of anonymous clinical study. Most importantly, the patient is guaranteed the right to revoke the granted access given to the doctors or institutions. The revocation of the base certificate will then in turn revoke the access of any associated researcher. This mechanism not only simplifies and improves transaction efficiency regionally and nationally in the long run but also reduces mistakes and duplication of effort and costs, while meeting the HIPAA regulation for reducing healthcare fraud and abuse.

## Security-Enabling Practices

There are well-known security-enabling practices from the system administration standpoint, such as to keep the database server behind a firewall, to limit the number of users with operating system accounts on the server host, and to harden the operation systems by disabling unnecessary network services such as Telnet and FTP. These are minimal means.[36-39] For mission-critical medical information management such as the WBgM

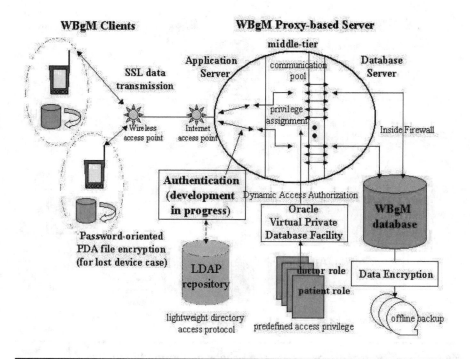

**Figure 13.5  WBgM security components.**

system, this kind of security is not good enough.[40-41] In general, security deployment for wireless applications roughly consists of three components: privacy protection, user authentication and authorization, and data integrity.[42-45] Part of WBgM's security implementation approach is indebted to Oracle's™ Virtual Private Database facility,[27] which enforces security policy on data tables and roles for operations on the database. As illustrated in Figure 13.5, there are specific technical issues considered for the WBgM implementation. These will be summarized in this section.

*Privacy Protection:* Privacy protection should be enforced for the data in the server storage, for the data stream transmitted between server and PDAs, and for the data presented and retained in PDAs. Privacy protection mechanisms at the server database will be discussed in the third item in this list. For communication protection, the WBgM system has taken advantage of the Internet Protocol Security (IPSec) standard being implemented as a Secure Sockets Layer (SSL) paradigm on top of a public-key encryption method.[46-48] SSL is implemented as a generic library that provides cryptographic functionality to applications. Any communications that flow between PDAs and the application server are encrypted as in the case of a commercial transaction and thus unreadable without the proper decoding. Both the Palm OS and the Java micro edition used in the WBgM implementation are SSL capable. Our development of WBgM has not considered the use of encryption in the wireless LAN at this moment because the standard Wireless Equivalent Privacy (WEP) is known to be in a relatively unsophisticated level of encryption.[49,50] It is designed to protect against unauthorized access to the (open) wireless network and to the information transmitted in the air, and has no effect after the data is received at the access point. However, under our system, encryption can be implemented in the application even with WLAN technology. We treat the wireless network in the same way as a wired network in that we provide security through SSL or, alternatively, through a private-key system. Encrypting private data while it is actually on the PDAs is also important. This is for the purpose of protecting vital data on lost or stolen PDAs. Third-party products, such as PDASecure™ by Trust Digital, will insure that data encrypted on the stolen devices cannot be copied or beamed to another system that could read it.[50-53]

*Authentication and Authorization:* Authentication consists of correlating an electronic identity to a real-world, legally binding identity; and authorization consists of assigning rights to the authenticated identifier. The WBgM system has adopted a sophisticated proxy-based, three-tier authentication mechanism that is introduced as a building block of the Oracle 9*i*'s virtual private database facility. The central elements are as follows: the function of the WBgM server is divided into an application server and a database server. The application server is responsible for authentication and maintains a pool of connections to the database server, each of which represents a single, abstract database user. When connecting, the WBgM client authenticates against the WBgM application server. Once authenticated, the client is assigned to one of the reserved connections to the database server and authorized with a predefined set of data access privileges. As illustrated in Figure 13.5, the WBgM server creates a lightweight user session to communicate with the database on behalf of the authenticated user. A lightweight user session uses a thread-oriented internal database communication process. This session consists of client/server activities between the authenticated and authorized user, and the server database. The connections in the pool are reused, thus eliminating the overhead of creating a separate connection for each user. The security of this system rests on the fact that an abstract user has no inherent privilege. User data access privileges are granted only when the authentication completes. Thus, even if the abstract user's account is compromised, there is no scope for an intruder to bypass the access control mechanism to access database information. The authentication adopted for the WBgM application server is based on the industry standard certificate authentication, which offers much greater security than a simple password. The proxy-based authentication is considered the industry-standard best security practice today. It has the strength that real identification and associated access privileges of the wireless user are dynamically established and preserved through the application context and the specific database role, such as patient, doctor, and researcher.

*Data Integrity:* The data integrity component of security in general ensures that system slipups and malicious intruders will not be able to read or alter information. For WBgM deployment, data integrity may be violated when the information leaves the medical servers for off-site storage. For instance, security vulnerabilities will arise during the database backup process and system and database migration processes. To further enforce data integrity and data privacy, all patient information including identity data and measurement data can be encrypted within the database. Standard algorithms such as message digest 5 (MD5) algorithm or the secure hash algorithm (SHA-1) could be used to protect against both data modification attack and replay attack. In addition, auditing helps deter unauthorized user behavior that may not otherwise be prevented. The WBgM database maintains a record of system activity to ensure that users are held accountable for their actions, such as unsuccessful select statements and exception errors for data the users are not privileged to see.

The structure and function of our system is such that we can tailor application development to the limitations of the device without sacrificing security. This application would not be appropriate, for instance, for the transfer of transcripts of lengthy written patient records, evaluations, or prescription information. It must be noted that we cannot determine what type of patient behavioral statistics may be evaluated by a third party monitoring this system, and one must therefore proceed with great caution. This application is most appropriate for recording time-stamped data. In the future development of WBgM, these elements may be capitalized on. For transmission of data between the mobile device and the secure server, a private-key system of minimal processing demand can be chosen. Further, we can explore the implications of separating identifying data from informational data, i.e., dates from glucose levels and time intervals from magnitudes.

## AN APPROACH TO COMMERCIAL DEPLOYMENT

If wireless telemedicine systems are to be fully realized in the clinical setting, a systematic approach to development is needed. The rapid growth of distributed processing has led to a need for a coordinating framework for the standardization of ODP. RM-ODP is a joint effort by ISO and ITU, and is supported by companies as diverse as IBM, Lockheed Martin, Lucent, DoD Agencies, and Verisign, as well as consortia such as OMG and the Distributed Systems Technology Centre in Australia.[14] The model describes an architecture within which support of distribution, interworking, interoperability, and portability can be integrated. An approach based on ODP will be very useful for WBgM systems to enhance significant portability of applications across heterogeneous platforms, interworking between ODP systems to realize exchange of information and convenient use of functionality throughout the distributed system, and to preserve distribution transparency to hide the consequences of distribution from both the applications programmer and user.

WBgM is a tool that offers communication between three distinct operations environments in the clinical setting: the DMT, the HCAT, and the CST. As such, the value-added components of WBgM are not as easy to define as one that is purely technology based. In general, the successful implementation of a healthcare aid must have evaluation criteria based on measurable outcomes to the health of the user. This is not to say that the system is a failure only if health improves. It is equally successful if the system can inform the DBT about deteriorating or stable health profiles. To guarantee the agreed service level, charge the customers correctly, and improve the service providers' products, it is necessary to have an SLM process to articulate SLAs, create products, monitor the services, measure their service level, and produce SLRs. An SLM system assists the service provider SLM process in measuring the monitored services quality, acting before an SLA is violated, and producing SLRs.

In this section we describe an ODP reference model-based method of SLM system architecture specification for WBgM. The RM-ODP viewpoints are also used to identify QoS characteristics relevant to monitored services.

### Basic Concepts of SLM and QoS

Service performance is critical in the deployment of WBgM. The overall satisfaction of the user will depend on myriad performance elements, all of which must be correctly monitored and evaluated. The collective effect of these performance criteria is termed QoS.[15] For an application such as this, which will be used as a tool to assist in the improvement of communication between the patient and the DMT, there are three critical areas of QoS: communication, clinical utility, and security. Each of these areas includes a number of relevant QoS parameters, which may be identified, monitored, and adjusted to facilitate the implementation. In fact, the commercial business standard offers a useful modeling tool in the elaboration of the SLA to fully specify the products and terms of a provider–user relationship. To allow accurate accounting and reporting of the SLA of a given product implementation, an SLM system must be employed. The SLM system must perform four basic tasks: (1) collect relevant information for identified QoS parameters, (2) analyze this collected information, (3) store the information, and (4) deliver formatted information and analysis to the appropriate parties.[16]

For some commercial wireless services, such as Web browsing, there is a gray area between utility and entertainment. Some aspects of web browsing have no parallel in pre-Web information gathering systems and are marketable on the basis of novelty. However, for a wireless biometric healthcare application like the WBgM system, utility is not a matter of interest only to the sales team. All vital biometric aspects of the system will have a previous non-wireless analogue, either electronic or paper-based. In order for the DMT to be induced to participate, there must be a compelling advantage to switch and reliable assurance that fault detection and performance evaluation of the WBgM system will be at least as good as that for the previous system. There is, thus, no question that the increasing complexity resulting from recent introductions to electronic biometric technology has created a demand for more convenient, reliable, and device-independent management systems. In addition, the WBgM system must include management systems that will detect and generate reports on element fault, performance, configuration, and accounting. An RM–ODP-based SLM system will meet this requirement.

## An ODP-Based Specification Method of WBgM SLM Architecture

We will specify a general SLM architecture for the WBgM system using a method recently introduced by Miyata and Becerra.[12] Using ODP viewpoints as defined in Reference 14, eight steps lead to the development of the specification. These are (1) business modeling, (2) service modeling, (3) QoS characteristic identification, (4) QoS and SLA modeling, (5) monitoring definition, (6) report generation definition, (7) accounting definition, and finally (8) SLM architecture specification. Recall from the previous section "Basic Concepts of SLM and QoS" that three separate human resource elements must work together in this development — the DMT, the HCAT, and the CST. Although each organization will differ in the composition of these teams, members of the DMT typically can include the patient, the primary care physician, an endocrinologist, nutritionist, physical therapist, ophthalmologist, cardiologist, and support staff. The HCAT can include office managers, database administrators, accounting personnel, insurance liaisons, and medical ethicists. The CST can include WBgM design, development, and technical support teams and datacenter operations.

In any given deployment of a WBgM, complexity will vary. Some roles may be filled by several people, whereas a single person may cover several others. Our aim here is not to present a management structure but merely to indicate the degree of flexibility that is required. For example, three possible system configurations exist for the wireless data transfer architecture: (1) end-to-end control, (2) commercial provider control, and (3) hybrid control. End-to-end control defines a system in which the CST is responsible for the design and interconnectivity of each element of WBgM: the glucose meter, the wireless transmission device, and the data and application server. Commercial provider control consists of a system in which existing technology is configured to perform the desired wireless functionality. A hybrid system is one in which the CST designs some elements and relies on some existing technology. Due to the advanced state of glucose testing devices, hospital database management systems, and the rapid pace of wireless development, it is likely that most configurations will fall into one of the latter two cases. This means that most CST must have a broad knowledge of the state of current technology of potential system components in addition to the ability to develop new ones. As in the design of the overall system, component design must adhere to the principle of open development. Components must be able to be replaced or improved separately, and subsystems joined to or severed from other systems without requiring a complete redesign.

## Business Modeling

The overall objectives of the DMT are to facilitate the regulation of patient BGL at or near normal levels by following the procedures such as (1) to exchange accurate BGL data between patient and DMT, (2) to increase timeliness and frequency of BGL data exchange, (3) to provide interim self-care diagnostic information to the patient between office visits, (4) to improve the dynamics of data distribution among members of the DMT, (5) to reduce complexity of equipment connectivity apparatus, and (6) to reduce complexity of data analysis software.

The overall objective of the HCAT is to ensure that WBgM adheres to accepted healthcare protocol and administrative practices. Recall from the subsection "Accessibility of Electronic Medical Data" that the data management functionality of WBgM is closely related to the ownership definition of the patient data. Patient profile and blood-glucose readings are solely owned by the patient. WBgMs database architecture allows the registered patient to grant access to his or her profile and blood-glucose readings to a particular set of members of the DMT. The DMT registration is verified by the system administrator. Furthermore, clinical researchers at authorized institutions can access the patient data once given permission from the DMT and patient. This allows the researchers to conduct studies on diabetes across patients, not bound by a particular doctor or hospital. A chain of trust will be established electronically for the clinical studies, and the patient is guaranteed the right to revoke the granted access to the doctors (which in turn revokes associated researchers access) at any time. This mechanism not only simplifies and improves transaction efficiency regionally and nationally in the long run but also reduces mistakes, duplication of effort, and costs. All in all, the objectives of the HCAT are: (1) to ensure the security of patient confidential data, (2) to facilitate the addition or removal of members of the DMT, (3) to facilitate the addition or removal of data-access privileges of members of the DMT, (4) to reduce costs associated with diabetes management, (5) to reduce the incidence of fraud and abuse related to patient records, (6) to facilitate interinstitutional data sharing for research purposes, and (7) to facilitate compliance with local, state, and federal laws.

The overall objective of the CST is to ensure that industry best practice methods are used in the implementation. The CST should address the same business objectives listed for the DMT and HCAT. The specific objectives of the CST are: (1) to measure services resources utilization and performance, (2) to measure service usage time, (3) to measure service outage time, (4) to deliver usage report to HCAT, and (5) to implement new designs and configurations based on requests from the DMT and the HCAT.

These specific objectives listed above refer to two sets of performance parameters: those of the WBgM system and those of the CST technical support and development teams. These parameters in turn directly impact the WBgM service level. Thus, an SLM system architecture can be specified that will measure the WBgM service level.

## Service Modeling and QoS Characteristic Identification

The WBgM service level can be measured by monitoring WBgM systems performance and the performance of CST technical support and development teams. To elaborate on an appropriate WBgM services model, it is necessary that CST identify what services that compose the WBgM service need to be monitored. Figure 13.6(a) shows a list of services and service resources needed to deliver our implementation of WBgM. Figure 13.6(b) shows a generic service model. Each service listed in Figure 13.6 (a) is a

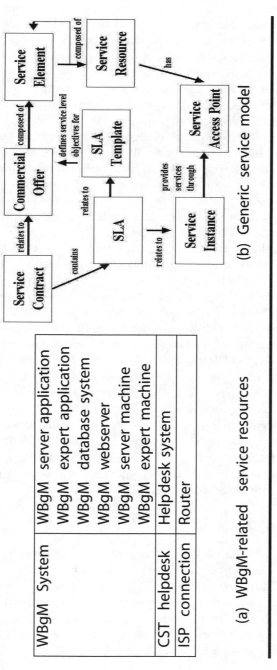

| WBgM System | WBgM | server application |
| | WBgM | expert application |
| | WBgM | database system |
| | WBgM | webserver |
| | WBgM | server machine |
| | WBgM | expert machine |
| CST helpdesk | Helpdesk system | |
| ISP connection | Router | |

(a) WBgM-related service resources

(b) Generic service model

**Figure 13.6  WBgM service modeling.**

service element that composes another service element called WBgM. Each service uses one or more service resources. Internet service provider (ISP) connection services and outsourced help-desk services are usually sold with SLAs. It is of interest to the CST that the SLM system also monitors these services to ensure that service providers are respecting the agreed service levels.

The DMT and HCAT business objectives identified in the previous subsection was divided into three areas: communication, clinical utility, and security. This classification is useful for the CST to address these objectives in the WBgM system design. The communication objectives consist of (1) to exchange accurate BGL data between the patient and the DMT, (2) to improve the dynamics of data distribution among members of the DMT, and (3) to reduce complexity of equipment connectivity apparatus. The clinical utility objectives consist of (1) to increase the timeliness and frequency of BGL data exchange, (2) to provide interim self-care diagnostic information to the patient between office visits, (3) to reduce the complexity of data analysis software, (4) to reduce the costs associated with diabetes management, (5) to facilitate interinstitutional data sharing for research purposes, and (6) to facilitate compliance with local, state, and federal laws. Finally, the security objectives consist of (1) to ensure the security of patient confidential data, (2) to facilitate the addition or removal of members of the DMT, (3) to facilitate the addition or removal of data-access privileges of members of the DMT, and (4) to reduce the incidence of fraud and abuse related to patient records. One way to evaluate how the implementation of the WBgM solution has achieved these objectives is to get subjective feedback from the DMT and HCAT, the end users of the system.

For each objective listed by the CST, it is necessary to identify relevant QoS characteristics that the SLM system needs to monitor. To measure the WBgM service performance, all of these QoS characteristics need to be monitored as detailed in Table 13.1. The resulting reports generated by the system are forwarded to the HCAT.

## QoS and SLA Modeling

Identifying QoS characteristics in terms of specific business goals allows two specific models to be generated by the WBgM teams. The first, the QoS model as illustrated in Figure 13.7(a), refers to the linkages between services and tunable quality aspects of the system. In this way, feedback with regard to one or a number of specific service elements may be directly related to a QoS characteristic that may be adjusted to improve user satisfaction. The second, the SLA model as illustrated in Figure 13.7(b),

**Table 13.1   QoS Characteristics for WBgM**

| Service Resource | QoS Characteristics |
|---|---|
| WBgM Server application | Availability<br>Service response time<br>Number of users connected<br>User connection time<br>Number of user operation |
| WBgM Expert application | Availability<br>Service response time |
| WBgM Database system | Availability<br>Hard disk utilization<br>Transactions rate |
| WBgM Web service system | Availability<br>Service response time |
| WBgM Server machine | Availability<br>CPU utilization<br>Hard disk utilization<br>Memory utilization |
| WBgM Expert machine | Availability<br>Service response time |
| Helpdesk system | Availability<br>Number of tickets opened<br>Average time to close a ticket |
| Router | Availability<br>CPU utilization<br>Bandwidth utilization<br>Memory utilization<br>Interface errors<br>Interface discards |

represents a configuration that relates services with users. The SLA model is necessary so that the effect on end users of any changes in services resulting from modification of QoS characteristics may be traced.

## Monitoring, Report Generation, and Accounting Definitions

Proceeding from the SLA model, the DMT, HCAT, and CST will then identify appropriate monitoring thresholds for the monitored characteristics. The CST will then design and develop appropriate monitoring systems

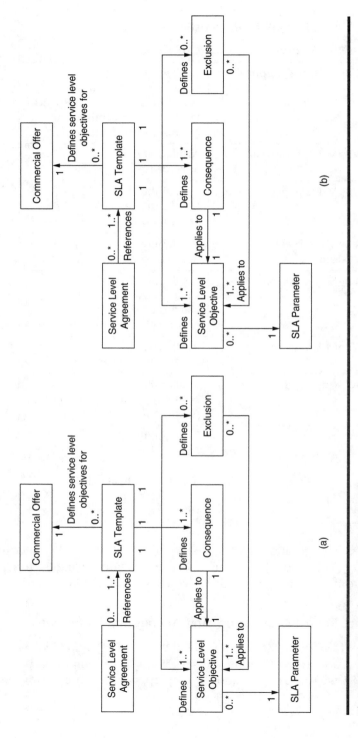

**Figure 13.7 Generic QoS and SLA models.**

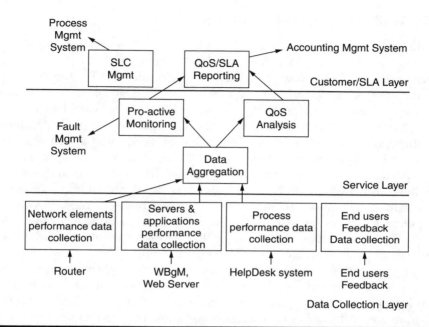

**Figure 13.8  WBgM SLM system architecture.**

that will record and report the status of the key characteristics. The ultimate purpose of this QoS-based monitoring system is to quantify user feedback into useful input. Because iterative design procedures must be followed for any clinical device, user experience is extremely valuable. This method allows precise identification of those aspects of the system that can actually be attributed to communication, clinical utility, and security issues, and which can be adjusted by the design teams. The section of QoS and SLA Modeling identifies the types of monitoring needed for this functionality. In this phase, the format and delivery schedule of generated reports is established. Ultimately, accounting refers to usage. Whether there is a financial element in terms of charging for services is specific to the implementation. However, all wireless biometric systems of this type will require an accounting method. This method is drawn directly from the Report Definition.

## SLM Architecture Specification

The cumulative result of the previous seven steps is an SLM architecture specification as illustrated in Figure 13.8. Beginning with the goal of adhering to RM-OPD principles, through identification of QoS character-istics and establishment of the SLA model, a clear specification for a WBgM is realized.

## SUMMARY AND CONCLUSIONS

In this chapter, we presented a detailed approach to realizing a WBgM system, the result of a joint research project begun in 2001 by the Bayer Corporation and the New York Center for Biomedical Engineering, the Department of Computer Science of The City College of New York, and the Department of Electrical Engineering Polytechnic School of São Paulo. Our research team has produced a robust WBgM system, with emphasis on the delivery of timely data to the DBT, and explored approaches for commercial deployment of this system. We applied an SLM approach to the specification of noncommercial components of system architecture. We have shown that QoS characteristics may be identified and monitored in a manner analogous to that of the standard business model even for a system in which the product is not a financial commodity in the strict sense. We hope our experiences will provide valuable resources for researchers and commercial vendors in this field.

When entering a new arena of technological application, users and designers inevitably try to achieve a standard method to specify application development. By borrowing from previous efforts in the wired computer environment, a flexible and comprehensive method of architecture specification can be established that will serve this purpose for wireless biometric systems. This is particularly important in the field of applied technology in medicine because so many parameters of successful healthcare practices are nonquantifiable. This method, taken directly from the modeling of commercial Internet systems, will provide a firm foundation for ODP-based architecture specification. Transactional systems by definition bear some similarity to the standard business model, and wireless computer information management systems are no exception. Wireless telemedicine and biometric management systems that adopt the approach herein outlined will both retain the flexibility of the ODP methodology and the integral robustness of the SLM approach.

## REFERENCES

1. B.A. Myers and B. Michael. Handheld computing. *IEEE Computers*, Vol. 36(9), September 2003.
2. Wireless Developer Network — Daily News eTForecasts: Smartphones Have Started to Impact PDA Sales. www.wirelessdevnet.com/news/2003/169/news7.html, June 17, 2003.
3. J.C. Lin. Applying Telecommunication Technology to Healthcare Delivery. *IEEE EMB Magazine*, 18(4), July/August 1999, 28–31.
4. S. Tacharka, R. Istepanian, and K. Bansitas. Mobile E-Health: The Unwired Evolution of Telemedicine. *Proceedings of HealthComB 2001, Italy*, July 2001.
5. C. Pattichis, E. Kyriacous, S. Voskarides, M. Pattichis, R. Istepanian, and C. Schizas Wireless telemedicine systems: an overview *IEEE Antennas Propag. Mag.*, 44(2), April 2003.

6. V. Stanford. Pervasive health care applications face tough security challenges. *IEEE Pervasive Comput.*, February 2002.
7. C. Berthelsen. Personal digital assistants: The healthcare matter at hand. *J. Am. Health Inf. Manage. Assoc.*, October 2001.
8. J. Cochrane. The Rise of Palmtop Technology in Medicine. *E-Healthcare-Connections* (www.ehealthcare-connections.com), 2001.
9. D. Chalmers and M. Sloman. A Survey of Quality of Service in Mobile Computing Environments. *IEEE Commun. Surv.* Retrieved February 9, 2004.
10. T.R. Henderson. Host Mobility for IP Networks: A Comparison. *IEEE Network*, November/December 2003, 18–26.
11. F. Hu and N.K. Sharma. A Priority-determined Multi-Class Handoff Scheme with Guaranteed Mobile-QoS in Wireless Multimedia Networks. *IEEE Trans. Vehicular Technol.* Retrieved February 9, 2004.
12. Center for the Advancement of Health, and Milbank Memorial Fund. Patients as Effective Collaborators in Managing Chronic Conditions. July 1999, Milbank Memorial Fund/Reports. http://www.milbank.org.
13. *New York Times*, January 30, 2003.
14. Diabetes Control and Complications Trial Research Group. Effect of intensive diabetes treatment on the development and progression of long-term complications in adolescents with insulin-dependent diabetes mellitus: Diabetes control and complications trial. *J. Pediatr.*, 125(2), 1995.
15. J. DeSonnaville, M. Bouma, L. Colly, W. Deville, D. Wijkel, and R. Heine. Sustained good glycemic control in NIDDM patients by implementation of structured care in general practice: 2-year follow-up study. *Diabetologia*, 40(11), 1334–1340, 1997.
16. Diabetes Control and Complications Trial Research Group. Implementation of treatment protocols in the Diabetes Control and Complications Trial. *Diabetes Care*, 18(3), 361–376, 1995.
17. S. Russell, A. Kawaguchi, and G. Qian. Development of a HIPAA Compliant Wireless Medical Data Application: Wireless Blood Glucose Monitor. *International Conference on Mathematics and Engineering Techniques in Medicine and Biological Sciences (METMBS'03)*, Las Vegas, Nevada, June 2003, pp. 210–215.
18. A. Kawaguchi, S. Russell, and G. Qian. Developing a Wireless Blood-Glucose Monitoring System: Concept and Practice. *The 7th World Multiconference on Systemics, Cybernetics, and Informatics (SCI '03), Orlando, Florida*, July 2003, pp. 345–350.
19. A. Kawaguchi, S. Russell, and G. Qian. Security Issues in the Development of a Wireless Blood-Glucose Monitoring System. *The 16th IEEE Symposium on Computer-Based Medical Systems (CBMS '03)*, New York, June 2003, pp. 102–107.
20. A. Kawaguchi, S. Russell, and G. Qian. Implementation Techniques for a wireless blood-glucose monitoring system. *Submitted for journal publication*.
21. S. Russell, C. Miyata, A. Kawaguchi, and J. Becerra. An ODP-based Method for Service Level Management Architecture Specification for Wireless Biometric Data Management Systems. *The 1st Hawaii International Conference on Computer Sciences*, Honolulu, Hawaii, January 2004, pp. 353–361.
22. C. Miyata, S. Russell, T. Carvalho, and A. Kawaguchi. Wireless telemedicine and service level management architecture specification. *The 1st International Conference on E-business and Telecommunications Networks,* Setubal, Portugal, August 2004, pp. 70–78.

23. J.R. Putnam. *Architecting with RM-ODP.* Prentice Hall PTR, Upper Saddle River, NJ, 2000.

24. C.M. Miyata and J.L.R. Bederra. An ODP Vision of QoS Management: An Application in the SLM Architecture. *International Conference on Software Engineering Research and Practice (SERP '03)*, Las Vegas, Nevada, June 2003.

25. A. Tanaka, Y. Nagase, Y. Kiryu, and K. Nakai. Applying ODP Enterprise Viewpoint Language to Hospital Information System. *Fifth IEEE International Enterprise Distributed Object Computing Conference*, Seattle, WA, September 2001.

26. A.S. Tanenbaum. *Computer Networks*, 4th ed., Prentice Hall, Upper Saddle River, NJ, 2003.

27. Oracle Corporation. Oracle9*i* R2 Database Security for E-Business. *Oracle White Paper*, January 2002.

28. Oracle Corporation. The Virtual Private Database in Oracle9*i* R2. *Oracle Technical White Paper*, January 2002.

29. A.S. Tanenbaum. *Modern Operating Systems*, 2nd ed., Prentice Hall, Upper Saddle River, N.J., 2001.

30. The Health Insurance Portability & Accountability Act of 1996. *Public Law* 104–191, August 21, 1996.

31. R.L. Mitchell. Wireless security: Good enough for medical records? *ComputerWorld*, July, 2001.

32. E. Grygo. Wireless health driven by HIPAA. *InfoWorld*, April 2002.

33. Alan Joch. Wireless Watchdogs. *Healthcare Info.*, July 2002.

34. M.E. Kabay. Identification, Authentication and Authorization on the World Wide Web. *WWW Security*, October 2002.

35. National Committee On Vital And Health Statistics, Subcommittee on Privacy and Confidentiality, *Minutes*. February 3–4, 1997.

36. E. Messmer and J. Cox. Making wireless LAN security air. *NetworkWorld*, December 2002.

37. The Pennsylvania State University Applied Research Laboratory. Wireless Handheld Electronic Devices Assisting Emergency Medical Field Personnel. *Final Report of Grant Agreement (Pennsylvania Department of Health Emergency Medical Services Office)*, August 2000.

38. A. Vigersky, E. Hanson, E. McDonough, T. Rapp, J. Pajak, and S. Galen. A wireless diabetes management and communication system. *Diabetes Technol. Ther.* 5(4), 2003, 695–702.

39. S. Plougmann, O.K. Hejlesen, and D.A. Cavan. DiasNet — a diabetes advisory system for communication and education via the Internet. *Int. J. Med. Inf.* 64(2–3), 2001, 319–330.

40. D.A. Cavan, J. Everett, S. Plougmann, and O.K. Hejlesen. Use of the Internet to optimize self-management of type 1 diabetes: preliminary experience with DiasNet. *J. Telemed. Telecare*, 9(3), 2003, 50–52.

41. E. Lou, M.V. Fedorak, D.L. Hill, J.V. Raso, M.J. Moreau, and J.K. Mahood. Bluetooth wireless database for scoliosis clinics. *Med. Biol. Eng. Comput.*, 41(3), May 2003, 346–349.

42. A.M. Albisser, J.B. Albisser, and L. Parker. Patient confidentiality, data security, and provider liabilities in diabetes management. *Diabetes Technol. Ther.*, 5(4), 2003, 631–640.

43. K. Hung and Y.T. Zhang. Implementation of a WAP-based telemedicine system for patient monitoring. *IEEE Trans. Inf. Technol. Biomed.*, 7(2), June 2003, 101–107.

44. B. Blobel. The European TrustHealth Project experiences with implementing a security infrastructure. *Int. J. Med. Inf.*, 60, 2000, 193–201.

45. L.G. Kun. Telehealth and the global health network in the 21st century. From homecare to public health informatics. *Comp. Methods Programs Biomed.*, 64, 2001, 155–167.

46. J.S. Kakalik and M.A. Wright. Privacy and security in wireless computing. *Network Security*, 2000, 2000, 12–15.

47. E. Schultz. Security Views. *Comp. Security*, 21, 2002, 11–12.

48. J. Allen and J. Wilson. Securing a wireless network. *Proceedings of the 30th Annual ACM SIGUCCS Conference on User Services*, ACM Press, 2002, pp. 213–215.

49. A.D. Rubin. Wireless networking security. *Comm. ACM*, 46(5), 2003, 28–30.

50. W.H. Diffie. New directions in cryptography. *IEEE Trans. Inf. Theory*, 22, 1976, pp. 644–654.

51. R.L. Rivest, A. Shamir, and L. Adleman. A method for obtaining digital signatures and public-key cryptosystems. *Comm. ACM*, 21, 1978, 120–126.

52. IETF PKIX Work Group, Public-Key Infrastructure (X.509). http://www.ietf.org/html.charters/pkixcharter.html.

53. National Cancer Institute. Confidentiality, data security, and cancer research: Perspectives from the National Cancer Institute. *Biotechnol. Law Rep.*, 18, 1999, 366–378.

# IV

## SECURITY I

# 14

---

# SECURITY SERVICES AND ISSUES IN WLANS

## M. Siyal and Fawad Ahmed

### SECURITY SERVICES AND ISSUES IN WIRELESS LOCAL AREA NETWORKS

Over the past several years, wireless LANs (WLANs) have gained tremendous popularity due to the user mobility and flexibility they offer in accessing information. In addition, the cost of wireless networking equipment has dropped, and there have been significant improvements in performance. The enormous flexibility offered by WLANs, however, presents a number of potential security challenges. A WLAN transmits data using radio waves, which makes the transmission signal available to anyone within the range of the transmitter, posing a high security risk to wireless networks. In this chapter, we discuss the mechanisms used to protect the security of WLANs, along with other related issues. The section titled "802.11 Wireless Local Area Network" gives a brief introduction to the IEEE 802.11 WLAN standard. In the section titled "Attacks against Wireless Networks," we describe different types of attacks against WLANs. In the section titled "WLAN Security Services," we present wireless network security services with particular emphasis on the IEEE 802.11 privacy and authentication mechanism. In the section titled "Security Weaknesses in WEP," we discuss a number of vulnerabilities that have been found in the Wired Equivalent Privacy (WEP) protocol. In the section titled "Enhancing WEP Security: The IEEE 802.1X Standard," we discuss the IEEE 802.1X standard, which enhances WLAN security. The section titled "Using Biometrics for Enhanced Wireless Authentication" gives a brief introduction to biometrics and illustrates how it can be augmented with wireless protocols to further increase the security of WLANs. In the section titled "Wi-Fi Protected Access (WPA) and the IEEE 802.11i Standard," we briefly describe the emerging WPA and IEEE 802.11i standards.

## 802.11 WIRELESS LOCAL AREA NETWORK

The development of WLAN standards started with IEEE 802.11, released by IEEE in 1997. The standard defines the medium access control (MAC) and physical (PHY) layer specifications for wireless connectivity for fixed, portable, and moving stations within a LAN.[1] Over the past few years, there have been many PHY layer specifications developed by IEEE. The original 802.11 standard specifies a 2.4-GHz operating frequency with data rates of up to 2 Mbps. In 1999, IEEE released the 802.11b standard, which works in the 2.4-GHz range and is capable of data rates of up to 11 Mbps. The data rate of 802.11b, however, can be scaled down to 1 Mbps. In 2001, 802.11a, which operates in the 5-GHz range and is capable of data rates of up to 54 Mbps, was released. The 802.11g standard is the fourth PHY specification released by IEEE that works in the 2.4-GHz range with a maximum data rate of up to 22 Mbps. The 802.11b standard is currently the most widely deployed PHY specification for WLANs. The MAC structure underlying the different PHY layer specifications is the same for all 802.11 wireless technologies. The 802.11 MAC provides several functions such as access to a wireless medium, joining and leaving the network, and security services. An 802.11 wireless network can be operated in one of two modes: *infrastructure* mode or *ad hoc* mode.[1] In the infrastructure mode, each wireless client relays all its communication to a central station called the access point (AP). The AP may relay traffic between two clients who are communicating with each other or may act as a bridge between a wireless network and a wired/wireless network as shown in Figure 14.1.

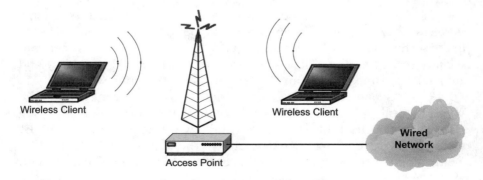

**Figure 14.1   Infrastructure mode.**

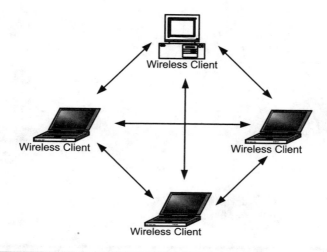

**Figure 14.2   Ad hoc mode.**

In the infrastructure mode, a wireless client intending to join the network must first *associate* with an AP, which involves identifying the AP and performing mutual authentication. The process involved is discussed in the section titled "Authentication." In the ad hoc mode, each client communicates directly with the other clients in the network, as shown in Figure 14.2. The participating clients associate with each other through the use of a common network identifier. Once associated, they can share resources exactly as they would in a wired peer-to-peer network. Though convenient to set up, ad hoc networks are difficult to manage for a large number of nodes and should be used only where security is not an issue.

## ATTACKS AGAINST WIRELESS NETWORKS

Unlike wired networks, wireless networks cannot be made physically secure, which makes them vulnerable to several kinds of attacks. These attacks can be broadly divided into passive and active attacks.

### Passive Attacks

In a passive attack, the attacker eavesdrops on the transmission to monitor the wireless session. In a wireless environment, anyone with a suitable transceiver in the range of the transmission can eavesdrop on the message. Figure 14.3 shows a typical scenario of passive eavesdropping between a wireless client and an AP. Passive eavesdropping enables the attacker to read and collect the data that is being transmitted in a wireless session.[2] The data collected can later be examined to get the source and destination

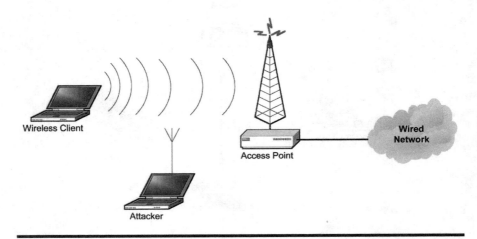

**Figure 14.3 Passive eavesdropping.**

addresses and other vital information about the communicating nodes. Passive eavesdropping is difficult to detect; the attacker can use powerful antennas to receive 802.11 transmissions even from hundreds of feet away.

## Active Attacks

Active attacks involve altering or destroying transmission data, or creating fraudulent packets. There are many types of active attacks that can be launched against 802.11 wireless networks. A brief summary of such attacks is presented in the following text; details can be found in Reference 2 to Reference 8.

### Attacks against the WEP

WEP is an encryption mechanism used to provide confidentiality and authentication to 802.11 wireless networks. This mechanism has several flaws, which make it vulnerable to different attacks.[2-5] The section titled "Security Weaknesses in WEP" discusses these attacks in detail.

### Denial-of-Service Attacks

Denial-of-Service (DoS) attacks are active attacks that are meant to cause disruption of network services. In a typical DoS attach, the attacker floods the network by saturating the 802.11 frequency bands with noise.[8] As a result, the network may either slow down or stop working.

## Man-in-the-Middle Attack

A man-in-the-middle attack can be launched either for the purpose of eavesdropping on a wireless session or to modify the contents of wireless packets before they reach the intended recipient. This attack is very powerful because it enables an attacker not only to read transmitted data but also to modify it. A man-in-the-middle attack can take several forms.[2] The Address Resolution Protocol (ARP) cache-poisoning attack[7] is one such form that requires the attacker and the target to be connected on the same local network. ARP is a mechanism that is used to determine the mapping between the Internet Protocol (IP) addresses and the MAC addresses in a network. When a host on a local network needs to send a message to another host, it relays an ARP request that is received by all the hosts on the network. The host that has the specific IP address sends an ARP reply, specifying its MAC address. The requesting host sends the packet to the supplied MAC address and updates its ARP cache. To reduce the number of ARP packets being broadcast, most ARP implementations update their ARP cache whenever an ARP reply is received. An attacker can exploit this mechanism to alter the IP-to-MAC-address mapping by sending forged ARP reply messages to reroute the traffic between two communicating nodes through his or her machine. This scenario is depicted in Figure 14.4.

In wireless networks in which the AP is directly connected to a hub or a switch, an attacker can launch an ARP cache-poisoning attack against all the hosts connected to that hub or switch.[7] This scenario is depicted in Figure 14.5.

## Session Hijacking

Through session hijacking, an attacker is able to take an authorized and authenticated session away from its legitimate user. To launch this attack, the attacker must be able to masquerade as a legitimate user to the wireless network. In addition, the attacker should also have the capability to stop the legitimate user from continuing the session.[2] The actual user may not even know that the session has been hijacked and attributes the disruption in the session to some fault in the wireless network. Once control over the session is gained, the attacker may use it for any desired purpose and is able to maintain the session for an extended period of time.

# WLAN SECURITY SERVICES

In this section, we present the details of various security mechanisms such as Service Set IDs (SSID), access control lists, and the IEEE 802.11 privacy

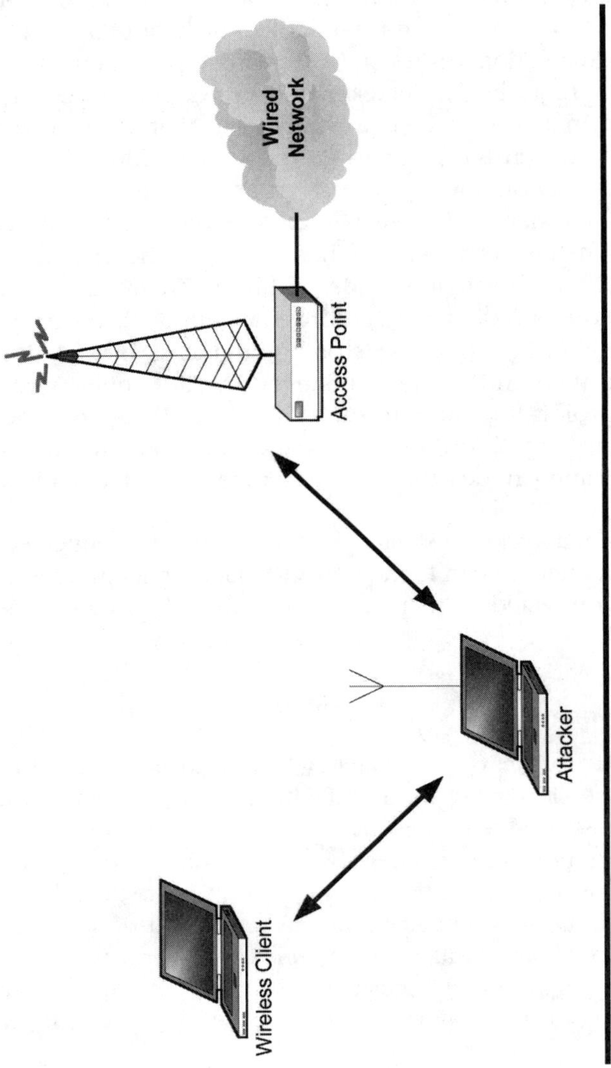

**Figure 14.4 Man-in-the-middle attack.**

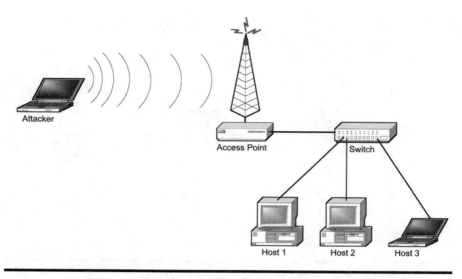

**Figure 14.5   A man-in-the-middle-attack used to attack wired hosts.**

and authentication schemes used to protect WLANs. SSIDs and access control lists provide a minimal level of security and are not adequate for wireless networks. To provide authentication and privacy to WLANs, the IEEE 802.11 standard defines the WEP protocol. The main goal of WEP is to provide WLANs with a level of privacy that is equivalent to the one provided by the physical security attributes inherent in a wired LAN.

## SSID

An SSID is a common network name that is used by all the APs and wireless clients participating in the same network. Only clients with a valid SSID are permitted to join the network. An AP broadcasts the SSID in its beacon by default. This feature is usually turned off, because it would enable an unauthorized user to read the SSID. Even if broadcasting of the SSID is turned off, it can be read by an attacker by eavesdropping on the communication between the AP and the client. Therefore, an SSID does not provide strong security to the wireless network.

## Access Control List

Every 802.11-enabled device is required to have a unique MAC address, which can be used to identify a genuine user to the network. This identification is accomplished by programming the AP with an access control list that consists of MAC addresses of clients who are allowed to access the network. A client whose MAC address is not listed in the access

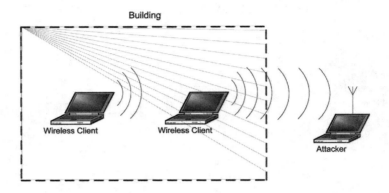

**Figure 14.6 Wireless signal traveling outside a building.**

control list will not be allowed to access the network. Similar to SSID, the access control list mechanism is weak from the security perspective, because MAC addresses can easily be discovered using sniffer software. An attacker can sniff the MAC address of a target user, reconfigure the MAC address of his or her wireless network card to the MAC address of the target user, and gain access to the network as a legitimate user.

## Privacy and Authentication

In wired LANs, the physical security of the cable can prevent unauthorized access to the network. This, however, is not possible in WLANs, where the physical medium has no precise bounds. The radio waves used by WLAN can pass through doors and walls, and even cross the outer boundaries of the building in which the wireless equipment is operated.

Wireless radio waves can easily be intercepted by an attacker, as shown in Figure 14.6. Because it is easy to capture radio waves in a wireless network, eavesdropping is a major concern. To prevent unauthorized users from eavesdropping and accessing WLANs, strong user authentication and confidentiality mechanisms are required. The IEEE 802.11 standard provides two basic services to bring the functionality of WLANs in line with that of wired LANs: authentication and privacy.[1] Authentication provides the ability to control WLAN access by requiring all wireless clients to establish their identities to the wireless station with which they wish to communicate. Privacy is used to keep the data confidential and to protect its contents while it is being transmitted between two communicating nodes.

## WEP

To provide privacy to the data traveling in air, IEEE 802.11 specifies an encryption scheme called WEP. WEP is also designed to provide data integrity and authentication services. The main goal of WEP is to provide, in WLANs, a level of privacy that is equivalent to the one provided by the physical security attributes inherent in wired LANs. The WEP protocol provides link-level protection by encrypting the 802.11 data frames transmitted between two WLAN devices. The WEP algorithm uses the RC4 symmetric cipher[12] along with a secret key that is shared between the communicating nodes. In a symmetric cipher, the same secret key is used for encryption and decryption. The key length specified in the initial 802.11 standard was 40 bits. However, to increase the security, a key length of 104 bits is also being used. It is important to note that the 802.11 standard does not define a service for the distribution of secret keys among the communicating nodes and assumes an external key-management service to perform this task. The WEP algorithm is used to encrypt only the MAC frame body. Figure 14.7 illustrates the WEP frame format. The MAC header and the frame check sequence (FCS) fields are not encrypted. The 802.11 MAC frame header consists of a 2-octet frame control field, which has a 1-bit WEP subfield. If the MAC frame body is encrypted using the WEP algorithm, the WEP subfield is set to 1. The 802.11 standard specifies the use of WEP for MAC data and authentication management frames.

The WEP frame consists of three major fields: the initialization vector field, the data field, and the integrity check value (ICV) field. The initialization vector field consists of a 24-bit initialization vector (IV), a 2-bit key ID field, and a 6-bit pad field. The purpose of IV is to produce a per-frame encryption key by selecting a new IV for each frame. This is accomplished by appending the IV to the shared key to form a family of $2^{24}$ keys. The WEP ICV is a 32-bit field containing the CRC-32 calculated over the data field. The purpose of ICV is to provide integrity of the data field. The WEP algorithm uses one of the $2^{24}$ keys for each frame transmission to encrypt the data and Cyclic Redundancy Check (CRC) value. The WEP algorithm starts by first calculating the CRC of the data to be transmitted. The algorithm then uses a new IV for each packet and concatenates it with the shared key to form a per-packet key. The RC4 algorithm is then used to get a keystream that is equal to the number of bits in the data and the CRC. The resulting keystream is then XORed with the data and CRC to get the ciphertext, which is appended with the IV to form the WEP frame. At the receiving end, the IV is extracted from the WEP frame, which is then appended with the shared key to generate the keystream using the RC4 stream cipher. The resulting keystream is then XORed with the encrypted payload to get the plaintext. The integrity of the transmitted

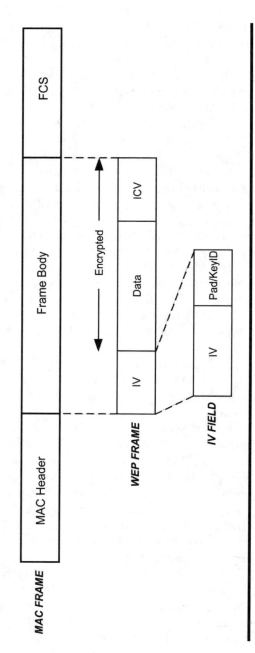

**Figure 14.7** WEP frame body.

message is verified by calculating the CRC of the decrypted data and comparing it with the CRC obtained from the decrypted WEP frame.

### Authentication

IEEE 802.11 specifies two types of authentication: open system authentication and shared-key authentication.[1] Open system authentication is the default authentication protocol, which authenticates anyone who requests authentication. Basically it provides a null authentication process. Shared-key authentication uses a challenge-and-response scheme along with a shared key. The use of this authentication mechanism requires the implementation of WEP. In shared-key authentication, the identity of the participating station is demonstrated by knowledge of the shared secret, i.e., the WEP encryption key. The station wishing to be authenticated (requester) sends an authentication request management frame to the authenticating station (responder). The responder in return sends the authentication management frame to the requester. This frame consists of a 128-octet challenge text that is generated using the WEP pseudorandom generator (PRNG). The requester then uses its WEP key to encrypt the challenge and sends it back to the responder in another management frame. Upon receiving this frame, the responder decrypts the encrypted challenge and compares it with the one sent to the requester. If the comparison is successful, then it is proved that the requester possesses the correct secret key. The responder then sends another authentication management frame to the requester, indicating authentication approval or failure. Figure 14.8 and Figure 14.9 show the open-key and shared-key authentication processes, respectively.

## SECURITY WEAKNESSES IN WEP

In recent years, researchers have found several security flaws in the WEP that severely undermine its encryption and authentication capabilities. In this section, we present some common attacks against the WEP protocol. More details can be found in Reference 2 to Reference 5.

### IV Reuse Problem

WEP provides the encryption mechanism using the RC4 stream cipher. WEP uses the secret key and IV to get a keystream of pseudorandom bits using the RC4 key schedule. Encryption is performed by XORing the bits of the plaintext with the keystream. Decryption is performed by XORing the ciphertext with the keystream. Let $C$ be the ciphertext, $P$ be the plaintext, and $K$ be the keystream. The encryption and decryption process is given by Equation 14.1 and Equation 14.2, respectively.

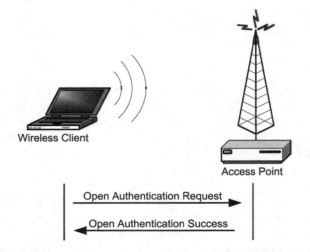

**Figure 14.8 Open system authentication.**

**Figure 14.9 Shared key authentication.**

$$C = P \oplus K \tag{14.1}$$

$$P = C \oplus K \tag{14.2}$$

In a stream cipher, it is required that the same keystream never be reused, because this can lead an attacker to launch known plaintext and chosen ciphertext attacks.[13] To illustrate this point, let us assume $P_1$ and $P_2$ to be two plaintext messages that have been encrypted using the same keystream $K$ to obtain the respective ciphertext $C_1$ and $C_2$. XORing the two ciphertexts would actually give the XOR of the two plaintexts. Moreover, the knowledge of one plaintext message would help an attacker to obtain the other.[3,4]

$$C_1 \oplus C_2 = ( P_1 \oplus K ) \oplus ( P_2 \oplus K ) = P_1 \oplus P_2 \tag{14.3}$$

To address this issue, WEP uses a different IV for each packet that is transmitted. This would ensure a different keystream, as the keystream generated is a function of the shared key and the IV. This, however, does not practically help in preventing the reuse of keystreams. WEP uses a 24-bit IV, which gives a total of $2^{24}$ possible keystreams. Thus, the number of possible keystreams is limited, and it may take about an hour for a single AP running at 11 Mbps to exhaust all the available keystream space.[4] The WEP standard also fails to specify how the IV is selected for each new frame. Many wireless network cards reset the IV to zero and then increment them by one for every IV, instead of selecting a random IV for each packet. Because the WEP frame contains the IV without encryption, any reuse of the old IV value exposes the system to keystream-reuse attacks.[3,4] The attacker can intercept the wireless traffic and find the WEP frames in which IV collision occurs. Once two such encrypted messages that use the same IV are found, Equation 14.3 can be used to get the XOR of the plaintext. If one of the plaintexts is found, getting the other is trivial. Due to the standard protocol structure used in IP traffic, the frame format is known in advance, making the job of recovering the plaintext quite easy for an attacker. Once the plaintext of one message is found, all the ciphertext that uses the same IV can be successfully decrypted using Equation 14.3. In addition, for a particular IV, knowledge of the plaintext and ciphertext of an intercepted message also reveals the keystream used to encrypt the message. This keystream can be used to decrypt any message that uses the same IV. By investing some time, it is possible for an attacker to intercept messages and find keystreams corresponding to different IVs. This would enable the attacker to build a decryption table that contains keystreams corresponding to different IVs.

The attacker can later use this table to decrypt any packet by simply matching the IV used in the packet to the corresponding keystream in the table.[3]

## Integrity Check Value Insecurity

To ensure that a WEP frame has not been modified in transit, a 32-bit ICV is used, as described in the section titled "Wired Equivalent Privacy." A CRC-32 checksum algorithm is used to calculate the ICV that forms part of the encrypted payload of the WEP frame. The CRC checksum is, however, not a cryptographically secure authentication code such as the SHA-1 or MD5. Because CRC-32 is a linear function, it becomes possible to flip arbitrary bits in an encrypted message and then correctly adjust the checksum so that the resulting message appears valid.[3] The WEP checksum, therefore, practically fails to provide data integrity. In addition, the WEP checksum is an unkeyed function of the message, allowing anyone to introduce arbitrary traffic into the network.[3] If an attacker is able to capture the complete plaintext of an encrypted frame, it is then easy to calculate the keystream for the specific IV.

$$P \oplus C = P \oplus ( P \oplus K ) = K \qquad (14.4)$$

Using the keystream $K$, an attacker can select an arbitrary message, calculate its CRC, encrypt it with the keystream $K$, and send it across the network with the corresponding IV. The transmitted message will be accepted as valid because the correct keystream was used with respect to the IV. This attack is possible because the CRC checksum is not a function of the secret key, and also because 802.11 allows the reuse of old IV values. The vulnerability in ICV can also be used to launch an IP redirection attack.[3] This attack is meant to fool the AP so that it can perform decryption of the ciphertext and send it to a wired IP address that is under the control of the attacker. The basic idea is to capture an encrypted packet and modify its destination IP address to the IP address of the machine that is under the attacker's control. This technique also requires fixing the IP checksum in the modified packet. Details of this attack can be found in Reference 3.

## Key Management

The WEP standard does not define any key-management protocol and presumes that secret keys are distributed to the wireless nodes by an external key-management service.[1] Most of the WEP implementations use a single WEP key that is shared between all wireless nodes. This, however, makes the overall network insecure because any station can intercept and decrypt traffic that was intended for another station. The situation worsens with increase in the number of users. In addition, APs and client stations must be programmed with the same WEP keys. As a result, the keys tend to remain in place for a long period of time, allowing attackers more time to use various attack methods to obtain the keys and decrypt the traffic. WEP also allows keys to be rotated by using a set of shared keys. This, however, does not help much to increase the security, as all the users in the network have knowledge of the shared keys.

### Weaknesses in RC4 Key-Scheduling Algorithm

In 2001, Scott Fluhrer, Itsik Mantin, and Adi Shamir published a paper titled "Weaknesses in the Key-Scheduling Algorithm of RC4." In this paper, they described weaknesses in the RC4 key-scheduling scheme that make it vulnerable the way it is implemented in the WEP protocol. Their observations consist of the fact that when the same secret part of a key is used with a number of different exposed values, it is possible for an attacker to rederive the secret part by analyzing the initial word of the keystreams.[9] This attack assumes that the attacker knows the first output byte of a large number of RC4 keystreams along with the corresponding IV that was used to generate them. In a typical WEP scenario, this assumption is realizable because the IV is sent in the clear. In addition, due to the payload format used with 802.11, the plaintext value of the first byte of the encrypted message, which is also the first byte of the SNAP header (AA in hexadecimal), is known.[10] The attacker can easily derive the first byte of the RC4 keystream by XORing the first byte of the plaintext and the ciphertext. Much worse is the fact that as the key length grows, the time it takes to break the key grows linearly and not exponentially. The attack described by Fluhrer, Mantin, and Shamir was practically implemented by Stubblefield, Ioannidis, and Rubin, using off-the-shelf hardware and software, to recover a 128-bit WEP key[11] by passively observing a real network. This attack demonstrated that RC4 is completely insecure the way it is implemented in the WEP, as pointed out by the authors in Reference 9.

## ENHANCING WEP SECURITY: THE IEEE 802.1X STANDARD

IEEE 802.1X is a port-based network access control standard that makes use of the physical access characteristics of the IEEE 802–LAN infrastructures for providing authentication and authorization services to devices attached to LAN ports.[14] 802.1X, when used in wireless networks, enhances security by supporting mutual authentication and providing centralized management and distribution of encryption keys. The problem of WEP key management is solved by generating dynamic, per-user, per-session encryption keys. 802.1X provides port-based access control, in which the network accessibility is provided through the concept of a port. A port is a single point of attachment to a LAN infrastructure. It can be a physical port, for example, a single LAN MAC attached to a physical LAN segment, or a logical port in the case of 802.11. In the latter case, the AP manages the logical ports, which are used for communication with wireless clients. The basic structure of 802.1X is shown in Figure 14.10. It consists of a supplicant, the authenticator, and the authentication server. A supplicant is a client machine attempting to gain access to the network through a port on the authenticator. The authenticator passes the supplicant's credentials to the authentication server and, accordingly, grants or denies network access to the supplicant. For a WLAN, the supplicant is the wireless client and the authenticator is the AP. The protocol functionality associated with the supplicant and the authenticator is provided by the protocol entity called port access entity (PAE). The supplicant PAE is responsible for responding to requests from an authenticator for information that will establish its credentials. The authenticator PAE is responsible for communicating with the supplicant and for passing the information received from the supplicant to an authentication server to verify the supplicant credentials for the purpose of authorization. The authentication information between the supplicant and authentication server is communicated using the Extensible Authentication Protocol (EAP).[15] EAP is a general protocol that supports multiple authentication mechanisms. Using EAP, a number of authentication schemes may be added including smart cards, Kerberos, public-key encryption, one-time passwords, etc. IEEE 802.1X defines an encapsulation format that allows EAP messages to be carried directly by a LAN MAC service. This encapsulated format is known as EAP over LAN or EAPOL and is used for all communication between the supplicant PAE and the authenticator PAE. The authenticator PAE further encapsulates the EAP packets for onward submission to the authentication server.

The 802.1X standard suggests the use of the Remote Authentication Dial-In User Service (RADIUS) protocol[16] for communication between the authenticator and the authentication server. RADIUS is a client/server protocol that enables remote access servers to communicate with a central server to authenticate dial-in users and authorize their access to the requested resource or service. RADIUS is used for providing authentication, authorization, and accounting services to a network. The EAP frames are carried out as message attributes in RADIUS. The authenticator basically acts as an EAP proxy between the supplicant and the authentication server. The authenticator accepts the supplicant's credentials in the form of EAPOL

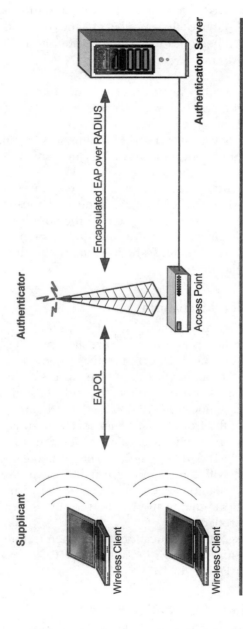

**Figure 14.10  Basic 802.1X structure.**

EAP packets and forwards them, using RADIUS, to the authentication server for approval. Similarly, the supplicant will also verify the authentication server's credentials to complete the process of mutual authentication. Once this process is complete, the authentication server issues a new temporary WEP key to the supplicant, and the authenticator will allow a WEP session to begin for the particular supplicant. The three main features of 802.1X/EAP are:

- Mutual authentication between client and authentication server
- Dynamic generation of WEP encryption keys after successful authentication
- Centralized policy control by performing session time-out triggers to facilitate reauthentication and new key generation

EAP messages consist of EAP requests, EAP responses, EAP successes, and EAP failures. EAP is basically a challenge–response authentication protocol. The EAP request message is sent to the supplicant, indicating a challenge. The supplicant replies using the EAP response message. The EAP success and failure messages are used to notify the authentication outcome to the supplicant.

## 802.1X Authentication Mechanism

In 802.1X architecture, the authenticator has two ports to access the network: the uncontrolled port and the controlled port. The uncontrolled port only allows authentication (EAP) messages to be exchanged between the supplicant and the authenticator. Once access is granted by the authenticator, the supplicant can use the network through the controlled port. Initially, the authenticator's PAE is set to the uncontrolled state, accepting only EAP traffic from the supplicant as shown in Figure 14.11.[14] The authenticator, upon receiving an EAP request from the supplicant, forwards the same as a RADIUS request to the authentication server. The authentication server will then send a RADIUS response back to the authenticator with permission to either allow or deny supplicant access to the network. If the authentication fails, the authenticator's port will remain in the uncontrolled state.

In case of a successful authentication, the authenticator's port switches to the controlled state, allowing the supplicant to access the network facilities as shown in Figure 14.12.[14]

## Authentication and Key-Distribution Process

Authentication can be initiated either by the supplicant's PAE or by the authenticator's PAE. A supplicant PAE may initiate the authentication

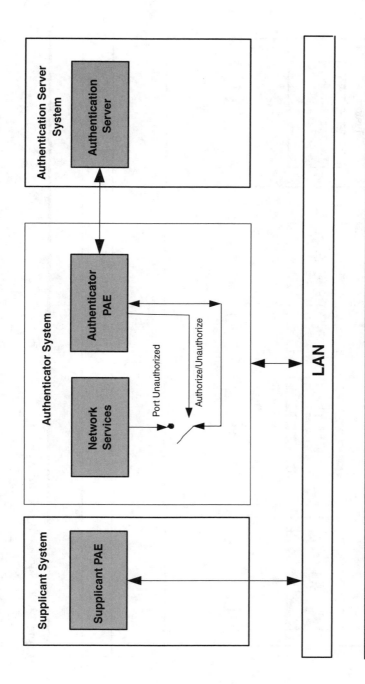

**Figure 14.11** Unauthenticated user.

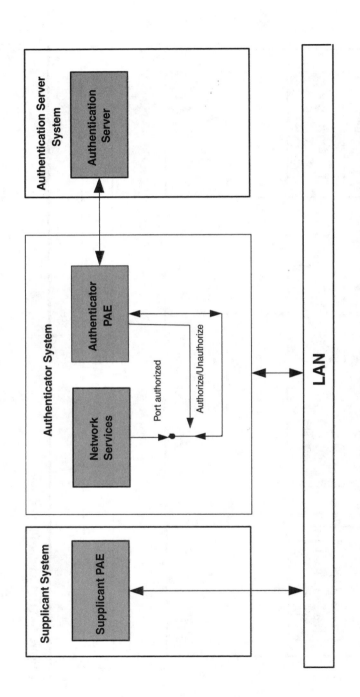

**Figure 14.12  Successful authentication.**

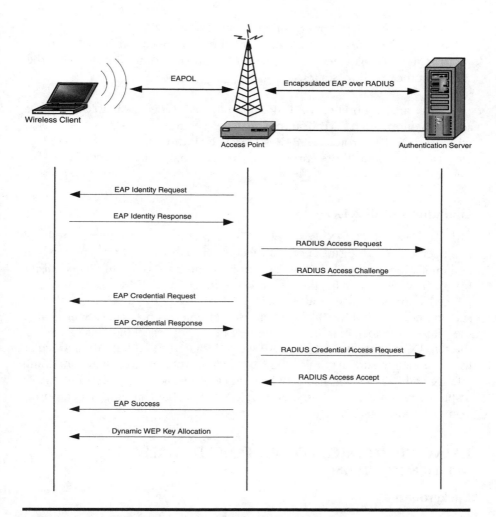

**Figure 14.13   EAP-enabled authentication process.**

sequence by sending an EAP-start frame. Authenticator PAE initiates the
authentication sequence by sending an EAP-request or identity frame.
Figure 14.13 shows the overall sequence of authentication and key distribution, which is summarized as below:

- AP initiates the authentication sequence by sending an EAP-request
  or identity frame to the wireless client.
- Client replies with EAP identity response.
- Authentication server sends a challenge to the client.
- Client provides the required authentication challenge credential to
  the authentication server.

- Authentication server authenticates the client.
- Client verifies the credentials of the authentication server.
- Client is allocated dynamic WEP keys to communicate with the authentication server.

There are a number of EAP-authenticated types available for user authentication, such as EAP-transport layer security (EAP-TLS), EAP-Cisco wireless (LEAP), protected EAP (PEAP), EAP-tunneled TLS (EAP-TTLS), etc. Details about these protocols can be found in Reference 17 to Reference 19.

## Limitations of 802.1X

The purpose of 802.1X is to provide mutual authentication, centralized management, and distribution of encryption keys. 802.1X is merely an authentication protocol that works around the existing WEP mechanism; however, it does not fix the weaknesses inherent in WEP itself. A recent paper by Arunesh Mishra and William A. Arbaugh[20] describes a number of security holes found in 802.1X when used in wireless networks. Specifically, the paper describes man-in-the-middle and session hijacking attacks against the 802.1X. Neither 802.1X nor EAP were designed to address the particular threat model present in wireless networks. The paper suggests that combining 802.1X with 802.11 will not provide sufficient levels of security until significant changes are made in the authentication mechanisms and synchronization of various state machines in these protocols.

## USING BIOMETRICS FOR ENHANCED WIRELESS AUTHENTICATION

### Background

The 802.1X/EAP protocol, discussed in the section titled "Enhancing WEP Security: The IEEE 802.1X Standard," provides a mechanism through which a wireless client can authenticate to an AP through an authentication (RADIUS) server using different types of credentials. In the case of password-based credentials, the security of the wireless client depends on the strength and safe custody of the password. However, password-based systems are weak due to three important reasons. First, users often choose simple passwords that may easily be guessed. Second, password-based systems are susceptible to dictionary attacks. Third, a password is not directly tied to a user, as it is only the knowledge that a user possesses. On the basis of a password, an authentication system cannot distinguish between a legitimate user and an impostor who fraudulently acquires that password. One way to make wireless authentication mechanisms more secure is to use biometrics-based identification. Biometrics refers to uniquely identifying an individual based on his or her physiological or behavioral characteristics. Physiological characteristics include fingerprints, hand or palm geometry, retina, iris, facial characteristics, etc. Behavioral characteristics include signature, voice, keystroke patterns, gait, etc.[21] Among all the biometric techniques, fingerprint technology is the most mature and proven technique. The cost of fingerprint sensors has gone down considerably, along with notable performance improvements in the verification algorithms. Fingerprint sensors are now integrated with keyboard and mouse for authentication purposes. A comparison of various fingerprint sensors can be found in Reference 22. In a wireless network scenario, each client station can be equipped with a fingerprint sensor that can capture fingerprint impressions for subsequent authentication. Besides fingerprints, other technologies such as iris pattern, face recognition, etc., can also be used. A detailed description of different types of biometric technologies can be found in Reference 21.

## Overview of Biometric-Based Verification System

A typical block diagram of a biometric verification system is shown in Figure 14.14.[23] The system has two basic modules: enrollment module

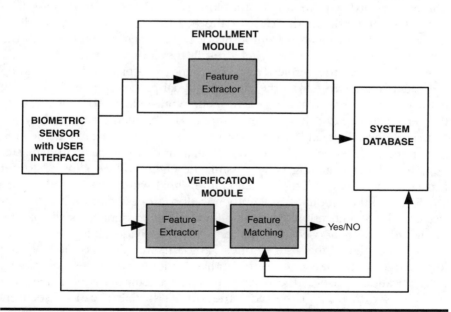

**Figure 14.14   Block diagram of a biometrics-based verification system.**

and verification module. The enrollment module scans the biometric sample of a person through a sensing device, extracts the relevant features, and then stores a representation (called template) in a database. In the verification phase, the user (whose claimed identity is to be verified) puts his or her biometric on the sensor, and the feature extractor in the verification module uses the same method to extract features as was done in the enrollment module. The features of the input biometric is then matched against the template of the claimed identity stored in the system database to give an accept response or a reject response.

A biometrics system may either work in the verification mode as discussed in the preceding text or in the identification mode in which all the biometric templates stored in the database are first matched with the input template. If a correct match is found, the identity of the user is established against the information of that template. In the case of wireless networks, the biometric system will be required to work in the verification mode because the user supplies his or her identity in the logon sessions. In this section, we will demonstrate how a biometrics-based authentication mechanism can be augmented with the 802.1X/EAP protocol to make the authentication mechanism more secure.

### Proposed Biometric-based 802.1X/EAP Authentication Scheme

There are many ways to augment a biometric authentication process with a wireless network. In the case of the 802.1X mechanism, every client using the network must first enroll his or her biometric template with the authentication server database. As a result, the authentication server database will have all the wireless network clients' master templates. To keep the client stations free from the complex signal processing involved while extracting biometric features, each client workstation should have only the sensor to capture the biometric input. The biometric captured by the sensor at the client end will be securely delivered to the authentication server for processing and subsequent matching with the stored master template. If the matching is successful, the client will be allowed to get access to the network, otherwise a failure signal will be generated. Figure 14.15 shows the modified EAP-authentication process employing biometrics-based authentication in addition to client credentials such as the password. The proposed EAP authentication is divided into two phases. In the first phase, the user verifies his or her credentials such as the knowledge of the password to the authentication server. After completing authentication phase 1, the user is assigned a temporary WEP key that will be used to encrypt the biometric sample so that it can be securely delivered to the authentication server. In the second phase of authentication, the user will be required to present his or her biometric to the sensor

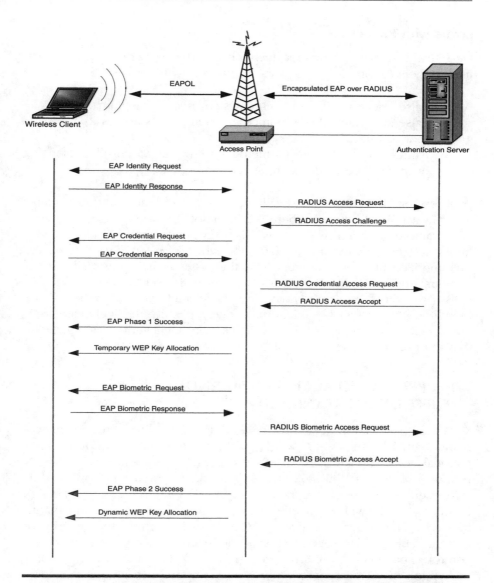

**Figure 14.15   A modified EAP sequence employing biometrics-based identification.**

attached to the machine. The sensor output will be encrypted using the temporary WEP key assigned in phase 1. The encrypted template will then be transmitted to the remote authentication server for subsequent authentication. If the biometric authentication is positive, the client will be assigned dynamic WEP keys and allowed to use the network resources. In case of an unsuccessful matching, access to the network will be denied, and the temporary WEP key assigned in phase 1 will be cancelled.

## Errors with Biometrics

Despite the fact that biometric technology offers a very strong form of user authentication, they are prone to the following errors:

■ False rejection: The biometric system rejects a legitimate user.
■ False acceptance: The biometric system grants access to an unauthorized user.
■ Failure to enroll: The biometric system fails to enroll a user; for example, due to dry skin, speech impediment, etc.

These parameters are statistically determined and measured in percentages. In a practical system, there is a trade-off between these quantities. For example, a system with a low false-acceptance rate will have higher false-reject rate, which means that although the probability of accepting unauthorized users will be very low, the system may occasionally reject a legitimate user. Using biometrics together with user credentials such as passwords, for example, provides a two-factor authentication scheme. The biometric system will be able to accurately identify the legitimate user, whereas the presentation of the correct password will verify the knowledge that the user possesses.

## WI-FI PROTECTED ACCESS (WPA) AND THE IEEE 802.11i STANDARD

WPA is a specification of standards-based, interoperable security enhancements issued by the Wi-Fi (Wireless Fidelity) Alliance, a nonprofit international association formed in 1999 to certify interoperability of WLAN products based on IEEE 802.11.[24] Any products tested and approved as Wi-Fi Certified (a registered trademark) by the Wi-Fi Alliance are certified as interoperable with each other, even if they are from different manufacturers. The major objective of WPA is to strengthen the security of current WEP standards by increasing the level of encryption and access-control capabilities for existing and future Wi-Fi WLAN systems. Data encryption in WPA is improved by using temporal key integrity protocol (TKIP). TKIP enhances the WEP data encryption mechanism by providing per-packet key-mixing function, a message integrity check (stronger than CRC used in WEP), an extended initialization vector, and a rekeying mechanism. To strengthen authentication, WPA uses IEEE 802.1X and EAP. WPA is a subset of the emerging IEEE 802.11i standard, which features 802.1X authentication protection and adds Advanced Encryption Standard (AES) for encryption along with other enhancements. In addition to WPA, 802.11i also includes Robust Security Network (RSN) for establishing secure communications. See Reference 25 and Reference 26 for an overview on 802.11i.

## SUMMARY

In this chapter, we presented an overview of various security mechanisms used in WLANs. Security considerations in WLANs are critical because the transmission media used is air, which allows anyone to intercept wireless transmissions. This poses a potential security risk to the network, as well as to the information that is being transmitted. The IEEE 802.11 standard proposed the WEP protocol as a method to protect wireless network access and data confidentiality and integrity issues. WEP protocol, however, failed to address the fundamental goals for which it was designed. To address the security flaws present in WEP, many organizations use 802.1X for improved access control and user authentication. 802.1X is merely an authentication protocol that works around the existing WEP mechanism and does not fix the weaknesses inherent in WEP itself. A recent paper by Arunesh Mishra and William A. Arbaugh[20] describes a number of security holes found in 802.1X when used in wireless networks. It is believed that the upcoming IEEE 802.11i standard, which is based on RSN, is expected to redress the significant security and scalability issues present in wireless networks. In this chapter, we have also proposed the use of biometrics as a means of enhancing the authentication mechanism in wireless networks. In recent years, the cost of biometric sensors has gone down considerably, although with significant performance improvements in the verification algorithms. This makes the biometrics technology a viable option for deployment in applications that require high-security access-control measures.

## REFERENCES

1. LAN MAN Standards Committee of the IEEE Computer Society, Wireless LAN medium access control (MAC) and physical layer (PHY) specifications, IEEE Standard 802.11, 1999.
2. Welch, D. and Lathrop, S., Wireless Security Threat Taxonomy, IEEE workshop on information assurance, 2003, pp. 76–83.
3. Borisov, N., Goldberg, I., and Wagner D., Intercepting Mobile Communications: The Insecurity of 802.11, *Proceedings of the 7th annual international conference on mobile computing and networking*, July 2001.
4. Walker, J.R., IEEE P802.11 Wireless LANs, Unsafe at Any Key Size; An Analysis of the WEP Encapsulation, 2000. http://citeseer.ist.psu.edu/558358.html
5. Arbaugh, W.A., Shankar, N., and Wan, Y.C.J., Your 802.11 Wireless Network Has No Clothes, Department of Computer Science, University of Maryland, College Park, MD, 2001.

6. Klaus, C.W., Wireless LAN Security FAQ, Version 1.7. http://www.iss.net/wireless/WLAN_FAQ.php

7. Fleck, B. and Dimov, J., Wireless Access Points and ARP Poisoning: Wireless Vulnerabilities that expose the Wired Network, Cigital, Inc. http://www.barbedwiretech.com/Technology/wp-pdf/BW-wifi_ARPPoison.pdf

8. Fleck, B. and Potter, B., *802.11 Security*, O'Reilly, 2002.

9. Fluhrer, S., Mantin, I., and Shamir, A., Weaknesses in the Key Scheduling Algorithm of RC4, *Proceedings of the 8th Annual International Workshop on Selected Areas in Cryptography*, 2001.

10. Gast, M., *802.11 Wireless Networks: The Definitive Guide*, O'Reilly, 2002.

11. Stubblefield, A., Ioannidis, J., and Rubin, A.D., Using the Fluhrer, Mantin and Shamir Attack to Break WEP, AT&T Labs Technical Report TD-4ZCPZZ, 2001.

12. Rivest, R.L., The RC4 Encryption Algorithm, RSA Data Security, Inc., San Mateo, CA.

13. Schneier, B., *Applied Cryptography,* Second edition, John Wiley & Sons, NY, 1996.

14. IEEE Standard for Local and Metropolitan Area Networks — Port-Based Network Access Control, IEEE 802.1X, 2001.

15. Blunk, L. and Vollbrecht, J., PPP Extensible Authentication Protocol (EAP), RFC 2284, March 1998. http://www.faqs.org/rfcs/rfc2284.html

16. Rigney, C., Willens, S., Rubens, A., and Simpson, W., Remote Authentication Dial-In User Service (RADIUS), RFC 2865, June 2000. http://www.ietf.org/rfc/rfc2865.txt

17. Aboba, B. and Simon, D., PPP EAP TLS Authentication Protocol, RFC 2716, 1999. http://www.faqs.org/rfcs/rfc2716.html

18. Funk, P. and Blake-Wilson, S., EAP Tunneled TLS Authentication Protocol (EAP-TTLS), April 2004. http://ietfreport.isoc.org/ids/draft-ietf-pppext-eap-ttls-05.txt

19. Cisco SAFE, Wireless LAN Security in Depth. http://www.cisco.com/warp/public/cc/so/cuso/epso/sqfr/safwl_wp.pdf

20. Mishra, A. and Arbaugh, W.A., An Initial Security Analysis of the IEEE 802.1X Standard, Department of Computer Science, University of Maryland, College Park, MD, 2002.

21. Jain, A.K., Bolle, R., and Pankanti, S., Eds., *Biometrics: Personal Identification in a Networked Society*, Kluwer Academic Publishers, Boston, MA, 1999.

22. Fingerprint Sensor Comparisons, Kinetic Sciences, http://www.kinetic.bc.ca/Biometrics/sensor-comparison.html

23. Jain, A.K., Hong, L., Pankanti, S., and Bolle, R., An Identity Authentication System Using Fingerprints, *Proceedings of IEEE*, Vol. 85, No. 9, pp. 1365–1388, 1997.

24. http://www.wi-fi.com/OpenSection/index.asp

25. Cam-Winget, N., Moore, Tim., Stanley, D., and Walker, J., IEEE 802.11i overview, presentation slides. http://csrc.nist.gov/wireless/S10_802.11i%20Overview-jw1.pdf

26. Edney, J. and Arbaugh, W.A., *Real 802.11 Security: Wi-Fi Protected Access and 802.11i*, Addison-Wesley, Boston, MA, 2003.

# 15

## WLAN SECURITY

*Salahuddin Qazi and Farhan A. Qazi*

### INTRODUCTION

Wireless local area networks (WLANs) have made it possible to connect mobile phones and handheld computing devices to critical business resources and applications on an anytime, anywhere access basis. This in turn has led to increased productivity and opportunities for businesses and consumers. However, a WLAN is subject to interference from other users because it operates in the unlicensed band, with a coverage range of about 30 to 300 m. It also suffers from lack of privacy due to flaws in the Wired Equivalency Protocol (WEP), which has weakened its security. WEP vulnerabilities arise from its 24-bit initialization vector (IV) and the absence of a cryptographic checksum. Additionally, it suffers from the Fluher–Mantini–Shamir (FMS) weakness, which allows a malicious attacker to discover the secret key and decrypt the data packets that are being passed along the exposed network.

The standards organizations have responded to WEP security vulnerabilities by working on a number of initiatives. One such initiative is currently addressed by Wi-Fi Protected Access (WPA) protocol, which is available as a firmware upgrade to the existing WEP devices. WPA is an interim solution and will be superseded by the full-blown IEEE 802.11i protocol when it is ratified. IEEE 802.11i uses a 48-bit IV, addresses the concerns of WEP by improving the method of RC4 implementation and adds automatic key settings. WPA also includes 802.1X server-based authentication using Remote Authentication Dial-In User Service (RADIUS) by supporting Extensible Authentication Protocol (EAP). This is helpful in stopping and tracking security breaches in larger Wi-Fi installations.

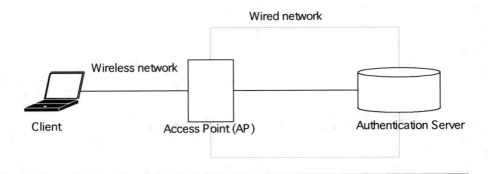

**Figure 15.1  WLAN topology.**

To further enhance the security of a WLAN, a new standard, IEEE 802.11i, is being developed that is expected to be ratified in 2004. IEEE 802.11i specifications are based on IEEE 802.1X for its scalable authentication, access control, and key-agreement framework. Its authentication and key-agreement functions can be implemented with a centralized authentication server, using RADIUS and EAP. The industry has also responded to the vulnerabilities in WEP and 802.11b WLAN security by developing a range of products to address these weaknesses. Numerous security tools and utilities are currently available for user authentication, intrusion detection, and generation of counterfeit 802.11b access points (APs) to confuse wardrivers, netstumblers, script kiddies, and other undesirable elements.

For greater security, an added layer of protection can be provided by biometrics, public key infrastructure (PKI), smart cards, and virtual public networks, when used on their own or along with another security solution.

## WLANS

WLANs are small-scale wireless networks with a range of about 100 m, used as a substitute for fixed LANs. Figure 15.1 shows a single-user model of WLAN topology, where a wireless client is attempting to access a stationary server via an AP, which is part of the wired network infrastructure. The most prevalent form of WLAN technology is called Wi-Fi, which includes 802.11a, 802.11b, and 802.11g standards. Wireless Internet via Wi-Fi offers high data speeds — 11 Mbps at the low end with 802.11b and 54 Mbps at the high end with 802.11a and 802.11g. Growth in wireless technologies for LANs has increased rapidly due to proliferation of mobile computing devices such as laptops, personal digital assistants (PDAs), handheld digital devices, and wearable computers. The age of PCs in which individual users utilize one computing device per person is moving to the age of ubiquitous computing in which individual users utilize, at the same time, several electronic platforms through which they can access all the required information whenever they need and wherever they may be. This "anytime anywhere" type of service is a very attractive technology for limited life-threatening situations in the areas of medical emergencies, nuclear power use, aviation, and defense. WLANs can provide connectivity to fixed, portable, and moving stations within a local area and also to a mobile device. WLANs can also provide ubiquitous connection to Internet information services. All these developments are giving rise to a wide-range usage of wireless communication technology in the public, private, and business sectors.

WLAN access can be achieved from points of communication in public hot spots, which is a relatively new phenomenon occurring in populated areas. Hot spots are scattered in places with a high demand for wireless connectivity, such as airports, hotels, and coffee houses. According to Gartner, Inc. (http://www3.gartner.com), the number of hot-spot users worldwide will total 30 million in 2004, and more than 50 percent of professional notebooks will have WLAN capability by the end of 2004. It is imperative that corporations take action in implementing a strategy that gives their employees secure access and control over the cost of access to hot spots.

The successful deployment of WLANs in recent years has yielded a demand to integrate WLANs with third-generation (3G) cellular networks such as GSM/GPRS, UMTS, CDMA2000, etc. Such integration is essential in making wireless multimedia and other high-data services a reality for large populations at strategic locations. The integration of 3G cellular and WLAN was started in 2003 by the Third-Generation Partnership Project (3GPP) and the Third-Generation Project 2 (3GPP2). The 3GPP is a joint initiative of the U.S., Japanese, European, and Korean telecommunications standardization organizations to produce global specifications for the Universal Mobile Telecommunication System (UMTS). 3GPP2 was set up to expedite the development of open, globally accepted technical specifications for the next generation cdma2000 wireless communications.[1,2]

New trends are emerging to migrate mobile networks to fully IP-based networks. The fourth-generation (4G) systems will utilize a new spectrum and emerging air interfaces to provide a bandwidth of over 10 Mbps and will use packet-switching technology. However, technical and political difficulties may ultimately prevent wireless technologies from reaching their full potential. The lack of universal industrywide standards, for example, holds back the technologies from delivering one of the true ideals of wireless, namely, the ubiquitous access to data.[3]

## IEEE 802.11

IEEE standardized WLANs in 1997 used the 802.11 standard, which is widely used for Wi-Fi WLANs. IEEE 802.11 specifies an over-the-air interface between a wireless client and a base station or between two wireless clients. IEEE 802.11 is limited to the physical layer (PHY) and medium access control (MAC) sublayer. The physical layer of IEEE 802.11 has the RF technologies — direct sequence spread spectrum (DSSS), frequency hopped spread spectrum (FHSS), and infrared. There are several specifications in the IEEE 802.11 WLAN family:

- *IEEE 802.11* provides 1 or 2 Mbps transmission in the 2.4-GHz band, using FHSS or DSSS.
- *IEEE 802.11a* is an extension of 802.11 that provides up to 54 Mbps in the 5-GHz band, using the orthogonal frequency division multiplexing (OFDM) encoding scheme.
- *IEEE 802.11b* is an extension of 802.11 that provides 11-Mbps transmission (with a fallback to 5.5, 2, and 1 Mbps) in the 2.4-GHz band. 802.11b uses only DSSS and was ratified in 1999.
- *IEEE 802.11g* provides 54 Mbps in the 2.4-GHz band by using OFDM.

Other IEEE 802.11 standards that address specific issues are given below:

- *IEEE 802.11d* addresses the spread of technology to countries not addressed by IEEE.
- *IEEE 802.11e* focuses on the quality of service (QoS) associated with wireless transmission of multimedia applications.
- *IEEE 802.11f* addresses roaming between APs and interoperability between different vendor groups.
- *IEEE 802.11h* focuses on frequency selection and power concerns on the 5-GHz band in European countries.
- *IEEE 802.11i* focuses on enhancing WLAN security and authentication for 802.11 that include incorporating RADIUS, Kerberos, and the IEEE 802.1X.
- *IEEE 802.11j* focuses on the Japanese equivalent of 802.11h. Japan has authorized frequency band (4.9–5 GHz) for 802.112-like functionality.
- *IEEE 802.11n* focuses on increasing the speed of 802.11a/g WLAN beyond 1000 Mbps. Completion in 2005 or 2006 is anticipated.

Other proposed IEEE 802.11 standards for future WLAN technology include 802.11p (wireless performance prediction), 802.11r (short-range communications), and 802.11s (mesh networking).

IEEE 802.11 WLANs consist of two types of modes: infrastructure mode and ad hoc mode. In the infrastructure mode, each client or station communicates through an AP by sending a packet to it, and the AP transmits the packet to the destination client, by way of a wired LAN. The AP is generally connected to the wired network and the setup of this WLAN provides wide-area mobility without losing connectivity. An ad hoc network is a collection of wireless mobile clients dynamically forming a temporary network without the use of an AP or existing network. In this mode, each client communicates directly with other clients in the RF range. Clients, such as laptop computers, communicate with each other using multi-hop wireless links in which each client in the network also acts as a router, forwarding data packets for other nodes.

Ad hoc networks are used when a network needs to be established for a short duration, whereas the infrastructure mode is used when the user wants to be connected to a wired network such as a local Ethernet or an Internet connection. This chapter will focus on the infrastructure mode because it is more commonly used and offers a much better platform for building security than ad hoc networks, which present a much more challenging task.[4,5]

## IEEE 802.11 WLAN IN INFRASTRUCTURE MODE

The working of a WLAN in the infrastructure mode can be explained by considering a classic 802.11 state machine, as shown in Figure 15.2. A state machine is a system with a set of unique states, which are connected by transitions. To create network connectivity in a WLAN, a client must establish a relationship with an AP, called an association. After the association is established, the client and the AP can exchange data. The association with an AP involves transition among the three following states:

1. Unauthenticated and unassociated
2. Authenticated and unassociated
3. Authenticated and associated

There are two basic types of 802.11 frames: a management frame and a data frame. A client moves (transitions) between the states, using specific management frames. The first step in establishing communication between the client and the network is to find an AP in the infrastructure mode or a basic service set (BSS). This takes place in the first state, unauthenticated and unassociated. To achieve this and to associate with an AP, a client listens for a beacon management frame, which is within range, that is transmitted by the APs at a fixed interval. To find an AP affiliated with a desired Service Set ID (SSID) a client may also send a probe request

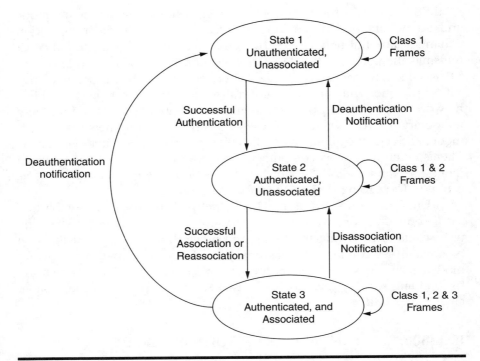

**Figure 15.2   802.11 state machine.**

management frame. After finding an AP, the client and AP perform mutual authentication by exchanging several management frames, using either an open systems authentication or a shared-key authentication mechanism. After successful authentication, the client moves to the second state, authenticated and unassociated. The client gains access to the wireless infrastructure network by moving from the second state to the third state, authenticated and associated. This process requires the client to send an association request frame, and the AP to respond with an association response frame.[6,7]

## SECURITY MECHANISM

With the popularity of WLANs growing, security concerns have also become an important issue. *Security* can be defined as the ability of a system or a network to withstand an attack. In a wireless network, unlike in a wired network, an attacker could be in an unsecured location such as a parking lot or a passing car. There are two types of security mechanisms, preventive and detective. Most of the preventive mechanisms use cryptography as a building block. Three steps are followed to accomplish

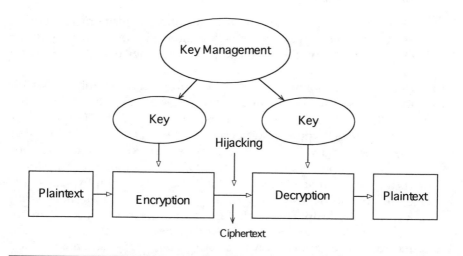

**Figure 15.3  Basic cryptographic system.**

security. The first step is identification, which names entities; the second step is authentication, which checks that an entity is who or what it claims to be; and the third step is authorization, which either grants or refuses access rights based on some security policies, which are a part of an organization's policy. Policies usually define access control rules and translate the trust that is placed on entities into access control decisions. Different cryptographic techniques for preventive security are discussed in the next section.

## Cryptographic Techniques

Cryptography is defined as the study of creating and using encryption and decryption for providing information security. A cryptosystem (for encrypting and decrypting data) usually involves an algorithm for combining the original data plaintext with one or more keys, which are numbers or strings of characters known only to the sender or the recipient. The resulting output is known as ciphertext, which is shown in Figure 15.3. A strong cryptosystem will produce a ciphertext that appears random in all standard statistical tests and resist all previously known methods (known as cryptanalysis) for breaking codes. Cryptographic algorithms are functions that transform information to hide it. There are three types of cryptographic techniques:[8]

1. *Secret-key cryptography:* This uses one secret-key cipher for encryption and decryption of a message. Secret-key cipher is also called *symmetric cipher* (or conventional cipher). The same key is used

to encipher a message and then to decipher the resulting ciphertext. Secret-key encryption provides confidentiality because only those entities that know the secret key can decipher the plaintext messages.

2. *Public-key cryptography:* This is a form of cryptography in which the key used for encrypting is not the same as the key used for decrypting. In other words, it uses a pair of keys (private and public) for encrypting and decrypting messages. This is also called *asymmetric cipher.* A public-key cipher uses one key to encipher plaintext (public key) into ciphertext and a different key to decipher (private key) that ciphertext. Public-key cryptography can overcome the problems of key distribution. Because the enciphering key cannot decipher the ciphertext, the enciphering key can be exposed (for example, on a Web page) without revealing the message. The exposed key is called the *public key*, and the retained hidden key is called the *private key*. The public key is distributed widely, so anyone can use it to encipher a message that presumably can only be deciphered by someone who has the hidden private key.

3. *Cryptographic hash function:* A hash algorithm is a function that transforms an input of variable size or a string of characters into a shorter fixed-size string value or a key that represents the original string. When used as cryptographic hash function, the algorithm must meet additional requirements that include the following:
   ■ The input can be of any length, but the output must be fixed length (usually smaller in size).
   ■ The hash value must be easy to compute.
   ■ The hash value must be one way, making it difficult to recreate the data based solely on the hash value.
   ■ The hash value must be collision free so that identical hash values for two random datasets must be difficult to find.

Cryptographic hash function is used for detecting integrity violations by first computing the hash value for a given dataset. A new hash value for the same data is then computed at a later time and compared to the previous value. If the two values are unequal, the data is said to be modified. It is also used to encrypt and decrypt digital signatures and to authenticate message senders and receivers. Hash algorithms do not use keys.[8]

## Key Management and Key Distribution

Key management is an essential part of any security based on cryptography. A cryptographic protocol uses keys to authenticate entities and grant access to guarded information to those who exhibit their knowledge of the keys. When private keys may have been exposed or just used for a long time, it is important to change those keys. In changing them, it is important that the keys be securely generated and distributed to appropriate entities.

Key distribution is fundamental to protocols that use cryptography for security. It is the process of ensuring that both the sender and receiver have access to the key required to encrypt and decrypt a message, while making sure that the key does not fall into enemy hands. Public keys are public knowledge and are distributed through certificates. A certificate binds a public key with an entity, and certificates are certified, stored, and distributed by one or more trusted parties. There are two ways to distribute secret keys: the preestablished secure channel and the open channel. Key distribution was a major problem in terms of logistics and security before the invention of public-key cryptography.

## THREATS AND ATTACKS

Attackers in WLANs, unlike those in wired networks, do not need physical access to the wires. For example, an attacker sitting in the parking lot of a building can attack the building's WLAN security. The attacks are typically divided into passive and active attacks, as discussed in the following text.

In a passive attack, an unauthorized party gains access to the network but does not modify its contents. The attacker does not have to be a part of the network to listen to whatever is going on in it. Furthermore, the attacker cannot fiddle with the network and such an attack has no risks. The two passive attacks are listed below:

■ Eavesdropping
■ Traffic analysis

In an active attack, an unauthorized party makes modifications to the information or the datafile. The attacker has to get into the network to modify or corrupt the data. The attackers can hijack, interrupt, and control the network at their will. There are four different types of such attacks:

■ Replay
■ Modification
■ Masquerading
■ Denial-of-Service

These attacks give rise to security risks and are threats against the 802.11 wireless networks. The consequences of these attacks include loss of network service and loss of proprietary information, resulting in user inconvenience and legal and recovery costs. Details of attacks for WLANs are discussed in the section titled "Weaknesses and Security of WEP."

## REQUIREMENTS FOR SECURITY

Secure communication based on encryption or decryption, authentication, and integrity are important features of network security. However, more security is needed in WLANs infrastructure to prevent denial-of-service attacks from malicious users and potential hackers. To achieve comprehensive security for a distributed wireless network system, additional countermeasures (discussed in the section titled "Countermeasures in WLANs") are needed that mitigate the security risks posed by some of these attacks. The important requirements for secure communications leading to network security are illustrated by the following:[9,10]

- Availability, which implies uninterrupted services. It ensures the continuity of network services even if the node of the network does not behave in an expected manner.
- Authentication, which means knowledge of the identity of a communicating party or the source of a piece of information. It helps to ensure that the node identifies the peer node that it is communicating with.
- Confidentiality, which deals with disclosure of information. It ensures that certain information is never divulged to unauthorized entities.
- Integrity, which means no unauthorized modification of resources. It guarantees that a message is not modified while transmitted.
- Non-repudiation, which means nondeniability of committed actions. It ensures that the originator of a message cannot deny having sent the message.
- Authorization, which establishes the rules that define what each network node is allowed or not allowed to do.

## FIRST-GENERATION SECURITY OF IEEE 802.11 STANDARD[11-13]

In comparison with a wired network, it is easier to gain access to the wireless AP for infrastructure networks by being in its proximity, such as in a parking lot adjacent to the building. Wireless networks, as a result, need secure access to the AP and the ability to isolate it from the internal private network prior to user authentication into the network domain. There are three basic methods to secure access to APs that are built into IEEE 802.11 networks, and these are best suited to small networks:

- SSID
- MAC address filtering
- WEP

## SSID

Network access control, a measure of security, can be implemented using an SSID associated with an AP or group of APs. This is a unique, case-sensitive, alphanumeric, 32-character long key used by WLANs to differentiate two networks logically. An SSID-based mechanism segments a wireless network into multiple networks serviced by one or more APs. Each AP is programmed with an SSID corresponding to a specific wireless network, and to access this network, the client must be configured with the correct SSID. Because a client computer must present the correct SSID to access the AP, the SSID acts as a simple password and thus provides a measure of security. The SSID, however, does not provide any data privacy function or a way to authenticate the client to the AP. In addition, this minimal security is further compromised if the AP is configured to broadcast its SSID. The SSIDs are widely known and easily shared when users configure their own client systems with the appropriate SSIDs. Even in the absence of SSID broadcast, the AP will answer a client requesting the identity of the SSID.

## MAC ADDRESS FILTERING

This is another way that has been used to secure networks over and above the IEEE 802.11b standards. A client computer can be identified by the unique MAC address of its IEEE 802.11 network card. It is a 12-digit hexadecimal number that is unique to each and every network card in the world. To increase the security of an IEEE 802.11 network, each AP can be programmed with a list of MAC addresses associated with the client computers that were allowed to access the AP. If a client's MAC address is not included in this list, the client is not allowed to associate with the AP.

The first problem with MAC address filtering is the management and upkeep of an updated database of MAC addresses, and the list of every device allowed by the LAN administrator to access the network. This database must be kept either on each AP individually or on a special RADIUS server that each AP looks at any time a device is added, lost, stolen, or changed in any way. The WLAN administrator must update the database of allowed devices. In an enterprise network with hundreds or thousands of devices, this is simply not a practical solution. As a result, MAC address filtering and SSID only improve the security and are best suited to small networks in which the MAC address list can be efficiently managed. The second problem is in the spoofing of MAC addresses by an attacker, who can capture them and gain unauthorized access to the network.

## WEP

To minimize the risk of being easily intercepted in wireless transmission, IEEE 802.11 specified WEP for encryption and authentication. It was included in the original IEEE 802.11 standard in 1997 and was widely deployed when IEEE 802.11b and Wi-Fi became established in 1999. WEP, however, suffers from several security weaknesses. Even with its weaknesses, WEP is the only choice until new security methods are established. WEP security as described in the standard has two parts. The first is the authentication phase, and the second is the encryption phase. The goals of WEP include access control and privacy, as defined in the following list:[11]

- Access control is achieved by preventing unauthorized users who lack a correct WEP key from gaining access to the network.
- Privacy is obtained by protecting WLAN data streams by encrypting them and allowing decryption only by the users with the correct WEP keys.

### Authentication

A client or an AP must be authenticated to communicate or participate in a WLAN. The process of authentication is a security measure that can prevent unauthorized IEEE 802.11 devices from joining the wireless network. The method of authentication must be set on each client, and the setting should match that of the AP with which the client wants to associate. The IEEE 802.11b standard defines two types of authentication: the open system and shared-key.

- Open system authentication is a basic form of authentication that consists of a simple authentication request containing the station ID and an authentication response containing success or failure. In open system authentication, any device can participate in the wireless network, provided the station's SSID matches the AP's SSID. It can also use any SSID option to associate with any available AP within range despite its configured SSID. The whole authentication process is performed in cleartext, and a client can associate with an AP even without supplying the correct WEP key. On

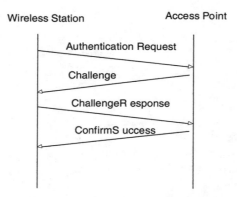

**Figure 15.4   Shared-key authentication.**

success, both stations are considered to be mutually authenticated. Open system authentication is usually used as default.

■ In shared-key authentication, the AP and station are required to have the same WEP "shared" key to authenticate. It is assumed that this key has been transmitted to both stations through some secure channel other than the IEEE 802.11 channel. In typical implementations, this might be set manually on the client station and the AP. In shared-key authentication, the AP sends the client a challenge text packet that must be encrypted by the client with the correct WEP key and returned to the AP. The one-way authentication will fail, and the client will not be allowed to associate with the AP if the client has the wrong key or no key. To obtain mutual authentication, the process is repeated in the opposite direction.

Shared-key authentication is a poor choice of authentication because it provides exactly the information needed to defeat WEP encryption. However, the shared-key authentication approach provides a better degree of authentication than the open system approach. To utilize shared-key authentication, a station must implement WEP to share the secret key, which is available only to the MAC coordinator. Figure 15.4 shows the operation of shared- key authentication.

## Encryption

WEP security is obtained by providing encrypted transmission between the client and an AP. This is achieved by two processes — first, to encrypt the data or the plaintext, and second, to protect it from tampering or

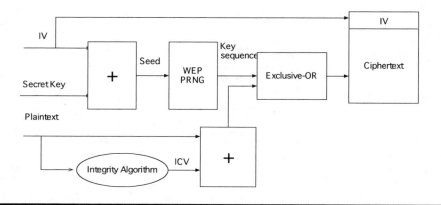

**Figure 15.5a    WEP encryption**.

unauthorized data modification. WEP uses Rivest Cipher 4 (RC4) as an encryption algorithm, which was designed by Ron Rivest in 1987. It remained a trade secret till 1994, when it was made public anonymously by reverse engineering.

WEP encryption process uses a 40-bit or 104-bit secret key that is used by the sender and the receiver. RC4 uses the same algorithm for encryption and decryption. An IV of 24 bits is combined with the secret key, which forms the seed for the input to the pseudorandom number generator (PRNG). The IV is different for every transmitted packet and its purpose is to confuse the attacker as to how the message is encrypted. The PRNG generates a pseudorandom key sequence based on the input key and is used to encrypt the data by performing a bit-by-bit exclusive OR (XOR) operation. The resulting sequence is ciphertext, which is transmitted to the receiver by attaching it with the IV. Each time the IV changes, the PRNG sequence also changes. Integrity check value (ICV) is produced by using an integrity algorithm CRC-32 on the plaintext for protection against unauthorized data modification. The plaintext, IV, and ICV form the actual data sent in the data frame. Figure 15.5a shows the WEP encryption process.

In decryption, shown in Figure 15.5b, the receiver uses the IV of the incoming messages to generate the same key sequence used by the sender to decrypt the message. The original plaintext and ICV is recovered by combining the ciphertext with the proper key sequence. The receiver verifies the decryption by performing the integrity check algorithm on the recovered plaintext and comparing the incoming ICV to the one at the output of the receiver. If the two are not equal, the receiver rejects the data, and an error indication is sent back to the sending station.

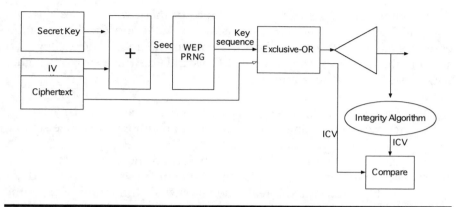

**Figure 15.5b   WEP encryption.**

## Weaknesses and Security of WEP

Many flaws and weaknesses have been reported[14,15] in the WEP algorithm. The first problem is due to the use of a 40-bit symmetric stream cipher (RC4) for WEP encryption, which used 24-bit IVs to generate different secret keys for different data packets. These IVs were supposed to be unique for each packet, but the space was too small to avoid duplications. As a result, the reuse of IV allowed the attacker access to the plaintext. The second problem was identified by Fluhrer, Mantini, and Shamir, who found a flaw in the key-scheduling algorithm of RC4 that made certain RC4 keys fundamentally weak and easy to breach. They designed an attack that enabled a passive listener to recover the secret key by simply collecting a sufficient number of frames encrypted with weak keys. Listed below are some of the problems and weaknesses of WEP and 802.11b WLAN:[16,17]

- Implementation of WEP is not mandatory, and security features in vendor products are frequently not enabled.
- IVs are short and are not protected from reuse.
- Cryptographic keys are short and shared, and they are susceptible to FMS attacks.
- Cryptographic keys have no built-in provisions to update the keys automatically and frequently.
- There is no effective detection of message tampering (or message integrity is poor).
- RC4 is inappropriately used and provides weak keys.
- There is no protection against message replay.

A number of attacks can result because of these weaknesses,[18] which apply to both 40-bit and 128-bit versions of WEP and can be mounted easily using inexpensive off-the-shelf equipment. In addition, they also apply to the networks using IEEE 802.11b standard, which supports higher data rates. These attacks are given below:

- Passive attacks to decrypt traffic based on statistical analysis
- Active attacks to inject traffic from an unauthorized mobile station, based on known plaintext
- Active attacks to decrypt traffic, based on tricking the AP
- Dictionary building attacks that, after analysis of about a day's worth of traffic, allow real-time automated decryption of all traffic

## NEW-GENERATION SECURITY STANDARDS OF WLANS[19–25]

To address the insecurities and weaknesses of IEEE 802.11 WLANs and improve WLAN security, standards organizations such as the Internet Engineering Task Force (IETF) and IEEE have been working with a number of new initiatives. The first initiative by the IEEE 802.11 Task Group i (TGi) to enhance the security of Wi-Fi involved an immediate software-upgradable security solution, a WPA standard, and a new addendum to the IEEE 802.11 standard called IEEE 802.11i. WPA is an interim standard introduced in 2003 to replace the easily crackable WEP. The second initiative for improving WLAN security is the short-term solution to address the problems of WEP by defining the Temporal Key Integrity Protocol (TKIP). Its goal is to remove well-known vulnerabilities by using firmware and software drivers on the existing Wi-Fi certified hardware before the completion of IEEE 802.11i standard. The third initiative is the introduction of a new standard, IEEE 802.1X, which defines a generic framework for port-based access control and key distribution. It also addresses the serious omission in the WEP standard by providing secure delivery of session keys.

## IEEE 802.1X

IEEE 802.1X is a new standard for access control for wireless and wired LANs and was introduced in 2001 by the IEEE 802.11 working group. It is designed to provide controlled port access between wireless client devices, APs, and servers. IEEE 802.1X uses dynamic keys, instead of static keys as used in WEP and EAP, for user authentication to connect with both the wired and WLAN media. There are a number of variations of EAPs, as follows:

- LEAP: Cisco proprietary EAP implementation
- EAP-TLS: IETF open standard with maximum vendor support

- EAP-MD5: IETF open standard with minimal security
- EAP-TTLS: Funk software's IETF open standard with high security
- PEAP: Cisco/Microsoft/RSA IETF open standard with high security

IEEE 802.1X authentication for WLANs has the following components:

- Compatible client device, which typically includes laptops and PDAs
- Supplicant software, which is usually the client software to present its credentials for joining the network as a client (A device that seeks to join a wireless network is called *supplicant*.)
- Authenticator, which is usually the AP that must verify the identity of a supplicant before granting network access to it
- Authentication server, usually a RADIUS server, although RADIUS is not specifically required by 802.1X
- User database, which is a list of valid users and their credentials that the authentication server consults to validate authentication requests

The authentication process in IEEE 802.11X starts when a client attempts to connect to the AP. The AP opens the restricted port and allows the client to pass EAP packets to the authentication server on the wired side of the AP. The port blocks all types of traffic. A central authentication server, RADIUS, is used, which also employs mutual authentication so that a wireless user does not accidentally join a rogue network that might steal its network credentials. IEEE 802.1X supports multiple authentication methods such as token cards, Kerberos, one-time passwords, certificates, and public-key authentication.

## TKIP

TKIP (pronounced tee kip) is a quick-fix method of overcoming the inherent weaknesses in WEP security, especially the reuse of encryption keys. It was introduced in 2001, after WEP was broken, to allow WEP systems to be upgraded and become secure. It is obtained by adding a per-packet key mixing, a message integrity check, and a rekeying mechanism, thus fixing the flaws of WEP. This is accomplished by adopting a security process with a 128-bit temporal key that is shared among clients and APs. TKIP combines the temporal key with the client machine's MAC address and then adds a relatively large 16-octet IV to produce the key that encrypts the data. This procedure ensures that each station uses a different keystream to encrypt the data. Similar to WEP, TKIP uses RC4 to perform the encryption. A major difference from WEP, however, is that

TKIP changes temporal keys every 10,000 packets. This provides a dynamic distribution method that significantly enhances the security of the network. TKIP solves the problem by not changing the major components, as is the way RC4 is implemented, but by adding a series of corrective tools around the existing hardware. The following changes are made to overcome the weaknesses of the WEP outlined in the section titled "Weaknesses and Security of WEP":[17]

■ Changing the encryption key for each frame
■ Increasing the size of IV to avoid reuse of the same IV
■ Changing the rules for how IV values are selected and reusing the IV as a replay counter
■ Adding a message integrity protocol to prevent tampering that can be implemented in software on a low-power microprocessor
■ Adding a mechanism to distribute and change the broadcast keys

## WPA

WPA is an industry-supported, prestandard version of 802.11i utilizing the TKIP, which fixes the problems of WEP, including use of dynamic keys. It was put together by the Wi-Fi Alliance as a data encryption method for 802.11 WLANs. WPA will serve until the 802.11i standard is ratified at the end of 2004. It was introduced in 2003 and is designed to improve the weak data encryption feature of WEP and provide user authentication, which was largely missing in WEP. It is a specification of standards-based, interoperable security enhancement that is applicable for both home and large-enterprise users. WPA is a subset of current 802.11i draft, taking already implemented pieces of the draft such as 802.1X and TKIP to the market. WPA utilizes TKIP to improve data encryption, which includes a prepacket key-mixing function, a message integrity check (MIC) named Michael, an extended IV with sequencing rules, and a rekeying mechanism. WPA uses an IV increased from 24 bits to 48 bits, which makes it harder to crack. It also uses a checksum security technique by checking the validity of an 8-bit message integrity code within the frame and also by testing the 802.11's frame 4-byte integrity check value. These enhancements help alleviate all the known vulnerabilities of WEP.

## IEEE 802.11I

IEEE 802.11i is a supplemental draft standard expected to be ratified in the Fall of 2004 to further improve the authentication and encryption gains implemented by WPA. It also defines the concept of a Robust Security

Network (RSN) by making extensive use of protocols above the IEEE 802.11 MAC layer to provide authentication and key management. There are three main areas in the security of IEEE 802.11b, which were lacking in the WEP used by it, that IEEE 802.11i intends to improve on:

- Authentication
- Key management
- Data transfer

In the 802.11i standard, unlike in WPA, key management and message integrity are handled by a single component CCMP (Counter mode/CBC-MAC Protocol) built around AES. The counter mode is used for data encryption and the CBC-MAC (cipher block chaining-message authentication code) ensures message integrity. It includes the authentication scheme of 802.1X and EAP in addition to enhanced security features such as a new encryption scheme and dynamic key distribution. EAP and RADIUS are used for new packet security methods of key distribution that replace the broken WEP algorithm. RADIUS is an authentication, authorization, and accounting (AAA) protocol that can function as an EAP transport mechanism for user authentication. However, RADIUS has several limitations in the areas of security, roaming support, robustness, and server-initiated operations.

To overcome these limitations, the IETF has specified DIAMETER as the AAA protocol successor to RADIUS. DIAMETER provides many functions in addition to those specified in RADIUS, making the former a good alternative to the AAA purposes in IEEE 802.11i once its standardization is ratified.[2] TKIP is used to produce a 128-bit temporal key to encrypt data using different keys by different stations. It introduced a novel key generation function for encrypting all data packets sent over the air, with its own encryption key, which greatly increases the complexity and difficulty of decoding the keys. These new standards are more complicated than the existing wireless security mechanisms but are more secure and are scalable to larger networks. The following are some of the improvements and potential problems:[23]

*Improvements:*

- Better security by way of encryption using AES
- Stronger key management
- Immune to man-in-the-middle attacks by use of two-way authentication
- Better message integrity performance by virtue of using keyed MICs

**Table 15.1   Comparison of IEEE 802.11i, WEP, and WPA Security Standards**

| Standard | WEP | WPA | 802.11i |
|---|---|---|---|
| Cipher algorithm | RC4 | RC4 | Rijndael (AES) |
| Encryption key length | 40 bit and 104 bit | 128 bit | 128 bit |
| IV length | 24 bit | 48 bit | 48 bit |
| Packet key | Concatenated | Mixing function | Not needed |
| Data integrity | CRC-32 | Michael | CCMP |
| Key management | Static | 802.1X(EAP) | 802.1X(EAP) |
| Header integrity | None | Michael | CCMP |
| Replay attack | None | IV sequence | IV sequence |

*Potential Problems:*

- Costly to implement on the preexisting networks because the new encryption CCMP. AES requires hardware upgrade.
- Vulnerable to security risk because of its heavy reliance on secrecy session keys.
- Needs additional hardware because of the use of two-way authentication.
- More complicated than the existing systems.

Table 15.1 compares IEEE 802.11i with WEP and WPA security standards. WPA fixes all the problems of WEP but does not include advanced features of the IEEE 802.11i.[24]

At the time of writing this chapter, the ratification of the 802.11i draft was in the final stages of approval and was reported to receive widespread vendor support and cooperation by multiple standard bodies including IEFT, IEEE 802, 3GPP, 3GPP2, and GSMA.[25]

## RSN

RSN is a new type of wireless network that makes use of protocols above the IEEE 802.11 MAC sublayer to provide authentication and key management. An RSN provides a number of additional security features, which are not present in the basic IEEE 802.11 architecture. The features include the following:

- Enhanced authentication mechanisms for both APs and wireless clients
- Enhanced data authentication mechanisms such as TKIP and WRAP
- Secure capabilities negotiation including ciphersuits and authentication methods
- Dynamic, association-specific cryptographic keys and key-management algorithms
- Protection of management and control frames

For subsequent upgrades, the IEEE 802.11i committee has also defined a transitional specification called Transitional Security Network (TSN), which allows RSN and older WEP systems to operate in parallel in the same WLAN. However, the wireless network will not be fully secure until it is completely an RSN. RSN and WPA have a lot in common. Whereas WPA has a subset capability focused on a single way to implement a network, RSN allows more flexibility in implementation. SRN and WPA use the same security architecture for upper-level authentication, key distribution, and key renewal. WPA is built around TKIP, however, which is available as a firmware upgrade to most legacy hardware. RSN is more comprehensive and includes support for AES, which is available only on the latest WLAN hardware.

## AES

This is an optional replacement for RC4 in the IEEE 802.11i. AES is a NIST-standard secret key cryptography method that officially replaced the Triple DES (Data Encryption Standard) method in 2001. It uses a mathematical ciphering algorithm employing variable key sizes of 128, 192, or 256 bits, making it more difficult to decipher than WEP. AES uses the Rijndael algorithm developed by Joan Daemen and Vincent Rijmen of Belgium. It can be encrypted in one pass instead of three, and its key size is greater than Triple DES's 168 bits. In addition to the increased security that comes with larger key sizes, AES can encrypt data much faster than Triple DES. This encryption protocol requires a new encryption and decryption chipset and AES; however, it is not readily compatible with today's WI-Fi-certified LAN devices.

### Counter-Mode with CBC-MAC Protocol (CCMP)

802.11i, including both RC4-based encryption and the AES-based algorithm, is called the *CCMP*. This is based on a FIPS (Federal Information Processing Standard)-approved cipher and is expected to provide network security at the link layer. CCMP's design criteria include wireless network confidentiality, integrity, and authentication. It uses the 128-bit AES for data protection, in which data is encrypted in 128-bit chunks using the block chaining (CBC) mode and provides data integrity checks via a MAC. CCMP uses a 48-bit IV to seed the initial key deviation process and to seed the MIC used in CCMP packets. The input to CCM includes three elements:

1. Authenticated and encrypted data, called the payload
2. Associated data such as a header, for authentication but not encryption
3. A unique value called a *nonce*, assigned to the payload and the associated data

There are two related pairs of processes pertaining to CCM. The first is "generation encryption" and the second is "decryption verification." These processes combine two cryptographic primitives, which consist of Counter-mode encryption and cipher block chaining-based authentication.

## COUNTERMEASURES IN WLANS[5,11,16,26–29]

The medium in which WLANs propagates is inherently insecure, and this will always be a security risk. Security in the existing IEEE 802.11 specifications have also come under intense scrutiny because of several vulnerabilities in the authentication, message integrity, and data privacy mechanisms. The risks WLANs face, however, can only be mitigated and not eliminated completely. IEEE and IETF have come up with several initiatives to address these vulnerabilities, as described in the previous section. There is, however, not one particular method that is a good solution for all the risks in a network security, but a combination of different methods may be the best solution. The following countermeasures may be applied to further address the threats and vulnerabilities against WLANs and to provide an added layer of protection when used on their own or along with another security solution. These countermeasures may be divided into the following categories:[16]

- Management
- Operational
- Software
- Hardware

## MANAGEMENT COUNTERMEASURE

Management and operational countermeasures are nontechnical security practices that could be used regardless of the security offering built into the standards. Management countermeasures deal with the implementation of a comprehensive policy for securing wireless networks. Many organizations either do not have a security policy or have a weak one. Implementation of a security policy and its compliance form the basis for other countermeasures. In implementing a comprehensive management countermeasure, it is important to ensure that all critical personnel are properly trained on the use of wireless technology and the security risks that WLANs and wireless devices pose. A WLAN security policy derived from NIST wireless network security draft specifies the following:[16]

- Identify the personnel who may use WLAN technology in an organization and whether Internet access is required.
- Describe the personnel who can install APs and other wireless equipment, the type of information that may be sent over wireless links, the conditions under which wireless devices are allowed, the hardware and software configurations of any access device, and the limitations on the use of wireless devices, such as location.
- Define standard security settings for APs and the frequency and scope of security assessments.
- Provide guidelines on reporting losses of wireless devices and security incidents and on the use of encryption and other security software. Provide also limitations on the location of and physical security of APs.

## OPERATIONAL COUNTERMEASURE

Operational countermeasure deals with physical security, which means that security risks can be minimized by giving physical access to only authorized users. It is the most basic measure to ensure security by protecting the WLANs from intruders. There are three types of physical security measures that are often used to mitigate risks. The first is access control, which ensures that only authorized users have access to the WLAN network. The second is personnel identification, which is required because physical access alone is not enough for WLAN security. Personal identifications such as ID cards, photo badges or biometric devices, and logins and passwords are also used to access the WLAN equipment, and they reduce the risk of unauthorized users. The third is the external boundary protection, which can be achieved by locking doors and installing video cameras for surveillance around the building where the WLAN equipment and APs are located. The location and range of APs is an important factor because in their vicinity RF radiation can be picked up easily. It is desirable to place the AP inside the building in such a way that its range does not extend beyond the boundary of the building to allow unauthorized personnel to eavesdrop.

## SOFTWARE COUNTERMEASURES

Software and hardware countermeasures are technical countermeasures as opposed to the two previous ones. These are dependent on the wireless industry and standards organizations to developing the products and helping to implement the standards for WLAN security.

Software countermeasures are achieved by configuring APs, implementing authentication and IDS solutions, performing security audits, adopting effective encryption, and by using software patches and upgrades. The implementation of software countermeasures involves the following:

- AP configuration
- Authentication
- Encryption
- Intrusion detection system
- Personal firewalls
- Software patches and upgrades
- Security assessments

## HARDWARE COUNTERMEASURES

Unlike software countermeasures, these require additional hardware, which makes the existing system more expensive and often more complicated. The need for such countermeasures should be based and determined from the level of security required in a certain application. Hardware countermeasure include virtual public networks (VPNs), the use of PKI, smart cards, and biometrics, which are described briefly in the following subsections:

### VPN

This is the extension of a private network that provides an encrypted connection between a user's remote sites distributed over a public network such as the Internet. In the remote-user applications, VPN provides a secure dedicated path (or tunnel) over an untrusted network such as the Internet. A virtual network can be used for secure wireless access where the untrusted network is the wireless network. Most of the VPNs currently use the IPSec protocol suite that was developed by the IETF. IETF is a framework of open standards for enabling private communications over IP networks, and it provides robust protection by the use of PKI, smart cards, and biometrics.

## PKI

This is the use and management of encryption technologies, software, and services that enable secure transmission and authentication of data across public networks. For applications requiring higher levels of security, WLANs can integrate PKI for authentication to secure network transactions because it provides strong authentication through user certificates. Furthermore, the user can use the same certificates in application-level security, such as signing and encrypting messages. However, for lower levels of security, the user should consider the cost of complexity and implementation. Wireless PKI, handsets, and smart cards that integrate with WLANs are available from third-party manufactures.

## Smart Card

Smart cards are portable media used in systems to store and maintain secure data. This data can be protected by the user via methods such as a personal identification number (PIN) or through biometrics. In wireless networks, smart cards provide the added feature of authentication, although they also add another layer of complexity. These devices are beneficial in situations that require authentication beyond simple usernames and passwords. Furthermore, they are portable, and hence users can securely access their networks from various locations. Organizations can use smart cards in a two-factor authentication by combining it with biometrics.

## Biometrics

*Biometric authentication* can be defined as the process of verifying human identity based on different physiological or behavioral characteristics or traits. For higher levels of security, biometrics can be integrated with wireless laptops, wireless smart card, or other wireless devices and used to authenticate the user to access the wireless network. Biometric authentication is considered more reliable than common forms of authentications such as passwords and access codes because it makes it necessary for the persons to be identified to physically authenticate themselves. It also helps to reduce forgery of passwords and to alleviate the problems of remembering a password and carrying around some authentication device. Some of the popular types of biometric systems use face recognition, fingerprinting, palm prints, and iris and retinal scanning.

Biometric authentication combined with encryption is being used to improve the security of encryption systems. To make brute force attacks obsolete, a biometric key with a person's unique personal identification can be added to or can replace the normal encryption key. Biometric encryption also makes key management unnecessary because the encryption key becomes a unique physical characteristic of a person and is hard to break. When a user wishes to access a secure key, that person will be prompted for a biometric sample. The key is released and can be used to encrypt or decrypt data if the verification sample matches the enrollment template.

## CONCLUSION

WLAN is a growing technology with many applications. However, WLANs based on the IEEE 802.11 standard suffer from security risks, weaknesses of WEP, and vulnerabilities due to the insecure medium in which it propagates. New and interim standards for enhancing the security of IEEE 802.11 WLANs on existing WEP devices using firmware and software are currently being used. The future of IEEE 802.11 WLANs security is, however, addressed by the development of 802.11i standard, which focuses strictly on security, and improvement of the protocols offered by the existing WLAN standards. IEEE 802.11i is designed to be more secure, scalable, and complicated, and will require additional hardware. Countermeasures against WLAN vulnerabilities and attacks help to mitigate the security risks. These countermeasures can be used as an added layer of security when used either alone or along with another security solution. New trends such as biometric encryption and integration of 3G cellular and wireless LANs are important developments toward the growth of ubiquitous high-speed wireless data with mobility and security.

## REFERENCES

1. V.K. Varma, K.D. Wong, K. Chua, F. Paint, Integration of 3G Wireless and Wireless LANs, *IEEE Communications Magazine*, November 2003.
2. K. Ahmavaara, H. Haverinen, and R. Pichna, Interworking Architecture Between 3GPP and WLAN Systems, *IEEE Communications Magazine*, November 2003.
3. M. Ilyas, *The Handbook of Ad Hoc Wireless Networks*, CRC Press, Boca Raton, FL, 2003.
4. S. Qazi, Implementation of Secure IEEE 802.11 Wireless Local Area Network, Final Report for Visiting Faculty Research Program, AFRL, Rome, 2001.
5. J.S. Park, A. Nanda, and J. Howison, Security Challenges and Countermeasures in WLANs, OCCCT 2003 Conference, Orlando, FL, July 2003.
6. A. Mishra, and W.A. Arbaugh, *An Initial Analysis of the IEEE 802.1X Standard*, University of Maryland, MD, February 2002.
7. W.A. Arbaugh, N. Shankar, and Y.C.J. Wan, Your 802.11 Wireless Network Has No Clothes, *IEEE Wireless Communications*, December 2002.

8. D. Zhou, Security Issues in Ad Hoc Networks, *The Handbook of Ad Hoc Wireless Networks*, CRC Press, Boca Raton, FL, 2003.

9. C. Kaufman, R.Perlman, and M. Spencer, *Network Security: Private Communication in a Public World*, Prentice Hall, Englewood Cliffs, NJ, 1995.

10. P. Papadimitratos, Z.J. Haas, Securing Mobile Ad Hoc Networks, *The Handbook of Ad Hoc Wireless Networks*, CRC Press, Boca Raton, FL, 2003.

11. Wireless LAN Security Solution for Large Enterprises, Wireless LAN security white paper, Cisco Systems, http://www.cisco.com.

12. Wireless LAN Security Roadmap, Intel Business/Enterprise Newsletter, http://www.intel.com.

13. Wireless Security in 802.11(Wi-Fi) Networks, white paper, Dell, Inc., January 2003, http://www.euro.dell.com.

14. S. Fluhrer, I. Mantin, and A. Shamir, Weaknesses in the Key Scheduling Algorithm of RC4, *8th Annual Workshop, Selected Areas in Cryptography*, August 2001.

15. V. Chang, Wireless Security Re-Invents Itself Again, *Wireless Systems Design*, November/December 2003.

16. Wireless Network Security, draft, NIST Special Publications 800-48, http://csrc.nist.gov/publications/drafts/draft-sp800-48.pdf.

17. J. Edney and W.A. Arbaugh, *Real 802.11 Security: Wi-Fi Protected Access and 802.11i*, Addison-Wesley, Reading, MA, 2004.

18. N. Borisov, I. Goldberg, and D. Wagner, Security of the WEP algorithm, http://www.issac.cs.berkley.edu/issac/wep-faq.htm, January/July 2001.

19. Overview of Wi-Fi Protected Access, Wi-Fi Alliance, http://www.wi-fi.org.

20. R. Puzmanova, Wireless Hot Spot Security, February 2003, http://www.computerbits.com/archive/2003/0200/hotspotsecurity.html.

21. Wireless Security Recommendations for Rutgers, The State University of New Jersey, NJ, November 2003, http://techdir.rutgers.edu/wireless.html.

22. G.C. Ou, Wireless LAN Security, June 2002, http://www.lanarchitect.net/Articles/Wireless/index.htm.

23. B. Brown, 802.11: The Security Difference between b and I, *IEEE Potentials*, October/November 2003.

24. A. Dornan, Emerging Technology: Wireless Security — Is Protected Access Enough? October 2003, http://www.networkmagazine.com/shared/article/showArticle.jhtml.

25. B. Aboba, IEEE 802.11i: A Retrospective, March 2004, http://www.drizzle.com~aboba/IEEE.

26. F.A. Qazi, A survey of Biometric Authentication Systems, accepted for presentation and publication at The 2004 International Conference on Security and Management, June 21–24, 2004, Las Vegas, NV.

27. R. Arun, S. Prabhakar and A. Jain, An Overview of Biometrics, Michigan State University, 2003. .http://biometrics.cse.msu.edu/info.html.

28. C. Soutar, A. Stoianov, R. Gilroy and B.V.K. Vijaya Kumar, Biometric Encryption, http://www.bioscrypt.com/technology/white_papers.shtml.

29. B. Steyn, An Introduction to Biometrics And Biometric Based Systems, May 2000, http://www.cs.uct.ac.za/courses/CS400W/NIS/papers00/bsteyn/report.html.

# 16

# MINIMIZING WLAN SECURITY RISKS

*Surinder M. Bhaskar and Syed A. Ahson*

## INTRODUCTION

Wireless Local Area Networks (WLANs) provide mobility to network users while maintaining the connectivity to the resources of their organization. The users are more prone to use laptops and handheld devices as their primary computing devices because they allow greater portability in meetings and conferences, and during travel. WLAN offers greater productivity per employee by providing constant connectivity. Many wireless Internet services are appearing in airports, hotels, malls, and convention centers. IEEE 802.11-based WLAN technologies have come to extensive use in government departments and public and private sectors. These deployments are causing security concerns due to the fact that 802.11, by its very nature, uses radio frequency (RF) for data transfer. The objective of this chapter is to discuss a variety of issues and processes that help in minimizing the risks associated with 802.11-based WLAN technologies.

## MAJOR SECURITY RISKS

The WLANs that are deployed without considering the security aspects often become targets for hackers. The rapid growth in the use of WLANs is due to the following reasons: (1) low cost of the devices, (2) ease of deployment, and (3) large productivity gains.

A major problem is the fact that most vendors sell devices in which all security features have been disabled. This type of deployment has attracted the attention of the hacker community. There are a number of Web sites that provide information on freely available wireless connections. The hackers use their connections as a means to get free Internet access or to hide their identity. Some hacker groups use these freely available connections to break into networks that might otherwise have been difficult to break into from the Internet. In WLANs, data is sent over the air and may be accessible outside the physical boundary of an organization. The unencrypted data packets can be accessed by anyone within the RF range. Any person with a laptop, a WLAN adapter, and a program like TCPDUMP can receive, view, and store all the packets that are transmitted on a given WLAN.

WLANs face the same threats as wired LANs, as well as those that are unique to wireless communications. Threats associated with WLANs are the following.[5,6,10,14]

- *RF interference and jamming:* It is easy to interfere with wireless communications. A jamming transmitter can make communications impossible, for example, by continuously attacking an access point with access requests, which, whether successful or not, will eventually exhaust the access point's available RF spectrum. Other wireless services that are operating in the same frequency range as a WLAN can reduce the range and available bandwidth of the WLAN. Bluetooth technology, used to communicate between handsets, employs the same 2.4-GHz RF. 802.11b devices operate at 2.4 GHz. The 2.4-GHz RF is an unlicensed frequency, and therefore other devices can interfere with the radio signals between wireless devices. This band was originally intended for industrial, scientific, and medical use, and also for use by cordless phones. This threat may cause a denial-of-service (DoS) like situation.
- *Wireless spoofing:* Access points can identify every wireless card ever manufactured by a vendor through the unique media access control (MAC) address that is burned into and printed on the card. However, the real scenario is much more complex because it requires that the cards be registered before becoming a part of the WLAN. Therefore, each access point in the WLAN should have this list. Even with this configuration, it cannot stop hackers who use cards with firmware that does not have a built-in MAC address, but a randomly chosen or deliberately spoofed address.
- *Wireless sniffing:* The default WLAN settings of various vendors are easily available on the Internet. If an access point has been

configured using default settings, anyone can connect to this access point. Hackers use wireless sniffers to gain access to WLANs with ease.

■ *DoS:* Denial-of-Service (DoS) attacks are common to both wireless and wired networks. DoS attacks are intentional interferences initiated by hackers to flood the 2.4-GHz frequency band with noise, such that devices cannot communicate. The IEEE 802.11-based management messages, including the beacon, probe request or response, association request or response, reassociation request or response, disassociation, and deauthentication, are not authenticated, which makes DoS attacks possible. Open source tools are available on the Internet, capable of performing such attacks.

■ *Poor configuration:* Most vendors set devices at default configuration for ease of deployment, helping users treat them as plug-and-play devices. However, such a configuration results in lack of security, making access points easy targets. Wireless clients can access each other in two modes: (1) infrastructure mode and (2) ad hoc mode. In infrastructure mode, most wireless clients connect through an access point for all communications. However, it is possible to configure devices to form a peer-to-peer network, which is more commonly called an ad hoc network. In ad hoc networks, two laptops or desktops within range of one another can share files directly without the use of an access point. The security impact of ad hoc networks is significant. Many wireless cards can be configured to form an ad hoc network. Any hacker with an adapter configured for ad hoc mode and using an identical setting as the other adapters may gain unauthorized access to the clients.

■ *Man-in-the-middle (MITM) attacks:* An MITM attack is a situation in which a malicious hacker uses an access point to effectively hijack mobile nodes by sending stronger signals than the legitimate access point. The mobile nodes then associate to this rogue access point, sending their data — possibly, sensitive data — into the wrong hands.

## MINIMIZING SECURITY RISKS

In the previous section, we have discussed some of the possible security risks in WLANs. In the following sections, we will describe the steps that should be taken to reduce the security risks associated with WLANs, the different security technologies available, and how these technologies work.[1–9,14,17]

## WLAN Security Policy

The security policy defines the acceptable configuration of the WLAN devices. The security policy deals with issues such as changing the default factory configuration of an access point (default administrator password, Simple Network Management Protocol [SNMP] community string, Wired Equivalent Privacy [WEP] setting, Service Set Identification [SSID]), identifying the associated security risks, and planning to mitigate security risks.

## Site Survey

The network administrator should conduct periodic surveys of the site, using sniffers or other tools, and perform the following:[4,5]

- Search for unsecured access points.
- Educate the user community about the security issues involved in reducing the risks.
- The power output of the access point should be limited to the desired area of coverage.
- If feasible, change the antenna from omnidirectional to unidirectional.
- Install the antenna as far away from an outside wall as possible. The problem of upper and lower floors in a multistorey facility should also be taken into account. The objective is to limit the signal so that it is available only to the desired users.

## Turning on SSID

SSID[9] is a unique 32-character identifier that is attached to the header of a data packet. SSID acts as a password when a device tries to connect to an access point. SSID differs from one WLAN to another; therefore, all access points and devices attempting to connect to a specific WLAN must use the same SSID. A device will not be permitted to join the network unless it can provide the unique SSID. However, because an SSID can be sniffed in plaintext from a packet, it does not provide any security to the network. An SSID is also referred to as a network name because essentially it is a name that identifies a wireless network. Each of the base stations supplied by various vendors have default SSIDs. Attackers can use these default SSIDs to attempt to penetrate base stations having default configuration. The following are some default SSIDs: "tsunami" (Cisco), "Compaq" (Compaq), and "Intel" (Intel).

SSID provides the first line of defense against intruders. It allows multiple wireless networks to operate in the same physical area without the threat of accessing the wrong network. The same SSID is assigned to both the access point and the mobile devices, and if the SSID of access point and the mobile device do not match, access to the network is denied.

## MAC Control List

MAC addresses can be used to prevent unwanted systems from becoming a part of the WLAN. The MAC addresses of all the desired systems should be added to the access list of the WLAN access point. This is also known as MAC level authentication.

## WEP

WEP[4,17] was introduced to counter the problem of eavesdropping on WLANs. The 802.11 standards define WEP as a mechanism to secure the over-the-air transmission between access points and network interface cards. WEP operates at the data-link layer and requires that all the communicating devices share the same secret key. In this case, the data between the access point and the clients are encrypted, thus making it difficult to eavesdrop on the data. WEP is available in 40-bit and 128-bit encryptions.

The traditional WLAN security mechanisms, which use open or shared-key authentication and static WEP keys, provide a basic level of access control and privacy, which can, however, easily be compromised.

Security in 802.11 networks can be broken into two primary elements: encryption and authentication. The implementation of authentication and encryption has been tested and documented as insecure by the security community at large. A simple configuration with a WEP key server and two access points is shown in Figure 16.1.

WEP consists of three main components: (1) frame encryption, (2) authentication, and (3) key management.

### Frame Encryption

Encryption is the process of taking a message (plaintext) and passing it through a mathematical algorithm to produce what is known as ciphertext. Decryption is the reverse of encryption; it is the process of extracting the message from ciphertext. Two major encryption algorithms currently in use are symmetric encryption and asymmetric encryption. The Institute of Electrical and Electronics Engineers (IEEE) has specified that WEP be used for protection of the 802.11 data frames. WEP employs the RC4[23] stream cipher algorithm for encryption, which was invented by Ron Rivest of RSA Data Security, Inc. In RC4 algorithm, the key length can be up to 256 bits. IEEE specifies that 802.11 devices support 40-bit keys, with the

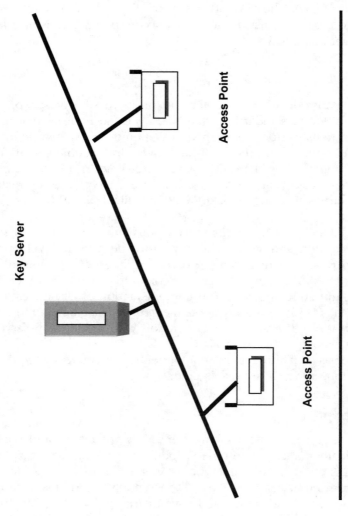

**Figure 16.1  Centralized encryption keyserver.**

provision to use longer key lengths. To avoid regeneration of the same ciphertext by the same plaintext, an initialization vector is concatenated with the symmetric key before generating the ciphertext. The initialization vector is a 24-bit value (ranging from 0 to 16777215) generated by the sender and must be sent to the receiver unencrypted in the header portion of the 802.11 data frame. The receiver can then concatenate the initialization vector with the WEP key to decrypt the data frame. Because the initialization vector is transmitted as plaintext and placed in the 802.11 headers, anyone sniffing a WLAN can access it.

As shown in Figure 16.2, the plaintext itself is not run through the RC4 algorithm; rather, the RC4 cipher is used to generate a unique keystream for a particular 802.11 frame, using the initialization vector and the WEP key. The resulting unique keystream is then combined with the plaintext and run through the eXclusive OR (XOR) mathematical function to produce the ciphertext.

The IEEE 802.11 standard specifies the use of the RC4 algorithm and key in WEP but does not describe the specific methods for key distribution. This may result in implementation problems due to the possibility of human error in key input and management.

## Authentication

The IEEE 802.11 standard supports two types of client authentication: (1) open-key authentication and (2) shared-key authentication. Open-key authentication involves supplying the correct SSID, and WEP encryption is used for transmitting and receiving data from an access point. Shared-key authentication is based on a challenge–response mechanism. An access point transmits the challenge text to the client device. The client encrypts the challenge text with the correct WEP key and sends it back to the access point. If the client has a wrong key or no key, authentication will fail, and the client will not be able to communicate with the access point. Shared-key authentication is not secure because a hacker can detect the challenge text and the WEP key.

## Key Management

Static WEP key is often used but not considered secure. Static WEP key is composed of either 40 or 128 bits and is defined on the access point and on all the clients associated with the access point. Static WEP key has to be entered on every device in the WLAN, and if any device is lost or stolen, the possessor of the stolen device can access the WLAN. It is difficult to detect the unauthorized access until the theft is reported. The

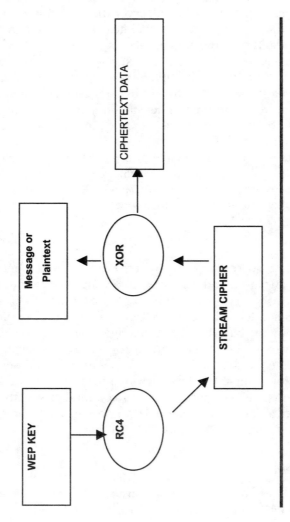

**Figure 16.2  WEP using RC4 algorithm.**

administrator must then change the WEP key on every device having the same WEP key as that of the missing device. It is more dangerous if the static WEP key is deciphered using the tools available on the Internet. Both the 40-bit and 128-bit variants of WEP can be easily cracked by using these off-the-shelf tools. The hacker community has also developed methods to break the 128-bit encryption by gathering enough traffic through eavesdropping, and then they can find the encryption key.

Key rotation is another issue with WLANs. WEP does not have a key manager, and there is no mechanism that allows for key distribution and key rotation automatically. Some vendors, however, offer products that feature dynamic WEP, in which the administrator can select the interval at which the keys are updated, thereby adding a layer of difficulty to those who may be trying to eavesdrop on a WLAN.

## SECURITY EXTENSIONS

In the previous section, we have explained the various techniques for enhancing the security of a WLAN. This section deals with the various security extensions that can be added to improve the security of a wireless network. These extensions can be (1) network layer encryption approach based on IP Security (IPSec), (2) a mutual authentication-based key-distribution method using 802.1X, and (3) improvement to WEP

### IPSec

IPSec[18] is a framework of open standards for ensuring secure private communications, and it provides security for transmission of critical and sensitive information over unprotected networks such as the Internet. IPSec operates at the network layer, protecting and authenticating IP packets between the participating IPSec devices. IPSec provides the following network security services:

- Data confidentiality — The IPSec sender can encrypt packets before transmitting them across a network.
- Data integrity — The receiver can authenticate packets sent by the IPSec sender to ensure that the data has not been altered during transmission.
- Data origin authentication — The IPSec receiver can authenticate the source of the IPSec packets sent. This service is dependent on the data integrity service.
- Antireplay — The IPSec receiver can detect and reject a replayed packet.

To deploy IPSec in the WLAN environment, an IPSec client software is loaded on every PC connected to the wireless network, and the user establishes an IPSec tunnel to route any traffic to the wired network. Filters are deployed to prevent any wireless traffic from reaching any destination other than the virtual private network (VPN) gateway and the Dynamic Host Configuration Protocol (DHCP) or Domain Name System (DNS) server. As discussed earlier, IPSec is capable of providing confidentiality, authenticity, and antireplay capabilities to IP traffic. Confidentiality is achieved through the application of encryption algorithms such as Data Encryption Standard (DES), Triple DES, or Advanced Encryption Standard (AES). Extension to IPSec allows for user authentication and authorization to occur as integral parts of the IPSec process.

## IEEE 802.1X and Extensible Authentication Protocol (EAP)

### IEEE 802.1X

Security for wireless networks can be broken down into three main components: (1) the authentication mechanism, (2) the authentication algorithm, and (3) data frame encryption.

802.1X focuses on the authentication mechanism or framework, taking advantage of EAP (RFC 2284),[20] which is developed around the point-to-point protocol (PPP), and ties it to the physical medium, be it Ethernet, Token Ring, or WLAN. EAP messages are encapsulated in 802.1X messages and referred to as EAPoL or EAP over LAN. 802.1X consists of three main components: (1) client software, (2) access point, and (3) authentication server.

When a client tries to connect to an access point, the access point detects the client, enables the client's port, and puts the port into an unauthorized state so that only 802.1X traffic is forwarded. The client then sends an EAP-start message. The access point then responds with an "EAP-request identify" message to get the client's identity. The client's response is forwarded to the authentication server for verification of the client's identity. The authentication server uses a specific authentication algorithm for the client's authentication. The authentication server sends an accept or reject packet to the access point, using the authentication algorithm. On receiving the accept packet, the access point will change the state of the client's port to an authorized state, and the traffic will be forwarded. 802.1X for WLANs does not provide a mechanism for key distribution or management. This is left for the vendor to implement. At log-off, the client sends an EAP-log-off message. On receiving this message, the access point changes the state of the client's port to an unauthorized state.

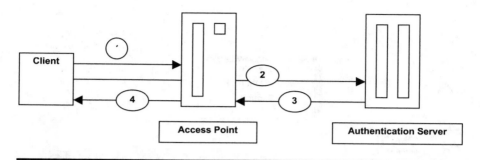

**Figure 16.3   802.1X authentication for WLANs.**

802.1X authentication for WLANs is depicted in Figure 16.3. It is basically a centralized server-based authentication of end users. The following steps are involved in 802.1X-based authentication.

1. A client sends a "start" message to an access point, which requests the identity of the client.
2. The client replies with a response packet containing an identity, and the access point forwards the packet to an authentication server.
3. The authentication server sends an "accept" packet to the access point.
4. The access point places the client port in an authorized state, and traffic is then allowed.

## *EAP*

This security approach focuses on providing centralized authentication and dynamic key distribution. This approach is based on the IEEE 802.11 Task Group "i" end-to-end framework using 802.1X and EAP to provide improved functionality. The main elements of 802.1X and the EAP approach are: (1) mutual authentication between client and authentication server Remote Access Dial-In User Service (RADIUS),[12] (2) encryption keys dynamically derived after authentication, and (3) centralized policy control.

In such an arrangement, a wireless client that associates with an access point cannot gain access to the network until the user performs a network log-on. The sequence of events is as follows (Figure 16.4):

1. A client associates with an access point.
2. The client supplies network log-in credentials (user ID and password, user ID and one-time password [OTP], or user ID and digital certificate) via an EAP supplicant.

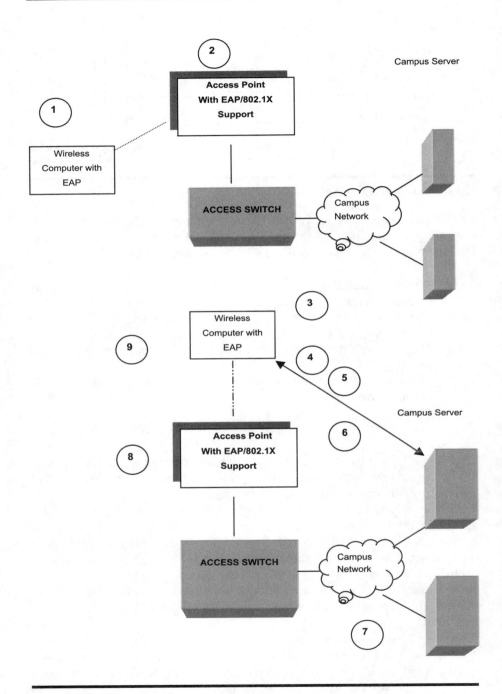

**Figure 16.4 EAP: sequence of events.**

3. Using 802.1X and EAP, the client and the RADIUS server perform mutual authentication through the access point.
4. After successful mutual authentication, the RADIUS server and client determine a WEP key corresponding to the client. The client loads this key.
5. The RADIUS server sends the WEP key (session key) to the access point.
6. The access point encrypts its broadcast key with the session key and sends the encrypted key to the client. The client uses the session key to decrypt the broadcast key.
7. The client and access point activate WEP and use the session and broadcast key for all communications during the session, or until a time-out has occurred and new WEP keys are generated.
8. Both the session key and broadcast key are changed at regular intervals.

EAP provides many advantages over basic 802.11 security:

1. Mutual authentication scheme — This scheme effectively eliminates MITM attacks due to rough access points and RADIUS servers.
2. Central management and distribution of encryption keys — This scheme eliminates the problem of distribution of static keys to all the access points and clients in the network. This scheme is of great help when a wireless device is lost and the network would need to be rekeyed to prevent the lost system from gaining unauthorized access.
3. Ability to define centralized policy control — This is useful when session time-out initiates reauthentication and new key generation.

EAP: sequence of events.

1. Client associates with access point.
2. Access point blocks all user requests to access LAN.
3. User provides login authentication credentials.
4. RADIUS server authenticates user.
5. User authenticates RADIUS server.
6. RADIUS server and client derive WEP key.
7. RADIUS server delivers unicast WEP key to access point.
8. Access point delivers broadcast WEP key, encrypted with unicast WEP key, to client.
9. Client and access point activate WEP and use unicast and broadcast WEP keys for transmission.

## EAP Types

Numerous EAP types are available today for user authentication over wired and wireless networks. Some of the EAP types are listed below:

- *LEAP-EAP:* LEAP supports all three of the 802.1X and EAP elements. The mutual authentication is based on a shared secret — the user's log-on password, which is known to the client and the network. The RADIUS server sends a challenge to the client. The user uses the hash value of the user-supplied password and sends the response back to the RADIUS server, which validates the response through its user information database. The mutual authentication process involves the RADIUS server authenticating the client and the client authenticating the RADIUS server. When this mutual authentication succeeds, an "EAP-success" message is sent to the client, and the client and the RADIUS server gets the dynamic WEP key.
- *EAP-Transport layer security (TLS):* This is an Internet Engineering Task Force (IETF) standard (RFC 2716)[21] and is based on the TLS protocol (RFC 2246).[22] This scheme uses digital certificates for both user and server authentication and supports the key elements of 802.1X and EAP, as discussed earlier. The client validates the RADIUS server certificate and the RADIUS server validates the client certificate. When this is complete, an "EAP-success" message is sent to the client, and both the client and the RADIUS server get the dynamic WEP key. The sequence of events is shown in Figure 16.5.

- EAP-TLS: sequence of events.

1. Client associates with access point.
2. Access point blocks all user requests to access LAN.
3. User authenticates RADIUS server (via digital certificate).
4. RADIUS server authenticates user (via digital certificate).
5. RADIUS server and client derive unicast WEP key.
6. RADIUS server delivers unicast WEP key to access point.
7. Access point delivers broadcast WEP key encrypted with unicast WEP keys for transmission.
8. Client and access point activate WEP and use unicast and broadcast WEP keys for transmission.

- *Protected EAP (PEAP):*[11,15] PEAP is an IETF draft RFC, authored by Cisco systems, Microsoft, and RSA Security. It uses a digital certificate for server authentication. The user is authenticated by various

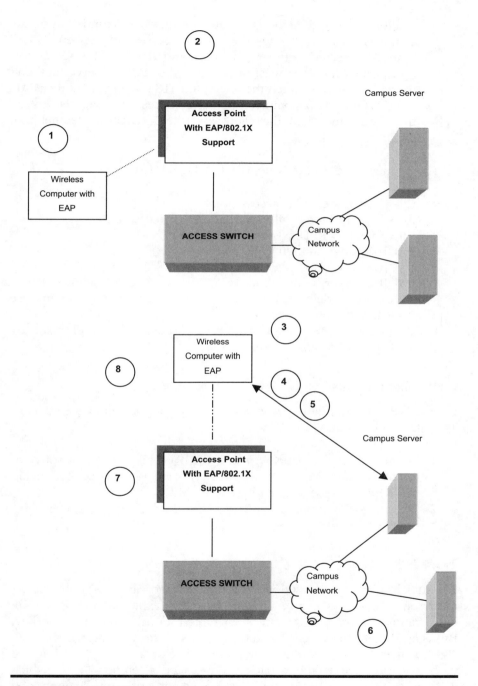

**Figure 16.5  EAP-TLS: sequence of events.**

EAP-encapsulated methods within a protected TLS tunnel. In phase 1, the client authenticates the RADIUS server, and an encrypted tunnel is created between the user and the RADIUS server for transporting the messages. In phase 2, the RADIUS server authenticates the client through the encrypted TLS tunnel via another EAP type. When this is complete, an "EAP-success" message is sent to the client, and both the client and the RADIUS server get the dynamic WEP key.

The sequence of events is depicted in Figure 16.6.

1. Client associates with access point
2. Access point blocks all user requests to access LAN
3. Client verifies RADIUS server digital certificate
4. RADIUS server authenticates user
5. RADIUS server and client derive unicast WEP key
6. RADIUS server delivers unicast WEP key to access point
7. Access point delivers broadcast WEP key encrypted with unicast WEP key to client
8. Client and access point activate WEP and use unicast and broadcast WEP keys for transmission

## WEP Enhancements

The IEEE 802.11i standard[7,14,17] includes two encryption enhancements in its draft standard 802.11 security:

1. Temporal Key Integrity Protocol (TKIP): This is a set of software enhancements for RC4-based WEP. This includes per-packet keying (PPK) and message integrity check (MIC).
2. Advanced Encryption Standard (AES),[13] which is a stronger alternative to RC4.

### TKIP: PPK

The initialization vector used in WEP is weak and can be exploited. The most common attacks against WEP are based on exploiting multiple weak initialization vectors in a stream of encrypted traffic using the same key. By using different keys per packet, it is possible to minimize this threat. As shown in Figure 16.7, the initialization vector and the WEP key are hashed to produce a unique packet key (called a temporal key), which is then combined with an initialization vector and run through one mathematical function XOR with plaintext.

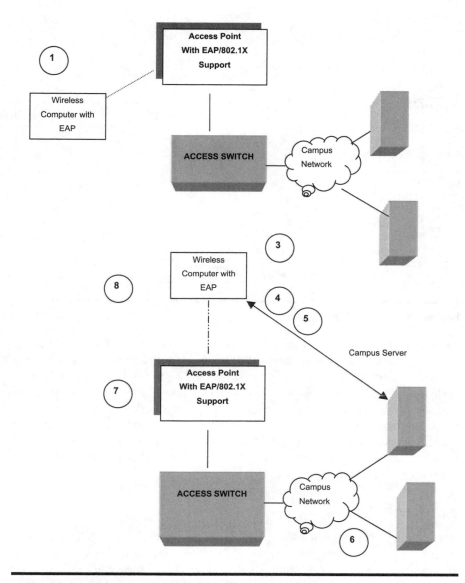

**Figure 16.6  PEAP: sequence of events.**

This mechanism prevents hackers from deriving the base WEP key by using the weak initialization vector. In this case, weak initialization vectors allow you to derive only the per-packet WEP key and not the base WEP key. To prevent attacks due to initialization vector collisions, the base key should be changed before the initialization vectors are repeated. Mechanisms

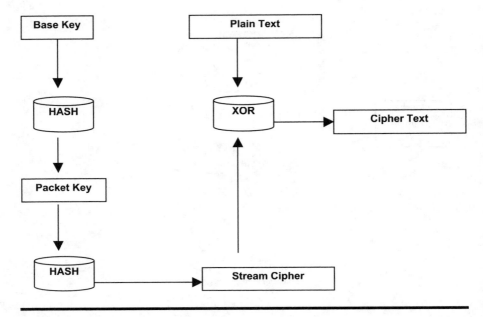

**Figure 16.7 TKIP-PPK.**

such as EAP or a similar protocol should be implemented to perform the rekey operation. Access points and clients use broadcast keys for layer 2 broadcast and multicast communication. Due to weak initialization vector collision, the broadcast key is prone to attacks. Some vendors provide support for broadcast key rotation in devices supplied by them. The access point dynamically calculates the broadcast WEP key using a random number generator. Broadcast WEP key rotation can be implemented with EAP protocols such as LEAP, EAP-TLS, and PEAP, which support dynamic derivation of encryption keys.

### TKIP-MIC

MIC protects WEP frames from being tampered with. MIC is based on a seed value, destination MAC, source MAC, and payload. It employs a hash algorithm to derive the resulting value. This is an improvement over the CRC-32 checksum function used in standard WEP. A comparison of various wireless encryption technologies is shown in Table 16.1.

### EAP-Tunneled TLS (EAP-TTLS)

EAP-TTLS is an extension of TLS and was developed to overcome the need created by TLS for client-side certificates.[6,8] In TTLS, server-side certificates are required. EAP-TTLS is a two-step authentication process:

Minimizing WLAN Security Risks ■ 415

**Table 16.1  Encryption Technology Comparison**

| Parameter | LEAP with TKP | EAP-TLS with TKIP | EAP-PEAP with TKIP | IPSec-based VPN |
|---|---|---|---|---|
| Key lengths (bits) | 128 | 128 | 128 | 168/128, 192, 256 |
| Encryption algorithm used | RC4 | RC4 | RC4 | 3DES or AES |
| Packet integrity check mechanism | CRC-32/MIC | CRC-32/MIC | CRC-32/MIC | MD5-HMAC SHA-HMAC |
| Device authentication | None | Certificate | None | Preshared secret or certificates |
| User authentication | Username/password | Certificate | User/password or OTP | Username/Password or OTP |
| Certificate requirements | None | RADIUS server WLAN client | RADIUS server | Optional |
| User differentiation | Group | Group | Group | User |
| Single sign-on support | Yes | Yes | No | No |
| ACL requirements | Optional | Optional | Optional | Required |
| Additional hardware | No | Certificate server | Certificate server | IPSec concentrator |
| Per-user keying | Yes | Yes | Yes | Yes |
| Protocol support | Any | Any | Any | IP unicast |
| Client OS support | Wide range | Wide range | Wide range | Wide range |
| Open standard | No | Yes | IETF draft RFC | Yes |

*Source:* With permission Cisco-Aironet Wireless LAN Security Overview, http://www.cisco.com/warp/public/cc/pd/witc/ao350ap/prodlit/a350w_ov.htm; Cisco SAFE: Wireless LAN Security in Depth, April 2003, http://www.cisco.com/warp/public/cc/so/cuso/epso/sqfr/safwl_wp.pdf.

(1) in step 1, an asymmetrical algorithm based on server keys is used to verify the server's identity, and on being successful, a symmetric encryption tunnel is created and (2) in step 2, the user's credentials, which are in the form of attribute–value pairs and not digital certificates, are validated by the server.[12]

The symmetric encryption tunnel of TTLS is used exclusively for protecting the client authentication method. After verification, the encryption tunnel is collapsed, and wireless devices create a WEP encryption tunnel for the purpose of data privacy. This approach is easy to implement because it provides one required security during authentication, besides accommodating existing end-user methods.

### EAP-Subscriber Identity Module (SIM)

EAP-SIM[6,8,14] is the current standard authentication method for cellular devices. This is similar to the smart card-based authentication method. Smart cards are small cards with microprocessors and memory that can store small programs such as encryption algorithms, device history, its usage, and information regarding the owner of the card. In cellular devices, the size of these cards has been downsized even further, and clients use these small cards, containing their credentials, to authenticate to the network. EAP-SIM provides identity privacy with a method called pseudonyms.

## WIRELESS TRANSPORT LAYER SECURITY (WTLS)

WTLS provides an enhanced level of security. WTLS consists of three operating modes: (1) Class 1 authentication is an anonymous authentication that offers no security, (2) Class 2 is a server-based authentication, and is susceptible to MITM attacks as well as session hijacking, and (3) Class 3 consists of both the client and the server. The keys for both the client and the server may be either private or public. In the case of a public key, a secure key-management infrastructure is required, whereas private keys require secure key distribution and storage.

## ADVANCED ENCRYPTION STANDARD (AES)

In the year 2001, the National Institute of Standards and Technology (NIST) published a new encryption algorithm called the Advanced Encryption Standard (AES).[1,2,8] This was chosen from a number of algorithms submitted in response to a request from NIST. The two researchers who developed and submitted the block cipher called Rijndael for AES are both cryptographers from Belgium: Dr. Joan Daemen of Proton World International and Dr. Vincent Rijmen, a postdoctoral researcher in the Electrical Engineering Department (ESAT) of Katholieke Universiteit Leuven. AES is gaining acceptance as a replacement for the RC4 algorithm used in WEP-based encryption. AES uses the Rijndale algorithm in the following specific key lengths: 128, 192, and 256 bits.

## FILTERING

Filtering[8] is a basic security mechanism that can be used in addition to WEP and AES. The filtering process works in the same way as an access control list (ACL) works on a router. There are three basic types of filtering that can be performed on a WLAN: (1) SSID filtering, (2) MAC address filtering, and (3) protocol filtering.

### SSID Filtering

SSID filtering[9,17] is a rudimentary method of filtering, and it is used for the most basic access control. The SSID of a client device must match that on the access point (in infrastructure mode) or of the other stations (in ad hoc mode) for communication. SSID is broadcasted in cleartext in every beacon that an access point transmits, and it is very easy to access using a sniffer. The network can be configured in such a way that the SSID is taken out from the beacon frame of an access point, and clients with matching SSID associate themselves with the access point. A network configured in this way is known as a "closed" system.

### MAC Address Filtering

Almost all the access points are equipped with the MAC filter functionality.[8,17,19] The network manager can maintain a list of allowable MAC addresses and program them into each access point. If a PC network interface card or other client with a MAC address that is not in the access point's MAC filter list tries to gain access to the WLAN, the MAC address filter functionality will not allow that client to associate with the access point. MAC filters can be implemented on a RADIUS server instead of maintaining a list of allowable MAC addresses on each access point.

Although MAC filters seem to be a good method of securing a WLAN, they have the following weaknesses that can compromise security: (1) theft of a network card whose MAC address is present on the MAC filter list of an access point (shown in Figure 16.8) and (2) sniffing the WLAN and then spoofing with the MAC address after office hours.

MAC filters are good for home and small office networks with a limited number of client stations. MAC addresses of wireless devices are transmitted

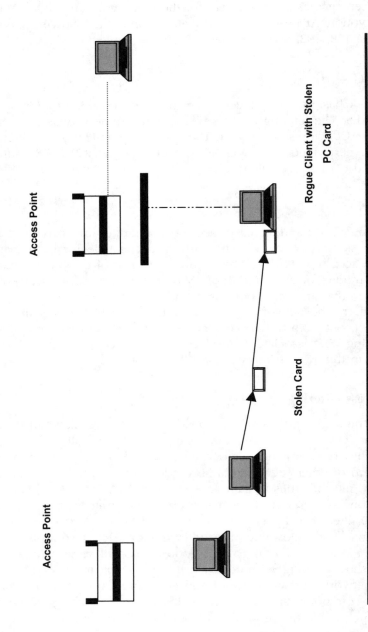

**Figure 16.8** MAC filters.

in clear text; hence, a hacker who can listen to the traffic on a network can find most MAC addresses available on the network. Some wireless PC network cards permit the changing of their MAC address through software or even operating system configuration changes. After getting a list of allowed MAC addresses, the hacker can simply change the PC network card's MAC address to match that of another PC network card on the target network and can gain instant access to the entire network.

## Protocol Filtering

WLANs can filter packets based on layer 2 to layer 7 protocols. Most of the vendors are providing protocol filters[19] independently configurable for both the wired segment and wireless segment of the access point. The protocol filtering mechanism is depicted in Figure 16.9.

# MONITORING WIRELESS NETWORKS

Implementing the steps described in the preceding text will help in minimizing the vulnerabilities associated with WLANs. The next important step is to monitor the wireless network using an intrusion detection system (IDS), which should be deployed to monitor (1) rogue access points, (2) unauthorized ad hoc networks, (3) unauthorized 802.11 network cards, and (4) DoS attacks.

These items help to identify how and when an intruder compromised the network and the origin of attack (external or internal).

# WIRELESS NETWORK DESIGN GUIDELINES

The following points may be useful for designing a WLAN. The design guidelines cover access point security recommendations and client security recommendations.[4,5]

Access point security recommendations are: (1) Enable a centralized user authentication server (e.g., RADIUS, TACACS+, etc.), (2) Choose strong community strings for SNMP and change them frequently, (3) Use SNMP in read-only mode if the network infrastructure allows it, (4) Disable any insecure and nonessential protocol provided by the manufacturer, (5) Use secure management protocols (SSH, etc.), (6) Limit management traffic to a dedicated wired subnet only, (7) Encrypt all management traffic and isolate it from user traffic, and (8) Enable wireless frame encryption.

The client security recommendations are: (1) Disable ad hoc mode and (2) Enable wireless frame encryption when available.

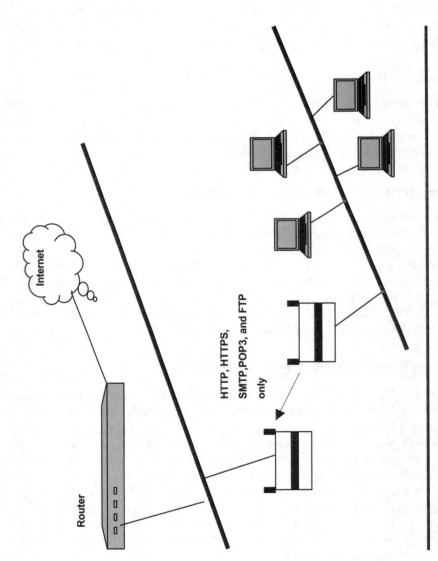

Figure 16.9  Protocol filtering.

## TIPS FOR SECURE WLAN NETWORKING

In the following, we briefly list a number of tips that may be used for deployment of secure WLANs.[4-6,8]

- Periodic surveys should be conducted to find rogue access points. All efforts to secure a wireless network could be wasted if a rogue access point were plugged into your network, behind your firewall.
- Take a notebook equipped with network tools and an external antenna outside your office building and see how far the office WLAN signal radiates.
- Apply MAC address-based authentication for associating a wireless device with an access point.
- Use an additional level of authentication, such as RADIUS, for association with access points.
- On wireless routers, assign static IP addresses to devices, turn off DHCP, and also change the IP subnet.
- Use access points and network cards with 128-bit (or higher) WEP capabilities.
- Make sure that you can upgrade your access points with new security enhancements.
- It is suggested that access points be installed in demilitarized zone (DMZ) and that users associate to access points using VPNs.
- Enable WEP encryption. Although WEP is not very secure, it provides a basic level of security to wireless communication.
- It is recommended that the default SSID of devices be changed. However, do not use SSIDs that reflect the organization's main names, divisions, products, street address, etc., because it could be easy to find and may attract hackers who, with little effort, can break the WEP encryption keys.
- Disable the "broadcast SSID" feature on each access point, because with the feature enabled, an access point will accept any SSID. By disabling the feature, the SSID configured in the client must match that of the access point for communication to ensue.
- Change the default passwords of access points and wireless routers, because tools are available to identify the manufacturer, based on the MAC address, and it is easy to find out what type of device is being used.
- Install access points or central devices in the center of the building rather than near the outer boundary. If access points are located near the outer boundary, a stronger signal will be radiated outside the building, making it easier for hackers to gain access to the target network.

## CONCLUSION

In wireless networks also, just as in the case of wired networks, no one can guarantee a completely secure networking environment that prevents all penetrations. Security is a dynamic issue, and network administrators and WLAN manufacturers need to keep one step ahead of hackers.

Security experts suggest several layers of defense across the network to mitigate the threats. Additional security components can be firewalls, intrusion detection systems (IDS), and virtual LANs (VLANs). Network administrators can lessen the risk by properly designing and installing their wireless networks and implementing proven security measures.

Research and development is still in progress on data encryption techniques to be applied for data security in wireless communication.

In this chapter, we have presented the various techniques that are currently employed for security of wireless networks.

## REFERENCES

1. Stallings, W., *Cryptography and Network Security: Principles and Practice*, Second Edition, Prentice Hall, 1999.
2. Schneier, B., *Applied Cryptography Protocols and Algorithms*, John Wiley & Sons, 1996.
3. IEEE working group for WLAN standards http://grouper.ieee.org/group/802/11/index.html.
4. Practically networked, Wireless Encryption Help, http://www.practicallynetworked.com/support/wireless_encrypt.htm.
5. Practically networked, Securing Your Wireless Network, http://www.practicallynetworked.com/support/wireless_secure.htm.
6. Cisco-Aironet Wireless LAN Security Overview, http://www.cisco.com/warp/public/cc/pd/witc/ao350ap/prodlit/a350w_ov.htm.
7. Intel Building Blocks for Wireless LAN Security, http://www.intel.com/network/connectivity/resources/doc_library/white_papers/WLAN_Security_WP.pdf?iid=ipp_prowireless+security_wp&.
8. Cisco SAFE: Wireless LAN Security in Depth, April 2003, http://www.cisco.com/warp/public/cc/so/cuso/epso/sqfr/safwl_wp.pdf.
9. What is SSID?, http://www.webopedia.com/TERM/S/SSID.html.
10. Wireless LAN Security FAQ, http://www.iss.net/wireless/WLAN_FAQ.php.
11. Mathew, G., A Technical Comparison of TTLS and PEAP, The O'Reilly Network white paper, October 17, 2002, http://www.oreillynet.com/pub/a/wireless/200210/17/peap.html.
12. Remote Authentication Dial-In User Service (RADIUS), RFC 2865, http://www.ietf.org/rfc/rfc2865.txt.
13. AES, Advanced Encryption Standard, available at http://www.nist.gov/aes.
14. Air Defense, Wireless LANs: Risks and Defenses, white paper available at http://www.airdefense.net/whitepapers/rd_request2.php4.
15. Andersson, H., Josefsson, S., Zorn, G., and Aboba, B., Protected Extensible Authentication Protocol (PEAP) IETF Internet-Draft.

16. RFC2865, http://www.faqs.org/rfcs/rfc2865.html.
17. SAFE-Blueprint from Cisco, http://www.cisco.com/en/US/netsol/ns340/ns394/ns171/ns128/networking_solutions_package.html.
18. What is IPSec Tunnelling?, http://www.microsoft.com/serviceproviders/columns/what_is_ipsec_tunneling_987.asp.
19. Wireless Home Networking, Part III — Wi-Fi Security, http://www.wi-fiplanet.com/tutorials/article.php/1495811.
20. RFC 2284, http://www.faqs.org/rfcs/rfc2284.html.
21. RFC 2716, http://www.faqs.org/rfcs/rfc2716.html.
22. RFC 2246, http://www.faqs.org/rfcs/rfc2246.html.
23. RSA Laboratories, Frequently Asked Questions About Today's Cryptography, available at http://www.hackpalace.com/encryption/general/rsa faq.html.

# 17

## SECURITY IN WLANS

*T. Andrew Yang and Yasir Zahur*

### INTRODUCTION

Following the widespread use of the Internet, especially the World Wide Web, since 1995, wireless networking has spread rapidly, becoming a buzzword at the beginning of the new millennium. New terms such as wireless communications, wireless local area networks (WLANs), wireless Web, wireless application protocols (WAP), wireless transactions, wireless multimedia applications, etc., have emerged and become common in the vocabulary of computer and information professionals. Among the emerging wireless technologies, WLANs have gained much popularity in various sectors, including business offices, government buildings, schools, and residential homes. The set of IEEE 802.11 protocols (especially 11a, 11b, and 11g), nicknamed *Wi-Fi*, has become the standard for WLANs since the late 1990s.

An increasing number of 802.11-based WLANs have been deployed in various types of locations, including homes, schools, airports, business offices, government buildings, military facilities, coffee shops, and bookstores, as well as many other venues. One of the primary advantages offered by WLAN is its ability to provide untethered connectivity to portable devices such as wireless laptops and PDAs. In some remote communities, WLANs are implemented as a viable last-mile technology,[1] which link homes and offices in isolated locations to the global Internet.

Further widespread deployment of WLANs, however, depends on whether secure networking can be achieved. To deliver critical data and services over WLANs, a reasonable level of security must be guaranteed. The Wired Equivalent Privacy (WEP) protocol, originally proposed as the security mechanism of 802.11 WLANs, is known to be easily cracked by commonly available hacking software. WLANs suffer from various security vulnerabilities such as eavesdropping, resource stealing, denial-of-service (DoS) attacks, static WEP keys, absence of mutual authentication, and session-hijack attack, etc. To deploy a secure WLAN, it is necessary to implement an alternative security mechanism, such as Secure Socket Layer (SSL), virtual private network (VPN), Wi-Fi Protected Access (WPA), or the IEEE 802.11i protocols that are being developed.

In this chapter, the security aspects of WLANs are studied. We first give an overview of the various types of WLANs and the respective vulnerabilities of various protocols, followed by a discussion of alternative security mechanisms that may be used to protect WLANs.

## TYPES OF WLANS

The 1999 version of the 802.11 standard[2] defines three types of wireless networks: independent basic service set (IBSS), basic service set (BSS), and extended service set (ESS).

1. *IBSS, i.e., ad hoc network:* "An ad hoc network is a network composed solely of stations within mutual communication range of each other via the wireless medium."[2]

   As shown in Figure 17.1, in an ad hoc WLAN, end nodes communicate without any AP and thus without any connection to a wired network. Node 2, for instance, may communicate directly with node 1 and node 3. Node 3 and node 1, however, cannot have direct communication with each other, because they are outside each other's communication range. An ad hoc network is typically created in a spontaneous manner, allowing nontechnical users of the wireless devices to create and dissolve the ad hoc network conveniently. It is useful in allowing quick setting up of a wireless network among end users.

   As noted in the 1999 edition of the 802.11 specifications,[2] "The principal distinguishing characteristic of an ad hoc network is its limited temporal and spatial extent." To achieve a more permanent wireless network with a larger communication range, infrastructure modes (see the following text) are often used.

2. *BSS:* A BSS is "a set of stations controlled by a single coordination function."[2]

   A BSS (commonly referred to as an *infrastructure network*) consists of a single access point (AP) and a number of end nodes as shown in Figure 17.2. All the communication between any two nodes has to pass through the AP. The coverage area is greatly increased as

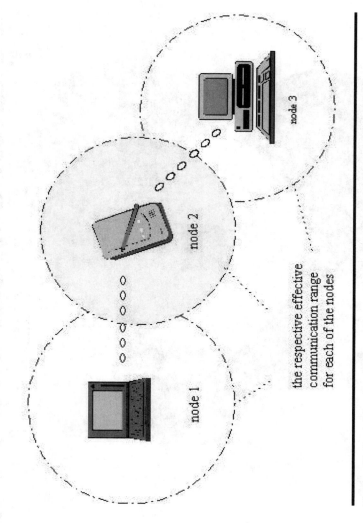

the respective effective
communication range
for each of the nodes

node 1

node 2

node 3

**Figure 17.1   Ad hoc mode.**

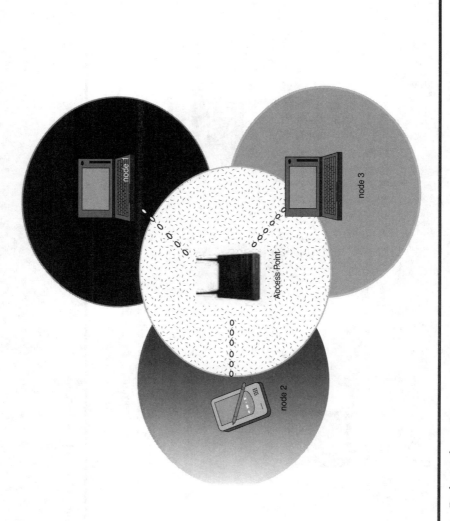

**Figure 17.2  Basic service set.**

compared to an IBSS. Mobile nodes A, B, and C, for example, cannot communicate with each other via ad hoc mode because they are outside each other's communication range. However, by communicating through the AP, they effectively form a WLAN for data communications. The AP behaves in a way similar to a hub in a star topology network.

3. *ESS:* An ESS is "a set of one or more interconnected basic service sets (BSSs) and integrated local area networks (LANs) that appears as a single BSS to the logical link control layer at any station associated with one of those BSSs."[2]

As shown in Figure 17.3, an ESS consists of multiple BSSs, each having a single AP. The AP in each BSS is connected to a distribution system that is usually a wired Ethernet network. An ESS is a hybrid of wireless and wired LANs, and extends a wireless station's connectivity beyond its local AP.

With the exception of ad hoc wireless networks, a WLAN typically consists of a central connection point called the AP, which transmits data between different nodes of a WLAN and, in most cases, serves as the only link between the WLAN and the wired networks.

## THE IEEE 802.11 PROTOCOLS AND BUILT-IN SECURITY FEATURES

Between 1997 and 2001, the IEEE had released a series of 802.11 WLAN standards,[3] some of which are summarized in Table 17.1. The first standard, 802.11, was released in 1997 and revised two years later. The 802.11a and 802.11b standards were released in September 1999. A new standard, 802.11g, was released near the end of 2001, as a high-rate extension to 802.11b.

Depending on the specific standards, the IEEE 802.11 WLAN may run as fast as 2 Mbps at 2.4 GHZ or 54 Mbps at 5-GHZ frequency. Newer wireless standards, currently being developed by various IEEE task groups, may bring the speed of WLANs up to the range of 200 to 300 Mbps.

In addition to the major 802.11 standards, as listed in Table 17.1, there exist other IEEE standards or recommended practices related to the 802.11 protocols, such as 802.11f and 802.11h. There also exist several ongoing task groups working on developing protocols related to the performance and security of 802.11 protocols, such as 802.11e and 802.11i. The standards and drafts are available online from the IEEE Web site.*

---

* http://standards.ieee.org/getieee802/802.11.html.

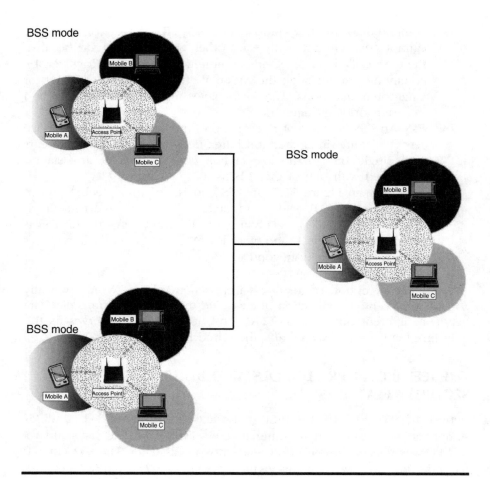

**Figure 17.3  Extended service set.**

■ *802.11e (QoS for 802.11)*: The goal of the IEEE 802.11e task group is to enhance the 802.11 medium access control (MAC) layer, to improve and manage QoS, and to enhance security and authentication mechanisms. These enhancements should provide the quality required for services such as IP telephony and video streaming.

■ *802.11f (Inter-Access Point Protocol [IAPP])*: IEEE 802.11f, released in July 2003, documents IEEE's recommended practice for multivendor AP interoperability via an IAPP across distribution systems supporting IEEE 802.11 operation.

■ *802.11h (power management extension for Europe)*: 802.11h is an amendment to the IEEE 802.11 protocol. Its goal is to provide a spectrum and transmit power management extensions in the 5-GHz band in Europe. The standard was released in October 2003.

**Table 17.1 The IEEE WLAN Standards**

| Standards | Description / Dates of Approval |
| --- | --- |
| 802.11 | Data rates up to 2 Mbps in the 2.4-GHz ISM band / July 1997: first release; 1999: current edition |
| 802.11a | Data rates up to 54 Mbps in the 5-GHz UNII band / September 1999 |
| 802.11b | Data rates up to 11 Mbps in the 2.4-GHz ISM band / September 1999 |
| 802.11g | High-rate extension to 802.11b allowing for data rates up to 54 Mbps in the 2.4-GHz ISM band / November 2001: draft standard adopted; June 2003: full ratification |

■ *802.11i (the forthcoming 802.11 security standard)*: The goal of the IEEE 802.11i task group is to enhance the 802.11 MAC layer with security and authentication mechanisms. The current status of that task group is available at IEEE grouper site.* As of April 2004, the group is working on draft 7.0 of the standard. We will discuss 802.11i in the section titled "The IEEE 802.11i Protocol."

## Built-In 802.11 Security Features

The security features provided in 802.11 are as follows:

1. *Service Set ID (SSID)*: The SSID acts as a WLAN identifier. Thus, all devices trying to connect to a particular WLAN must be configured with the same SSID. It is added to the header of each packet sent over the WLAN and verified by the AP. A client device** cannot communicate with an AP unless it is configured with the same SSID.
2. *Wired Equivalent Privacy (WEP) protocol:* According to the 802.11 standard, "Wired equivalent privacy is defined as protecting authorized users of a wireless LAN from casual eavesdropping. This service is intended to provide functionality for the wireless LAN equivalent to that provided by the physical security attributes inherent to a wired medium."[2] IEEE specifications for wired LANs do not include data encryption as a requirement. This is because approximately all of these LANs are secured by physical means

---

* http://grouper.ieee.org/groups/802/11/Reports/tgi_update.htm.
** Throughout the text, the word *client* is used interchangeably with the word *station* and the word *node*. All of these refer to the wireless device used to connect to a WLAN.

such as walled structures, controlled entrance to buildings, etc. However, no such physical boundaries can be provided in the case of WLANs, thus justifying the need for an encryption mechanism such as WEP.

3. *MAC address filtering*: In this scheme, the AP is configured to accept association and connection requests from only those nodes with MAC addresses that are registered with the AP. Association and connection requests sent by other wireless devices will be rejected. Although an unrealistic protection method in an enterprise network environment, MAC address filtering can be an effective method in smaller networks at homes or small businesses.

## WLAN VULNERABILITIES

Ubiquitous network access without wires is the main attraction underlying wireless network deployment. Although this seems to be enough of an attraction, there is another side to the picture. In this section, we discuss how WLANs could be vulnerable to a myriad of intrusion methods.

### General Wireless Network Vulnerabilities

All wireless networks share a unique difference from their wired counterparts, i.e., their use of radio as transmission medium, which contributes to a unique vulnerability — lack of physical security. Besides, wireless networks may suffer other vulnerabilities, some of which they share with wired networks, such as invasion and resource stealing and DoS. Other vulnerabilities such as rogue APs are associated only with wireless networks. We discuss these vulnerabilities below:

■ *Lack of physical security:* Unlike wired networks, the signals of a wireless network are broadcasted among the communicating nodes. A hacker with a compatible wireless device can intercept the signals when the intercepting device is within the broadcasting range of the communication paths. A hacker with a wireless laptop, for example, may be physically outside a building but can still intercept and then decrypt wireless communications among devices within the building.

■ *Invasion and resource stealing:* Resources in a network include access to various devices (such as printers and servers) and services (such as connectivity to an intranet or the Internet). To invade a network, the attacker will first try to determine the access parameters for that particular network. Hacking techniques such as MAC spoofing may be used to attack a WLAN.[4,5] For example, if the underlying network uses MAC address-based filtering of clients, all an intruder has to do is to find out the MAC address and the assigned IP address for a particular client. The intruder will wait until that client goes off the network and then start using the network and its resources, appearing as a valid user.

■ *Traffic redirection:* An intruder can change the route of the traffic, causing packets destined for a particular computer to be redirected to the attacking station.

■ *DoS:* Two types of DoS attacks against a WLAN can exist. In the first case, the intruder tries to bring the network to its knees by causing excessive interference. An example could be excessive radio interference caused by 2.4-GHz cordless phones.[6] A more focused DoS attack would be an attacking station sending an 802.11 "disassociate" message or replaying a previously captured 802.1X "EAPoL-log-off" message* to the target station and effectively disconnecting it (as in *session-hijack attacks*). The latter type of DoS attack is described in more detail in the section titled "The IEEE 802.11i Protocol."

■ *Rogue AP:* A rogue AP is one that is installed by an attacker (usually in public areas such as shared office space, airports, etc.) to accept traffic from wireless clients, to which it appears as a valid authenticator. Packets thus captured can be used to extract sensitive information or for launching further attacks by, for example, modifying the content of the captured packet and reinserting it into the network.

## IEEE 802.11 Vulnerabilities

The previously stated concerns relate to wireless networks in general. Some of the security concerns raised specifically against IEEE 802.11 networks are as follows.[7]

■ *MAC address authentication:* Such authentication establishes the identity of the physical machine, not its human user. Thus, an attacker who manages to steal a laptop with a registered MAC address will appear to the network as a legitimate user.

■ *One-way authentication:* WEP authentication is client centered or one-way only. This means that the client has to prove its identity to the AP but not vice versa. Thus, a rogue AP may successfully authenticate the client station and then subsequently will be able to capture all the packets sent by that station through it.

---

* *EAP over LAN (EAPoL)* is a standard for encapsulating EAP messages.

- *Static WEP keys:* There is no concept of dynamic or per-session WEP keys in 802.11 specifications. Moreover, the same WEP key has to be manually entered at all the stations in the WLAN, causing key-management issues.
- *SSID:* Because SSID is usually provided in the message header and transmitted as cleartexts, it provides little security.
- *WEP key vulnerability:* Many concerns have been raised regarding the usefulness of WEP in securing 802.11 WLANs. Some of them are as follows:
  1. *Manual key management* — Keys need to be entered manually on all the clients and APs. Such overhead may result in infrequently changed WEP keys.
  2. *Key size* — The IEEE 802.11 design community blames 40-bit RC4 keys for WEP vulnerability and recommends the use of 104- or 128-bit RC4 keys instead. Although using larger key size does increase the work of an intruder, it does not provide a completely secure solution.[8]
  3. *Initialization vector (IV)* — IV is used to avoid encrypting two identical plaintexts with the same keystream, thus resulting in the same ciphertext. By combining a randomly generated IV with the key, the probability of two identical plaintexts being encrypted into identical ciphertexts is minimized. In WEP encryption, the secret WEP key is combined with a 24-bit IV to create the key. RC4 takes this key as input and generates a key sequence equal to the total length of the plaintext and the IV. The key sequence is then eXclusive-ORed (XORed) with the plaintext and the IV to generate the ciphertext. According to the findings reported in Reference 8, the vulnerability of WEP stems from its IV and not from its smaller key size. WEP is based on RC4 algorithm. Two frames that use the same IV almost certainly use the same secret key and keystream. Moreover, because the IV space is very small, repetition is guaranteed in busy networks.
  4. *Decryption dictionaries* — Infrequent rekeying and frames with the same IV result in a large collection of frames encrypted with same keystreams. These are called *decryption dictionaries*.[9,10] Therefore, even if the secret key is not known, more information is gathered about the unencrypted frames and may eventually lead to the exposure of the secret key.

With vulnerabilities outlined in the preceding text, it is reasonable to assume that an 802.11 WLAN protected by WEP alone can be easily cracked by using readily available tools such as AirSnort and WEPCrack. Alternative security solutions are apparently needed.

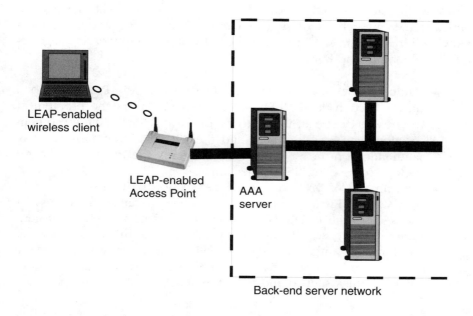

LEAP-enabled
wireless client

LEAP-enabled
Access Point

AAA
server

Back-end server network

**Figure 17.4   Wireless security via LEAP.**

## ALTERNATIVE SOLUTIONS FOR WLAN SECURITY

To secure 802.11 WLANs for critical applications, several alternative solutions have been adopted. Some of the common solutions are discussed in this section, including Cisco's proprietary LEAP protocol, SSL, VPN, the upcoming IEEE 802.11i protocol, and the WPA protocol.

### The Cisco LEAP Protocol

Cisco *Lightweight EAP\** supports mutual authentication between a client and a RADIUS\*\* server. LEAP was introduced by Cisco in December 2000 as a way to quickly improve the overall security of WLAN authentication.

As shown in Figure 17.4, both the wireless client and the AP must be LEAP enabled. An authentication server such as RADIUS is present in the server network to provide authentication service to the remote user.

---

\* Extensible Authentication Protocol (EAP) is a method of conducting an authentication conversation between a user and an authentication server.

\*\* Remote Authentication Dial-In User Service (RADIUS) is a protocol that provides authentication, authorization, and accounting (AAA) services to a network.

Cisco has addressed the above described WEP vulnerabilities with WEP enhancements, such as message integrity check (MIC) and per-packet keying.[11] In addition, LEAP provides the following countermeasures against WEP vulnerability in 802.11:

- Mutual authentication between client station and AP: As described in the section titled "IEEE 802.11 Vulnerabilities," the problem of rogue APs can be attributed to the one-way, client-centered authentication between the client and the AP. LEAP requires two-way authentication, i.e., a client can also verify the identity of the AP before completing the connection.
- Distribution of WEP keys on a per-session basis: As opposed to the static WEP keys in 802.11, LEAP supports dynamic session keys. Both the RADIUS server and the client independently generate this key, so it is not transmitted through the air. An attacker posing as an authenticated client will not have access to the keying material and will not be able to replicate the session key, without which frames sent to and from the attacker will be dropped.

## SSL

SSL is an application-level protocol that enables end-to-end security between two communicating processes. As shown in Figure 17.5, in a WLAN environment, the SSL client runs on the wireless station, and the SSL server runs on the target application or Web server. When communicating with an AP, a user is not able to access resources over the wireless connection until properly authenticated. This authentication is accomplished via the additional level of SSL security encryption. Once an SSL client is authenticated with an SSL-enabled server, subsequent data transmissions between them are encrypted.

Being an application level protocol, SSL provides the system implementers selective authentication for some of the back-end applications or servers behind the APs. In comparison, most other wireless security solutions, including LEAP, VPN, 802.11i, and WPA, are network or lower-level protocols, which typically enforce across-the-board implementation of secure access to the network behind the AP.

## VPN

VPN technology provides the means to securely transmit data between two network devices over an insecure data transport medium.[12] VPN has been used successfully in wired networks, especially when using an insecure network such as the Internet as a communication medium. The

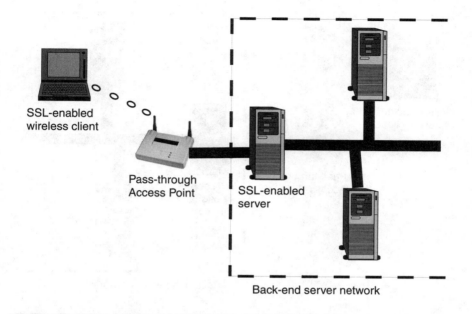

**Figure 17.5    Wireless security via SSL.**

success of VPN in wired networks and the Internet has prompted developers and administrators to deploy VPN to secure WLANs. As shown in Figure 17.6, when used to secure a WLAN, the VPN client software runs on the wireless client machine, whereas the VPN server runs on one of the back-end servers. An encrypted tunnel is formed between the VPN client and the VPN server, thus ensuring the confidential data transmission over the wireless network.

VPN works by creating a tunnel on top of a protocol such as IP. VPN technology provides three levels of security:[12]

- *Authentication*: A VPN server should authorize every user who logs on at a particular wireless station, trying to connect to the WLAN using a VPN client. Thus, authentication is user based instead of machine based.
- *Encryption*: VPN provides a secure tunnel on top of an inherently insecure medium like the Internet. To provide another level of data confidentiality, the traffic passing through the tunnel is also encrypted.
- *Data authentication*: It guarantees that all traffic is from authenticated devices.

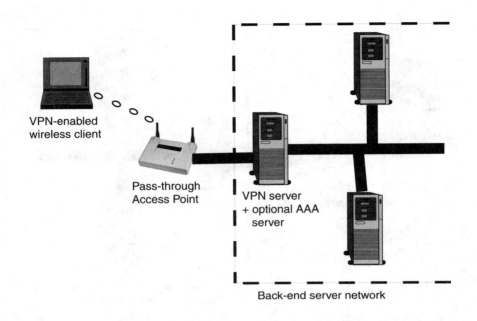

**Figure 17.6    Wireless security via VPN.**

## The IEEE 802.11i Protocol

As stated in a page* of the IEEE 802.11i task group, "The purpose of task group I is to: Enhance the current 802.11 MAC to provide improvements in security." To reach that goal, the IEEE 802.11i task group proposed a new protocol called Robust Security Network (RSN).

### *RSN in 802.11i*

RSN uses the IEEE 802.1X port-authentication standard to authenticate wireless devices to the network and to provide the dynamic keys it requires.[13] RSN consists of two basic subsystems:[14,15]

1. *Data privacy mechanism*: The Temporal Key Integrity Protocol (TKIP) is used to patch WEP for legacy hardware based on RC4, whereas the AES-based protocol is used as a long-term security solution.

---

\*    http://grouper.ieee.org/groups/802/11/Reports/tgi_update.htm (accessed 4/22/04).

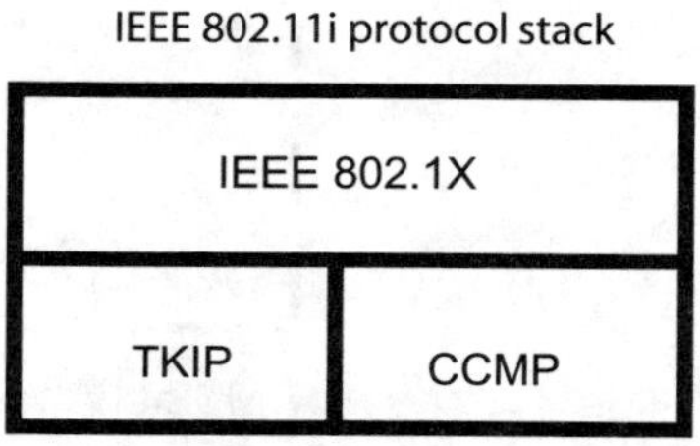

IEEE 802.11i protocol stack

| IEEE 802.1X | |
| :---: | :---: |
| TKIP | CCMP |

**Figure 17.7   IEEE 802.11i.**

2. *Security association management*: It adopts IEEE 802.1X authentication to replace IEEE 802.11 authentication, and it uses IEEE 802.1X key management to provide cryptographic keys.

As shown in Figure 17.7, the 802.11i protocol consists of three underlying protocols organized into two layers. On the bottom layer are TKIP and CCMP, the Counter Mode CBC-MAC Protocol.* Both encryption protocols provide improved data integrity over WEP. CCMP is an AES-based protocol, chosen to replace the old RC4 protocol in future 802.11 devices. Although TKIP is optional in 802.11i, CCMP is mandatory for anyone implementing 802.11i.[16]

Placed on top of TKIP and CCMP is the 802.1X protocol, which provides user-level authentication and encryption key distribution.

## The IEEE 802.1X Protocol

IEEE 802.1X is a port-based authentication protocol, which may be used in wired or wireless networks. Figure 17.8 is an illustration of 802.1X used in a wireless network. There exist three different types of entities in a typical 802.1X network, including a supplicant, an authenticator, and an authentication server.[17] The authenticator is the port that enforces the authentication process and routes the traffic to appropriate entities on the network. The supplicant is the port requesting access to the network. The authentication server authenticates the supplicant, based on the supplied credentials and is typically a separate entity on the wired side of the network but could also reside directly in the authenticator.[16]

To permit the EAP traffic before the authentication succeeds, a dual-port model is used in IEEE 802.1X specifications. In an unauthorized

---

* *Counter-mode cipher block chaining message authentication code* is an encryption mechanism for data on packet-based networks.

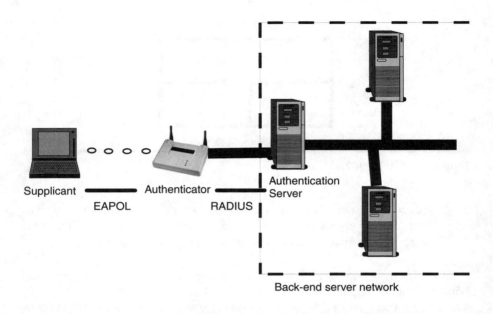

Back-end server network

**Figure 17.8   IEEE 802.1X in 802.11 WLANs.**

(uncontrolled) state, the port allows only DHCP and EAP traffic to pass through.

When applied to 802.11, the 802.1X specification includes two main features: (1) logical ports and (2) key management.[18] In the rest of this section, we first discuss these two features, followed by discussions of vulnerabilities unveiled by some researchers.

- *Logical ports:* Unlike wired networks, wireless stations are not connected to the network by physical means. They must have some sort of association relation with an AP in order to use the WLAN. This association is established by allowing the clients and the AP to know each other's MAC address. This combination of MAC address of the AP and that of the station acts as a logical port. This then acts as a destination address in EAPoL protocol exchanges.
- *Key management:* In IEEE 802.1X, key information is passed from an AP to a station, using EAPoL-key message. Keys are generated dynamically, at a per-session basis.

A typical configuration of a WLAN using IEEE 802.1X is shown in Figure 17.8. The supplicant is the wireless station, which authenticates with the authentication server by using EAPoL to communicate with the AP, which acts as the authenticator. Messages are exchanged between the supplicant and the authenticator to establish the supplicant's identity. The authenticator then transfers the supplicant's information to the authentication server, using RADIUS. All communications between the authentication server and the supplicant pass through the authenticator, using EAP over LAN (i.e., EAPoL) and EAP over RADIUS, respectively. This creates an end-to-end EAP conversation between the supplicant and the authentication server.

## Vulnerabilities of 802.11i

The design goal of the IEEE 802.11i protocol is to achieve enhanced MAC level security by integrating the 802.1X protocol with 802.11. Recent findings by Mishra and Arbaugh,[19] however, have unveiled design flaws and the resulting vulnerabilities in such an integration. Two of the vulnerabilities identified are absence of mutual authentication and session hijacking, of which we provide an overview in the following text. (For more technical details about the design flaws and other vulnerabilities, consult the original publication.)

### Absence of Mutual Authentication

According to 802.1X specifications, a supplicant always trusts the authenticator but not vice versa. Consider Figure 17.9. There is no "EAP request" message originating from the supplicant (the client). It only responds to the requests sent by the authenticator (the AP). This one-way authentication opens the door for man-in-the-middle attacks. The "EAP success" message sent from the authenticator to the supplicant contains no integrity-preserving information. As shown in Figure 17.9, an attacker can forge this packet to start the attack.

### Session Hijacking

With IEEE 802.1X, RSN association has to take place before any higher-layer authentication. Thus, we have two state machines in 802.11i, one being the classic 802.11 and the other the 802.1X-based RSN state machine. Their combined action should dictate the state of authentication. However, due to a lack of clear communication between these two state machines and message authenticity, a session-hijacking attack becomes possible.

The following list shows how a session may be hijacked in 802.11i (refer to Figure 17.10):

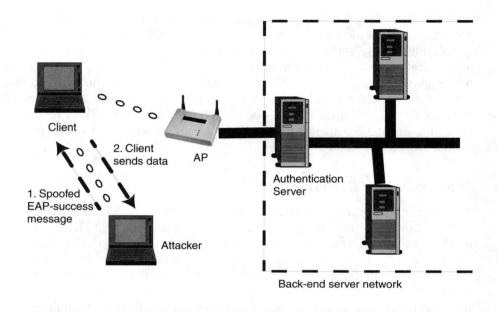

**Figure 17.9 Man-in-the-middle attack in 802.11i.**

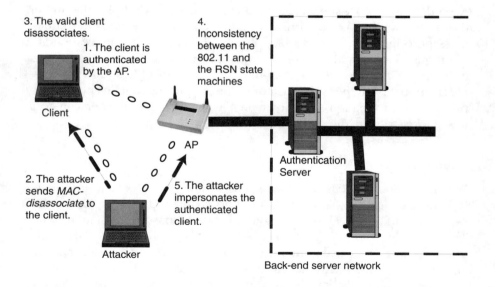

**Figure 17.10 Session-hijack attack in 802.11i.**

1. First, the supplicant and the authenticator (the AP) engage in the authentication process, which results in the supplicant being authenticated.
2. An attacker then sends a "MAC-disassociate" message, using the AP's MAC address.
3. The valid supplicant will disassociate when receiving the "MAC-disassociate" message.
4. This causes the RSN state machine to transfer to the unassociated state. However, because this disassociate message was sent by the attacker (impersonating the real AP), the real AP does not know about it. Thus, the 802.11 state machine remains in authenticated state for that particular client in the real AP.
5. The attacker then gains network access, using the MAC address of the authenticated supplicant (which is disassociated by now).

## The WPA Protocol

As anticipated by most Wi-Fi manufacturers, the forthcoming IEEE 802.11i protocol, currently still in draft mode, is the "ultimate" security mechanism for 802.11 WLANs. The WPA protocol, backed by the Wi-Fi Alliance*, is considered as a temporary solution to the weakness of WEP.

WPA is designed to secure all versions of 802.11 devices, including 802.11b, 802.11a, and 802.11g, multiband and multimode.[21] As a temporary solution prior to the ratification of the 802.11i standard, WPA is forward compatible with the 802.11i specification and is actually a subset of the current 802.11i draft, taking certain pieces of the 802.11i draft that are ready to bring to market today, such as its implementation of 802.1X and TKIP.

WPA offers two primary security enhancements over WEP: an improved data encryption, which was weak in WEP, and a user authentication, which was largely missing in WEP.[20]

1. *Enhanced data encryption through TKIP*: To address WEP's known vulnerabilities (as discussed in the section titled "IEEE 802.11 Vulnerabilities"), TKIP provides important data encryption enhancements including a per-packet key-mixing function, a message integrity check (MIC) nicknamed "Michael," an extended IV with sequencing rules, and a rekeying mechanism.
2. *Enterprise-level user authentication via 802.1X and EAP*: To strengthen user authentication, WPA implements 802.1X and EAP. Similar to LEAP, WPA's user authentication framework uses a central

---

* Wi-Fi Alliance: http://www.wifialliance.com/OpenSection/index.asp.

authentication server such as RADIUS to authenticate each user before allowing them to connect to the network. It also supports mutual authentication to allow a user to authenticate an AP, preventing rogue APs.

The main pieces of the 802.11i draft that are not included in WPA are secure IBSS, secure fast handoff, and secure deauthentication and disassociation, as well as enhanced encryption protocols such as AES-CCMP.[20]

## SUMMARY

Following an introduction to WLAN, we gave an overview of the various types of WLANs. Vulnerabilities of WLANs and the IEEE 802.11 protocols were then discussed, followed by a discussion of alternative security solutions that may be used to protect WLANs, including Cisco's LEAP, SSL, VPN, IEEE 802.11i, and WPA. It should be noted that IEEE 802.11i protocols are yet to be standardized, and vulnerabilities discussed in this chapter are associated with the current draft, which will be further revised.

An important issue related to WLANs is the potential impact a security measure may have on the performance of a given WLAN. Performance of WLANs can be optimized by tweaking various factors like encryption methods, fragmentation threshold, RTS/CTS, etc. It is also dependent upon external factors including distance, other devices operating in the same frequency range such as microwave ovens, cordless phones, etc. Readers interested in this line of work may see Reference 22.

## ACKNOWLEDGMENT

This material is based upon work supported by the National Science Foundation under Grant No. 0311592.

This chapter was adapted from one of the earlier publications of the authors: Wireless LAN Security and Laboratory Designs. *The Journal of Computing Sciences in Colleges,* Volume 19, Issue 3, January 2004.

## REFERENCES

1. John, C. (2002). Report Forecasts WLAN "Last-Mile" boom. Network World Fusion, August 5, 2002. http://www.nwfusion.com/news/2002/0805alex.html.
2. IEEE 802.11 (1999). Wireless LAN Medium Access Control (MAC) and Physical Layer (PHY) Specifications. http://standards.ieee.org/getieee802/download/802.11-1999.pdf.
3. WLAN Association (2002). Wireless Networking Standards and Organizations, WLANA Resource Center, April 17, 2002. http://www.wlana.org/pdf/wlan_standards_orgs.pdf.

4. KLC Consulting (2003). Change MAC Addresses on Windows 2000 and XP. http://www.klcconsulting.net/Change_MAC_w2k.htm.

5. Wright, J. (2003). Detecting Wireless LAN MAC Address Spoofing. http://home. jwu.edu/jwright/papers/wlan-mac-spoof.pdf.

6. Vollbrecht, J., David, R., and Robert, M. (2001). Wireless LAN Access Control and Authentication, a white paper from Interlink Networks Resource Library, 2001. http://www.interlinknetworks.com/images/resource/WLAN_Access_ Control.pdf.

7. Interlink Networks (2002a). Wireless LAN Security Using Interlink Networks RAD Series AAA Server and Cisco EAP-LEAP, application notes at Interlink Networks Resource Library, 2002. http://interlinknetworks.com/images/ resource/wireless_lan_security.pdf.

8. Walker, J.R. (2000). Unsafe at Any Key Size: an Analysis of the WEP Encapsulation, 802.11 security papers at NetSys.com, October 27, 2000. http://www.netsys. com/library/papers/walker-2000-10-27.pdf.

9. Sangram, G. and Vetha Manickam, S.A. Wireless LAN Security: Today and Tomorrow, an online report, the Center for Information and Network Security, Dune University. http://hyatus.dune2.info/Wireless_802.11/wireless-lan-security.pdf.

10. Borisov, N., Goldberg, I., and Wagner, D. (2001). Intercepting mobile communication: the insecurity of 802.11. In *Proceedings of the 7th Annual Conference on Mobile Computing and Networking*, pp. 180–188, 2001.

11. Cisco Networks (2002). Cisco Aironet Response to University of Maryland's paper. http://www.cisco.com/warp/public/cc/pd/witc/ao350ap/prodlit/1680_pp. pdf.

12. Trudeau, P. (2001). Building Secure Wireless Local Area Networks, a white paper at Colubris.com, 2001. http://download.colubris.com/library/whitepapers/ WP-010712-EN-01-00.pdf.

13. Moskowitz, R. (2003). The WLAN's Weakest Link. Network Computing, March 5, 2003.

14. Philip, C. Robust Security Network: The Future of Wireless Security, System Experts Corporation. http://www.systemexperts.com/win2k/SecureWorldExpo-RSN. ppt.

15. Mansfield, B. (2002). WLAN and 802.11 SECURITY, Internet Developers Group, Netscape Communications, June 18, 2002. http://www.inetdevgrp.org/ 20020618/WLANSecurity.pdf.

16 Eaton, D. (2002). Diving into the 802.11i Spec: A Tutorial, CommsDesign.com, November 26, 2002. http://www.commsdesign.com/design_center/wireless/ design_corner/OEG20021126S0003?_loopback=1.

17. Open Source Implementation of IEEE 802.1X. http://www.open1x.org/.

18. Interlink Networks (2002b). Introduction to 802.1X for Wireless Local Area Networks, a white paper at Interlink Networks Resource Library, 2002. http:// www.interlinknetworks.com/images/resource/802_1X_for_Wireless_LAN.pdf.

19. Mishra, A. and William, A.A. (2002). An Initial Security Analysis of the IEEE 802.1X Standard, *CS-TR-4328*, Department Of Computer Science, University of Maryland, http://www.cs.umd.edu/~waa/1x.pdf, February 6, 2002.

20. Wi-Fi Alliance Overview at http://www.wi-fi.org/OpenSection/pdf/Wi-Fi_ Protected_Access_Overview.pdf.

21. Wi-Fi Protected Access: Strong, Standards-Based, Interoperable Security for Today's Wi-Fi Networks (2003). http://www.wifialliance.com/OpenSection/ pdf/Whitepaper_Wi-Fi_Security4-29-03.pdf, April 29, 2003.

22. Zahur, Y., Doctor, M., Davari, S., and Yang, T.A. (2003). 802.11b performance evaluation. *The Proceedings of the 2nd IASTED International Conference on Communications, Internet, and Information Technology (CIIT 2003)*. http://sce.cl.uh.edu/yang/research/CIIT03finalPaper.pdf.

# V

---

# SECURITY II

# 18

## WLAN SECURITY: ISSUES AND SOLUTIONS

*M. Farrukh Khan*

### INTRODUCTION

One of the key requirements for a robust IT infrastructure is the provision of extensive facilities that allow safe and secure dissemination, exchange, and archiving of critical information. Wireless systems have recently matured to a degree where they are complementing or even replacing the traditional wireline systems. As far as technological feasibility is concerned, communication bandwidth and computational prowess have advanced sufficiently to bring many E-commerce and E-government applications into the realm of possible deployment. This is particularly true in the case of static and mobile end users as well as smaller systems.

Today's wireless computer and communication systems are highly complex. Consequently, the successful and productive deployment of security facilities, whether small or large, requires a multifaceted approach that balances the issues of efficiency, cost-effectiveness, robustness, scalability, stability, interoperability, etc. Many of these issues are common with wireline technologies, but most problems are exacerbated in the wireless environments. As we describe wireless and WLAN security problems and solutions, special attention will be drawn to those distinctive issues.

In the next section, we review the basics of practical cryptographic technologies and describe a few applications in E-business and E-government that are enabled by these technologies. We then proceed to examine the underlying reasons for the movement to use special problems introduced in wireless security domains. In the subsequent sections, we describe challenges that need to be attended to so as to benefit practically from wireless networking technologies. There is strong hope that emerging technologies will mature to form a strong basis for a safe wireless security infrastructure.[7]

## ENABLING SECURITY TECHNOLOGIES AND SAMPLE APPLICATIONS

Currently, there is an exponential growth in the deployment of wireless systems. We are witnessing a period computational history when advances in technology coupled with advances in cryptographic methodologies have made an exciting array of secure services that were never before practical on a large scale.

At a very basic level, cryptography utilizes a key to allow the transformation of data into unintelligible material that cannot be reprocessed to obtain the original data without knowing the key. The available cryptography techniques may be broadly divided into secret-key cryptography and public-key cryptography. In secret-key cryptography, a secret (or key) is shared between the sender and receiver. This shared secret is used to process the data at each end of the communication. Without possession of the secret, it should be computationally infeasible to encrypt or decrypt data. In public-key cryptography, a key for user x consists of a pair (public(x) and private(x)). For the purpose of encryption, the owner of the key uses private(x), whereas all others use public(x). In addition, it should be computationally infeasible to discover private(x) given only the value of public(x). With a slight modification, this system can also be used to provide nonforgeable digital signatures[2].

Given these cryptographic functions, it should be relatively simple to achieve all of the following security objectives, at least in theory. These security objectives include the following:

*Confidentiality:* Keeping secret the communication between the sender and receiver, one or both of which exist on a WLAN

*Integrity:* During any transaction using wireless systems, ensuring that data is not modified maliciously

*Authentication:* Verifying the identity of a principal in wireless communication

*Non-repudiation:* Inability of a principal to disown an earlier communication that took place using a wireless system

*Availability:* Ensuring that a wireless service is operational and usable whenever a principal wishes to use it

Each of the preceding services may be constructed using cryptographic techniques. A few examples follow. A basic requirement of governmental and business communication is that it be confidential. For example, business customers involved in online shopping using credit cards do not wish to divulge their credit card numbers to third parties. In the realm of E-government, this could be video conferences between ministers, with all data being transported over a WLAN. Cryptography solves the problem of transporting such communication over public channels because wireless communications can be easily monitored by a third party to observe what data travels over it.

Data integrity has implications for the accuracy of data being exchanged. This is particularly relevant in terms of financial transactions, quantitative data, textual data, and sensitive data such as that meant for real-time embedded systems in transportation and industrial control. Unfortunately, wireless systems are notorious for introducing much higher error rates than comparative wireline systems. Robust encoding techniques such as Code Division Multiple Access (CDMA), and forward error correction techniques have mitigated these problems to some extent.

Authentication of the source is essential to avoid impersonation attacks, where a malicious intruder masquerades as a genuine principal. In today's highly connected world, it would be desirable to have two-way mutual authentication, i.e., both the source and the destination authenticate each other before full communication can be established between them. The problem is compounded in wireless systems because one can no longer assume the identity of the party at the other end of the secure wire; indeed, there is no wire that can be secured. Therefore, cryptographic authentication becomes mandatory in sensitive wireless applications.

Non-repudiation is essential to prevent a principal from denying a previous commitment with another principal. Thus, in the case of a dispute, a third party such as a judge in a court of law should be able to ascertain and affirm the identities of concerned parties in the context of a given transaction. The issues in WLAN environments are very similar to those in wireline systems.

Suppose that citizens are uniquely identified through a combination of cryptographic information on a live smart card that is part of a secure personal area network (PAN) such as some type of modified Bluetooth. Secure, authenticated communication in different domains of business or government is an obvious application of this. A potential application could be secure and the voting verifiable using public-key cryptography. Each principal could vote after encrypting the vote with the voter's private key. For the purposes of verification, each particular vote is readily verified by using the respective voter's public key. This process could be repeated as many times as necessary by different entities.

On the other hand, a large number of votes (such as 100 million) could be decomposed into much smaller jobs, and these jobs could be processed in parallel. In this manner, a government can potentially accelerate the vote counting process to any desirable speed. We ensure that we have integrity (tampering with the voting information will effectively destroy it), non-repudiation (only the concerned voter knows the private key), and verifiability (anyone can use the public key and verify a particular vote). Wireless PANs add to the mobility of the users and hence to the speed of the entire process. Thus, they give a strong enough underlying system; citizens need not line up to get desired services. As far as confidentiality is concerned, it is a policy issue. If needed, confidentiality could be incorporated by either using secret key encryption or by not publishing public keys except to the concerned parties, such as election commissions and judges. Further elaboration may be found in standard texts such as in Reference 1.

## THE NEED FOR STRONG CRYPTOGRAPHY IN WIRELESS APPLICATIONS

Over the last five decades, wireline systems evolved concurrently with security techniques and protocols. This meant that proliferation of weak cryptography was abetted in large part by the first-mover advantage. Weak cryptographic techniques were hard to get rid of once they had been embedded in the systems. This problem continues to this day in wireless systems. For example, shortly after the development of WEP (the Wired Equivalency Protocol), exploitable flaws were discovered by researchers at Berkeley and other places.[8] However, the lack of alternatives and commercial inertia has meant continued heavy deployment of WEP in WLANs and other wireless systems.[1]

For almost three decades, DES has been the mainstay of conventional cryptography, and probably more bits have been encrypted with it than any other cipher. Data Encryption Standard (DES), developed and standardized by the National Security Agency (NSA) along the lines of an IBM proposal, uses a 56-bit key. There have been strong concerns regarding DES. One is that the U.S. government may have deliberately left a trapdoor in the standard that would allow it decipher encrypted data with relative ease. The U.S. government denies such a design loophole, but unease about DES somehow persists.

The other, and probably more practical, concern is that a key length of 56 bits is not sufficient for the current adversaries' computational prowess. Thus, anyone who is well funded (such as many governments and corporations) may be able to listen in to DES-encrypted communication. For example, distributed.net and EFF responded to an RSA Labs challenge designed to emphasize the need for strong crypto by breaking a DES key in less than 24 hours.

DES was the most famous but not the only datapoint in events pointing the future toward strong cryptography. For example, some versions of RC4 with 40-bit key lengths were duly hacked by individuals or small teams utilizing server farms or a small number of easily available super-computers. A stopgap measure to address the key length concern is the usage of Triple DES, a standard that effectively increases the key length to 168 bits and makes cracking it several orders of magnitude harder. Although this is strong enough for today's adversaries, the problem with triple-DES is that it is too slow for the level of security that it provides.

To address the aging problems and distrust related to DES, and to counter competing strong cryptographic technologies from the EU and Japan, the NSA has sanctioned a new recommended standard for secret-key encryption — the Advanced Encryption Standard (AES). AES has several things going in its favor, including its being an open standard, unencumbered by patent protection and close scrutiny by the international security community. AES-based technologies allow key lengths of 128, 192, or 256 bits. This gives users ample flexibility in terms of choosing to use strong encryption for high security or weaker encryption for the sake of efficiency. This is particularly relevant in WLANs in which small footprint devices lack sufficient resources for larger key processing.

In addition, there are a few other usable international standards available that fit the bill of strong cryptography, such as IDEA, which have been well documented.[2] In some of there standards, there are a myriad of problems such as patent protection, efficiency, etc. These may be the determining factors in deciding the specific strong cryptography technologies that permeate future wireless systems. Currently, the IEEE 802.11i group is evolving a comprehensive draft security standard for wireless communications. The Wi-Fi Alliance has already floated a standard largely compatible with parts of 802.11i and named it Wi-Fi Protected Access (WPA).

## FURTHER ISSUES FOR WIRELESS SECURITY

For the past decade, a lot of research has taken place in developing strong cryptography. These efforts have been highlighted with the development of such standards as Rijndael, IDEA, etc. Strong cryptography provides an essential ingredient of privacy and security in the emerging wireless IT infrastructures. Technologies for strong cryptography are also readily available. However, as is true in the maturing of most new technologies, the technical feasibility of a product or service is but one part of the wider picture. In the case of strong cryptography, there are a variety of issues that need to be addressed to complete the picture. These include infra-structure development, security policies, and appropriate use of technology. These dimensions need to be incorporated in consumer education and cost–benefit analyses of the security infrastructure. We elaborate in the following text on each of these dimensions related with viable provision of security services.

## Infrastructure Support for Wireless IT

Secure communication is one of the largest consumers of available bandwidth among many applications. Sometimes the communication overhead could be as high as 3000 percent. In addition, wireless infrastructure planners need to ensure widespread access to security services, including authentication facilities through a robust, diverse, and secure public key infrastructure (PKI). Conventional PKI is based on certificates issued by certification authorities (CAs). PKI can be prohibitively expensive or extremely challenging in wireless domains with limited I/O bandwidths and low processing capabilities. In some applications such as secure online voting, one can anticipate millions of encrypted transactions from an extremely wide geographic area and with stringent timing requirements. This means that adequate infrastructure facilities are required for the provision of practical and sustainable security-enhanced goods and services based on the wireless infrastructure.[3] A large amount of communication bandwidth should be provided among all centers of cryptographic data exchange. This is at odds with the inherent limitations in wireless devices with regard to bandwidths, processing power, and memories.

There is large potential for rapid growth of E-business and, as such, encryption of wireless communication needs to become widely prevalent rather than an exception. There are also things that, at first sight, may seem outside the purview of policy makers and IT administrators of non-IT-savvy countries. However, informed and intelligent choices, in many instances, are required to avoid security debacles. Wireless device operating systems need to incorporate and support current open security standards. Application-level software needs to pay particular attention to robustness in the face of malicious attackers. Dependence on products from a single company should be avoided, and consumers should be empowered with choices. Products not conforming to open international wireless standards should be avoided but tolerated to an extent.

## Robust IT Policies for Wireless Security

Whatever regulatory policies are implemented will introduce overheads for wireless systems. A major policy dichotomy is how to balance between people's right to privacy and anonymity on the one hand and the need and desire to know everything on the part of various law enforcement agencies. The government may stipulate extreme policies. For example, there may be a requirement that all cryptographic keys be deposited with (or even generated by) the big brother before they are used for encryption. The big brother could masquerade as a third party. This will be an obvious overhead for already stretched wireless systems.

In either case, there will be a detrimental effect on the growth of E-business and E-government. Such policies will largely ensure that a country's security infrastructure in the business and service sectors will not be compatible and competitive with international peers, again stunting potential growth in E-business and IT-based service sectors.

On the flip side, situations that demand transparency or accountability require that unconditional privacy or anonymity be tempered with key-recovery or key-disclosure mechanisms. These mechanisms should be well documented and consistent, and applied with proper safeguards (such as search warrants from law enforcement agencies). All the concerned parties should be clearly aware of the defined scopes, rights, and responsibilities. This also points to a need for a competent, fair, and stable conflict resolution environment for all the concerned parties, including those inside the system.[4] We can see that wireless systems will have substantial communication and processing loads if enhanced security facilities are deployed.

In short, regulation should be well thought out and carefully measured, and follow the minimalist credo. Finding the optimal balance between consumer freedom and governmental oversight will require careful calculation of risks and benefits. To avoid stunting the growth potential of the wireless IT sector, any policy must consider the resource limitations of wireless systems.

## Wireless Systems Implementation Issues

The security debacles that arose in the early versions of WEP deployment in wireless systems have taught us important lessons. Sometimes, current technology has full potential to provide an envisaged security service. However, the way a particular service is implemented and deployed may well compromise its security value by introducing dangerous vulnerabilities.[5] This is not an easily addressed problem, as experts in software engineering and theory of computation will attest. However, preventive measures can be employed to ameliorate the problems. These measures include formal protocol testing and verification methodologies, sound design methodologies, extensive prototype testing, and adopting the open systems paradigm.

Unlike consumer transactions themselves, protocols and specific implementations must be placed under the microscope of rigorous scrutiny to assess potential weaknesses, scalability, robustness, etc. The bulk of this must be done before the implementations are effected. Any flaws that are discovered in the production phase should be expeditiously communicated to all the concerned parties, along with a recommended course of action to combat the flaws or minimize potential damage. To this end, standards bodies are of great service to the wireless community, including the aforementioned IEEE 802.11i Working Group, the Wi-Fi Alliance, etc.

## CONCLUSIONS AND FUTURE DIRECTIONS

Wireless systems suffer from a relative scarcity of transmission bandwidths, processing power, memory buffers, bus widths, etc. Security mechanisms based on cryptography push the envelope of resource utilization even in wireline systems. Wireless systems, however, make it hard to implement many useful security functions because of their inherent limitations.[6]

Cryptographic techniques and protocols are important ingredients in a robust security service menu. However, they alone are not sufficient as the basis of a viable wireless security infrastructure. Policymakers and researchers need to address other issues before practical deployable solutions are found. These issues include a strong, efficient, and interoperable (with the wireline PKI) PKI, perhaps based on ECC rather than RSA;[9] a robust conflict resolution mechanism; and transparency and accountability in the deployment of communication, hardware, software primitives, etc.

A few relevant issues that are beyond the scope of this chapter but should be mentioned include physical security (a much greater threat than wireline systems because of the broadcast nature of wireless); stability and some level of continuity in terms of national security policies; international interoperability among products from various standards bodies; prioritization of wireless IT security in terms of budgeting and manpower; and dissemination of general and technical knowledge related to the limitations, potential, and utilization of the aforementioned technologies to all concerned. As with any other engineering solution, it is difficult or impossible to bulletproof the systems against determined adversaries: they will probably find holes in reliable systems built out of unreliable components. This is truer for nascent wireless systems. This is where each individual or community will have to make choices after weighing the potential benefits of secure wireless systems against potential costs and potential risks. If the potential risks are greater than the benefits, then it would be wise to stick to alternative systems providing sufficient levels of security.

Finally, education about the availability, limitations, capabilities, idiosyncrasies, liability limits, etc., of all goods and services related to secure wireless systems is essential before the community at large can adopt and fully benefit from the wireless infrastructure and concomitant facilities in the domains of business, government, entertainment, etc.

## REFERENCES

1. Kaufman, C., Perlman, R., and Speciner, M. *Network Security: Private Communication in a Public World.* Prentice Hall PTR, Englewood Cliffs, NJ, 2000.
2. Menezes, A., van Oorschot, P., and Vanstone, S. *Handbook of Applied Cryptography.* CRC Press, Boca Raton, FL, 1997.
3. Madron, T. *Network Security in the 90s: Issues and Solutions for Managers.* John Wiley & Sons, NY, 1992.
4. Christianson, B. et al. (Eds.) *Security Protocols.* LNCS 1550. Springer-Verlag, NY, 1998.
5. Anderson, R. Why Cryptosystems Fail. In *Proceedings of the 1st Conference of Computer and Communication Security,* 1993, pp. 215–227.
6. Potter, B. Wireless security's future. *IEEE Security and Privacy,* July/August 2003, pp. 68–72.
7. Miller, S. Facing the challenge of wireless security. *IEEE Computer,* July 2001, pp. 16–18.
8. Patel, S. Weakness of North American wireless authentication protocol. *IEEE Personal Communications,* June 1997, pp. 40–44.
9. Lauter, K. The advantages of ECC for wireless security. *IEEE Wireless Communications,* February 2004, pp. 62–67.

# 19

## GENERAL SECURITY ANALYSIS OF AN IT CONFIGURATION BASED ON WLAN

*Thijs Veugen and Sander Degen*

### INTRODUCTION

This chapter presents a structured overview of security threats and attacks for WLAN and security services that can be used to overcome or reduce these threats. When analyzing the security of a specific technology, it is important to consider a complete system that uses this technology because it is not only the technology that causes security threats but how it is used in practice within a complete system. One can try to perfectly secure the WLAN, but when the network and applications that use WLAN are not sufficiently secure, the complete system will be insecure. In general, security is just as strong as the weakest link in the communication chain. For example, what is the use of encrypting WLAN messages when they are transmitted in plaintext over the Internet through an access point? Hence, to achieve end-to-end security, the communication link has to be secured up to the application level.

Therefore, when analyzing the security of WLAN, it is important to consider a complete IT configuration in which IEEE 802.11 technology is used. Such an IT configuration is depicted in Figure 19.1.

Figure 19.1 shows two computers (wireless clients) that can communicate with each other through a WLAN. The WLAN consists of two stations, one for each computer, and an access point. Through the access

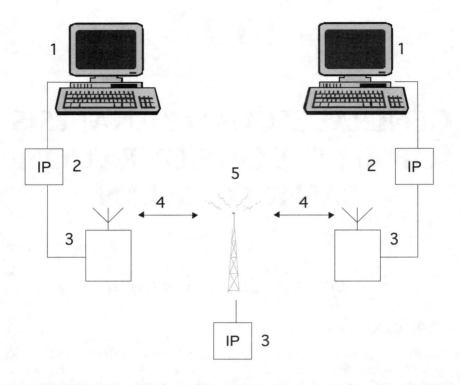

**Figure 19.1   General IT configuration when using WLAN.**

point, the computers can establish a connection to wired networks such as the Internet.

When analyzing the security of the general IT configuration, as depicted in Figure 19.1, five components are considered crucial:

1. Applications running at a wireless client such as e-mail and Internet browsers. Only those applications that involve communication over the WLAN are considered.
2. The wired network technology (usually IP) used to communicate between applications, and between an application and wired networks such as the Internet.
3. The station responsible for the wireless communication with the access point or other stations.
4. The wireless communication link based on the IEEE 802.11 protocol.
5. The access point that enables stations to communicate with each other and to wired networks such as the Internet.

Further analysis of security threats and services in this chapter will be based on the five main components of Figure 19.1. Each security threat can be seen as a threat to one or more security criteria. The following security criteria are considered:

■ Authentication: The source of transmitted data should be absolutely clear. The entity that claims to have sent this piece of data should be the entity that did indeed send it.

■ Confidentiality: People who are not authorized to read specific data should not be able to read this data. The most common way to protect the confidentiality of data is to encrypt it.

■ Integrity: People who are not authorized to change specific data should not be able to change it. For example, it should be possible to detect when the content of a message was manipulated during transmission.

■ Availability: Information (systems) should be available whenever they are supposed to be. In the current world of the Internet, the so-called Denial-of-Service attacks are threats to the availability of Web servers.

The following section contains an overview of security threats and attacks to WLAN, based on the configuration in Figure 19.1. For each threat, which security criteria are involved is explained. Next, an overview of security services that are available for WLAN is provided. These services will be positioned according to the configuration in Figure 19.1 and, for each service, it is explained which security criteria are protected by it. After the first two sections, we will discuss specific WLAN security protocols. A detailed explanation of the Wireless Equivalent Privacy (WEP) protocol, which forms the basic security protocol within WLAN technologies, is given. Some weaknesses of WEP are shown, and suggestions are made on how to overcome these weaknesses.

## SECURITY THREATS AND ATTACKS

In this section an overview of the security threats and attacks to WLAN, based on the configuration in Figure 19.1, is given. Each attack can be seen as a threat to one or more security criteria of one of the five crucial components, as described in the previous section.

1. Threats to applications running on the wireless client: We consider applications that are involved in communication over WLAN such as an Internet browser or an e-mail application. Each application has its own threats and vulnerabilities. It goes beyond the scope

of this chapter to describe all possible threats to these applications, but there are, roughly speaking, three kinds of threats.

- Software exploits: These threats, which evolve from bugs in the software, are well known and published on the Internet. That is the reason for software manufacturers publishing patches to their applications — to fix these bugs. The most common type of software errors is unforeseen buffer overflows.
- Misconfiguration: Although many applications have functionality for security protection, these components have to be configured properly to obtain suitable protection.
- Importing malicious software by e-mail: Besides application weaknesses, there is the risk of malware (malicious software) and spyware (software spying on your network) that can be imported by e-mail.

2. Threats at the level of the wired network technology: It goes beyond the scope of this chapter to describe all possible threats at the network level, but three important threats are:

- Exploits in software and hardware network components: Similar to applications, all network components (routers, switches, servers) are vulnerable due to weaknesses in software and hardware. This also includes the communication protocol at network level, which usually is IP.
- Denial-of-Service by IP: A common threat in the wired world also holds for the wireless network. By flooding a network with IP packets, some network components may not be able to process regular traffic anymore.
- File sharing: By using IP services such as file sharing, one could obtain access to confidential files on a wireless client. This threat could come from other clients within the WLAN or from networks connected to the access point.

3. Threats to the station for wireless communication at the wireless client:

- Insertion of an unauthorized station: An attacker can try to insert a station that is not authorized into the wireless network. This is easy when the access point does not require a password or other kind of authentication method. Otherwise, a brute force dictionary attack could be used to compromise the password.
- Obstruction of legitimate traffic to the station by signal jamming: Because all wireless communication in a WLAN uses the 2.4-GHz frequency, one could easily disrupt the network by overwhelming the legitimate traffic with large amounts of illegitimate traffic. This threat is comparable to the DoS attack at the IP level. It could disable a station from receiving legitimate traffic.

4. Threats to the wireless communication link: These threats are considered in more detail later in this chapter. Here we mention the following threats:
   - Unauthorized interception and monitoring of wireless traffic: Because WLAN communication is transmitted over the air, unauthorized persons could easily intercept the signal. Because access points transmit their signals in a circular pattern, the transmission area extends beyond the boundaries of the WLAN stations.
   - Cryptanalysis of the encrypted messages: Even when the intercepted signal is encrypted, it could be possible to decrypt it by means of cryptanalysis.
5. Threats to the WLAN access point:
   - Insertion of an unauthorized access point: An attacker could try to fool legitimate wireless clients by placing an unauthorized access point with a stronger signal in close proximity to wireless clients. Users attempt to log into the substitute servers and unknowingly give away passwords and similar sensitive data.
   - Obstruction of legitimate traffic to the access point by signal jamming: Because all wireless communication in a WLAN uses the 2.4-GHz frequency, one could easily disrupt the network by overwhelming the legitimate traffic with large amounts of illegitimate traffic. This threat is comparable to the DoS attack at the IP level. It could disable an access point for receiving legitimate traffic.
   - Misconfiguration of the access point: Just like applications, access points can be misconfigured such that they remain at a high risk for attack or misuse.

Each of the above-mentioned attacks forms a threat to one or more security criteria. Table 19.1 shows which security criteria are possibly involved in which threats.

A check (✓) in Table 19.1 means that this particular attack forms a threat to this component with respect to this specific security criterion. Note that threats with respect to one security criterion could lead to threats with respect to other security criteria. For example, an attacker gaining unauthorized access to a station (authentication threat) could consequently be able to read confidential files (confidentiality threat) or alter certain files (integrity threat). In Table 19.1, only the initial threat is considered.

## SECURITY SERVICES

In the previous section a general overview of all security threats to an IT configuration using WLAN was given. This section deals with possible

**Table 19.1  An Overview of Security Threats**

| Threat | Authentication | Confidentiality | Integrity | Availability |
|---|---|---|---|---|
| **1. Threats to applications** | | | | |
| Software exploits | ✔ | ✔ | ✔ | ✔ |
| Misconfiguration of the application | ✔ | ✔ | ✔ | |
| Malicious software | ✔ | ✔ | ✔ | |
| **2. Threats at the network level** | | | | |
| Exploits in network components | ✔ | ✔ | ✔ | ✔ |
| Denial-of-Service | | | | ✔ |
| File sharing | | ✔ | ✔ | |
| **3. Threats to the station** | | | | |
| Insertion of an unauthorized station | ✔ | | | |
| Station signal jamming | | | | ✔ |
| **4. Threats to the wireless communication link** | | | | |
| Unauthorized interception | | ✔ | | |
| Cryptanalysis of encrypted messages | | ✔ | | |
| **5. Threats to the access point** | | | | |
| Insertion of an unauthorized access point | ✔ | ✔ | | |
| Access point signal jamming | | | | ✔ |
| Misconfiguration of the access point | ✔ | | | |

solutions (security services) to overcome these threats. Some of these security services are already available within the technologies used in the IT configuration, other security services are additional.

Just as in the previous section, we use the five different components of Figure 19.1 as a basis for positioning these services. Each described security service can be considered as a way to reduce the security threat with respect to one or more security criteria.

1. Security services in client applications:
   - Public key infrastructure (PKI): This is an additional infrastructure that manages the distribution of cryptographic keys. By using PKI, each client could be provided with an asymmetric key pair. The private key of this pair can be used to authenticate messages that are sent to other clients. Other clients are able to verify these messages by using the public key of the sender. It is also possible to protect confidential messages by encrypting them with the public key of the receiver. Authentication and encryption at the application level is essential for providing end-to-end security.
   - Virtual private network (VPN): This is a means for achieving a "virtual" confidential network while actually transmitting data over an untrusted network (such as a WLAN). This is realized by encrypting at the application level all communication to specific clients. All participating clients together form the actual VPN.
   - Secure e-mail: Some e-mail applications are capable of securing their messages by using, for example, S/MIME. In this way the authentication, confidentiality, and integrity of e-mail messages can be protected.
2. Security services in the wired network technology:
   - TCP/IP security services: The IP protocol has several security services known as IPSec. These security services include authentication of the IP header, exchange of cryptographic keys over IP, and encapsulation of encrypted IP messages.
   - Firewall: A firewall is an additional component that is able to control network traffic. It can filter or block network ports, e.g., FTP, HTTP, SMTP, or Telnet. A firewall can be used to separate wireless networks from the Internet or internal networks, or to lock down access to clients.
   - Router Access Control List (ACL): A router is a specific network component that routes network traffic. An intruder could use a router to obtain unauthorized access to confidential clients. Access to routers can be restricted by using an ACL that specifies who is authorized for access to the router.

■ Intrusion Detection System (IDS): An IDS is an additional component that is used to detect intrusions. A network IDS scans the network traffic for possible attacks, and a host IDS guards all communication at a client.

3. Security services at the station for wireless communication:
   ■ Server Set ID (SSID): An SSID can be seen as a single shared password between the access point and the stations. Only stations with the correct SSID are able to communicate with the access point. This security service is available in most WLAN, but the default (commonly known) value should be changed.

4. Security services in the WLAN technology:
   ■ Wireless Equivalent Privacy (WEP): The main security service in WLAN technology is WEP. WEP uses encryption to protect confidentiality, and a cyclic redundancy check for integrity protection. WEP is explained in more detail in the next section.

5. Security services at the access point:
   ■ Server Set Identifier (SSID): An SSID can be seen as a single shared password between the access point and the stations. Only stations with the correct SSID are able to communicate with the access point. This security service is available in most WLAN, but the default (commonly known) value should be changed.
   ■ SNMP community passwords: Many access points run SNMP agents to provide an interface for SNMP to external networks. A community password is available to restrict access to the SNMP agent, but the default (commonly known) value should be changed.
   ■ Web interface password for configuration: Some access points have a Web interface to configure the access point from a distance. Although from a security point of view such an interface is risky, anyone using this interface should at least use a password to restrict access to the interface and change the default value.

A check (✓) in Table 19.2 means that this particular security service reduces the threat to this component with respect to this specific security criterion. Note that services with respect to one security criterion could lead to reductions with respect to other security criteria. For example, increasing access control such that an attacker is less likely to gain unauthorized access to a station (authentication threat reduction) could consequently reduce the risk of reading confidential files (confidentiality threat reduction), altering certain files (integrity threat reduction), or executing a DoS attack (availability threat reduction). In Table 19.2, only the initial reduction is considered.

**Table 19.2  An Overview of Security Services**

| Security Service | Authentication | Confidentiality | Integrity | Availability |
|---|---|---|---|---|
| **1. Security services at applications** | | | | |
| PKI | ✔ | ✔ | ✔ | |
| VPN | | ✔ | ✔ | |
| Secure e-mail | ✔ | ✔ | ✔ | |
| **2. Security services at the network level** | | | | |
| IPsec | ✔ | ✔ | ✔ | |
| Firewall | ✔ | ✔ | ✔ | ✔ |
| Router ACL | ✔ | | | |
| IDS | ✔ | ✔ | ✔ | ✔ |
| **3. Security services at the station** | | | | |
| SSID | ✔ | | | |
| **4. Security services within WLAN** | | | | |
| WEP | ✔ | ✔ | ✔ | |
| **5. Security services at the access point** | | | | |
| SSID | ✔ | | | |
| SNMP community password | ✔ | | | |
| Web interface password for configuration | ✔ | | | |

In the following two sections we will focus on the WLAN technologies within our general IT configuration. This part of the IT configuration of Figure 19.1 is illustrated in Figure 19.2.

The three components shown in Figure 19.2 are the access point, the wireless communication link, and the station. The names depicted are commonly used names within cryptography and usually denote the sender

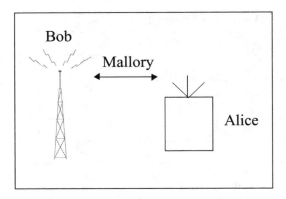

**Figure 19.2   WLAN technologies within the general IT configuration.**

(Alice), the receiver (Bob), and the evil eavesdropper (Mallory). These names are used in the next section.

## WIRED EQUIVALENT PRIVACY

Wired Equivalent Privacy (WEP) is the security module within the IEEE 802.11 protocol. It is designed to make IEEE 802.11 approximately as secure as regular, wired networks. Features of WEP are:

- Required authentication before gaining access to the WLAN network
- Encryption of data through RC4
- Monitoring the data integrity during transmission by means of a Cyclic Redundancy Check (CRC)

Although WEP is a first step in securing a WLAN, it is disabled by default for almost all WLAN products. This decision was made when WLAN producers had to choose between ease of installation and security, which has led to serious problems ever since.

Shortly after the IEEE 802.11 protocol was released, weaknesses were found in WEP implementation. These weaknesses will be discussed in this section, and the solutions to counter them in the following section. But first, it is important to know how WEP was designed.

## WHAT IS WEP?

WEP employs the shared-key principle, so everybody that is connected to a specific access point uses the same password. This password is used for authentication as well as encryption, as will be discussed later in this chapter.

When client Alice connects to access point Bob, Bob has two options:

1. Open authentication
2. Shared-key authentication

With open authentication, Alice immediately has free access to the network. Shared-key authentication, however, requires her to encrypt a challenge text from Bob. Bob then decrypts Alice's reply, using the secret key. If the result matches the original challenge text, it has been proven that Alice knows the secret key and Bob grants her network access.

If WEP encryption is enabled, all the data sent by Alice is encrypted using the secret key.

For example:

Alice performs:
<pre>
        Plaintext:    11001010110
        Keystream:    01011011010
                      ---------------- (XOR)
        Encrypted:    10010001100
</pre>

Bob translates:
<pre>
        Encrypted:    10010001100
        Keystream:    01011011010
                      ---------------- (XOR)
        Plaintext:    11001010110
</pre>

When the same key is used too often, it is possible for attackers to "crack" the key. This cracking can be done by guessing, analyzing, or forcing the plaintext. An example is that the attacker (Mallory), who is able to eavesdrop on the victim's (Alice) network traffic, sends Alice an e-mail. When Alice retrieves that e-mail, Mallory knows the contents of the data, the so-called plaintext, and she can also pick up the WEP-encrypted packets. The exclusive-or (XOR) of the plaintext and the encrypted text returns the keystream; with that, Mallory is capable of decrypting all other packets that were encrypted with the same key.

To counter that threat, WEP employs a so-called initialization vector (IV). This is used as an addition to the secret key, and it is supposed to limit the number of times that the same key is used. The IV is 24 bits, so after $2^{24} = 16,777,216$ differently encrypted packets, the same keys will be used again. Unfortunately, this number of packets is reached after just 1 hr, when an 11-MB connection is used. Second, the "birthday paradox" is also applicable, meaning that there is a 50 percent chance that a key was already used after 4823 (~$2^{12}$) packets. Third, because the packets have a maximum size of about 1500 bytes, an attacker is able to create a keystream database for every IV + secret key combination with a hard disk of (1,500 * $2^{24}$) 26 GB. This is not as ridiculously big as it was in 1997, when 802.11 was approved.

Apart from the problems for the IV, the key itself also has several security issues. Due to a faulty implementation of the RC4 algorithm, a number of so-called "weak keys" can reveal parts of the secret key.

The integrity check that is used in WEP is a CRC of the data in the packet. Alice computes this checksum over the payload, and the recipient Bob will also calculate this checksum and compare it with the one that Alice sent. If there is a difference between the two checksums, the data must have been changed during transit. CRC is adequate for detecting transmission errors but offers very little protection against deliberate attacks on the data.

## WEP IN DETAIL

The schematics that describe the functionality of WEP encryption are shown in Figure 19.3. The secret key and the IV, either a random value or otherwise, are concatenated to form the actual encryption key. This is fed into the pseudorandom number generator (PRNG) to generate a keystream. This keystream is equal to the size of the plaintext message + 4 bytes for the CRC value, and the next step is to XOR the keystream with the message and the concatenated CRC value. This results in the encrypted message, also known as the ciphertext. This ciphertext, together with the IV, is sent to the receiver, who could be either a client or an access point.

The encryption process is then executed backward (Figure 19.4). The recipient uses his or her secret key together with the IV from the sender to form the decryption key. This is used as the seed for the PRNG and should produce the exact same keystream as the sender's. This keystream

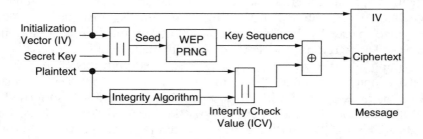

**Figure 19.3 WEP encryption schematics.**

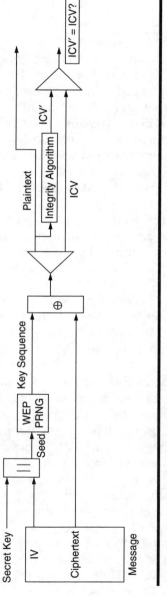

**Figure 19.4  WEP decryption schematics.**

and the ciphertext are XORed, which results in the plaintext message and the CRC value. This CRC value is compared with a CRC value that the recipient calculates over the plaintext, and if they match, the data has been successfully received.

## IEEE 802.1X

The basis for IEEE 802.1X is the Point-to-Point Protocol (PPP), a protocol that is often used in dial-up connections to an Internet service provider. Besides such functionality as redialling in case of a dropped connection, it also grants a user Internet access based on his or her credentials. This latter part is known as the Extensible Authentication Protocol (EAP). IEEE 802.1X is a port-based authentication protocol standard that uses EAP from PPP for handling authentication and key exchange. This specific protocol is called EAPoL (EAP over LAN).

There are three important parties during the IEEE 802.1X authentication:

- Supplicant: the party requesting access to a dial-in server
- Authenticator: the party requesting authentication of the supplicant
- Authentication server: the dial-in server that could grant access to the Internet

In WLANs, usually the client (component 1 in Figure 19.1) is the supplicant, the access point (component 5 in Figure 19.1) is the authenticator, and the Remote Authentication Dial-In User Service (RADIUS) server (positioned between component 5 and its IP component 2) is the authentication server.

The IEEE 802.1X authentication protocol is depicted in Figure 19.5.

When a client associates with an access point, the access point sends an "EAP-Request/Identity" packet to the client. It, in turn, replies with a "EAP-Response/Identity" packet. The access point forwards this to the RADIUS server, which generates an "EAP-Request/Challenge" and sends it back to the client through the access point. The client now generates a response that matches the challenge, and it sends this "EAP-Response/Challenge" to the access point. The access point checks the validity of the response and grants the client network access when the user is correctly authenticated.

IEEE 802.1X supports different authentication algorithms such as the MD-5 challenge authentication, but it also supports one-time passwords, Transport Layer Security (TLS), Tunnelled TLS (TTLS), etc. In Figure 19.5, the MD-5 challenge authentication is used. One big disadvantage of this authentication protocol is that it only authenticates the client and not the server. TLS and TTLS do require the authentication server to prove his or

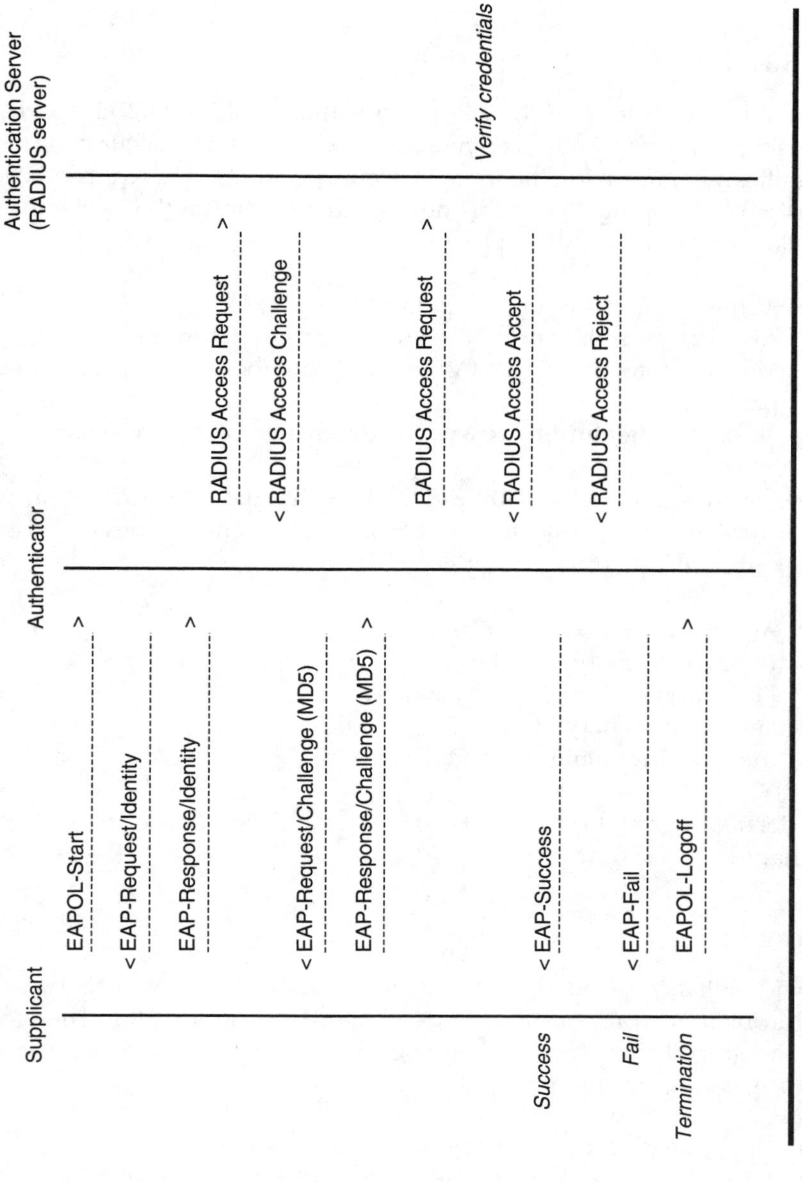

Supplicant

Authenticator

Authentication Server
(RADIUS server)

EAPOL-Start >

< EAP-Request/Identity

EAP-Response/Identity >

RADIUS Access Request >

< RADIUS Access Challenge

< EAP-Request/Challenge (MD5)

EAP-Response/Challenge (MD5) >

RADIUS Access Request >

*Verify credentials*

< RADIUS Access Accept

*Success*

< EAP-Success

< RADIUS Access Reject

*Fail*

< EAP-Fail

*Termination*

EAPOL-Logoff >

**Figure 19.5   IEEE 802.1X authentication.**

her identity by using a certificate. EAP-TLS authentication is also supported in the Windows 2000+ operating systems.

Many vendors of WLAN equipment support IEEE 802.1X authentication combined with WEP.

## IEEE 802.11i

IEEE 802.11i refers to the i-th task group within the IEEE 802.11 standard. This task group is currently reengineering the IEEE 802.11 security module by significantly improving the level of security and offering real protection against eavesdropping. When this new standard is finished, it will support the following:

- Authentication through EAP, like IEEE 802.1X
- AES, the current state-of-the-art encryption algorithm
- Message Integrity Check (MIC), protecting both the header and the payload
- 48-bit IV, significantly lowering the chance of key reusage

One of the drawbacks of IEEE 802.11i is that the AES encryption will require new hardware due to processor requirements. It offers a lot of benefits, though, such as:

- Authentication of server and client
- Per-packet encryption keys
- Protection against integrity attacks
- Strong AES encryption
- Improved roaming support

A detailed description of IEEE 802.11i would be outside the scope of this chapter.

## WPA

The Wi-Fi alliance decided in 2002 that the insecurity of WLANs was too big a problem to wait for the IEEE 802.11i task group to solve. Therefore, they announced that they were going to design Wi-Fi Protected Access (WPA). It had a few requirements:

- Upgrade WLAN security to an acceptable level.
- Be firmware-upgradable for current WLAN equipment.
- Be forward compatible with IEEE 802.11i.

WPA makes use of some of the IEEE 802.11i solutions, namely IEEE 802.1X, per-packet encryption keys, MIC integrity protection, and a 48-bit IV.

WPA supports two different modes, Enterprise mode and Pre-Shared Key (PSK) mode. Enterprise mode requires an authentication server like RADIUS or LDAP, whereas the PSK mode is based on a shared key in both the client and the access point. Enterprise mode offers much more protection and is better scalable to multiple users, but in home or small office environments the PSK mode is easier to set up and does not require extra servers.

Although it was planned to be released in Q1 2003, it was not until Q1 2004 that WPA-enabled WLAN products were shipped.

## HOW TO IMPROVE WEP SECURITY?

There are several settings in WEP itself that can raise the level of security.

## SSID NAME

To connect to a WLAN, you need to know the Service Set ID (SSID). Almost all manufacturers have default SSID names for their hardware, which makes it easy to guess an SSID name and try to log in on that network. By changing the SSID to a hard-to-guess name, preferably a random value, it becomes very hard to guess.

It is also noteworthy that an access point periodically sends out beacon messages containing (among others) this SSID name. Most vendors have built in an option to remove the SSID name from the beacons. Together with choosing a hard-to-guess SSID name, this will make one's network "invisible" from normal users. However, the SSID name is also sent during normal communication when a legitimate user sends a probe response, so real hackers will not be stopped with these actions.

## SHARED-KEY AUTHENTICATION

As mentioned earlier, WEP supports two types of authentication: open and shared-key. Using open authentication means that everybody who knows the SSID name can access the WLAN network and send network traffic. Shared authentication, however, requires the connecting client to prove it has the appropriate WEP key in its possession. This prevents "guests" from being able to access the network, something that is desirable from a security point of view.

But, as discussed in a previous section, WEP alone is inadequate for security. It is outdated and has been made obsolete by the newer WPA standard, and the soon-to-be-released IEEE 802.11i (rumored to be named WPA2). If upgrading to WPA is not possible, using IEEE 802.1X authentication and key exchange, or VPN software, are the only other real alternatives for link-layer security.

## CONCLUSION

When considering security threats to WLAN, it is important to look at the complete IT configuration wherein WLAN technology is used. Having a perfectly secure WLAN does not imply having a secure IT configuration because the security is just as strong as the weakest link in the chain.

In a general IT configuration using WLAN, five components can be considered that play an important role in securing the configuration:

- The applications at the wireless client that are used for communication
- The wired network technology that is used to form the link between the application and the WLAN technology
- The station at the wireless client responsible for sending and receiving wireless signals
- The wireless link itself that uses WLAN technology
- The access point responsible for the wireless network and communication to other wired networks

Each of these components are vulnerable to security threats and attacks, of which a general overview was given. Each attack forms a threat to one or more security criteria of the specific component. We have already considered the security criteria: authentication, confidentiality, integrity, and availability.

To reduce the security threats, some security services are mentioned. Some of these are available within the used technologies, others are additional. Each security service can reduce the threat to a specific component with respect to one or more security criteria.

We focused on the security technologies that are used in WLAN. The WEP protocol, the main security protocol of WLAN, was explained in detail. Threats to WEP and possible improvements were considered. In Table 19.3, an overview of all the security services that are, or will be, available within WLAN standards is given. For each security service, it is indicated which security criterion is protected.

Although these standards have different means for authentication, their levels are very different. The authentication services in Table 19.3 are mentioned in increasing order of authentication protection.

**Table 19.3  Protection by the Different WLAN Security Services**

| WLAN Security Services | Authentication | Confidentiality | Integrity | Availability |
|---|---|---|---|---|
| **1. WEP** | | | | |
| Open authentication | ✔ | | | |
| Shared-key authentication | ✔ | | | |
| 802.1X | ✔ | | | |
| RC4 | | ✔ | | |
| CRC | | | ✔ | |
| **2. 802.11i** | | | | |
| EAP | ✔ | | | |
| AES | | ✔ | | |
| MIC | | | ✔ | |
| **3. WPA** | | | | |
| Pre-shared key mode | ✔ | | | |
| Enterprise mode and 802.1X | ✔ | | | |
| Packet encryption | | ✔ | | |
| MIC | | | ✔ | |

Note that the security analysis given in this chapter is that of a general IT configuration. Although this chapter gives a good insight into the security problems and solutions when using WLAN, to secure a specific IT configuration, a more detailed risk and security analysis has to be made that is fine-tuned to that IT configuration. To complete the security analysis, procedural, legal, and organizational elements have to be incorporated.

# 20

## AN OVERVIEW OF NETWORK SECURITY IN WLANS

*Jiejun Kong, Mario Gerla, B.S. Prabhu, and Rajit Gadh*

### INTRODUCTION

IEEE 802.11 Wi-Fi (Wireless Fidelity) is a suite of specifications for wireless local area networks (WLANs), one among the numerous wireless data networks that coexist today. 802.11 is also referred to as wireless Ethernet because at the link layer it uses frame definitions and random Medium Access Control (MAC) protocols (CSMA or CSMA/CA) similar to the ones used in 802.3 Ethernet (CSMA/CD). Wireless Ethernet can serve as a wireless extension of an internal Ethernet in an institution that may require strong user authentication to prevent unauthorized users, as well as strong privacy protection to keep information confidential. Unfortunately, the first set of 802.11's security design has been proved to be insecure. Several research groups have independently discovered effective attacks exploring security flaws in the approved 802.11 standards. Open-source implementations of these attacks are now widely available.[4,28] This chapter serves as a survey of security problems in 802.11 wireless Ethernets. In this survey we study the state-of-the-art attacks against the 802.11 security protocol suite. We present an overview of common attacks, proposed countermeasures, and some known vulnerabilities of these proposals. Compared to many other articles discussing 802.11 security, we will present more technical details to show how the attacks work, as well as pros and cons of the corresponding countermeasures.

The rest of the chapter is organized as follows. In the section titled "802.11 WEP and Related Cryptology," we first describe 802.11 WEP and related cryptological concepts. We then present an overview of well-known attacks against WEP in the section titled "A Brief Overview of Common Attacks." In the section titled "New WLAN Security Protocols," we further describe a series of proposals that may help provide adequate protection to WLANs.

## 802.11 WEP AND RELATED CRYPTOLOGY

### Basic Cryptology

In a cryptosystem, the protected message is called *a plaintext*. The processed message that can be seen by all parties is called *ciphertext*. A cryptosystem is usually a collection of algorithms that translate plaintext into ciphertext by using secret *keys*, and vice versa. In modern cryptosystem design, a principle known as "Kerckhoff Desiderata" states that security should not be achieved via obscurity. Instead, all algorithms and protocol design should be made public by default, or the system's scalability and deployment could be impeded. To satisfy this requirement, the best option probably would be to protect the system using just the secrecy of cryptographic keys. As a result, in a cryptosystem based just on the secrecy of cryptographic keys, if the key gets compromised, all services protected by the key will be compromised.

The term *cryptology* refers to the study of both cryptography and cryptanalysis. *Cryptography* is the study of creating and using cryptosystems to protect messages so that only authorized parties can use their cryptographic keys or authorization tokens to access critical information. *Cryptanalysis* is the study of breaking such cryptosystems so that unauthorized cryptanalysts can pay reasonable costs to gain illegal access to critical information.

Depending on the resources available to the potential attackers, the following classes of the cryptanalytic attacks are related to 802.11 security design discussed in this chapter, and are ordered by their increased potential to compromise a cryptosystem:

*Ciphertext-only attack:* The attacker only has the encrypted ciphertext from which to determine the secret plaintext, with no knowledge whatsoever of the latter.

*Known-plaintext attack:* The attacker has a certain amount of plaintext and corresponding ciphertext, but out of his or her choice. This data is said to be *compromised*.

*Chosen-plaintext attack:* The attacker can obtain the ciphertext corresponding to an arbitrary plaintext data of his or her choice. If the attacker can determine the ciphertext of chosen plaintexts in an interactive or iterative process based on previous results, the attack is called *adaptive chosen-plaintext attack*.

The feasibility of mounting these attacks in practice is very high in wireless networks. In wired networks, the adversary has to pass several lines of defense, such as firewalls and security gateways, to gain access to the transmitted data. But in wireless networks there is no clear line of defense inside the wireless part of the network. The adversary can easily intercept needed ciphertexts. Besides, standard network protocols typically use fixed-packet formats, hence a wireless adversary has lots of opportunities to know the original plaintext of a specific ciphertext, in particular those data fields in packet headers. Finally, chosen-plaintext attacks are feasible if the adversary can trigger encryption upon plaintext it has chosen. In the 802.11 environment, this can be achieved by two collaborating adversaries: one communicates with a victim station with chosen plaintext, and the other intercepts corresponding ciphertexts.

## 802.11 WEP

In the 802.11 protocol suite, Wired Equivalent Privacy (WEP) defines the cryptographic algorithm that is designed to provide WLANs with a level of security and privacy that is equivalent to that of the wired LAN and prevent illegal access and protect authorized users from eavesdropping. WEP assumes that a secret key $k$ has already been securely delivered to all communicating parties. Each sender follows the following procedures to protect every payload message $M$:

- *Adding Integrity Check Value (ICV):* The 802.11 unit computes a CRC-32 checksum $c(M)$ on the payload $M$. The encryption key plaintext is the concatenation* $P = (M \mid\mid c(M))$.
- *Selecting Initialization Vector (IV):* The 802.11 unit randomly selects an IV $v$. The encryption key is $K = (v \mid\mid k)$.
- *Encryption:* The RC4 device accepts the encryption key $K$ and outputs a *keystream* — i.e., a long sequence pf pseudorandom bits — denoted by $RC4(K)$. The ciphertext $E$ is obtained by Vernam cipher $E = P \oplus RC4(K)$.
- *Transmission:* The unit prepends original 802.11 packet header and IV $v$ to the ciphertext, then transmits the packet over the radio link.

After a receiver recovers plaintext $P$, it verifies whether $c(M)$ is the CRC-32 checksum of $M$, then forwards the packet to upper layers if the checksum is valid. Figure 20.1 shows how WEP encrypts and decrypts 802.11 packets.

---

* In the chapter "$\mid\mid$" denotes concatenation, and "$\oplus$" denotes bitwise XOR.

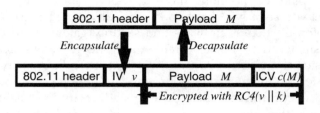

**Figure 20.1    Encrypting and decrypting 802.11 packets with WEP.**

## A BRIEF OVERVIEW OF COMMON ATTACKS

Since the beginning of the millennium, attacks on WEP have attracted critical attention in both academia and industry. Several network security research groups have published cryptanalysis against WEP.[9,24,25] They have successfully explored the broadcasting nature of wireless channels to reveal protocol design flaws. Given the capability of overhearing large amounts of ciphertext transmitted in wireless channels, Borisov and co-workers[9] revealed several insecurities in WEP design including short IV and linear message authentication code. Adversaries can easily launch data confidentiality, message modification, and message injection attacks against the wireless access point and gain unauthorized access to the network. In another research group, the Fluhrer–Mantin–Shamir attack[13] realized by Stubblefield and colleagues[24,25] can successfully reveal a 128-bit WEP secret key by eavesdropping about 5,000,000 packets, or less than 2,000,000 packets after the attack is improved. These two sets of attacks represent the most significant vulnerabilities discovered so far. Related details are described in the following text. Interested parties may see to other references for discussions on other vulnerabilities.[26,5]

### Basic Insecurities and Vulnerabilities of WEP

At the link layer, an adversary is capable of launching ciphertext-only, or known-plaintext, or even chosen-plaintext attacks against security protocols. The design of WEP does not address these threats properly.[5,9]

- *IV reuse:* When two WEP packets reuse the same IV $v$, adversaries can launch attacks against the wireless application's data privacy and data integrity. Suppose the plaintexts in the two WEP packets are $P_X = X \| c(X)$ and $P_Y = Y \| c(Y)$, and the corresponding ciphertexts are $E_X$ and $E_Y$, respectively, then

- When IV $v$ is reused, RC4 device will produce exactly the same pseudorandom keystream $RC4(K)$ up to the bit length* of $X||c(X)$ or $Y||c(Y)$. If an adversary is capable of launching known-plaintext attack and knows $P_X$, the adversary can reveal the keystream $RC4(K) = P_X \oplus E_X$ and immediately reveal $P_Y = RC4(K) \oplus E_Y = P_X \oplus E_X \oplus E_Y$. Even when neither $P_X$ nor $P_Y$ is known, the adversary can do further analysis by knowing the bitwise XOR of the plaintexts from the ciphertexts $P_X \oplus P_Y = E_X \oplus E_Y$.

- WEP does not use keyed message authentication code[16] to generate cryptographic checksums. In contrast, its integrity checksums are produced by linear CRC-32 function $c$. For two messages $X$ and $Y$,

$$c(X) \oplus c(Y) = c(X \oplus Y).$$

- Thus an adversary can generate a new "valid" packet from two existing packets. By WEP's design the receiver cannot detect the compromise of data integrity.

$$(X||c(X)) \oplus (Y||c(Y)) = (X \oplus Y)||(c(X) \oplus c(Y)) = (X \oplus Y)|| (c(X \oplus Y)).$$

■ *Short IV field:* The chance of IV reuse is very large because the IV field used by WEP is only 24 bits wide. In addition, because the size of IV space is only $2^{24}$, an adversary can build a decryption dictionary for all keystreams. As the upper bound of 802.11 packet size is about 2000 bytes, the dictionary size is less than $2^{24} * 2000$ bytes = 32 GB, a storage space available to most modern PCs.

■ *Revealed keystream:* An adversary can use a revealed keystream $RC4(K)$ to inject random messages and launch other attacks. For example, because WEP's access control is based on a symmetric key challenge–response scheme, an adversary can easily use the revealed keystream to produce a valid response, then gain access to the network.

## Fluhrer–Mantin–Shamir Attack

RC4 was arguably the most widely used stream cipher when WEP was designed. Unfortunately, Fluhrer and co-workers[13] discovered that the initial output in the RC4 keystream is disproportionally affected by a small

---

* In the CRC-32 checksum system, the shorter one of $X$ and $Y$ is identical to an elongated version with the zero-bit padded. Thus $X$ and $Y$ can be regarded as being of the same length.

```
KSA(K):                                    PRGA(S):

Initialization:                            Initialization:

  For i = 0..255                             i = 0

    S[i] = i                                 j = 0

  j = 0

                                           Generation Loop:

Scrambling:                                  i = (i + 1) mod 256

  For i = 0..255                             j = (j + S[i]) mod 256

    j = (j + S[i] + K[i mod l]) mod 256      Swap(S[i], S[j])

  Swap(S[i], S[j])                           Output z = S[(S[i] + S[j]) mod 256]
```

**Figure 20.2    RC4's Key Scheduling Algorithm and Pseudorandom Generation Algorithm.**

number of key bits.* Moreover, the secrecy of RC4 key is vulnerable to related key cryptanalysis.[7] It is used on the observation that adversaries may know the difference between two cipher keys though both keys are not completely revealed yet. In WEP, the cipher key used by the RC4 device is the concatenation $v||k$ where $v$ is the exposed IV and $k$ is the shared WEP secret key. As the result, related key cryptanalysis against RC4 is enabled by the design. The Fluhrer–Mantin–Shamir attack is practical for any key size and for any modifier size, including the 24 bit recommended in the original WEP and the 128 bit recommended in the revised version of WEP2. After numerous different exposed IVs are used, an attacker can derive the WEP secret key by analyzing the initial word of the keystreams.

As depicted in the following text, the RC4 algorithm comprises two modules: key scheduling algorithm (KSA) and pseudorandom generation algorithm (PRGA) (Figure 20.2).

KSA turns an *l*-byte random key (whose typical size is 40–256 bits) into a scrambled initial permutation $S$ of $\{0, ..., N\ 1\}$, then PRGA uses this permutation to generate a pseudorandom output sequence that can be used in stream cipher.

In this section we briefly describe the *known IV attack* (as named in Reference 13), that is, how related key cryptanalysis can reveal the secret part (i.e., the WEP secret key) upon knowing the exposed IV prefix. Once

---

* Earlier credits go to Ian Farquhar, who posted a warning to sci.crypt in 1996 that there was an approximately 35 percent probability that the first byte of the keystream generated by RC4 would be the same as the first byte of the key.

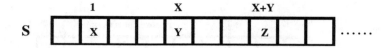

**Figure 20.3   RC4's** *Resolved Condition* **(after KSA).**

the WEP secret key is revealed, cryptographic protection provided by the key is no longer effective.

The first PRGA keystream output word depends only on three specific permutation elements. If the state of the permutation immediately after KSA is as shown in Figure 20.3, then the value labeled $Z$ will be output as the first word. In addition, let $S_i$ denote the snapshot permutation of $S$ at round $i$. If the key setup reaches a stage where $i$ is greater than equal to 1, $X = S_i[1]$ and $Y = S_i[X]$, then (if we model the remaining swaps in the key setup as in a truly random process) with probability greater than $e^3$ ($\approx$5 percent), none of the elements referenced by these three values will participate in any further random swaps. This is because

$$\left(1 - \frac{3}{N}\right)^N \geq e^{-3}$$

for $N$ rounds. The probability that none of the three positions is chosen for swapping each round is $\left(1 - \frac{3}{N}\right)$.

If after KSA $S$ is in a resolved condition, then the value $S[S[1] + S[S[1]]]$ will be output as the first word by PRGA. With probability less than 1 $e^3$ ($\approx$95 percent), at least one of the three values will participate in a swap and be set to an effectively random value, which will make the output value effectively random. However, the chance of 5 percent is non-negligible.

Now we analyze the weakness of KSA and see how a resolved condition can be set up. Consider the WEP scenario where the IV is prepended to the secret key. In this circumstance, assuming we have an $I$-byte exposed IV and an $(l\ I)$-byte secret key ($K[I + 0]$, $K[I + 1]$, ..., $K[l\ 1]$). These values can be mapped into real values used in 802.11. In WEP, $I = 3$ and $l = 8$. In WEP2, $I = 16$ and $l = 32$. Our goal is to derive information on a particular byte $B$ of the secret key (i.e., to find out $K[I + B]$, $0 \leq B \leq l\ I\ 1$).

Let us see a simple example. In WEP, $I = 3$ and $l = 8$. Suppose the cryptanalyst has inductively revealed the sequence secret key bytes ($K[3]$, ..., $K[A + 2]$), and now he wants to reveal $K[A + 3]$. The entire ($l\ I$)-byte secret key ($K[I + 0]$, $K[I + 1]$, ..., $K[l\ 1]$) can be inductively revealed in

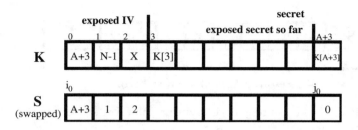

**Figure 20.4   KSA round 1.**

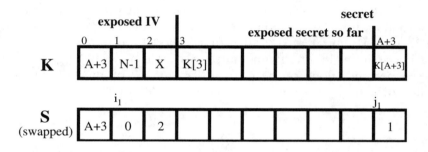

**Figure 20.5   KSA round 2.**

this way as the process starts from the scratch line where $A = 0$ and $K[3]$ is the target. The cryptanalyst examines a series of IVs of the form $(A + 3, N\ 1, X)$. At the first KSA round, $j$ is advanced by $A + 3$, and the $S[i]$ and $S[j]$ are swapped, resulting in the key setup state shown in Figure 20.4.

In the next round, $i$ is advanced, and then the advance on $j$ is computed, which happens to be $(N\ 1) + 1 \bmod N = 0$. Then $S[i]$ and $S[j]$ are swapped, resulting in the state shown in Figure 20.5.

Then, in the next round, $j$ is advanced by $X + 2$, which implies that each distinct IV assigns a different value to $j$, and thus, beyond this point, each IV acts differently. Unfortunately, because the cryptanalyst knows the value of $X$ and $(K[3], \ldots, K[A + 2])$, he or she can compute the exact behavior of the key setup until he or she reaches round $A + 3$. At this point, he or she knows the value of $j_{A + 2}$ and the exact values of the permutation $S_{A + 2}$. If $S_{A + 2}[0]$ or $S_{A + 2}[1]$ has been disturbed (i.e., $S_{A + 2}[0] \neq A + 3$ or $S_{A + 2}[1] \neq 0$), the cryptanalyst rests the case and the $K[A + 3]$ is not revealed. Otherwise, $j$ is advanced by $S_{A + 2}[i = A + 3] + K[A + 3]$, and then the swap is done, resulting in the state shown in Figure 20.6.

We are looking for a resolved condition where $(S_{A + 3}[1] + S_{A + 3}[S_{A + 3}[1]]) = 0 + A + 3 = A + 3$. Then with more than 5 percent probability $S_{A + 3}[A + 3]$ will be output as the first byte in RC4's keystream. As described

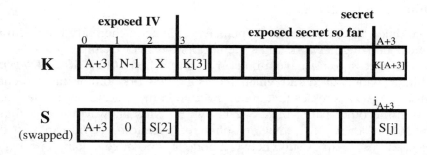

**Figure 20.6   KSA round (A + 3).**

previously, in particular in Reference 9, there are many ways to recover this keystream and to know $S_{A + 3}[A + 3]$.

At KSA round $(A + 3)$, the cryptanalyst knows the permutation $S_{A + 2}$ and the value of $j_{A + 2}$. Now he or she also knows the value of of $S_{A + 3}[A + 3]$, and then he or she knows its location in $S_{A + 2}$, which is the value of $j_{A + 3}$. He or she can use the known information to compute $K[A + 3]$:

$$K[A + 3] = j_{A + 3} - j_{A + 2} - S_{A + 2}[A + 3].$$

The case described above is a special case. In general, by searching for IV values such that, after the first $I$ KSA rounds, $S_I[1] < I$ and $S_I[1] + S_I[S_I[1]] = I + B$. Then, with high likelihood ($\approx e^{2B/N}$ if we model the intermediate swaps as in a truly random process), we will be in a resolved condition after KSA round $(I + B)$, and then the most probable first byte of keystream output will be

$$keystream = S_{I + B\ 1}[j_{I + B}] = S_{I + B\ 1}[j_{I + B\ 1} + K[B] + S_{I + B\ 1}[I + B].$$

Or, in other words, if we know the value of $j_{I + B\ 1}\ S_{I + B\ 1}$, then given the first byte of keystream output, we can predict the value

$$K[I + B] = S_{I+B-1}^{-1}[keystream] - j_{I+B-1} - S_{I+B-1}[I + B]$$

where $S^1[X]$ denotes the location within the permutation $S$ where the value $X$ appears. This prediction is accurate more than 5 percent of the time, and effectively random less than 95 percent of the time. By collecting sufficiently many values from different IVs, we can reconstruct $K[I + B]$.

Because the known IV attack only uses the first byte of keystream output from any given secret key and IV, the attack is applicable to secret key and IV of *any size*. This means the elongated key design adopted by WEP2 is not an effective countermeasure. The substitutes of WEP and WEP2 are being developed to address the new challenges. They will be reviewed in later sections of this chapter.

## Other Simple Attacks

In addition to these serious cryptanalytical attacks, there are other simple attacks utilizing network and system loopholes. For example, most 802.11 (Wi-Fi) access points have a built-in feature called *MAC Address Filtering*, which is widely used and which allows the network administrator to enter a list of MAC addresses that are allowed to communicate on the network.

On the other hand, most 802.11 (Wi-Fi) NICs allow you to configure the MAC address of the NIC in software. Therefore, if you can sniff the MAC address of an existing network node, it is possible to join the network using the MAC address of the victim node. This shows that anonymity protection is needed in WLANs as well.

## NEW WLAN SECURITY PROTOCOLS

### WPA

Knowing the vulnerabilities of Wired Equivalent Privacy (WEP), the Wi-Fi Alliance, in conjunction with the IEEE, has begun an effort to bring strongly enhanced, interoperable Wi-Fi security to replace WEP. The result of this effort is Wi-Fi Protected Access (WPA), which is a specification of standards-based, interoperable security enhancements that strongly increase the level of data protection and access control for existing and future WLAN systems. Designed to run on existing hardware as a software upgrade, WPA is derived from and will be forward-compatible with the upcoming IEEE 802.11i standard.

WPA protects WLANs using technology such as IEEE 802.1X, RADIUS, and TKIP. The 802.1X technology provides authentication services, likely from a RADIUS server. EAP provides a variety of algorithms for authenticating the client with the RADIUS server. TKIP ensures confidentiality and integrity for wireless communications. WPA operates in two modes: PSK (Pre-shared Key) and Enterprise. In the PSK mode, a shared password is used in authentication. It is essentially the same key-management scheme assumed in 802.11 WEP, so that an authentication server (e.g., RADIUS) is not required. In the Enterprise mode, an authentication server is needed in service provisioning.

802.1X, RADIUS, and TKIP are described in detail in the following text.

### *802.1X*

802.1X leverages existing standards such as EAP,[8] which stands for Extensible Authentication Protocol. EAP lets the network use a variety of algorithms for authenticating the client with an authentication server.

802.1X defines three roles: (1) *supplicant* is a user or client requesting authentication, (2) *authentication server* is the server providing authentication service, and (3) *authenticator* is the being-requested device to be accessed by the supplicant. The conversation among the three roles is depicted in Figure 20.7. In a typical scenario, an authenticator (i.e., the wireless access point) is configured to allow open and shared authentication. The initial client authentication is open because dynamically derived encryption keys are not yet available at this stage. Once the client associates with the authenticator, the depicted 802.1X authentication conversation (between the access blocked and access allowed) grants access to supplicants with valid credentials. Prior to EAP-Success, the authenticator filters all non-EAPoL (EAP over LAN) traffic from the supplicants. Upon EAP-Success, the authenticator removes the filter.

802.1X can be run on top of both wired LAN (802.3) and WLAN 802 (802.11) with minimal changes. It is particularly well suited for WLAN applications because it requires very little processing power on the part of the authenticator (access point). In addition, 802.1X also supports user-based identification based on a Network Access Identifier,[1] which enables support for roaming access in public domains.[3]

## RADIUS Authentication

RADIUS[20,21,31] is not part of the WPA or 802.11i standard, but it is the *de facto* back-end authentication protocol. A RADIUS server customized for 802.11 WLAN is designed on the same core architecture as a wired RADIUS server. The related service includes user-specific check, session time-out, idle time-out, and framed IP address, etc. However, protocols such as SLIP, PPTP, and L2TP are particularly designed for wired dial-ups; they are not suitable in a wireless environment. Therefore, RADIUS servers customized for WLANs need not implement these features.

For backward compatibility, RADIUS servers customized for WLAN may implement proxy supports, which let the network maintain accounting information and user data in an existing database. This database holds information pertaining to conventional RADIUS usage, such as dial-up, VPN, and firewall-access accounts.

A RADIUS server is essentially a centralized database of authentic IDs and associated attributes. It can perform a lookup authentication or a log-in authentication. In the lookup mode, the server authenticates the user against the ID store and returns user-specific attributes from the ID store with the access-accept packet. In the log-in mode, the user's credentials are used to log in to the authentication system. If the credentials are validated, the user receives an access-accept packet.

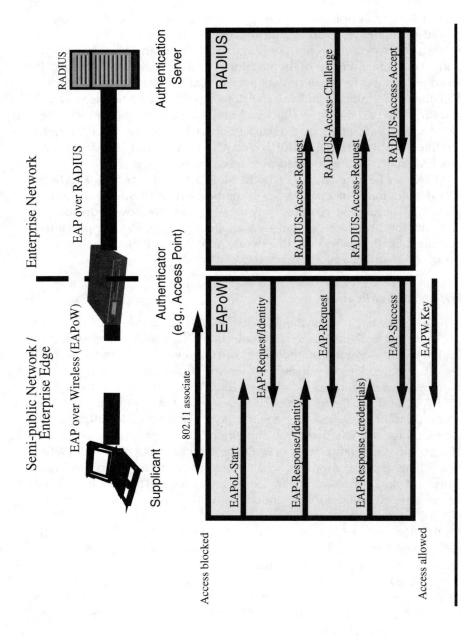

**Figure 20.7  Roles and control flows in 802.1X.**

## TKIP

TKIP (Temporal Key Integrity Protocol) is an interim standard allowing for backward compatibility. It will be replaced by standards with better security supports, such as CCMP (Counter Mode with Cipher Block Chaining Message Authentication Code Protocol). However, the backward compatibility with TKIP is necessary because most legacy Wi-Fi (802.11) hardware does not support the AES (Advanced Encryption Standard[19]) algorithm used by CCMP.

TKIP enforces a policy that the RC4 cipher key must be refreshed every 10,000 packets. This prevents a Fluhrer–Mantin–Shamir attacker from accumulating enough information to reveal a constant cipher key. In addition, the RC4 cipher key is not a constant. As depicted in Figure 20.8, TKIP has a key-mixing function to generate the RC4 cipher key per packet. Finally, TKIP uses a 48-bit IV (versus the 24-bit IV in WEP). At least a 44-bit chunk of the IV field is used as a counter to avoid IV reuse and replay attacks. Any packet with an out-of-sequence IV counter will be discarded. Now two IVs collide after about $2^{44}$ frame transmissions, which would typically take thousands of years.

To ensure message integrity, a keyed hashing for message authentication (HMAC) algorithm named Michael[11] is used in each direction of wireless transmission, and a 64-bit key is used in Michael to generate a 64-bit cryptographic checksum (also known as message integrity code, MIC). In WEP, CRC-32 is linear and is only applied on the payload. In TKIP, Michael is not only cryptographically strong but also applied on both the payload and header. This further decreases the chance of checksum collision and enlarges the covered range of integrity protection.

## 802.11i

WPA is viewed by many merely as an interim measure. IEEE 802.11i is the draft security standard for 802.11 wireless network protection. Nevertheless, 802.11i relies heavily on many existing WPA components.

802.11i introduced the RSN (Robust Secure Network) protocol for establishing secure communications. Like WPA, RSN uses 802.1X to authenticate wireless devices to the network. Two new protocols based on AES, namely CCMP and WRAP (Wireless Robust Authenticated Protocol), are introduced to protect data privacy and integrity.

### 802.11i Key Management

Key management is a critical service in any security protocol suite using cryptography to protect the system. The key-management module of 802.11i is far more sophisticated than the flawed 802.11 WEP and 802.1X.

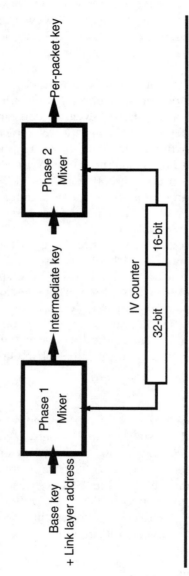

**Figure 20.8  TKIP's key mixing.**

There are two classes of keys: pairwise keys and group keys. In pairwise key management, a hierarchy of three types of keys protects the network:

1. Master Key (MK), which is established upon appropriate authentication.
2. Pairwise Master Key (PMK), which is derived from MK per connection or session. The difference between MK and PMK is due to security policy. Because the authentication server is the appropriate place to verify a wireless station's identity and grant associated privileges, MK is the fresh token representing the relation between the authentication server and an authenticated wireless station. On the other hand, PMK is the fresh token representing the relation between an access point and the authenticated station. In terms of PMK, the authentication server can be regarded as a trusted third party.
3. Pairwise Transient Key (PTK) is the collection of operational keys including Key Confirmation Key (KCK), Key Encryption Key (KEK), and Temporal Key (TK). They are derived from PMK. The entire derivation process uses random nonces selected by both access point and the wireless station; thus a man-in-the-middle attack is not feasible. KCK is used to prove the possession of PMK and to bind the wireless station to the access point via the PMK. KEK is used to distribute Group Transient Key (GTK), which belongs to group key management described in the following text. TK is used as a cipher key to secure data traffic.

In group key management, the access point selects a GTK and encrypts it with KEK. The wireless station uses KEK, which is derived from the shared PMK, to decrypt and know the GTK. Messages of this GTK key exchange handshake are authenticated by KCK, which is also derived from the shared PMK. GTK can be shared among a number of wireless stations, and thus the access point can be used to protect broadcast and multicast traffic.

## CCMP

CCMP (Counter Mode with Cipher Block Chaining Message Authentication Code Protocol) is the mandatory encryption protocol in the 802.11i standard. It uses CCM mode[29] and 128-bit keys, with a 48-bit IV for replay detection.

In CCMP, each sender follows the following procedure to protect every payload message $M$:

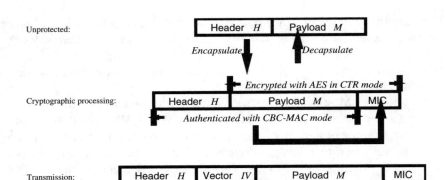

**Figure 20.9   Encrypting and decrypting 802.11 packets with CCMP.**

- *Adding MIC:* The 802.11i unit computes a cryptographic checksum $MIC(H||M)$ on the header $H$ and payload $M$. In CCMP, the cipher block chaining message authentication code (CBC-MAC) mode is used to produce the *MIC*.
- *Encryption:* The unit then encrypts payload $M$ and *MIC* separately. Payload $M$ is encrypted in CTR mode with counter values counting from 1. *MIC* is encrypted in CTR mode with counter value 0.
- *Transmission:* The unit prepends the IV before the payload $M$ and transmits the packet over the radio link.

After a receiver recovers $M$ and *MIC*, it verifies whether *MIC* is the CBC-MAC checksum of $M$, then forwards the packet to upper layers if the checksum is valid. Figure 20.9 shows how CCMP encrypts and decrypts 802.11 packets.

### WRAP

WRAP (Wireless Robust Authenticated Protocol) is an encryption protocol in the 802.11i standard. WRAP is based upon the Offset Codebook (OCB) mode proposed by Rogaway and coworkers.[22]

OCB is a composite mode ensuring data privacy and integrity at the same time. It is efficient. The overhead of OCB's composite operations is about the same workload of encryption alone — given a message of $n$ 128-bit blocks, encryption alone needs $n$ AES cipher calls, and the OCB mode only needs $n + 2$ AES cipher calls (versus $2n + 2$ calls in the CCM mode).

The major drawback of WRAP is not a technical one. The OCB mode is proven to be provably secure by its designers and has so far survived cryptanalysis. Unfortunately, three different parties have filed for patents on WRAP. These intellectual property issues caused the IEEE to adopt CCMP as the standard. The WRAP protocol becomes an optional component because some vendors have already implemented WRAP-conforming hardware.

## Known Vulnerabilities

### 802.1X Design Trade-Offs and Vulnerabilities

802.1X recommends that an implementation use RADIUS.[21,20] But this is not mandatory. An implementation can employ a variety of authentication methods, such as PKI certification based on X.509,[15] Kerberos,[23] etc. This design choice gives WLAN administrators certain flexibility in their management. However, there are trade-offs. This design is suitable for a closed network but not for an open wireless network for mobile users. For example, even if a mobile supplicant has installed and enabled its 802.1X support in its home WLAN, it cannot connect to a WLAN with another type of 802.1X support.

Mishra and Arbaugh[18] reported two potential vulnerabilities of 802.1X: (1) 802.1X should mandate mutual identity authentication (or class 3 authentication named in WTLS[27]) in all scenarios, so that an adversary cannot launch a man-in-the-middle attack and insert itself between a supplicant and an authenticator. Fortunately, EAP-TLS[2] does provide mutual identity authentication. Also, this man-in-the-middle attack is detectable. In other words, the supplicant and the authenticator can detect anomalous communications, such as duplicated identities, if they are within one-hop radio transmission range. (2) Not all 802.11 frames and EAP frames are appropriately authenticated. This contributes not only to the feasibility of a man-in-the-middle attack but also to session hijacking, where an adversary — the man in the middle — deassociates the supplicant and acts as the supplicant itself. The key to addressing these vulnerabilities is to implement appropriate per-connection identity authentication and per-frame message authentication. As described previously, the design of the 802.11i key-management scheme can be considered as a related effort along this track.

### Michael Vulnerabilities

Michael is an efficient message authentication algorithm, but it is much weaker than other counterparts such as MD5 and SHA1. Due to the "birthday paradox," it is expected that one MIC collision will happen for every $2^{32}$ packets (versus $2^{64}$ in MD5 and $2^{80}$ in SHA1). It is shown in the original document[11] that this number can be decreased to $2^{29}$. Nevertheless, this value is considered to be big enough and is unlikely to affect a wireless session.

Recently there has been an unpublished technical report[30] showing that Michael can easily be inverted by a known-plaintext attacker given a single known plaintext message and its known MIC. The designer of Michael seemed to be aware of this vulnerability because the need of keeping the MIC values secret is stated several times in the original document.[11] The technical report further shows a related message attack that utilizes the invert function as a subroutine to reveal Michael's secret key with relaxed conditions.

## AES Vulnerabilities

No successful cryptanalysis against AES/Rijndael is publicly known at this time. But some interesting results have been unveiled. Ferguson and colleagues[12] pointed out certain interesting algebraic properties of AES/Rijndael, which implies that the security of AES/Rijndael relies on a new computational hardness assumption. But it is unknown whether effective attacks against the hardness assumption are feasible in the future. Fuller and Millan[14] also discovered an interesting mathematical property of AES/Rijndael. Again, it is not even an attack.

A more recent result published by Courtois and Pieprzyk[10] illustrated an attack against AES/Rijndael and Serpent. The authors described an XSL attack, which is a better-than-brute-force attack against Serpent, and possibly one against Rijndael as well.* AES is a newly proposed standard. Studies on related key cryptanalysis and other cryptanalytic attacks against AES are inconclusive and may potentially reveal more vulnerability and even insecurity.

## Composition of Authentication and Encryption

Security protocols and applications, such as SSL/TLS/WTLS, IPSec, and SSH, apply both symmetric authentication (such as HMAC) and encryption to the transmitted data. However, these popular protocols have chosen a different method to combine authentication and encryption. The three methods used by SSL, IPSec, and SSH are referred to as *authenticate-then-encrypt (AtE)*, *encrypt-then-authenticate (EtA)*, and *encrypt-and-authenticate (Et&A)*, respectively.

---

* This is a controversial issue. Interested parties may refer to http://www.cryptosystem.net/aes/ for details.

SSL:    $a = Auth(x)$, $C = Enc(x||a)$, transmit $C$

IPSec:  $C = Enc(x)$, $a = Auth(C)$, transmit $(C||a)$

SSH:    $C = Enc(x)$, $a = Auth(x)$, transmit $(C||a)$

where $x$ is a message, $Enc(\cdot)$ is a symmetric encryption function, $Auth(\cdot)$ is a message authentication code, and '$||$' denotes concatenation — in this notation the secret keys to the algorithms are implicit.

Under the generic assumptions, HMAC functions are secure against chosen-message attacks and symmetric encryption functions are secure against chosen-plaintext attacks. Bellare and Namprempre[6] and Krawczyk[17] proved that IPSec's Et&A method is a generically secure method for implementing secure channels.

In contrast, the Et&A method used in SSH is vulnerable to privacy attacks even if we use a secure HMAC function. The AtE method used in SSL has the same problem unless specific encryption modes are used, for example, two very common forms of encryption: CBC mode and stream ciphers.

The 802.11i CCMP protocol adopts the AtE approach. It is considered secure if the CTR mode effectively realizes a strong pseudorandom generator as the CBC mode does. The 802.11i WRAP protocol utilizes the OCB mode, which achieves message privacy and message authentication simultaneously. Thus composition vulnerability is not applicable to WRAP.

## SUMMARY

Compared to deploying an 802.11 WLAN, protecting a deployed WLAN is a relatively more difficult job. In this survey we limit our scope to 802.11 security attacks and the proposed countermeasures. We have illustrated in detail the design flaws of the original 802.11 WEP standard. We have also shown that countermeasures are feasible. Currently, WPA/TKIP is considered an inexpensive and effective means to address wireless cryptographic attacks. In the near future, proposed solutions such as CCMP and WRAP will provide better protection to wireless users. We will continuously watch the ongoing game of cryptanalysts versus cryptographers (or attackers versus counterattackers). Though we believe that a more secure WLAN environment will be realized in the foreseeable future, the final conclusion should not be made today because more investment, both in terms of resources and time, are needed, and there will be more security surprises waiting for us ahead.

# REFERENCES

1. B. Aboba and M. Beadles. The Network Access Identifier. http://www.ietf.org/rfc/rfc2486.txt, 1999.
2. B. Aboba and D. Simon. PPP EAP-TLS Authentication Protocol. http://www.ietf.org/rfc/rfc2716.txt, October 1999.
3. B. Aboba and J. Vollbrecht. Proxy Chaining and Policy Implementation in Roaming. http://www.ietf.org/rfc/rfc2607.txt, 1999.
4. AirSnort Project. http://airsnort.shmoo.com/, 2002.
5. W. A. Arbaugh, N. Shankar, and Y. J. Wan. Your 802.11 Wireless Network has No Clothes. http://www.cs.umd.edu/~waa/wireless.pdf, 2000.
6. M. Bellare and C. Namprempre. Authenticated Encryption: Relations among Notions and Analysis of the Generic Composition Paradigm. In T. Okamoto, Ed., *ASIACRYPT'00, Lecture Notes in Computer Science 1976*, pp. 531–545, 2000.
7. E. Biham. New Types of Cryptanalytic Attacks Using Related Keys. In *Advances in Cryptology-EUROCRYPT'93*, pp. 487–496. Springer-Verlag, Heidelberg, 1994.
8. L. Blunk and J. Vollbrecht. PPP Extensible Authentication Protocol (EAP). http://www.ietf.org/rfc/rfc2284.txt, 1998.
9. N. Borisov, I. Goldberg, and D. Wagner. Intercepting Mobile Communications: The Insecurity of 802.11. In *ACM MOBICOM*, 2001.
10. N. Courtois and J. Pieprzyk. Cryptanalysis of Block Ciphers with Overdefined Systems of Equations (or the XSL attack on block ciphers). In Y. Zheng, Ed., *ASIACRYPT'02, Lecture Notes in Computer Science 2501*, pp. 267–287, 2002.
11. N. Ferguson. Michael: an improved MIC for 802.11 WEP. IEEE standard 802.11-02/020r0. Available from http://grouper.ieee.org/groups/802/11/Documents/DocumentHolder/2-020.zi%p, 2002.
12. N. Ferguson, R. Schroeppel, and D. Whiting. A Simple Algebraic Representation of Rijndael. In *8th Annual Workshop on Selected Areas in Cryptography*, pp. 103–111, 2001.
13. S. Fluhrer, I. Mantin, and A. Shamir. Weakness in the Key Scheduling Algorithm of RC4. In *8th Annual Workshop on Selected Areas in Cryptography*, pp. 1–24, 2001.
14. J. Fuller and W. Millan. On Linear Redundancy in the AES S-Box. http://eprint.iacr.org/2002/111/, 2002.
15. International Telecommunication Union. Recommendation X.509(11/93), The Directory: Authentication Framework.
16. H. Krawcyzk, M. Bellare, and R. Canetti. HMAC: Keyed-Hashing for Message Authentication. http://www.ietf.org/rfc/rfc2104.txt, 1997.
17. H. Krawczyk. The Order of Encryption and Authentication for Protecting Communications (or: How Secure Is SSL?). In J. Kilian, Ed., *CRYPTO'01, Lecture Notes in Computer Science 2139*, pp. 310–331, 2001.
18. A. Mishra and W. Arbaugh. An Initial Security Analysis of the IEEE 802.1X Standard. http://www.cs.umd.edu/~waa/1x.pdf, 2002.
19. National Institute of Standards and Technology. Advanced Encryption Standard. http://csrc.nist.gov/encryption/aes/, 2001.
20. C. Rigney. RADIUS Accounting. http://www.ietf.org/rfc/rfc2139.txt, 1997.
21. C. Rigney, A. Rubens, W. Simpson, and S. Willens. Remote Authentication Dial-In User Service (RADIUS). http://www.ietf.org/rfc/rfc2138.txt, 1997.

22. P. Rogaway, M. Bellare, J. Black, and T. Krovetz. OCB: a Block-cipher Mode of Operation for Efficient Authenticated Encryption. In *CCS*, pp. 196–205, 2001.
23. G. Steiner, B. C. Neuman, and J. I. Schiller. Kerberos: An Authentication Service for Open Network Systems. In *USENIX Winter*, pp. 191–202, January 1998.
24. A. Stubblefield, J. Ioannidis, and A. D. Rubin. Using the Fluhrer, Mantin, and Shamir Attack to Break WEP. Technical Report TD-4ZCPZZ, AT&T Labs, August 2001.
25. A. Stubblefield, J. Ioannidis, and A. D. Rubin. Using the Fluhrer, Mantin, and Shamir Attack to break WEP. In *Network and Distributed System Security Symposium (NDSS'02)*, 2002.
26. J. R. Walker. Unsafe at any key size: An analysis of the WEP encapsulation. IEEE document 802.11-00/362, October 2000.
27. WAP Forum. Wireless Transport Layer Security Specification. http://www1.wap-forum.org/tech/documents/WAP-261-WTLS-20010406-a.pdf.
28. WEPCrack Project. http://wepcrack.sourceforge.net, 2002.
29. D. Whiting, R. Housley, and N. Ferguson. Counter with CBC-MAC (CCM). http://www.ietf.org/rfc/rfc3610.txt, September 2003.
30. A. Wool. A Note on the Fragility of the "Michael" Message Integrity Code. Technical Report EES2003-2, Department of Electrical Engineering Systems, Tel Aviv University, April 2003.
31. G. Zorn. Microsoft Vendor-specific RADIUS Attributes. http://www.ietf.org/rfc/rfc2548.txt, March 1999.

# 21

---

# SECURITY
# IN WIRELESS NETWORKS

## Partha Dasgupta and Tom Boyd

## INTRODUCTION TO WIRELESS NETWORKS

Security has often taken a backseat in the design and deployment of networking technologies. The growth of the Internet was spurred by the availability of low-cost networking access points; provision of security did not appear to be greatly needed and thus was not a major design consideration. This mistake has been felt repeatedly since the late 1990s when, after the explosion of Internet usage, exploitation of weak security emerged as one of the biggest threats to the long-term viability of the global networking medium. The same problem may plague wireless networks. In most of the protocols that use wireless networking (local area or wide area), the security mechanism is more of an afterthought. Attacks on wireless networks are quite uncommon even though they are deployed quite widely. However, the extent of deployment of wireless networks is a small fraction of the reach of wired networks. As the penetration of the wireless medium increases, the problems with wireless security are bound to become more significant. Even today, we are aware of many weak points that exist in current wireless network designs, and we can be certain that many more will be discovered.

Since the early 1990s, wireless communication has been used for data networking, mostly using proprietary technologies. One of the first data communication networks without wires was the ALOHA network in Hawaii (circa 1970). The emergence and acceptance of standards around 2000 has exploded the use of wireless access, and currently (year 2004) several forms of wireless communication are widely used by the mainstream computing community. These forms include, among others, the IEEE 802.11 series of wireless products, various forms of data access provided by cellular providers, and an emerging technology for short-range communication called Bluetooth.

Before wireless data communication could emerge into the mainstream, it had to cross a few political and technological barriers. These barriers to wireless communication in the early 1990s were many. Spectrum was in short supply, which was later resolved by the Federal Communications Commission (FCC) opening up several large bands in the 2-GHz and 5-GHz ranges for unlicensed use. The price of producing hardware that operates at the multi-gigahertz range fell sharply due to advances in miniaturization and innovative production techniques. Even with falling prices and availability of spectrum, the barrier was interoperability; that is, signaling protocols and frequencies used by a manufacturer of wireless hardware were not compatible with those used by another vendor, causing customers to get "locked in" to a particular provider. This was enough of a customer disincentive to stifle the wireless market.

Several simultaneous occurrences finally pushed wireless access to the foreground of consumer products in the 2003 time frame. These are the decline in the price of laptop computers and personal digital assistants (PDAs), the perceived need and allure of untethered Internet access, and the emergence of standards, notably IEEE 802.11b, which allowed products from any vendor to seamlessly interact with products of other vendors.

Along with the emergence of almost ubiquitous low-cost wireless access, we are now saddled with risks, vulnerabilities, and a general lack of security at the data transmission level. This chapter discusses the vulnerabilities of wireless access and presents the industry standard solutions that can in some cases correct, mitigate, or at least provide some level of confidence in wireless communication.

## VULNERABILITIES IN COMPUTER NETWORKS

A vulnerability is a flaw in any hardware or software system that is the result of either oversight or poor design. Sometimes, the basic nature of a particular system can be exploited to disrupt the intended operation of the system. The disruption may be in the form of the introduction of a malfunction or the gaining of unauthorized access, as well as the theft of some or part of the information stored or in transit in the system. A complete description of vulnerabilities in computer systems and network systems is not within the scope of this chapter.

There are two major causes for the existence of vulnerabilities. The first is poor design. To ensure secure networking, the protocol and the security substrate need to be designed together. Historically, all networking protocols are designed to handle data transmission needs, and later, security mechanisms are added. This approach has repeatedly proven to be flawed because security schemes are an additional layer that does not integrate well, or that can be disabled or circumvented. The second cause is the presence of bugs in software. Because bug-free software is practically impossible, these overlooked bugs will always be around for exploitation by attackers.

Until around the early 1990s, vulnerabilities in computing and networking systems were not well understood and were generally ignored. There was a naïve and ultimately flawed assumption that operating systems are secure and computer users are largely honest, and any miscreants who were around lacked access to adequate equipment or technical knowledge to exploit any vulnerability that may have existed in computer and networking systems. As the emergence of the Internet brought computer networks into the mainstream, this flawed assumption gave rise to a painful environment of exploits, hacks, spoofs, spam, and many such destructive and often expensive intrusions upon the Internet infrastructure and upon the hosts that are connected to the Internet.

Today we understand the nature and cause of a large number of these vulnerabilities (and many more are discovered almost daily). We are clearly cognizant of the need for identifying vulnerabilities and protecting users.

## Wired Networks and Vulnerabilities

In a wired network, data packets traverse from one host (sender) to another (receiver) over a collection of wires, terminated by switches, routers, and gateways. In the majority of the current Internet, the data travels as "clear text," that is, the data packets contain the data in native form. Even if the data is encrypted, the "headers" of the data packets with routing information are visible in the clear. Anyone with physical access to the wire or the switches and routers can physically siphon off the data and record, and look and analyze the contents. In addition, the intruder may change, tamper with, or destroy the data. In the extreme, the intruder may simply cut the cable or disable the router. We classify the vulnerabilities into several loose classes:

*Eavesdropping:* Anytime two (or more) computers communicate over a network, the data packets can be intercepted, copied, stored, or analyzed. This is a passive form of a security vulnerability that does not disrupt the communication and is almost always undetectable but may cause leakage of data and activity information. Because every packet has source and destination addresses, the attacker can gain information about who communicates with whom and how much data is being exchanged.

The content of the data is also visible if the data is sent unencrypted. Encryption converts the data into a nonreadable form that requires effort to reconvert back to its readable form. If encryption is used properly, the contents may remain secret, but the routing and the size of communication still remain detectable, raising privacy concerns.

*Man-in-the-middle (MITM):* The MITM attack is one step beyond the eavesdropping attack and is an active form of attack. In this form, the attacker intercepts the data traveling, for instance, from Alice to Bob and alters the contents intelligently. The intruder may drop packets, or replay or modify them, or even completely change the contents. For example, if Alice is surfing www.cnn.com, the intruder might keep most of the CNN site's look and feel intact but insert spurious or false news stories. If Alice is downloading a security patch from Microsoft.com, the intruder may substitute the downloaded file with an executable one of his or her own that infects Alice's computer with a virus. Some forms of MITM attacks can use even more nefarious tactics to convince Alice to part with sensitive or damaging information. The assumption in the analysis of MITM attacks is that the intruder has computers with unbounded power and ample time to attack, spoof, and fool Alice for any gain or purpose. Making MITM attacks infeasible on communication networks is one of the harder problems in network security.

*Infrastructure vulnerabilities:* The networking fabric itself uses computers, servers, and switches. For example, the Domain Name System (DNS) servers provide name translation facilities, the Border Gateway Protocol (BGP) servers provide routing information, the Dynamic Host Configuration Protocol (DHCP) servers provide addressing at the edge, and programmable routers provide flexible routing as well as "ingress and egress" filtering. All of these components can be subject to attacks using one or more vulnerabilities, including spoofing, MITM, and buffer overflow. These attacks can cripple a network by hitting at the core engines that run the data communication.

*Denial-of-service (DoS):* The DoS attack is a generalization of a large number of different attacks, which essentially leads to one participant in a communication link being unable to communicate. Cutting the wire between a client and a server is a physical DoS attack. Similarly, sending a large number of fake but complex queries to a server, causing the server to spend an inordinate amount of time performing useless computations and not being able to do genuine work, is a form of a host-based DoS attack. In a well-designed network, attacking the network via DoS is not feasible, but the distributed denial-of-service (DDoS) attack is a form of DoS that is quite effective on the Internet backbone.

*DDoS:* The DDoS attack is a variant of the DoS, in which the attacker floods the network fabric with traffic, causing congestion that essentially stops major parts of the network from operating properly. The Internet has many backbone links that carry rather large quantities of data between very heavily accessed sites. Saturating one or more such backbones has the effect of crippling the Internet. Such attacks are launched by first recruiting a very large number of zombies — i.e., random machines scattered around the Internet, which are programmed via a virus to become nonconsensual accomplices. Then, these zombies send data to selected attack targets such that the aggregate traffic congests the Internet backbones. The DDoS attack is particularly nasty because it is almost impossible to prevent or contain, and has been used very effectively to cripple the Internet.

*Buffer overflows:* The buffer overflow attack is a technically sophisticated attack and exploits unknown bugs in network software to attack a victim computer. In simple terms, an attacker sends a large amount of data that is nonconforming to the data format expected by the receiver. This data is crafted in some very particular and ingenious way so that the receiving software, in the process of copying the data to its internal buffers, overwrites some key address elements in the stack or heap of the receiving machine. Then, there is a snowball effect in which the receiving computer actually starts executing some code that was part of the message it received — in effect, an executable object code provided by the attacker. This object code opens up listening ports, installs backdoors, and makes the victim a slave of the attacker. Due to difficulties in designing a good buffer overflow attack, it is not too common. However, there have been plenty of such attacks in the past that have caused significant damage (e.g., MSBlaster), and new ones are continuously being discovered.

*Trojans, viruses, and other hacks:* All of these are variants of a basic attack that download executable code to a victim computer to use it to launch other attacks or to steal information on it. The techniques range from e-mailed attachments, social engineering, doctored Web pages, and phony advertisements to a grab bag of other tricks that form the hacker's playground. Informed and educated users can "almost always" ensure that they do not fall prey to these tricks. However, because the majority of the computer users of today are relatively unsophisticated, Trojans and viruses are a significant source of concern. These hacks can utilize the unsuspecting victim's computers to perform DDoS attacks, generate SPAM, and perpetrate other misdeeds that affect the networked community as a whole.

## Wireless Networks and Vulnerabilities

Wired and wireless networks share many properties and yet are inherently different. Vulnerabilities that exist in wired networks also exist in wireless networks. However, attacks such as MITM and eavesdropping need physical access to the wired media by the attacker and hence are not very common in wired networks. Consider the typical home computer user who connects to the Internet via an Internet Service Provider (ISP), using dial-up, cable modem, or Digital Subscriber Line (DSL), all of which are physical connections. In each of these cases, he or she is using the ISP-owned (or telco-owned) equipment to connect to a corporate access point (AP). Thus, his or her communications through the Internet are "almost always" safe from eavesdropping and MITM attacks. Similarly, an office user is protected from random sniffing or hijacking attacks.

In wireless networks, such inherent physical protection is absent. Because all the signaling is "over the air," attacks such as MITM and eavesdropping are easier to launch. Of course, the attacker has to be in some physical proximity or be using high-gain antennas and more powerful radios. Remote attacks on wireless networks can be launched using the same methods as remote attacks on wired networks, and the wireless characteristic of the network is hence not exploited.

Furthermore, in most of the current designs of wireless networking protocols, the security system is not integrated into the wireless protocol or has not been carefully designed. Because the standards are already deployed, there is not much scope to redo the designs, and thus the vulnerabilities are already baked in. Some of the notable attacks on wireless networks are documented in the following text.

*Rogue APs:* The rogue AP is a simple but effective technique to steal credentials and perform a variety of attacks on wireless networks, particularly 802.11b networks. Consider a coffee shop that provides wireless service to its customers. Valid users have a username and password. An intruder, Ivan, sets up another AP near the shop with the same Service Set ID (SSID) as that of the coffee shop. When a customer, Alice, tries to connect to the coffee shop's network, she inadvertently connects to Ivan's network, which retrieves her username and password, and then, using her account, logs on to the coffee shop's network and provides Internet access to Alice. However, Ivan can not only steal Alice's credentials but can also be an MITM for all customers of the coffee shop who stray into Ivan's wireless signal. From this point, Ivan can launch all kinds of MITM attacks on unsuspecting users (who will not see anything amiss) with impunity.

*Rogue clients:* A wireless client can gain unauthorized access to a wireless network by stealing credentials, sniffing wireless signals, cloning medium access control (MAC) addresses, etc. These clients can thus gain unauthorized entry into a corporate network, sniff traffic, access protected resources, and often be "behind" the firewall or, in other words, be part of the trusted intranet.

*Open APs:* Many wireless networks are set up as "open," that is, without keys or authentication mechanisms, as this is often the factory default. For such networks, anyone with a wireless client can gain access to the Internet. Also, the intruder can gain access to the network (often inside a firewall) and can access data and other resources available on the network.

*WEP key attack:* The encryption used in Wired Equivalent Privacy (WEP), the standard encryption system for 802.11b (discussed in later sections), is particularly weak and can be compromised by eavesdropping. Attacks against the WEP protocol allow intruders to gain access to the network and steal encrypted traffic.

*Jamming:* All wireless networks are prone to jamming, that is, emission of radio signals in the frequencies used by the network, making communications impossible. This is a form of the attack, but it is easier to launch on wireless networks. Jamming is illegal, but so are most attacks.

*High-gain antennas:* Low-power wireless networks such as 802.11b seem to be secure from any intruder not in the vicinity. However, this has been shown not to be true. It has been demonstrated, that using high-gain antennas, an intruder can access a 802.11b network from up to 15 mi away, even though the network is designed for a maximum operational range of about 300 ft.

*Software vulnerability exploits:* Wireless networking devices use software to perform many of the protocol functions. Inadvertent bugs in the software cause vulnerabilities that can be exploited. Such vulnerabilities are characteristic of the particular version and manufacturer of the software. Not many such bugs have been discovered, but as the adoption of wireless networking grows, the discovery and exploitation of such flaws are inevitable.

## Countermeasures for Wireless Networks

Much of the countermeasures for attack avoidance on wireless networks are the same as that in wired networks. Because a redesign of protocols cannot feasibly be done, we have to resort to "higher-level" security methods such as application-level encryption and the use of digital certificates and authentication. Encryption, when used properly, defeats all eavesdropping attacks. However, the protocols for encryption and key exchange are often flawed and provide holes that an intruder can exploit. Digitally signed certificates, if used by the client and the server, provide immunity to MITM attacks. Digital certificates are, however, not often used. Running a virtual private network (VPN) from the wireless client to some trusted wired Internet proxy is the most secure form of wireless communication and should be used for all Internet communications (even casual

Web browsing). This guarantees wireless clients that data will not be stolen and MITM attacks will not be launched. Yet, VPNs are not widely used for two reasons: (1) they are inconvenient and need considerable installation, configuration, and expertise at the client end, and (2) they add significant visible overhead and slow down communications. Both these problems could have been avoided if the networking protocol used VPN-like methods at the data transport layer.

Alternative to the use of VPN, all wireless networking technologies provide security mechanisms. The 802.11 family provides a plethora of protocols such as WEP, WPA, LEAP, etc. Other protocols have similar features. As we explain later, most of these techniques have been shown to be severely flawed or have not been adequately tested. As of now, VPN is considered to be the only protocol that is relatively free from exploits and vulnerabilities.

## WIRELESS COMMUNICATION TECHNOLOGIES

This section will discuss briefly the actual specifications and standards for the 802.11 family: Code Division Multiple Access (CDMA), General Packet Radio Service (GPRS), and Bluetooth. We discuss the protocols and their operational properties, and provide insights into the relative security of each of these technologies. More details of the security methodologies used in these protocols are described in the section titled "Security in Wireless Networks."

### 802.11 Wireless LAN Standard

The IEEE 802.11 consists of a group or family of Wireless LAN (WLAN) standards. They are designed for use with wireless data access devices such as laptops and PDAs. Each member of the family builds upon the 802.11 base and is identified by a single-letter suffix to the standard. This leads to an alphabet soup of protocols (802.11a, 802.11b, 802.11c, 802.11d, etc.).

The 802.11 base or "legacy" standard set specifies the lower portion of the data-link layer's MAC and the physical layer's (PHY) operations. The most efficient way to communicate using the wireless environment is through the use of a Carrier Sense Multiple Access (CSMA) broadcast approach, which forms the basis for the Ethernet standards employed in most wired LANs. However, due to the wireless medium idiosyncrasies, Collision Avoidance (CSMA/CA) is used instead of Collision Detection (CSMA/CD). In the lower physical layer, there are three specifications defined for the transmission of data, frequency hopping spread spectrum (FHSS), direct sequence spread spectrum (DSSS), and infrared (IR). Most vendors choose to use the DSSS method, which uses two different phase shift keying or modulation approaches to achieve 1-Mbps (Differential Binary Phase Shift Keying [DBPSK]) or 2-Mbps (Differential Quadrature Phase Shift Keying [DQPSK]) data transmission rates. The data sent using these methods is first modulated using a specific pattern of ones and zeros referred to as the *chipping sequence.*

One of the issues that arises with this standard is the usage of the 2.4-GHz band. Many other devices such as microwaves and cordless phones are also using this band. In congested areas such as a large city with a large number of closely packed and tall buildings, signals may not be clear, and there is the possibility that differing signal types and strengths may cause wireless stations to select an AP other than the desired target.

The following variants of the 802.11 family of protocols are built upon the basic or legacy standard defined in the preceding text. Each one was developed to either extend the protocol using newer technologies or to solve a set of problems identified by the community. Even though they are built on the same base, they are not all interoperable within themselves. However, their position within the protocol stack is guaranteed so that they may be used with the existing MAC and PHY standards.

*802.11a:* This is a recently developed standard and is beginning to appear in the market. It is specified as using the unlicensed 5-GHz band and operating at data rates of 6, 9, 12, 18, 24, 36, 48, and 54 Mbps. However, interestingly, to meet the standard, data rates of only 6, 12, and 24 are required. It introduces a new PHY layer protocol: orthogonal frequency division multiplexing (OFDM). By using the 5-GHz band, it is assumed that interference with other devices would be less of an issue. However, the trade-off is that the signal attenuates more quickly (travels less distance) than with the 2.4-GHz band defined for the base standard.

OFDM uses a form of spread spectrum in that it transmits signals separated by a set of precise frequencies, alternately transmitting on these frequencies in a set time and order pattern. The spacing of the frequencies and the modulation of the signal within these frequencies is what provides the orthogonality. This approach is touted as having benefits including a high spectral efficiency, resilience to RF interference, and a lower multipath interference, all of which are benefits in dealing with closely packed buildings with multiple transmission sources.

*802.11b:* Most of the current wireless networks and network infrastructure are based on this standard. 802.11b uses the original 802.11-defined 2.4-GHz frequency range along with a fast DSSS encoding scheme. This standard is what the Wireless-Fidelity (Wi-Fi) Alliance has chosen as its initial standard. It uses a 1-, 2-, 5.5-, and 11-Mbps transmission rate set. The speed of transmission selected determines the type of modulation used. The 802.11 standard uses the 11-bit Bark code to create chips, but this code does not lend itself to supporting the higher data rate of this standard. To achieve this, Complementary Code Keying (CCK) is utilized in replacement for the Barker code. CCK has 64 unique code words, so it can encode up to 6 bits in a chip rather than 1 bit per chip that 802.11 defines. Security is provided in 802.11b using WEP, which is a shared key system and is known to be weak (see section titled "WEP").

*802.11c/d/e/f:* These variants are essentially variants of 802.11b with some changes. The "c" variant provides procedures for bridging on a network AP. The "d" variant, also referred to as the Global Harmonization standard, has other usages and practices in countries where the spectrum allocation is different. The "e" variant provides quality-of-service (QoS) optimizations in support of payloads such as voice and video. The "f" variant provides a uniform set of standards to support features and capabilities, such as roaming, between APs of different vendors.

*802.11g:* This is a recently ratified standard (June 2003) that is a high-speed superset of 802.11b. This standard achieves the higher data rates attributed to 802.11a only, using the 2.4-GHz band rather than 5 GHz. This standard allows for an intermixing of both 802.11b and 802.11g stations accessing the same APs. It uses the OFDM technology defined from the 802.11a standard in combination with 802.11b's DSSS-CCK to provide for compatibility. It has defined data rates of 1, 2, 5.5, and 11 Mbps using DSSS-CCK and 6, 9, 12, 18, 24, 36, 48, and 54 Mbps using OFDM. The mandatory rates that must be supported are 1, 2, 5.5, 11, 6, 12, and 24 Mbps.

The 802.11b and 802.11g standards are defined to be used predominately indoors, so their performance ranges are defined in terms of feet with a distinct trade-off between distance and bandwidth. In a nominal setting, their performances can be characterized as being fully capable of achieving their rated throughputs within 150 to 175 ft (45 to 54 m) and as being comparable in performance beyond that distance.

The 802.11g specification allows for the specification of data rate values that exceed the defined performance standards and allowances within the software control structures and fields. Therefore there are a number of vendor-provided extensions to the 802.11g that utilize these field options. These extensions allow for speeds, such as 108 Mbps, that exceed the 802.11g specification, still using the control fields and frames of the 802.11g standard. Because the extensions are vendor specific in semantic if not syntax, they are, or at least should be, interoperable within the vendor's environments and do not, as a general rule, play well with another vendor's environment (APs and wireless cards).

*802.11h:* This is a supplement to the MAC layer to support European compliance with 5-GHz WLANs.

*802.11i:* Due to security issues (identified in the following text) relating to the security of 802.11 WLAN protocols, this standard attempts to address these problems. It adds new encryption key protocols such as Temporal Key Integrity Protocol (TKIP) and Advanced Encryption Standard (AES). It is estimated that it could take 100 years of continuous key generation to exhaust the keyspace values. 802.11i also defines additional standards such as secure IBSS, secure fast handoff, and secure deauthentication and disassociation. Whereas some of the standards proposed are considered easy software upgrades to existing products, some of the new standards, such as this one, will require hardware changes to support them.

## WiMAX

Intel has introduced a wireless Metropolitan Area Network implementation based on the IEEE 802.16d wireless broadband standard and the European ETSI High-Performance Radio Metropolitan Area Network (HiperMAN) broadband wireless MAN (WMAN) standards. The intention of WiMAX is to act as a high-speed (75 Mbps), "last mile" connection technology for up to 30 mi (48.3 km). It operates in the 2 to 11-GHz range, which forms the basis for the higher aggregate data throughput and provides non-line-of-sight implementations (as compared to the original 802.16 implementation). It is not considered to be a replacement for the 802.11 implementations but to actually coexist as a primary approach to wireless delivery of T1/E1 level services to a corporation or a hot-spot deployment of 802.11.

Being a protocol developed after the problems with the 802.11 family was apparent, the WiMAX protocol integrates the security into the wireless transport. WiMAX uses DES in cipher block chaining mode to encrypt the payload. It uses Privacy and Key Management (PKM) protocol, which provides a certificate-based authorization with RSA public-key methods. The payload includes the control information for fragmentation, packaging, and control information. This means that little information can be deduced about the payload and the data from the information that remains in the clear. The protocol is new, and its vulnerabilities have not been adequately studied.

## GPRS

This is a standard for a value-added service to be used with the cellular telephone infrastructure using GSM and TDMA. Security was not a design consideration when GPRS was conceived. It augments the voice capability of cellular phones, providing digital wireless communication data rates of up to 171.2 kbps theoretically and 114 kbps practically. For this scheme to work over the existing network, it requires support from components placed at strategic points between the end user and the data sources. A Gateway GPRS Support Node (GGSN) acts as a back end and is used as the interface to other networks such as the Internet. The mobile device communicates directly with a Serving GPRS Support Node (SGSN), which acts to maintain the attached state for the mobile devices as it moves through the network. The GGSN and SGSN work together to maintain the virtual connections to the resource and to deliver the data.

Though GPRS seems to provide a good match for other packet-based backbones, it also brings with it a few downsides. One of these is that GPRS must share bandwidth with the GSM/TDMA voice channel. This means the data and voice transmissions interfere with each other, causing a quality versus bandwidth trade-off to be made. There is no inherently defined security protocol within the GPRS standards, so other existing protocols must be used. For example, to securely send data between two GPRS stations over the Internet, VPN could be employed at the higher layers to provide the least data encryption.

## CDMA

After the success of digital cellular systems such as GSM in most parts of the world, and TDMA in the United States, Qualcomm developed a competing standard — CDMA protocol. The original technology definition is currently also known as cdmaOne and is defined as the IS-95B standard. CDMA works by transmitting a digitally encoded analog signal using spread-spectrum technology combined with a special coding scheme over a 1.25-MHz channel. To significantly reduce the odds of a collision, it encodes the data using a set of 64-bit Walsh codes. In theory at least, up to 64 users could simultaneously use the same channel. This approach makes it difficult for a receiver not having the correct code to even receive any intelligible signal (the signal is equivalent to noise). For communications, CDMA is used for data transfer and control. The CDMA signaling technique has a side effect that casual eavesdropping on the transmitted signal is thought to be not possible.

## Bluetooth

Bluetooth is a standard developed as a short-range wireless link between devices. The maximum link distance is about 33 ft (10 m). The transmitter operates on the 2.4-GHz ISM band and uses a fast acknowledgment frequency–hopping scheme that improves the robustness of the signal in a noisy frequency environment. The technology is capable of frequent, quick hops (changes in frequency) due in part to its use of smaller data packets than any other device that uses the band, such as 802.11b. It uses low power for its transmissions, which requires the units to be relatively close to each other. Bluetooth was intended for use as a communications link between small, potentially low-cost devices. It is not intended for both mobility and distance. It has a data rate of between 300 and 400 kbps, which is sufficient for devices such as a mouse or keyboard communications with a PC or other devices. It is also beginning to be used to communicate between PDAs and cellular phones for local data exchange. In a phone or PDA setting, users agree to exchange data by providing each other with an access code or PIN, which enables access to perform a local and limited set of functions such as messaging or data exchange.

## SECURITY IN WIRELESS NETWORKS

WLANs are significantly less secure than wired LANs. Signals can be more easily captured from a number of stations by the simple choice of capture location. To prevent this type of data capture from being successful, the WLAN standards have included a set of protocols and facilities. However, as with anything new and interesting, it turns out that the initial protocols were not as robust in performing their task as expected.

Optionally, security for 802.11 and 802.11a, b, and g was initially defined to be based on the use of SSID and WEP to provide for both authentication and privacy through the encryption of data over the radio waves. Each WLAN has the option of specifying an SSID that can be exchanged at the initiation of communication between a system and an AP. The SSID in use must be the same between both sides before further communications can commence. Unfortunately, this exchange is in cleartext, so it is relatively easy to capture the SSID of the network, thus rendering the SSID little more than a convenient label for the AP.

### WEP

This is a key-based security protocol intended to prevent "casual eaves-dropping" of the data being transmitted over the wireless network. The key is used to encrypt or decrypt the data portion of a packet. The key that is defined in the original standards is a single 40-bit key, although larger keys, up to 128 bits, are defined by a follow-on standard sometimes referred to as WEP2. The same key is used at all stations that communicate over a particular WLAN. The entire key is never exchanged over the wireless network, so it is not directly captured. In principle, the WEP methodology is strong; however, the implementation is flawed and allows the key to be determined relatively easily. Another problem with the WEP approach is that it is a shared key, shared by all users of the network, and hence the key is subject to being leaked manually.

The WEP key generation is based on the RC4 stream cipher algorithm. The algorithm depends upon a permutation of all the possible $n$ bit words, a pair of indices, and the initial value of a variable key. RC4 defines the output of a key scheduling algorithm (KSA), which uses the variable key as input to drive the subsequent permutations of the algorithm. As it turns out, the keys used for this initial value can be based on a series of "weak" keys, which are keys for a small number of bits; the remaining bits can be easily generated given the original key generation methodology. This makes generating certain keys easy to do in a short period. Because time is a code breaker's enemy, any reduction in time improves the ability to break into the key and data.[1]

The second WEP key vulnerability is the ease with which the exposed or available portion of the encryption key, the initialization vector (IV), and data portion of the packet can be analyzed to determine a pattern that an attacker can readily use to generate the unseen or private portion of the key. An attacker can spot a pattern that can be concatenated with a guess about the secret part of the key and obtain a relatively quick and small set of choices to try to guess the key.

This WEP weakness has been well documented and exploited. In general, in an environment with large quantities of data in the air, all of the keys can be captured within a matter of a few hours. In a home network, with little or small bursts of airborne data, the time to capture the key is significantly longer but still relatively easy. Larger key sizes tend to increase the key-cracking time, but in most cases this appears to be only less than twice the time required to derive the 40-bit key.

"War Drivers" are hackers who actively travel around, and using standard hardware such as a high-gain antenna (sometimes made from a Pringles potato chip can) and some specialized software such as Network Stumbler, can actually map the locations and various characteristics of WLANs they detect. The software will indicate whether the WLAN is using any form of security. If not, then it is a trivial exercise to gain access to the WLAN. If it does use security, then it is not uncommon for a user to use the factory default SSID, again making access to the WLAN trivial. For those WLANs in which the user has set the SSID value, it becomes a matter of trying a relatively short list of values to determine the key.

To improve on this security issue, a new replacement standard was developed called Wi-Fi Protected Access (WPA). This was developed as a cooperative effort between the WFA and IEEE, and as a stopgap of sorts until the 802.11i implementation is in place. WPA is derived from the 802.11i standard, and as such is considered to be forward compatible with 802.11i. WPA is a software replacement for the WEP standard, so no hardware changes are required on most of the existing equipment. It improves on the WEP implementation by incorporating elements of the 802.11i standard's TKIP. Specifically, WPA improves encryption by including a per-packet key-mixing function, a message integrity check (MIC), an extended 48-bit IV with sequencing rules, and a rekeying mechanism. Also included in WPA is 802.1X/EAP network authentication. The last feature is required as part of the certification for WPA.

WPA provides a key that, unlike the WEP ASCII to HEX conversion, creates a cryptographically stronger key with more possible values in the key domain. This improves the randomality factor for the key, which should decrease its susceptibility to cracking. However, if keys are chosen with a small size and using dictionary words, then it would still be a relatively easy target for cracking.[4]

For the most part, most of the newer devices based on the 802.11g standard use WPA as the security base. In some cases, upgrades are available for the older 802.11b devices. Most of the APs and network cards will support only one of the security protocols that create a problem in a mixed WEP/WPA environment.

One of the drawbacks of WPA is the 802.1X/EAP implementation. Even though it is optional, those environments in which network authentication is important will not be able to use this nonsecure protocol. To fix this problem, companies such as Microsoft began to make available an EAP-TLS (Transport Layer Security) version of the protocol. There are some distinct advantages over non-TLS EAP such as mutual authentication, but to achieve this security level, client certificates must be installed on each system that wants to participate in the network. Unfortunately, EAP-TLS does have an issue with displaying some data in the clear which is its main weakness. There is a tunneled version (TTLS) that first establishes the session before sending any data using encryption. Microsoft and Cisco have been pushing a Protected EAP (PEAP). PEAP does not tunnel data but instead incorporates the EAP messages into a secure exchange.

WPA is considered to be an interim solution to the WEP security problem. A newer standard, referred to as WPA 2, is expected to be made available in a final form sometime in the 2004 time frame. WPA 2 is essentially a fully certified version of the IEEE 802.11i standard.

## 802.1X

The IEEE appears to recognize that the security introduced in the original and subsequent 802.11 standards was not as robust as it should be for an effective deployment. To this end, working committees have been formed to propose a variety of extensions, enhancements, and replacements. In addition to the work on the 802.11i standards, there is a Robust

Security Network (RSN) proposal, referred to as 802.1X, focused on a longer-term solution to the 802.11 problems.

RSN provides security by adding a third-party authentication server service to the authentication process. The way this works is that a station (referred to as the supplicant) that wishes to connect to the network by way of the AP (referred to as the authenticator) makes a connection request. The authenticator then contacts an authentication server, usually a RADIUS-type server, which either validates or rejects the request. EAP is used to make the authentication request. In this way, a high degree of secure flexibility exists as to the actual service that can be deployed. EAP is considered to be an effective approach in that it is a challenge–response model; however, as was noted in the preceding text, it is not a secure protocol. To this end, enhancements such at TTLS and PEAP are being proposed.

However, as with most new protocols, the shortcomings of EAP are being discovered. Mishra and Arbaugh[2] believe that they have identified weaknesses that allow for both a MITM attack and a session hijacking to occur. The first attack focuses on the relationship between the authenticator and the authenticating server. No explicit mutual authentication is specified in the standard, and thus someone sitting between these two entities could gain access and assume either role. The second attack relies on the wireless operational environment and the ability of an attacker to use certain management frames to change the supplicant's and authenticator's connection to a different supplicant while remaining in an authenticated state. Proposals are made by Mishra and Arbaugh as to how to correct this issue. In addition, Cisco makes proposals for how to resolve these security problems.[3] This dialogue is not complete; but when it is, it is expected to result in a much stronger standard.

## MAC Filtering

One approach that has been used to help resolve issues with WEP security is to use a list of valid MAC addresses (also known as a White List) at the AP control access. If your MAC address is not listed on the AP, you are not granted access. This sounds good in theory, but in actuality, valid or acceptable MACs can be captured or sniffed using standard wireless cards and software. Once this information is obtained, it is relatively easy for the wireless card to be set to use the valid MAC (MAC cloning). Once this occurs, the AP will no longer be able to provide network security.

The converse of the White List is the Black List, which is a list of those who cannot access the network. This approach suffers from the same type of attack as that of the White List. All that is required is a valid or

AP-acceptable MAC address. This can be obtained using the same tools and procedure as that of the White List. Once a valid MAC is obtained, the attacker simply changes the MAC address of his or her station's card and then gains access to the network by way of the AP.

## VPN

VPN is a transport methodology that allows for data to be moved between two networks by providing a local network–addressing service for both networks, combined with authentication and data encryption. This allows for a remote computer to have a connection to a network by way of a server that acts both as local host and data translator for the remote computer. The use of VPN over WLAN provides the data-level encryption and end-to-end authentication depending upon what services are provided at the wired network side. VPN can be used to augment existing protocols such as WEP and WPA, and can be used in lieu of the 802.11i or 802.1X standards.

Point-to-Point Tunneling Protocol (PPTP) was developed by Microsoft and several other vendors to connect and move data between networks using a Point-to-Point Protocol (PPP). IPSec and L2TP are newer standards that offer a number of improvements over PPTP, while providing the same VPN functionality. IPSec can either act as a transport of IP data between two networks, or it can perform tunneling of IP data inside another IP packet. Either approach can be made secure through the use of Encapsulated Security Payload (ESP) or Authentication Headers (AH). IPSec is viewed as a server-to-server protocol, which is why data traffic can be effectively delivered between networks without the need of a client *per se*.

L2TP's job is more of the client-to-server function. It is used to encapsulate and transport multi-protocol packets and is an extension of PPP, which allows the layer 2 to exist on a different device than the PPP endpoint. This approach allows for packets to be sent over X.25, ATM, Frame-Relay, or other non-IP-based protocols such as IPX, as well as across the IP cloud. L2TP does not provide security services, so a good matchup is with IPSec.

The VPN environment is considered by most to be very secure, and a variety of standards are being codified in support of a more widespread interoperability. It can also be used in a number of mixed or diverse transport environments. There are no defined vulnerabilities with VPN. Most of the problems arise with either vendor implementations or access to data and security information from either end of the VPN connection. This latter problem exists in any setting and certainly is not solvable through the usage of VPN.

One thing to note, however, is that because VPN will have an address on a remote network, it is possible that a hacker on the remote network could discover the address through normal discovery techniques and attempt to use or send known exploits to attack ports or other OS vulnerabilities, using the VPN remote address. Depending upon security implemented with the OS that is in use and the implementation of the VPN, as well as other factors such as the use of firewalls, these types of attacks can be effectively defended against.

## Cellular Network Security

The security of cellular networks has been studied but not as rigorously as other forms of wireless networks. This is due in part to the perception that the use of cellular networks is for noncritical data transmission. Cellular networks are being used for short messages, quick specialized Web browsing, and for sending pictures, video, as well as audio. Hackers have not been interested in eavesdropping on such activities. Thus, much of what is known of these vulnerabilities (or lack thereof) is somewhat speculative and not established.

The security of GPRS networks depend upon the A3, A5, and A8 algorithms used by the GSM system to authenticate the user and the base station, and to cipher all data and voice traffic between them. Although on the surface GPRS seems to be secure, many security holes have been discovered. The smart card used in the GSM system utilizes an authentication system in which a challenge response (CR) is performed with the mobile unit's ESN (Electronic Serial Number). The encoding used in this CR scheme has been shown to be vulnerable, and smartcards can thus be cloned.

The A5 cipher is used to encrypt all data communications. Researchers believe that A5 is not as strong as its 114-bit key length but can be broken using hardware-based cryptanalysis. However, such attacks are not prevalent because the importance of user data transmitted by GPRS networks is still quite small.

The CDMA systems are believed to be more secure than the GPRS network, mainly due to the nature of the radio frequency signaling. Although it is possible to listen in on a GPRS transmission using TDMA receivers, it is not possible with CDMA. A CDMA receiver has to be coded with the correct 64-bit code to receive a channel of CDMA traffic, and without this code, or with a wrong code, the received signal is noise. A brute force attack to find a correct code is not feasible. The code is exchanged between the sender and the receiver at the handshake, which happens over an encrypted channel.

In spite of the difficulty in tuning into a CDMA transmission, the data (or voice) transmission is further encrypted. This double layer of ciphering makes CDMA security possibly quite strong.

All cellular networks are, however, vulnerable to location-finding by triangulation or directional antennas. That is, an attacker can find the location of a mobile station with the use of radio-monitoring equipment. This does not compromise the privacy of the data but the privacy of the operator's location.

## Bluetooth Security

There are four elements in Bluetooth devices that are used to maintain link-level security. The first is the Bluetooth device address, which is a 48-bit value unique to each Bluetooth device and defined by IEEE. The second is a private authentication key, which is a 128-bit random number. Third, there is an 8- to 128-bit private encryption key. The last is a pseudorandomly generated, 128-bit number that the device creates. The use of these elements, to some extent, depends on the mode of the security-level setting of the Bluetooth device.

The choices for modes are 1 to 3, where Mode 1 is the nonsecure mode, Mode 2 is service-level enforced mode, and Mode 3 is link-level enforced mode. Mode 3 security begins the security prior to a communications channel being established. Devices can also be tagged as trusted and untrusted with service levels that include requiring both authorization and authentication, authentication only, and open to all.

Bluetooth uses RF data transmission, which makes it more vulnerable to attack. One area that may be a weakness, given certain circumstances, is the use of a *divide and conquer* approach against the 128-bit key, which might allow the key to be broken in $O(2^{64})$ time.[5] However, this may not be a large issue because the actual key usage time is so short that it is doubtful if this could actually be effectively implemented. One other possible area of weakness is that it is possible that after two devices are authenticated and a shared key is used, a third device could be introduced that uses the shared key. Given the knowledge of the shared key, the third device could potentially (1) use a phony device address to calculate the encryption key and therefore listen to the traffic, and (2) authenticate itself as one of the other two devices.

Another security flaw appears to be specific to certain cellular phones. Flaws in the implementation appear to allow a hacker to connect to a user's phone without the user's awareness. This is sometimes referred to as "bluesnarfing." Once connected, any data present on the user's phone can be downloaded and potentially altered. This type of flaw is not an inherent problem with the Bluetooth technology or its defined security as much as a poor implementation of the existing security.

## PROTECTING WIRELESS NETWORKS

Given the plethora of broken security protocols that are used in wireless networks, the best method of protecting the wireless network is probably not to rely on any one of the proposed or deployed standards but to rely on the time-tested VPN technique. VPN is a high-level protocol that does not depend upon any of the underlying transport protocols. It creates a secret encrypted tunnel from a client station to a designated VPN server, which then forwards the client's packets to the destinations via a wired network on an intranet. If used on top of an insecure wireless network, it ensures that no data packets are visible to an attacker and an MITM attacker cannot tamper or spoof the data exchange (as the endpoints are authenticated using digital certificates).

However, a VPN is not a panacea. First, it needs end-user expertise to deploy and use. Some applications fail when used over a VPN and have to be reconfigured. The throughput is lower as more hops and encryption overhead are introduced. Finally, although the VPN protocols are secure, they are implemented on top of insecure wireless protocol stacks, and attacks on these stacks may compromise the client machine using other trapdoors and vulnerabilities (though not many have been discovered yet).

The alternative to VPN is to use some form of well-designed, well-tested wireless security protocols. Newer alternatives to WEP and WiMAX protocol are possible candidates. The plethora of such new, improved protocols are, however, both confusing and worrisome. In the security arena, new does not mean good; it means unknown and untested. Given that so many standards are being proposed, it is highly likely that deployment of each will be minimal (or spread thin), and thus most of them will not receive adequate exposure and stress testing. This makes all protocols potentially untrustworthy.

## CONCLUSIONS

As wireless devices are gaining in popularity, their built-in security systems are beginning to show definable and exploitable weaknesses. The 802.11b and WEP scheme is the most popular but the most maligned due to its inadvertently poor design. Other systems are most probably better, or their weaknesses have not yet been discovered.

All wireless (and wired) systems are capable of supporting application-level security methods such as VPN, SSL, SSH, etc. Conventional wisdom states that the security provided by these higher-level protocols is much superior. Barring implementation flaws, these protocols provide as close to guaranteed security as we can achieve today. Because the underlying transport-level security does not affect the security of VPN-like protocols, they can be safely used over insecure wireless networks.

Wireless networking is insecure due to its lack of integrated security at the transport layer. This design deficiency is not expected to change. Integrating security at the transport layer would significantly increase the complexity of deploying wireless networks and thus is not considered feasible from a usability and marketing perspective. The lack of security is not considered a major issue, and thus the community largely chooses to ignore its implications.

## REFERENCES

1. Fluhrer, S.R., Mantin, I., and Shamir, A., Weaknesses in the Key Scheduling Algorithm of RC4, *Lecture Notes in Computer Science, Revised Papers from the 8th Annual International Workshop on Selected Areas in Cryptography*, Springer-Verlag, New York, 2001.
2. Mishra, A. and Arbaugh, W.A., An Initial Security Analysis of the IEEE 802.1X Standard, Technical Report, University of Maryland, Department of Computer Science, CS-TR-4328, UMIACS-TR-2002-10, Feburary 2001.
3. http://www.cisco.com/warp/public/cc/pd/witc/ao350ap/prodlit/1680_pp.htm, accessed February 29, 2004, posted August 22, 06:32:08 PDT 2002, Cisco Systems.
4. Moskowitz, R., Weakness in Passphrase Choice in WPA Interface, *Wi-Fi Networking News*, November 4, 2003, http://wifinetnews.com/archives/002452.html, accessed May 29, 2004.
5. Vainio, J.T., Bluetooth Security, http://www.niksula.cs.hut.fi/~jiitv/bluesec.html, accessed May 29, 2004.

# VI

STANDARDS

# 22

## MEDIUM ACCESS CONTROL: TECHNIQUES AND PROTOCOLS

*Zhihui Chen and Ashfaq Khokhar*

## INTRODUCTION

A wireless local area network (WLAN) consists of multiple stations that coexist within a limited geographic area and share a common wireless channel to communicate with each other. All the stations in a WLAN can access this channel by first reserving it and then transmitting or receiving data frames. However, due to the absence of a physical medium and on account of the limited communication range of transmitters and receivers, the medium access control (MAC) issues in wireless channels are significantly different from those in their wired counterparts. In this chapter, we review medium access control issues related to WLANs and provide an overview of different MAC protocols designed for such networks. We also review the popular IEEE 802.11[15] standard for the MAC layer in WLANs.

Depending on the configuration, a WLAN can be categorized into one of the following two types: ad hoc network or infrastructure network. An ad hoc network is a dynamic network formed by a group of wireless stations without the assistance of any preexisting network infrastructure. To create an ad hoc network, one only needs to have participating stations within the communication range of each other, and these stations then coordinate with each other to provide correct data communication service among them. In such networks, all the stations have the same functionality;

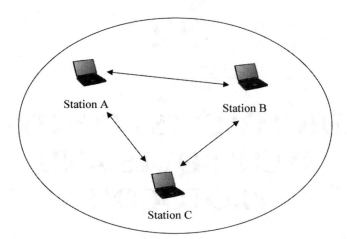

**Figure 22.1 An example of ad hoc network configuration without any explicit router.**

therefore, the network control is simple and completely distributed. A sample ad hoc WLAN is shown in Figure 22.1. In this example, there are three stations, and each can communicate with any other station within transmitting or receiving range.

The infrastructure network, on the other hand, contains a special station referred to as the access point (AP), which may also be connected to the outside network with a wired or wireless link. The communication range of an AP may be larger than the regular station. The AP acts as a router and, in this way, two stations that are out of radio coverage of each other can communicate via an AP. An example of the infrastructure network is shown in Figure 22.2. Here, there are two APs, AP1 and AP2, and four stations: A, B, C, and D. The stations A and D cannot communicate directly via the radio channel due to their limited range. However, station A can send data to AP1, and AP1 forwards the data to AP2. Then AP2 delivers the data to station D. In an infrastructure network, the AP coordinates access to the wireless channel.

## MAC PROTOCOLS

Regardless of the configuration of the WLAN, all stations in the network share a single wireless channel. So, at any point in time only one transmitter–receiver pair can communicate. If more than one station transmits in the same area, the transmitted data frames collide with each other, so that neither is received. For data frames be received successfully, stations must obey some rules to decide which station should access the

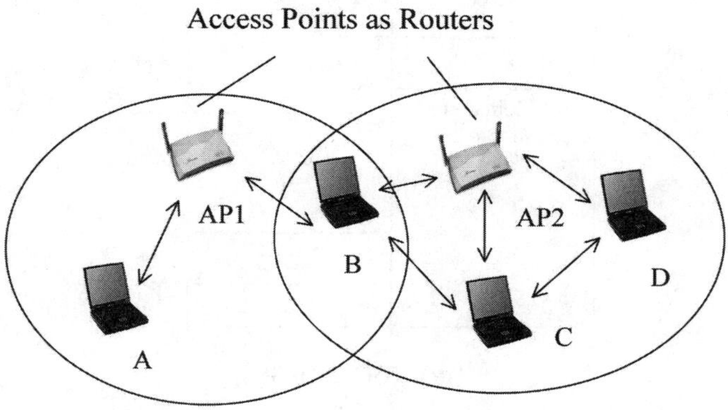

**Figure 22.2 An example of infrastructure network configuration.**

medium and at what time the station should start transmitting. These rules are referred to as MAC protocols. There are two ways of achieving medium access control: (1) let stations negotiate by themselves, which is called distributed control and (2) assign a station to be the central controller, which arbitrates access to the channel. In ad hoc networks, because all the stations perform the same role, no one can be assigned the role of central controller and therefore, distributed MAC protocols are more applicable. In infrastructure networks, the APs serve as the central control units and centrally controlled MAC protocols are applicable. In the following text, we first outline generic approaches to MAC protocols and then identify problems related to wireless channels.

The first MAC protocol, ALOHA,[1,7] can be dated back to the Fall Joint Computer Conference of 1970 at the University of Hawaii. The ALOHA scheme is a simple distributed control scheme in which a station starts transmission as soon as it has data to send. If two data frames collide, both frames are discarded. After the frame loss, transmitters wait for a random time duration and retransmit the data frame.

In this scheme, it is supposed that all wireless stations in the system can "hear" each other. It is simple and easy to implement but yields only moderate performance. The performance of a MAC protocol is usually measured as the maximum throughput, i.e., the maximum number of data frames that can be sent in unit time. The performance of the ALOHA protocol can be analyzed as follows: first, we suppose that data frames have a fixed length, and we normalize the time required to transmit a frame to be one time unit; second, we assume that the data arrival processes for all stations are Poisson. That is, the data arrival process is history independent. For a station with data arrival rate of frames per time unit, the probability of $k$ frames arriving in $T$ time units is:

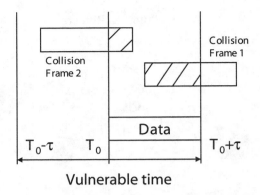

**Figure 22.3  Vulnerable time for a data frame collision in the ALOHA protocol.**

$$Pr(k) = \frac{(\lambda T)^k e^{-\lambda T}}{k!}$$

In the ALOHA protocol, data frames are sent out as soon as they arrive. For a frame to be received successfully, no collision should happen during the transmission of the data frame. Suppose a station starts sending data frame at time $T_0$. According to our assumption, the frame transmission occurs from $T_0$ to $T_0 + 1$, as shown in Figure 22.3. If any other station starts transmitting its data frame between $T_0$ and $T_0 + 1$, the frame will collide with the tail of the original data frame. On the other hand, frames starting between $T_0$ 1 and $T_0$ will collide with the head of the original data frame. So the vulnerable time for a packet is from $T_0$ 1 to $T_0 + 1$. The probability for a station with data arrival rate having no data frames arriving in this period is $Pr(0) = e^{-2\lambda}$. Suppose there are $N$ stations in the network and all of them have the same data arrival rate $\lambda$, then the probability of successful reception is:

$$P(success) = \left(Pr(0)\right)^{N-1} = e^{-2\lambda(N-1)}$$

Because the probability of a station transmitting in unit time is equal to the average data arrival rate, $\lambda$, the throughput is

$$T_{ALOHA} = N\lambda e^{-2\lambda(N-1)}$$

For a large number of stations, $T_{ALOHA} \approx N\lambda e^{-2N\lambda} = Ge^{-2G}$, where $G + N\lambda$ is the average data arrival rate of all the stations, or the traffic load of the system. By computing the first derivative, we can determine that throughput is maximum at $G = 0.5$, which corresponds to a throughput of about 18.4 percent. To improve the performance of the ALOHA protocol, several modifications have been proposed. These proposals are discussed in the following text:

■ *Slotted ALOHA:*[2] This scheme divides the time into slots and aligns frame transmission with slot boundaries. The benefit of Slotted ALOHA is that it reduces the vulnerable time. Because the frames are aligned, the vulnerable time for a data frame is reduced to one time unit; so the throughput of the Slotted ALOHA system can be formulated as:

$$T_{Slotted-ALOHA} = Ge^{-G}$$

The throughput of the Slotted ALOHA system is maximum when $G = 1.0$, corresponding to a throughput of 36.8 percent, which is 100 percent more than that of the ALOHA system.

■ *Carrier Sense Multiple Access (CSMA):*[16] This scheme lets the stations sense the medium first and start transmission only if the medium is clear or idle, so that the chance of collision is further reduced.

■ *p-Persistent CSMA:* Stations applying the p-Persistent CSMA protocol start transmitting with a certain probability when the medium is clear. This transmitting probability is called persistent. For both ALOHA and Slotted ALOHA systems, when the traffic load $G$ increases, the throughput also increases at first. However, after $G$ exceeds the maximum throughput point (0.5 for pure ALOHA and 1.0 for Slotted ALOHA), throughput drops when G keeps increasing, as shown in Figure 22.4. Finally, the medium is totally wasted due to collisions and throughput approaches zero. This phenomenon is called congestion. A good MAC protocol design is supposed to relieve congestion and keep the throughput at a certain level after the traffic load exceeds the maximum throughput point. p-Persistent CSMA is such a MAC protocol. It relieves congestion by blocking some stations that have data to send. The cost is lower network throughput when traffic load is low and additional traffic delay.

■ *CSMA with Collision Detection (CSMA/CD):*[17] Stations applying this protocol listen to the medium while transmitting, and if collision happens, the stations abort the transmission immediately. This scheme improves the throughput because the time wasted on transmitting collided frames is reduced.

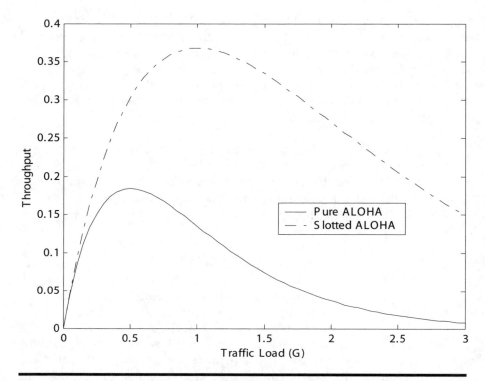

**Figure 22.4    Throughput comparison of pure and Slotted ALOHA protocols.**

## Characteristics of WLANs

Although the ALOHA family of protocols was originally designed for radio networks, it also found widespread use in wired LANs. In their original design, these protocols fail to work properly in WLAN environments. This is because the access control issues in a WLAN are significantly different from its wired counter part.

First, in the protocols discussed above, it is assumed that all stations in the system can communicate with each other, so the medium occupation status viewed by all the stations is identical. However, WLAN stations have transmission power and coverage constraints, and distant stations may be "invisible." So the medium may be busy at one location but free at another location. A naive approach to solving this problem is to increase the transmission power and let all the stations hear each other. This approach is not desirable for two reasons. First, increasing transmission power increases interference in other wireless systems and reduces the radio frequency utilization. The second reason is that some of the WLAN devices are battery powered (e.g., laptop, PDA), so raising transmission power leads to shorter battery life.

Also, in a wireless environment, listening to the medium and transmitting simultaneously requires two independent transceivers, so it is expensive to implement the CSMA/CD protocol. Also, keeping precise time synchronization is difficult because the propagation delay is time variant and unpredictable. The straightforward realization of wired LAN protocols in wireless environments increases the complexity of station equipment and thus is not cost effective.

### *Location-Dependent Medium Status — Hidden- and Exposed-Station Problem*[9]

As pointed out earlier, the medium status viewed by each station in a WLAN is location dependent. This poses a problem in assigning the medium to stations. Consider a CSMA/CD-based WLAN consisting of four stations: A, B, C, and D, as shown in Figure 22.5. Let us first consider the following scenario: Both stations A and C have data frames to send to stations B and D, respectively. Assume that stations A and C cannot hear each other. Now, suppose station A starts transmitting first, as depicted in Figure 22.5a. At this time, if station C senses the medium, it will not hear the packet from A because it is out of the range of A. So C thinks the medium is clear and decides to start the transmission of the packet to D. Unfortunately, the data frames from stations A and C collide at station B, so station B cannot receive the frame from A successfully. This problem is called the *hidden-station problem* which means a sender (station C) is hidden from another sender (station A), so station A has no knowledge of the existence of station C.

Second, let us consider the same network but reverse the transmission direction from A to B, i.e., station B and C have a data frame to be transmitted to A and D, respectively. When station B starts to transmit, station C senses the medium and finds the medium occupied. So station C decides to defer transmission, although the medium around station D is clear and the data frame can be received successfully. This transmission deferral is unnecessary and introduces additional delays. This situation is called the *exposed-station problem*, which means a sender (station C) is exposed to the range of another sender (station B).

From the previous examples, we learn that a sender can only sense the medium around itself. However, successful reception depends on the medium status at the receiver side as well. The medium statuses at the two sides may be different; therefore, estimation of the status at the receiver side by the sender is not a trivial matter.

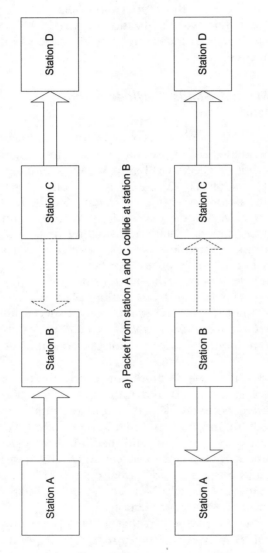

a) Packet from station A and C collide at station B

b) Packet from station B to A block transmission from station C to D

**Figure 22.5 Hidden- and exposed-station problems in WLAN. (a) Packet from station A and C collide at station B. (b) Packet from station B to A blocks transmission from station C to D.**

**Table 22.1  Categories of WLAN MAC Protocols**

|  | Sender-Initiated | Receiver-Initiated |
|---|---|---|
| Single-channel | MACA, IEEE802.11 | MACA-BI |
| Split-channel | BTMA | Receiver-initiated BTMA |

# DESIGN OF MAC PROTOCOL FOR WLAN

Several MAC protocols have been proposed to solve the hidden- or exposed-station problem.[3-6,8,13,14] Depending on who initiates the communication dialogue, the proposed protocol can be categorized as a sender-initiated protocol or receiver-initiated protocol. By using in-channel signaling or dedicated channel signaling, the MAC protocol can be grouped into single-channel protocols and split-channel protocols. All the in-between combinations are also possible; for example, we can have a sender-initiated single-channel MAC protocol or a receiver-initiated split-channel protocol, and the representatives of each category are listed in Table 22.1.

## Sender-Initiated MAC Protocols

Most of the MAC protocols proposed for WLANs are single-channel sender-initiated protocols, the idea of which is quite straightforward. Both data and control frames are sent via the same channel medium. A station tries to access the medium as soon as it receives the transmission request, provided the medium is clear. The transmission is deferred if the medium is busy. Multiple Access with Collision Avoidance (MACA) is an example of this category.[6] The basic idea of MACA is that transmitter and receiver exchange medium status information before actually sending out the data frame to avoid collision. The nearby stations can hear these control frames and learn that a transmission is about to start and thus avoid accessing the channel during the transmission.

Figure 22.6 illustrates an example of how the MACA protocol works. The wireless network consists of five stations. The left circle and right circle show the radio coverage of stations A and B, respectively. Suppose station A has a data frame to be transmitted to station B. First station A sends out a request-to-send (RTS) control frame to station B. This RTS frame contains the time duration for transmitting the data frame. Station B then responds with a clear-to-send (CTS) control frame to station A. The CTS frame also contains time duration information, which is copied from the RTS frame. After receiving the CTS frame, station A starts transmitting the data frame. All the stations that hear the RTS frame

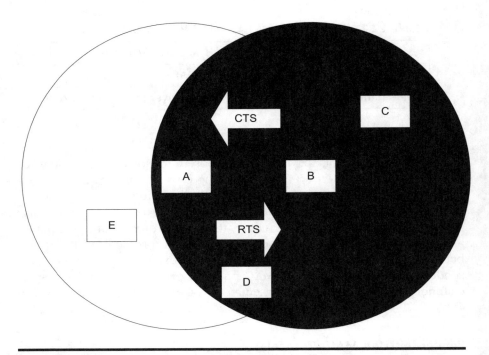

**Figure 22.6 The RTS–CTS exchange procedure in the MACA protocol.**

postpone their transmission until the CTS frame can be received success-fully, whereas those stations that receive the CTS frame (irrespective of whether they have heard the RTS frame) should postpone their transmission until the data frame is received successfully. In the system shown in Figure 22.6, stations C and D need to remain silent until the end of data frame transmission, whereas station E needs to wait only until the CTS frame is received.

However, the MACA protocol fails in the presence of the exposed-station problem. Consider the previous example, in which station E also applies the MACA protocol and sends out an RTS frame to B during the data frame transmission. Now the corresponding CTS frame cannot be heard by station E because it is exposed to station A, so the RTS–CTS exchange cannot be accomplished successfully. This implies that all the stations in the range of either station A or station B should remain silent during the whole communication duration. The channel resources inside the two circles are reserved for the usage of stations A and B during transmission.

Although the RTS–CTS exchange procedure can notify the hidden stations and avoid collisions during data transmission, it cannot avoid collisions among the control frames. MACA uses a backoff procedure to solve these collisions. First, after sending out an RTS frame, a station listens to the channel and determines that a collision has occurred if it does not receive the corresponding CTS frame. The stations involved in the collision wait for a random period of time and try to access the channel again. The random backoff time should be chosen carefully. If the range of the backoff time is too narrow, several stations may pick up the same value and simultaneously start retransmission, which again will trigger collisions; if the range of the backoff time is too broad, stations are more likely to choose larger numbers so the backoff time may be unnecessarily long.

### Binary Exponential Backoff

To find an optimal backoff time range, let us consider a WLAN consisting of $N$ stations. When a station needs to back off, it picks up a random number $k$ in the range from 0 to $k$ 1 time units. Because the number $k$ is randomly chosen, the probability $p$ of a station starting to transmit at any time between 0 and $k$ 1 time units is $1/k$. The condition for successful reception is that only one station transmits in any given slot and all other stations keep silent. So the probability of successful transmission can be formulated as

$$\Pr = N \times p \times (1 - p)^{n-1}$$

By computing the first derivative of this formula, we can easily find that the probability is maximized when $p = 1/N$. So the stations should adjust the backoff range to the number of stations in the system. But how a station can know the total number of stations in the system is nontrivial, particularly in ad hoc networks. However, in an infrastructure network, the AP may learn the number of stations in the network and then broadcast this information to all the stations. However, if it is an ad hoc network and the medium access control scheme is distributed, then there is no explicit way for stations to learn the exact number of stations in the system. One approach to setting the backoff period is to let each station estimate the number of nearby stations by listening to the media. If a lot of collisions occur, then the stations know the network is in congestion, and they should expand the backoff range. If the medium is idle most of the time, the station can reduce the backoff range. A special procedure called *binary exponential backoff* (BEB) is usually applied to adapt the backoff time to the number of stations. Initially, the BEB procedure sets the maximum backoff value $k$ to be one. Each time a station has a data frame to send, it starts a countdown timer with a random value picked from 0 to $k$ 1. When the timer expires, the station is triggered to send an RTS frame. If the CTS frame is received without collision, then the RTS–CTS exchange is complete, and the station starts transmitting its data frame. If the CTS frame is not received or is corrupted by collision, the station doubles the backoff range, and this step is repeated until the CTS frame is received. Because the average backoff period is $k/2$, the average backoff time increases exponentially with the number of collisions experienced. The maximum backoff value is reset to its initial value after any successful data frame transmission. With this BEB procedure, stations can respond to congestion very quickly with the backoff range increasing exponentially and decreasing sharply. The disadvantage of the BEB procedure is that it has a severe fairness problem. By applying the BEB procedure, stations in the same network may have different backoff ranges depending on their medium access history. The station acquiring the medium during the previous time unit resets its range to a minimum, and those stations that cannot access the medium usually have large backoff ranges. So the station that acquires the medium always starts sending out the RTS frame earlier than other stations and thus is likely to acquire the channel more often. On the other hand, if a station loses the medium, it needs to wait a long time for the chance to transmit again, which at times causes drastic frame delays.

### Improvements over MACA

Some enhancements based on simulation results have been proposed for MACA in Reference 3, and the new protocol is called MACA for Wireless (MACAW). MACAW has introduced the transmission of an acknowledge frame after the transmission of each data frame so that the error in the transmission data frame can be recovered by retransmission immediately after the error, more quickly than by waiting for the notification from a higher layer. In this case, the transmission pattern turns to be a four-step handshake of RTS-CTS-DATA-ACK. To deal with the fairness problem, MACAW has modified the BEB procedure with exponential increase and linear decrease. That is, the backoff range decreases by one after each successful transmission. Moreover, the MACAW protocol suggests that the backoff range should be copied from cell to cell so that all the stations in the network share the same range. However, broadcasting in the backoff range may increase the traffic load and worsen congestion. Another problem of the MACAW protocol is that congestion may not be equally distributed in the network. MACAW has also addressed the exposed-station problem. A preamble frame called data send (DS) is transmitted before the actual data frame to confirm the channel reservation. Exposed stations can learn about the successful exchange of the RTS-CTS frames by listening

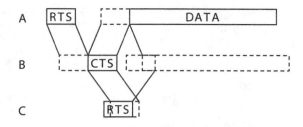

**Figure 22.7 Possible collision scenario after successful RTS–CTS exchange in the FAMA protocol.**

to the DS frame and then setting their medium status to busy until the end of the transmission. Multiple traffic stream support and multicast and broadcast issues have also been discussed in MACAW.

Floor Acquisition Multiple Access (FAMA)[4] is another improvement over the MACA protocol. The designers of FAMA argue that the transmitter and receiver should reserve the entire channel around them in advance to completely avoid frame collision. The process of acquiring the channel is called floor acquisition. It is pointed out that although MACA is designed to avoid collision, collision may still occur in some situations. Consider the scenario depicted in Figure 22.7. In a system of three stations, namely A, B, and C, both stations A and C can communicate with station B, while A and C are hidden from each other. Station A first sends out an RTS frame to B, and it is successfully received by B, after which B responds with a CTS frame. At the same time, station C also wishes to transmit a frame to station B; it senses the medium before the CTS frame from B reaches C and finds the medium to be clear. So station C starts sending out RTS frames to B. Now station A receives the CTS frame from B without any collision and "thinks" that the channel is reserved successfully, so it starts sending the data frame. Unfortunately, the data frame from A collides with the RTS frame from C.

The FAMA protocol suggests that carrier sensing and longer CTS frames can solve this problem. First, stations need to sense the carrier before trying to access the medium. Second, the transmitter should wait until all the packets that may cause collision have been successfully delivered. Consider the worst-case scenario in which the transmission delay from A to B is zero and that from C to B is $\tau$, which corresponds to the maximum transmission delay in the network. Suppose at time $T0$, as shown in Figure 22.8, station B sends out the CTS frame and it reaches station A immediately because there is no transmission delay. This frame is broadcast in the network and after the maximum transmission delay $\tau$, at time $T1 = T0 + \tau$, all the stations in the radio range of station B have heard the CTS from

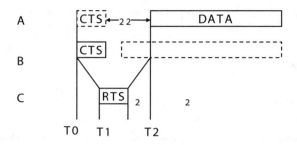

**Figure 22.8   Timing relationship between control and data frames in the FAMA protocol.**

B and learned that the medium is busy. However, if station C has sent an RTS frame to B during this time, it will be heard by all the stations in its coverage area latest by time $T2 = T1 + \tau$. So station A should defer transmission until time $T2$ and then start transmitting the data frame. The time difference between time point $T2$ and $T0$ is the sum of $2\tau$ and the time duration of an RTS frame. This is the minimum time for the source station to wait before data transmission.

## Receiver-Initiated Protocols

Unlike sender-initiated protocols, receiver-initiated MAC protocols let the receiver start the data transmission dialogue. When a station wishes to receive a data frame, it senses the medium and sends out a request-to-receive (RTR) frame provided the medium is clear. When the source station receives the RTR frame, it replies with the data frame; the receiver may send out an ACK frame after the data frame has been successfully received. The typical dialogue pattern is RTR-DATA-ACK. Compared with the RTS-CTS-DATA-ACK dialogue pattern in sender-initiated protocols, one control frame is saved and channel utilization may be improved. Because successful data reception depends on the medium status at the receiver side and receivers know this information better than the transmitters, receiver-initiated data transmission may have a better chance to be successful than sender-initiated transmission. However, the main problem in receiver-initiated protocols is how a receiver can know that a sender has pending data frames to send. In other words, when should a receiver poll the potential senders? If a receiver polls the senders too frequently, the cost of sending control frames will be too high; if the receiver's polling rate is too low, senders may have to wait for longer times before they have a chance to send out their data. In general, there are two methods of polling, independent polling and data-driven polling. With independent

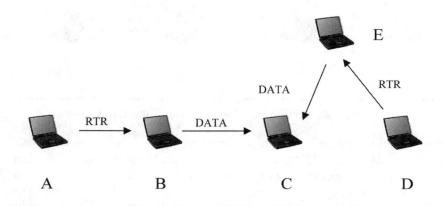

**Figure 22.9 Possible data frame collision scenario in the MACA-BI protocol.**

polling, a station polls with an independent rate. On the other hand, with data-driven polling, a station adapts its polling rate to the data rate it hears in the air. Most receiver-initiated protocols apply data-driven polling because it is hard to derive a proper independent polling rate.

MACA-BI (MACA by Invitation)[10] is an early version of a receiver-initiated protocol. According to this protocol, when a station is ready to receive, it sends out an RTR frame inviting a potential station to send the data. If the polled station receives the RTR correctly, it then sends out a data frame. In this protocol, polling station and receiver can be two different entities. All the stations that hear the RTR frame and do not have data to send back off to avoid collision. Unfortunately, this protocol only works well when no hidden stations exist. If there are hidden stations, data frames may still collide. Consider the network shown in Figure 22.9 with five stations. Suppose station A polls station B, and station B sends a data frame to station C. Meanwhile, station D polls station E, and station E also sends a frame to station C; then the two data frames collide and no frame can be received successfully.

To solve this problem, a protocol called Receiver-Initiated Multiple Access (RIMA) has been proposed.[11] There are two major modifications proposed in RIMA. The first modification is that a polled station can only send data frames to the polling station. This avoids the frame collision scenario described in the preceding text. However, collision may still happen in other scenarios, for example, the scenario depicted in Figure 22.10. If station A polls station B, station B senses the medium and finds that the medium is clear, so it starts to transmit a data frame to station A.

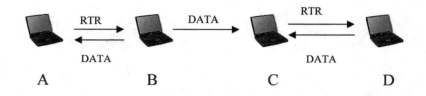

**Figure 22.10  Possible data frame collision scenario in the RIMA protocol.**

In the meantime, before the data frames from B to A arrive at station C, station C sends out an RTR frame to D, and station D receives the invitation without any error and then starts transmitting data frame to C; unfortunately, these two data packets will collide at station C. The problem here is that station C has no idea that a frame has been transmitted from station B because the receiver-initiated protocol only reserves the medium in one direction. To solve this problem, RIMA has added one more type of control frame called the no-transmission-request (NTR) frame. The polled station does not start data frame transmission directly after receiving an RTR; instead, it backs off for a certain period of time. If no NTR frame is received in this period, the transmission starts; otherwise the transmission is postponed. In the meantime, the polling station again senses the medium after sending out the RTR frame; if there is any interference in the medium, it sends the NTR frame to notify the polled station. According to the RIMA protocol, the waiting time $\varepsilon$ should be the sum of $2\tau$, the maximum round trip delay in the network, and the receiver-to-transmitter conversion time. Suppose the RTR frame from the polling station ends at time $T0$, then at time $T0 + \tau$, all the stations in the network should receive the RTR frame and should back off. So the interference frame can only be sent out after $T0 + \tau$ and should arrive at the polling station no later than $T0 + 2\tau$, and then the polling station can send out the NTR frame to notify the polled station. If the channel propagation delay is time invariant, then the maximum interval between consecutive RTR and NTR frames should be no more than $2\tau$, which is the waiting time for the polled station before data transmission can start.

However, collision still cannot be eliminated completely. Again, consider the case in which stations A and C poll stations B and D, respectively, as shown in Figure 22.10. If these two RTR frames are sent simultaneously, they would collide at station B and only station D would receive the RTR frame. After transmitting the RTR frame, station C senses the medium and finds that the medium is clear because station B is silent, so no NTR is sent out and station D starts data transmission after proper backoff. Meanwhile, station A gets no response. So station A retransmits the RTR after a random backoff and it is received by station B correctly; as a result, station B starts data transmission after a while. If the data transmission from D to C is long enough, it would be corrupted by the data transmission from B to A. The problem here is that station B has lost the information contained in the RTR frame from C to D, so it cannot sense the medium status correctly. The solution to this problem is to either let station A back off long enough, or let station B send out an NTR frame after detecting a collision. Then station C can notify station D to abort data transmissions.

Hybrid protocols combining sender-initiated and receiver-initiated approaches have also been proposed.[8] According to these protocols, stations poll when they have a data frame to transmit. If the polled station has no data to send, it replies with a CTS frame and invites the polling station to transmit; then the dialogue pattern is a typical sender-initiated pattern.

## Split-Channel MAC Protocols

Both sender-initiated and receiver-initiated protocols discussed in the preceding text are single-channel protocols. That is, both control and data frames are transmitted using the same channel. However, some protocols use one or two dedicated signal channels besides the data channel.[5,12] The Dual Busy Tone Multiple Access (DBTMA)[5] protocol falls in this category. According to the DBTMA protocol, two narrow-band channels $BTr$ and $BTt$ are separated from the main data channel. The senders set a busy tone in the $BTt$ channel, in addition to participating in the RTS/CTS dialogue, when transmitting data frames; the receivers set a busy tone in the $BTr$ channel starting from the sending of the CTS frame to the end of transmission of data frames. So the data transmission is protected continuously throughout the procedure. Neighboring stations hearing $BTr$ should postpone any transmission, whereas stations hearing $BTt$ can transmit but cannot receive data. The benefit of the DBTMA protocol is that stations can correctly judge the medium status by listening to the busy tone signals, so the protocol is robust with respect to RTS/CTS frame errors. However, there are several disadvantages. First, to receive narrow-band busy tone signals, stations should be equipped with highly selective bandpass filters, thus increasing the hardware complexity and power consumption. Second, this protocol cannot avoid collision completely due to transmission delay. The reason is the same as for the MACA protocol. Third, it cannot solve the exposed-station problem because stations exposed to $BTt$ cannot receive a CTS frame after sending out an RTS frame. Obviously the split-channel protocol can also be combined with receiver-initiated protocols. The Receiver-Initiated Busy-Tone Multiple Access (RIBTMA) protocol[13] belongs to this category.

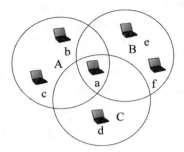

**Figure 22.11 An ad hoc network with six stations organized in three groups.**

## Collision-Free WLAN

The protocols discussed above are contention-based protocols. That is, all the stations compete for the medium when the medium is free. Although collisions during data frame transmission are avoided by some schemes, collisions still occur during the medium acquisition stage. For example, in MACA we have RTS–RTS collisions. Another approach to medium access control is to let a scheduler assign the channel to stations in a certain order, thus eliminating unsolicited transmissions and preventing collisions. Because there are a couple of collision-free wired LAN protocols, e.g., the IEEE 802.5 Token Ring protocol, one may try to implement those collision-free protocols in the WLAN environment as well. However, this approach is not applicable due to the existence of hidden stations. Consider a network with six stations, as shown in Figure 22.11. Stations can be divided into three groups: A, B, and C. Group A contains stations *a*, *b*, and *c*; Group B contains stations *a* and *d*; and Group C contains stations *a*, *e*, and *f*. Stations in the same group can hear each other, but stations from different groups are hidden from each other.

Consider applying the token ring protocol to this network. It seems that stations in each group should construct a subnetwork and generate a token, so we need to maintain three tokens, one for each group. A station can access the medium only if it possesses the token for that group, and it passes it to the next station after it has completed its transmission. Suppose that station *a* has a frame to send to station *b* and that it has the token of group A. However, if station *a* transmits, the frame would probably collide with frames in group B and/or in group C. To avoid collision, station *a* has to possess all three tokens before transmission. But the probability that all three tokens arrive at the same time is very low. If station *a* keeps the token from group A while waiting for the tokens from the other groups, stations in group A may assume that the token is lost and regenerate a token, which would disrupt the entire token management scheme. So, it is not practical for an ad hoc network to implement a distributed collision-free scheduling protocol.

For infrastructure networks, however, a collision-free protocol may work because there are center stations, namely APs, which can manage the medium assignment. The AP can maintain a station list and then poll stations one by one; a station can only access the channel when it is polled. However, if there is more than one AP and there is some overlapping area in the network, e.g., as in the network shown in Figure 22.2, hazards still exist. Consider a station in the overlapped area that can communicate with both APs. It can only access the channel when polled by both APs because of the same reason mentioned in the previous example. Coordination between APs is required to let the collision-free protocol work properly. The IEEE 802.11 MAC protocol defines a polling-based collision-free protocol called Point Coordinate Function (PCF).

## Channel-Aware MAC Protocol

In all the above MAC protocol analysis, the wireless channel itself is assumed to be perfect. That is, the channel is error free and the only reason for data loss is frame collisions. However, this model is too ideal. The behavior of wireless channels is quite different from that of wired channels. Wired channels have very low error rates, and the errors are independently and identically distributed. In contrast, wireless signals propagate in multiple paths, and each path has different path loss and delay characteristics. So, the signals from different paths may cancel each other out at some places and add up at other places. So the combined signal may have significant fluctuations. This phenomenon is called *channel fading*. Moreover, the path between the transmitter and receiver may be blocked by all types of obstacles such as buildings, walls, etc., thus increasing the path loss. This phenomenon is called *shadowing*. Both channel fading and shadowing are location dependent and time variant. So the signal-to-noise ratio at the receiver side, and subsequently the channel error rate are also location dependent and time variant. In fact, modeling wireless channel behavior is a complicated and extremely challenging task. In general, errors on the channel are numerous, time variant, and bursty. "Bursty" means that errors happen in sequence; that is, if an error occurs, it is probable that the error will occur again in the next bit or frame, which means that the errors are highly correlated and that error correction by retransmission may not work.

The correlated or bursty errors pose difficult challenges for error-resilient channel-coding techniques. To overcome bursty errors, more redundant information needs to be added into the frame, thus decreasing

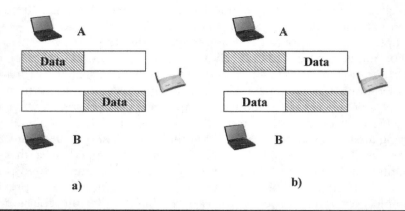

**Figure 22.12    Possible data transmission scenarios over time-variant fading channels.**

the transmission efficiency. Another method of ensuring error resilience is to interleave the bit stream among multiple frames such that bursty errors can be dispersed and then rectified by simple correcting codes. But interleaving also increases decoding complexity, reduces compression ratios, and introduces additional delay.

On the other hand, correlated channel behavior is also a good feature in a sense because it can help in predicting the future channel status by observing the current and past channel status. This channel-status information can be used in allocating channels to stations more efficiently. To see this, consider an infrastructure network with one AP and two stations, A and B. Suppose both stations have one data frame each to be sent to the AP in two time units. There are two sending sequences: station B follows station A or station A follows station B, as shown in Figure 22.12a and Figure 22.12b. Now consider the channel status; suppose the channel between station A and the AP suffers bursty errors in time unit one and the channel between station B and the AP suffers errors in time unit two. Then, if the data frames are transmitted in the order of Figure 22.12a, both frames are lost; however, if they are transmitted in the order of Figure 22.12b, both frames can be received successfully. Note that in both cases there are no frame collisions, and the frame errors depend only on the channel status. Based on this observation, the MAC protocol can schedule the channel access using channel status information and improve the overall channel utilization. Such MAC protocols are called channel-aware MAC protocols. The major challenge in the design of channel-aware MAC protocols is how to collect channel information, because a lot of factors contribute to channel loss. The easiest approach to determining the channel status is to send a probe frame via the channel and measure the receiving signal strength. Because the channel information turns to be invalid very quickly, probe frames should be sent out periodically with very high frequencies. But if too many probe frames are sent out, these frames will congest the network, and the MAC protocol will be cost inefficient.

Handshake-based Channel Aware (HCA)[14] is one of the channel-aware MAC protocols for infrastructure networks that can collect channel information at low cost. The basic idea is that only stations with good channel conditions are allowed to take part in the channel reservation. For the sake of clarity, let us assume that communication only tales place between the AP and the stations; so no direct station-to-station communication occurs, and the AP controls the channel allocation. The channel reservation is performed in three steps: (1) the qualify step, (2) eliminate step, and (3) rehandshake step. The data transmission procedure is described in the following text.

First of all, in step one or the qualify step, the AP sends a probe frame (the RTS frame) and a threshold; all stations measure the signal level of this beacon and compare it with the threshold. The stations in the network having channel status greater than the threshold are qualified, and they send CTS frames, whereas other stations block themselves. If only one CTS frame is received, then the channel is successfully reserved, and the data frame is sent out. If no CTS frame is heard, the AP sends the RTS frame again with a lower threshold, and step one is repeated. If more than one station qualifies, then step 2, called the eliminate step, is performed.

In step two, a BEB procedure is applied to relieve congestion. That is, stations hear the collision backoff for a random period before retransmitting the CTS frame. If a collision occurs again, the range of the random value chosen is doubled. At the end of step two, only one station wins the channel by sending the CTS frame and all other stations are blocked. After successful reservation, the AP sends out the data frame, and the destination acknowledges the successful transmission by sending an ACK frame.

Step three is performed optionally after data transmission. In this step, the AP measures the signal strength of the ACK frame and compares it with the rehandshake threshold. If the channel status is satisfactory, the AP still retains the channel for that destination; otherwise, the AP releases the channel and starts a new channel reservation round, which is called rehandshake. The purpose of this step is to keep using the channel as long as possible so that the handshake overhead of each data frame can be reduced. To be fair, if a station keeps using the channel long enough, the channel is released even if the channel status is still good.

HCA also relieves network congestion by blocking stations with bad channel status. Analytical calculation shows that the channel reservation overhead has a bounded average, no matter how many stations are in the network. As a comparison, overhead of BEB protocols increases logarithmically with the number of stations in the system.

## Threshold Computation

The threshold for step one in the HCA protocol is chosen to maximize the probability of a successful handshake. The condition for a successful handshake is only one station sending an RTS. As discussed in the preceding text, for a system with $N$ users, the probability of channel utilization is maximized if every station transmits with a probability of $1/N$. So the threshold should be chosen such that the probability of a station having channel status above the threshold is $1/N$.

If there is no station that satisfies this threshold, the threshold should be reduced such that in the next slot the transmission probability is still $1/N$. Because the probability of a station having channel status less than the first threshold is $1 \; 1/N$, the second threshold should be set such that the probability of having channel status above the threshold is $(1/N) \times (1-1/N) = 1 - (1-1/N)^2$. By induction, the $k$th iteration of step one would have a threshold such that the chance of a station having channel status above that threshold is $1 - (1-1/N)^k$.

The threshold for rehandshake should also be carefully chosen. If the threshold is set too high, then rehandshake would happen too often. In this case the overhead will be significant, and the communication efficiency will be reduced. On the other hand, if the threshold is too low, the channel status is likely to be bad in the next slot and the transmitted frame may be lost, which also reduces the communication efficiency.

## Channel Status Measurement

Another important issue in the design of channel-aware protocols is how to measure the channel status. Because low path loss usually means low error rate, path loss is a good candidate for measuring channel status. However, because the stations are randomly distributed, some stations may be close to the APs and others far away from the stations. Those close to the AP will always have comparatively low path losses and have a better chance to acquire the channel. The distant stations may be totally blocked too often, which introduces the fairness problem. A fair MAC protocol should guarantee that all the stations have equal chances to win the channel. To achieve this purpose, the HCA protocol measures the channel status by the ratio of current path loss over the mean path loss. The mean path loss is the average of the path loss in several recent channel slots. Thus, the channel status only reflects the fast-changing part of the channel status and the slow-changing part is eliminated. So, stations close to as well as those far away from the AP can have equal chances to win the channel.

## IEEE 802.11

In the preceding text, we discussed several MAC protocols available in the literature. However, only a few of them have been adopted as standards. This section introduces the most widely implemented WLAN protocol, the IEEE 802.11 MAC protocol. IEEE 802.11 is a part of the IEEE Local and Metropolitan Area Networks (LAN/MAN) standards. It specifies the physical (PHY) layer and MAC layer requirements and mechanisms for WLANs at several unlicensed wireless bandwidths. The IEEE 802.11 series of protocols are designed to support multiple PHY layer techniques to provide high-speed wireless data communication services. The first version of the IEEE 802.11 protocol was approved in 1997, in which three PHY layer units were specified. Unit one is the direct sequence spread spectrum (DSSS) PHY layer working at the 2.4-GHz band; unit two is the frequency-hopping PHY layer that also works at the 2.4-GHz band; and unit three is a baseband infrared unit. The maximum data rate supported by these PHY layer specifications was 2 Mbps.

In recent years, several amendments have been approved to meet the increasing high-bandwidth wireless data transmission demands. The specifications IEEE 802.11a and b were approved in 1999. IEEE 802.11a specifies a PHY layer working at the 5-GHz band that supports data rates up to 54 Mbps. To enable this high speed, new multiplexing techniques called orthogonal frequency division multiplexing (OFDM) are applied. IEEE 802.11b improves the DSSS PHY layer of 802.11 to support up to 11 Mbps data rate at the 2.4-GHz band. The newest PHY layer amendment, IEEE802.11g, was approved in 2003 and supports the same data rate as IEEE 802.11a with OFDM technologies at the 2.4-GHz band. Besides introducing new PHY layer specifications with higher bandwidth, other issues such as network management and security are also under consideration. IEEE 802.11f, which was approved in 2003, defines the coordination of APs to support roaming in a distributed system. The forthcoming IEEE 802.11i protocol would provide more secure authentication schemes. Several other standards have also been proposed, for instance, the Bluetooth protocol[18] and HIPERLAN/2.[19] However, the IEEE 802.11 series of standards are the most popular protocols for their high speed and simplicity. The major features of these protocols are summarized in Table 22.2.

The next-generation WLAN protocol is also being considered by the IEEE 802.11 designers. A new task group, the IEEE 802.11n Task Group, was formed in 2003. The target of this task group is to provide up to 100-Mbps

**Table 22.2  Major Features of WLAN Protocols**

| Protocol | Physical Layer Modulation Method | Frequency Band (GHz) | Max. Speed (Mbps) |
|---|---|---|---|
| 802.11 | Direct sequence spread spectrum Frequency hopping spread spectrum Infrared | 2.4 | 2 |
| 802.11a | OFDM | 5 | 54 |
| 802.11b | DSSS | 2.4 | 11/22 |
| 802.11g | OFDM | 2.4 | 54 |
| Bluetooth[18] | FHSS | 2.4 | <1 |
| HIPERLAN/2 | OFDM | 5 | 54 |

data service at the interface between the data-link layer and the MAC layer. To achieve this aim, new technologies such as multiple antenna arrays have been introduced at the PHY layer. The MAC layer protocol will also be revised to introduce QoS management to support real-time applications such as Voice-over-IP (VoIP). The task group expects that the new protocol can be approved in 2005.

In this section, we concentrate on the current versions of the IEEE 802.11 MAC protocol, which supports both ad hoc and infrastructure network configuration. The basic network block is a basic service set (BSS). The BSS is a coverage area in which member stations can talk to each other. An ad hoc network is composed of an independent BSS. BSSs can overlap with each other to form larger ad hoc networks. A station can join or associate with a BSS when it powers on or moves within the range of the BSS. An infrastructure network, on the other hand, consists of several BSSs and a distributed system (DS). A DS is a wired network to interconnect BSS. The interface between the DS and BSS is the AP. The stations in different BSSs can communicate via a DS relayed by APs. A station can also access wired networks via the AP.

## MAC Layer Services and Functions

The IEEE 802.11 MAC layer protocol defines the services and functions to be implemented by MAC layer entities. Services are to be used by upper-layer users and are provided through the service access point (SAP) of MAC layer entities. On the other hand, functions are transparent to upper-layer users. These are the proper operations to be executed among several MAC layer entities to enable the defined services. In other words, services are the language used between the MAC layer and the upper layers, whereas functions are the language used between peer MAC layer entities in different stations. The IEEE 802.11 protocol defines three types of services and two types of functions. The three types of services are asynchronous data service, security services, and MAC Service Data Unit (MSDU) ordering. The two types of functions are Point Coordination Function (PCF) and Distributed Coordination Function (DCF).

The asynchronous data service is the most important service provided by the IEEE 802.11 MAC protocol; it uses the PHY layer service to provide data transmission between two stations. This service can use any PHY layer service with varying data rates and different medium types. The security services provide confidentiality and authentication services. Because IEEE 802.11 uses the unlicensed band for communication and because everybody can listen to communications on these bands, confidentiality and authentication services are important to keep malicious users from reading the messages. The MSDU ordering service reorders MSDUs so that the upper layer receives the data in the correct order, thus improving the likelihood of successful delivery.

The layer above the MAC layer accesses MAC layer services by exchanging service primitives. Possible service primitives for data service are MA-UNITDATA.request, MA-UNITDATA.indication, and MA-UNITDATA-STATUS.indication. The upper layer passes a package of data to the MAC layer by sending an MA-UNITDATA.request. If the MAC layer receives a data packet from the PHY layer, it calls MA-UNITDATA.indication to notify upper layer for reading new incoming data. The parameters of these primitives include source address, destination address, routing information, data, priority, and service class. The destination address can either be an individual or a group MAC address, which can be used to support multicast and broadcast. The routing information parameter is used for source routines. The priority parameter specifies the priority desired for the data unit transfer. IEEE 802.11 allows two values, Contention or ContentionFree. Contention indicates to MAC layer entities to use DCF, whereas ContentionFree indicates to MAC layer entities to use PCF to provide the service. For those stations that do not support PCF, this parameter would be ignored. The service-class parameter specifies the service class desired for the data unit transfer. IEEE 802.11 allows two values, ReorderableMulticast or StrictlyOrdered.

For the authentication services, IEEE 802.11 defines two subtypes of authentication service, open system and shared-key. These services are realized by sending and receiving management frames. Open system authentication is the simplest authentication algorithm and is the default service. Actually, it does nothing and allows every station to use the data service. It involves a two-step authentication transaction sequence. First,

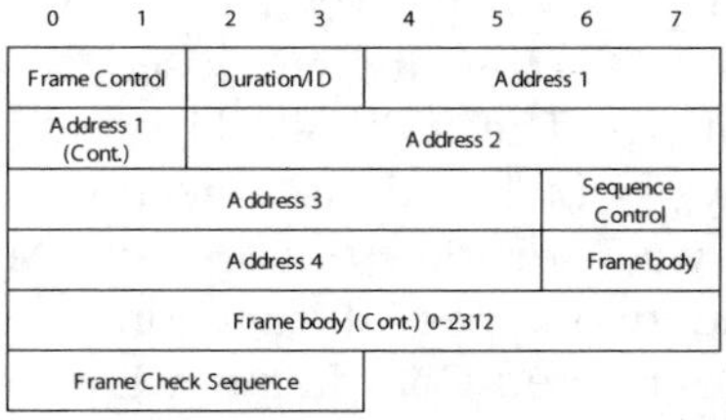

**Figure 22.13  IEEE 802.11 MAC frame format.**

a station sends a request for authentication to its destination (wireless station or AP). If the destination replies with an authentication result having the value "successful," then they are mutually authenticated. Shared-key authentication requires the support of the Wired Equivalent Privacy (WEP) standard. The communication pair needs to exchange four packets to accomplish shared-key authentication. More powerful authentication algorithms are under development and will be specified in IEEE 802.11i. For details of shared-key authentication, please refer to Section IV and Section V of this handbook, which discuss security issues.

Regarding the functions of the IEEE 802.11 MAC layer, DCF is the fundamental access method, which is based on MACA with Collision Avoidance (MACA/CS). All stations, both in ad hoc networks and infrastructure networks, implement this function. PCF is optional and only applicable in the case of infrastructure network configurations. PCF uses a point coordinator, which operates in an AP to control medium access. DCF and PCF can coexist in the same BSS. When a PC is operating in a BSS, it performs these two functions alternately, with a contention-free period (CFP) for PCF followed by a contention period (CP) for DCF.

## Frame Format

To be understood by the communication peer, data is organized into predefined formats. The IEEE 802.11 MAC protocol defines a uniform format for all types of frames. The frame format is shown in Figure 22.13. The length of each field is specified in bytes. A MAC frame of IEEE 802.11 contains up to 9 fields, namely Frame Control, Duration/ID field, Address 1, 2, and 3, Sequence Control, Address 4, Frame Body, and Frame Check Sequence (FCS), in which fields Address 2, Address 3, Sequence control, Address 4, and Frame Body are optional.

The Frame Control field contains 16 bits and its structure is shown in Figure 22.14. There are three fields and eight one-bit flags. The Protocol Version field is two bits in length and specifies the version of the standard to be applied; it should be set to 00 for the current version. The Type

| b0 | 1 | 2 | 3 | 4 | 5 | 6 | b7 |
|---|---|---|---|---|---|---|---|
| Protocol Version | | Type | | Subtype | | | |
| To DS | From DS | More Frag. | Retry | Power Mgt. | More Data | WEP | Order |

**Figure 22.14   IEEE 802.11 MAC frame control field format.**

field is also a two-bit field and indicates whether the frame is a management, control, or data frame. The Subtype field contains four bits that are used to specify the function of the frame, and following the subtype field are eight one-bit flags. To DS and From DS bits are set if the destination or source of this frame is in the distribution system. These two bits help APs to correctly relay data frames between the BSS and outer networks. The More Fragment bit indicates whether this frame is the final fragment of the current MAC data unit. The Retry bit indicates whether the frame is a retransmission frame to avoid redundant reception. The Power management bit indicates the power management mode of a station. A value of 1 specifies that the station will be in power-save mode and a value of 0 stands for the active mode. The More Data bit is used in PCF mode. A station sends out a data frame with this bit set to indicate to the AP that it has more data to send. The WEP bit is set to 1 if the frame body contains encrypted information for the WEP algorithm. The Order flag is set to 1 to indicate that this frame is using the StrictlyOrdered service class.

The Duration/ID is a 16-bit field in the MAC frame. It is retrieved as a 14-bit association identity (AID) in a Power Save Poll control frame. Otherwise, the Duration/ID field contains a duration value that is used for the virtual mechanism of all hearing stations to learn the communication duration of this dialogue. The maximum duration value is 32,767.

Following the Duration/ID field are four address fields. These fields are used to indicate the BSS ID, source address, destination address, transmitting station address, and receiving station address. The usage of these addresses depends on the frame type. Each address field contains 48 bits. The address could be an individual address, multicast address, or broadcast address. The address space is also partitioned into locally administered and universally administered addresses.

Sequence Control is a 16-bit-long field that is partitioned into two parts. The first four bits are the Fragment Number field and the next 12 bits represent the Sequence Number field. Each MAC data unit is assigned a sequence number from 0 to 4095, which increases by one after sending one frame. Retransmitted frames have the same sequence number. Fragments of the same data unit also have the same sequence number, but their fragment numbers are different. Next to the MAC header is the Frame

Body field, which is a variable-length field that contains frame data. Control or management frames may not contain any data.

At the end of the data frame is the FCS field. It is a 32-bit field containing a 32-bit cyclic redundancy code (CRC), which is computed using a standard 32-degree polynomial generator. The frame, including both the MAC header and the frame body, is treated as the coefficient of a polynomial and divided by the generator polynomial in modulo 2. The remainder is put in the FCS field. At the receiver side, the decoder can detect bit errors from FCS.

The control frames for DCF are RTS, CTS, and ACK. The RTS frame is 20 bytes in length, which contains the frame control, duration, receiving address, transmitting address, and FCS fields. The duration field contains the time in microseconds for the whole transmission dialogue, which is the time for transmitting CTS, data frame, ACK, and three SIFS. The CTS frame is 14 bits in length; it contains only the receiving address, and the other fields are same as in the RTS frame. The receiving address of the CTS frame is copied from the transmission field of the corresponding RTS frame. The duration field is also copied from RTS, but the time for CTS and one SIFS is deducted. The ACK frame has the exact format of CTS. PCF.

The control frames for PCF are CF-End and CF-End+CF-Ack. The CF-End frame is 20 bytes in length, which contains frame control, duration, receiving address, BSSID, and FCS. The duration field is set to zero, and the receiving address is the broadcast group address. The CF-End+CF-Ack frame has the same format as the CF-End frame.

## Distributed Coordination Function (DCF)

DCF is based on the CSMA/CA protocol (a variation of MACA/CA) with a random backoff time policy if the medium is busy. After data transmission, an ACK frame is sent to confirm the successful reception. If no ACK is received, the source station immediately schedules a retransmission. A threshold called *dot11RTSThreshold*, which is individually set by each station, assists the use of RTS/CTS mechanism. If the length of a data frame exceeds the threshold, RTS/CTS packets are exchanged to reserve the channel first. This is to avoid unnecessary protocol overhead for short data frames. The successful data frame transmission procedure can be described as follows:

$$\text{Carrier Sense} \rightarrow \text{random backoff} \rightarrow \text{RTS} \rightarrow \text{CTS} \rightarrow \text{Data} \rightarrow \text{ACK}$$
(for long frames)

or

$$\text{Carrier Sense} \rightarrow \text{random backoff} \rightarrow \text{Data} \rightarrow \text{ACK}$$
(for short frames)

There are two ways for a station to sense the carrier, through the PHY layer and through a virtual mechanism. A station should apply both methods. The virtual mechanism is called the *network allocation vector* (NAV). The NAV maintains a prediction of future traffic on the medium based on the duration information contained in RTS/CTS or other frames used by PCF. Either noise detected by the PHY layer or a busy NAV can set the medium status to busy. If a station receives an RTS or CTS, it copies the Duration/ID fields and starts a countdown counter. When the counter is zero, NAV indicates that the medium is idle; when the counter is nonzero, NAV indicates that the medium is busy.

## Interframe Spaces

As discussed earlier, stations need to wait long enough after the medium becomes idle to avoid collisions. The time depends on the communication scenario. In DCF, these waiting intervals are called *interframe spaces* (IFSs). The IFSs defined in IEEE 802.11 include: SIFS (short interframe space), PIFS (PCF interframe space), DIFS (DCF interframe space), and EIFS (extended interframe space).

SIFS is the shortest interframe space, which consists of the PHY layer processing time, the MAC layer processing time, and the time for hardware to switch from receiving mode to transmission mode. This is the time for a station to process the incoming frame and respond correctly. SIFS is used when stations have already acquired the medium, so they need not worry about collision frames and should respond as soon as possible. SIFS could be inserted before a CTS or ACK frame in DCF. Data packets following CTS frames also use SIFS. If the station responds too late, other stations may sense the medium and incorrectly think that the medium is idle.

PIFS is the second-shortest interframe space. The duration of PIFS is equal to the sum of an SIFS and a slot time. The slot time is the time unit for random backoff. It consists of the time for the PHY layer to sense the medium and judge whether it is busy or idle, the maximum air propagation time, and the MAC protocol processing time. PIFS is used for PCF to acquire the medium when a contention-free period starts.

DIFS is longer than PIFS and is used before the first frame of a DCF data transmission sequence. If RTS/CTS is used, then the first frame is an RTS frame; otherwise the first frame is a Data frame. The duration of DIFS is equal to the sum of SIFS and two slot times. DIFS is applied before a station starts the data transmission dialogue in the contention period. During DIFS, the station senses the medium twice. After SIFS, at the beginning of the first slot time, if the station senses the medium and finds the medium is idle, it knows that the previous communication has ended.

Next, at the beginning of the second slot time, the station senses the medium again; if the medium is idle, it knows no PCF communication has occurred; therefore, it can bid for the channel. After DIFS, the station still needs to back off for a random period of time before it actually sends out the first frame. DIFS is longer than PIFS, which guarantees that the AP has the priority to acquire the medium.

EIFS is applied when a station receives a bad frame. A frame is determined to be bad if the verification over its 32-bit CRC fails. EIFS is defined to provide enough time for the station to transmit one SIFS, one DIFS, and one ACK frame.

## Backoff Procedure

A station may transmit a pending data frame when it is operating under the DCF access method either in an ad hoc network or in the contention period of an infrastructure network. When a station determines that the medium is idle — both by the PHY layer probe and by the virtual mechanism — for longer than or equal to a DIFS period (or an EIFS period if a data frame with an erroneous FCS value is received), it performs the backoff procedure. A random number uniformly distributed between 0 and CW is generated, where CW is the contention window parameter, and *BackoffTime = Random* () × *SlotTime*. In each backoff slot, the station senses the carrier; if the medium is detected to be busy, it suspends the backoff procedure. The backoff procedure resumes when the medium is idle for longer than DIFS or EIFS. If the medium is idle when the backoff procedure countdown has reached zero, the station can start transmitting the pending data frame. This is to avoid collision among multiple stations. Figure 22.15 illustrates the scenario of sharing the common medium among three stations.

In Figure 22.15, there are three stations A, B, and C. All have a frame to send and the contention window is set to 7. After the medium is clear for one DIFS period, the stations start the backoff procedure. Station A picks number 7, station B picks number 2, and station C picks number 6. In the first two time slots, all three stations sense the medium and find the medium is idle, so they reduce their backoff counters by 2. At the beginning of the third slot, station B starts transmitting the pending data frame because its backoff counter has reached zero. Meanwhile, station A and C listen to the air and hear the data frame from B, so they suspend their backoff procedure and set the medium status to busy. The remaining backoff time in the counters of station A and C are 5 and 4, respectively. After station B successfully transmits the frame and gets the ACK response, the medium status is set to free again. Stations A and C then resume their

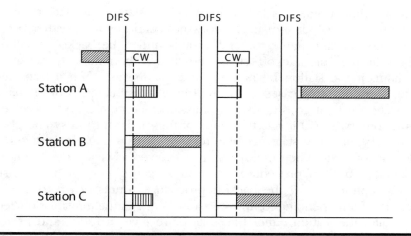

**Figure 22.15  Illustration of IEEE 802.11 backoff procedure.**

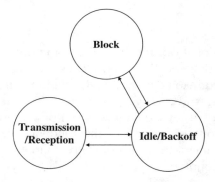

**Figure 22.16  FSM representation of IEEE 802.11 DCF.**

backoff procedures after one DIFS. After four slots, station C starts transmitting and station A suspends the backoff procedure with the remaining backoff time equal to one slot. Finally, station A starts transmitting after station C completes its frame transmission, and the air is free for one DIFS period. The backoff procedure of station A then counts down to zero.

DCF of a station can be implemented as a finite state machine (FSM). The station is always in one of the states and transfers from one state to another when a certain event is triggered. The FSM of a station is described in detail in the IEEE 802.11 protocol, annex C. However, a simple model is shown in Figure 22.16 for abstract understanding. This machine has three states, idle/backoff, block, and transmission/reception. When the station is initialized, it is in the idle/backoff state. This state means the medium is clear, but the station does not send anything, either because it has no pending frame to send or it is under the backoff procedure. When in the idle/backoff state, the station continuously senses the medium. If the station hears any frame on the air, then it transfers its state. There are two cases: (1) if the frame's destination address matches the station's ID, it goes to the transmission/reception state to receive the frame, otherwise (2) it sets the status of the medium to busy and goes to the block state. The state machine turns back from the block state to the idle/backoff state when the medium is detected as clear and resumes the suspended backoff procedure. The station can also go to the transmission/reception state if the backoff procedure counts down to zero and there is a frame pending for transmission. While in the transmission/reception state, the station either transmits or receives a frame and performs the corresponding RTS/CTS/ACK procedure. The state machine goes back to the idle state after the transmission/reception process completes, regardless of the success or failure of the process.

## Error Recovery

Transmission errors may occur during the communication. Error recovery is the responsibility of the station that initiates the data exchange sequence. At the frame level, for each data frame, the station maintains two counters, a short retry counter and a long retry counter. The station also maintains two station-level global retry counters, a station short-retry counter and a station long-retry counter. In any of the following cases of frame transmission, the station increases the short-retry counter of that frame and the station short-retry counter:

1. If an RTS is sent and no corresponding CTS is received.
2. If a direct data frame is sent and no corresponding ACK frame is received.
3. Any other short frame is sent, and no response is received.

If transmission of a short data frame succeeds or CTS response is received, both the frame level and the station level short-retry counters are reset, and the station increments the long-retry counter of that frame. The station long-retry counter is increased if the channel was reserved by RTS/CTS but the corresponding ACK was not received after transmitting the data frame. The long-retry counter is reset after a successful retransmission of a long data frame. Both long-retry counters are reset after any successful long data transmission.

The transmission failure is recovered by retransmissions, until the data frame is received successfully or either of the short- and long-retry counters has reached their upper threshold values. If a counter reaches its threshold value, the corresponding frame is discarded.

There may be several reasons for a station not receiving the corresponding CTS or ACK frames. For a missing CTS frame after an RTS was sent out, the possible reasons include: RTS–RTS collision, RTS–CTS collision, the destination's virtual carrier sense mechanism set to busy by other RTS or CTS, and either RTS or CTS corrupted by channel noise. For a missing ACK frame, it is possible that the data frame is corrupted by channel noise because the channel is already reserved by RTS–CTS dialogue. However, unexpected channel collision may also cause data frame corruption. Even though the data frame itself may be received correctly, the ACK frame may be lost or corrupted. Because the loss of control frames such as ACK may cause the failure of the whole transmission, and recovery by retransmission is extremely expensive, these control frames need to be protected. Special modulation methods or more powerful error-correcting codes are applied when transmitting control frames, so that their transmission is more robust to channel noise. For example, in the IEEE 802.11b protocol, although data frames can be transmitted at a speed of 11 Mbps, control frames are still transmitted at a basic rate of 1 Mbps.

Because ACK frames may be lost, it is possible that some data frames are received more than once. The destination station is responsible for filtering the duplicate frames. So, in order to perform the frame filtering, each station maintains a cache to store information on recently received frames. Each frame is uniquely represented by a tuple of <destination address, sequence number, fragment number>. If a frame with a retry bit is set in the MAC header, the destination station checks these three numbers in the cache and rejects the frame if it is a duplicate frame. The sequence number has 12 bits, so it repeats after every 4096 frames. Hence, the duration between two frames sharing the same sequence number is long enough and possibility of improper rejection is extremely low.

The error recovery process also influences the random number generation for backoff. As mentioned in the preceding text, before accessing the channel the station chooses a random number, uniformly distributed in the interval [0–CW], where CW is the contention window parameter. The CW value is initially set to CWmin. After every retry counter increment, CW takes the next value in the series, until it reaches CWmax, where it remains. The CW value is reset to CWmin when either station short- or long-retry count is reset. The values in the CW series increases exponentially. In IEEE 802.11, CWmin is set to 7, and CWmax is set to 255. The whole series is [7,15,31,63,127,255]. This is a modification of the BEB scheme. Because the increment of the CW value has a limitation, it is called truncated BEB.

## PCF

Besides DCF, IEEE 802.11 also defines a polling-based function, PCF, for infrastructure network configurations. PCF provides contention-free communications in the system and a point coordinator (PC) that resides in the AP and controls channel access. In a network where both PCF and DCF functions are activated, they are performed alternately with a certain frequency, so that stations can have a chance to transmit/receive data frames periodically. The aim of PCF is to guarantee a bounded time delay variance for frame transmission from a certain source. This is to satisfy the QoS requirement of those delay-sensitive applications such as audio and video communications.

The PCF is embedded into the DCF scheme. That is, the whole contention-free period is treated as a long frame or beacon under DCF. The AP needs to compete for the medium to reserve the channel for the contention-free period. However, the AP has priority to win the channel because it can claim the medium after PIFS, but other stations have to wait for DIFS.

### *CFP Beacon*

The AP generates the contention-free period beacon after every CFP repetition interval. The duration of a CFP repetition interval is defined as a number of delivery traffic indication message (DTIM) intervals. DTIM is a type of management message that is used to maintain system synchronization. When an AP generates a PCF beacon, the system is operating in the contention period. So the AP needs to sense the carrier, and if the medium is busy, it should defer any transmission until the current frame transmission is complete. After the medium is idle for an SIFS period, the AP senses the channel again to ensure that the medium is free. Then it starts sending the CFP beacon, so the total waiting time is equal to a PIFS period.

The beacon sent out by the AP contains a CF parameter set and a DTIM. The CF parameter set contains the fields of CFP Count, CFP Period, CFP MaxDuration, and CFP DurRemaining. The CFP count indicates how many DTIMs appear before the next contention-free period. The CFP period indicates the CPF repetition interval in terms of the number of DTIMs. CFP MaxDuration is the maximum duration of this contention-free period. After receiving the CFP beacon, stations set their virtual mechanism of carrier sense to busy according to this value. CFP DurRemaining records the maximum time remaining in this CFP. Both CFP MaxDuration and CFP DurRemaining are in terms of the number of time units, so they are not constrained to be an integer multiple of the DTIM interval. When a station hears this beacon, it uses the information to set its virtual carrier sense mechanism to be busy for the CFP.

## Data Transmission

After receiving the beacon, the CFP starts and the AP controls all channel accesses in the system. Frame transfers under PCF usually consists of an alternating pattern of transfer of frames from the AP to stations and from stations to the AP. RTS/CTS channel reservation schemes are not applied under PCF; instead, the AP polls the stations with CF-poll control frames. These CF-poll frames can be embedded into data frames by setting the subtype of data frames so that the interframe intervals can be saved. The subtypes of data frames include: Data, Data + CF-Ack, Data + CF-Ack + CF-Poll, Null, CF-Ack (no data), CF-Poll (no data), and CF-Ack + CF-Poll (no data). Among these subtypes, those subtypes containing CF-Poll can only be sent from the AP, and all stations can send other subtype frames. A station can only access the channel when it receives a frame containing CF-Poll with a destination address identical to its ID. When polled, the station should ignore the busy state indicated by the virtual mechanism and transmit the pending data frame. The CF-Ack to acknowledge the preceding frames is also embedded in the data frame.

Not all stations have the ability to respond to CF-Poll; a station that can respond to CF-Poll is called a *CF-Pollable station*. The AP should maintain a CF-Pollable station list and poll only the stations on the list periodically. A CF-Pollable station can be added or removed from the list by sending association and reassociation management messages. A station may remove itself from the polling list so that the AP will not poll it any more. Then the station can shut down its receiving circuit to save power.

The error recovery procedure of PCF and DCF are also different. Because the AP controls the channel accesses, if a station sends out a data frame and does not receive ACK, it cannot retransmit the frame until the AP polls it again. A station can always choose retransmission of the unacknowledged frame during the contention period. The AP can decide whether to retransmit the data frame immediately or to store the frame in the buffer for later transmission.

## Possible PCF Scenario

Figure 22.17 illustrates a possible PCF scenario in which the network consists of an AP and three stations. The CFP repetition interval is equal to six DTIM intervals and the CFP maximum duration is about 4.9 DTIM intervals. At the end of the contention period, the AP sends out a beacon that claims the starting of the CFP, and which contains the maximum duration of this CFP. After receiving this beacon, all stations set their NAV to busy according to the CFP MaxDuration value. After one SIFS of the CFP beacon, the AP sends out the first data frame (data frame 1) to station

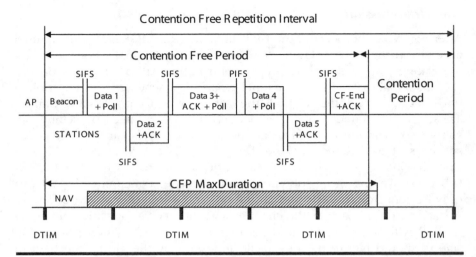

**Figure 22.17 Illustration of control and data frame coordination in IEEE 802.11 PCF.**

1 and polls the station as well. Station 1 receives this data frame, sends out a pending data frame, data frame 2, to the AP, and acknowledges data frame 1. After the AP receives data frame 2 and waits for one SIFS, the AP sends out data frame 3 to station 2 while acknowledging data frame 2 and polling station 2. Unfortunately, station 2 loses this data frame and keeps silent. The AP senses the medium after one SIFS and realizes that the data frame is lost. So it skips station 2 and sends out data frame 4 to station 3. Station 3 receives the data frame successfully and replies with data frame 5 together with acknowledgement of data frame 4. The AP sends out a CF-End+ACK frame to acknowledge data frame 5 and claim the end of this contention-free period. All stations receiving this frame then reset their NAV to free and enter the contention period.

## SUMMARY

WLANs share a single wireless channel among multiple stations. MAC protocols are designed to assign this single channel to competing stations that wish to access the channel. WLANs may operate in ad hoc or infrastructure configurations. One of the early MAC protocols for WLANs was ALOHA. Later on, Slotted ALOHA and carrier sense protocols were proposed. Compared with wired LANs, WLANs have a special problem called hidden- and exposure-station problem that needs to be addressed in the MAC protocol. To avoid frame collisions among data frames, collision-based protocols reserve the channel by exchanging control frames. Collision-based protocols can be categorized into sender-initiated and receiver-initiated protocols. They can also be grouped into single-channel and split-channel protocols. Channel-aware protocols can improve channel efficiency by carefully assigning the medium to stations, depending on their channel status. Collision-free MAC protocols can only be applied to infrastructure networks. The IEEE 802.11 standard for WLANs is widely implemented in existing WLAN network products. It defines a contention-based function called DCF and a polling-based function called PCF. In this chapter, we have outlined generic as well as state-of-the-art MAC protocol designs for WLANs. We have also provided a detailed description of the IEEE 802.11 standard for WLANs.

# REFERENCES

1. N. Abramson, The ALOHA System — Another Alternative for Computer Communications, *Proceedings of the Fall Joint Computer Conference*, 1970.
2. L. Roberts, Extensions of Packet Communication Technology to a Hand Held Personal Terminal, *Proceedings of the Spring Joint Computer Conference*, 1972.
3. V. Bharghavan, A. Demers, S. Shenker, and L. Zhang, MACAW: A Media Access Protocol for Wireless LANs, *Proceedings of ACM SIGCOMM '94*, 1994.
4. C.L. Fullmer and J.J. Garcia-Luna-Aceves, Floor Acquisition Multiple Access (FAMA) for Packet-Radio Networks, *Proceedings of ACM SIGCOMM*, 1995.
5. Z.J. Haas and J. Deng, Dual Busy Tone Multiple Access (DBTMA) — Performance Results, WCNC'99, 1999.
6. P. Karn, MACA — A New Channel Access Method for Packet Radio, ARRL/CRRL Amateur Radio 9th Computer Networking Conference, 1990.
7. S. Tanenbaum, *Computer Networks*, 3rd ed., Prentice Hall, 1996.
8. Y. Wang and J.J. Garcia-Luna-Aceves, Throughput and Fairness in a Hybrid Channel Access Scheme for Ad Hoc Networks, WCNC, 2003.
9. F.A. Tobagi and L. Kleinrock, Packet Switching in Radio Channels: Part II — the Hidden Terminal Problem in Carrier Sense Multiple-Access Modes and the Busy-Tone Solution, *IEEE Transactions on Communications*, Vol. 23, No. 12, pp. 1417–1433, 1975.
10. F. Talucci and M. Gerla, MACA-BI (MACA by Invitation): A Receiver Oriented Access Protocol for Wireless Multihop Networks, in *Proceedings of PIMRC '97*, 1997.
11. J.J. Garcia-Luna-Aceves and A. Tzamaloukas, Receiver-initiated Collision Avoidance in Wireless Networks, *ACM Wireless Networks*, Vol. 8, pp. 249–263, 2002.
12. F.A. Tobagi and L. Kleinrock, Packet Switching in Radio Channels: Part III — Polling and (Dynamic) Split-Channel Reservation Multiple Access, *IEEE Transactions on Communication*, Vol. 24, No. 8, August, pp. 832–845, 1976.
13. C. Wu and V. Li, Receiver-Initiated Busy-Tone Multiple Access in Packet Radio Networks, *Proceedings of the ACM workshop on Frontiers in Computer Communications Technology*, 1987.
14. Z. Chen, and A. Khokhar, An Improved MAC Protocol for Wireless Local Access Network, IEEE Conference of Local Computer Networks (LCN'03), October 2003.
15. ANSI/IEEE 802.11 Standard Part 11: Wireless LAN Medium Access Control (MAC) and Physical Layer (PHY) Specifications, 1999 ed., 1999.

16. L. Kleinrock and F.A. Tobagi, Packet Switching in Radio Channels: Part I — Carrier Sense Multiple-Access Modes and their Throughput-Delay Characteristics, *IEEE Transactions on Communications*, Vol. 23, No. 12, pp. 1417–1433, 1975.

17. ANSI/IEEE Standards for Local Area Networks: Carrier Sense Multiple Access with Collision Detection (CSMA/CD) Access Method and Physical Layer Specifications, 1985.

18. IEEE 802.15.1: Wireless Medium Access Control (MAC) and Physical Layer (PHY) Specifications for Wireless Personal Area Networks (WPANs), 2002.

19. ETSI TR 101 683 Broadband Radio Access Networks (BRAN); HIPERLAN Type 2; System Overview, February 2000.

# 23

---

# 3G UMTS–IEEE 802.11B WLAN INTERNETWORKING

*Syed A. Ahson*

## INTRODUCTION

Seamless wireless data and voice communication is fast becoming a reality. IEEE 802.11b wireless local area network (WLAN) is a high-speed network[1] designed as a wireless extension of the Ethernet.[2] IEEE 802.11b WLAN networks have been widely deployed in offices, homes, and public hot spots such as coffee shops and hotels. IEEE 802.11b offers a number of advantages such as low operational cost, ease of deployment, and low equipment cost. However, it is limited by its small coverage area (100 to 300 ft). 3G UMTS (Universal Mobile Telecommunications Service) networks offer higher speeds and more capacity than existing 2G networks.[3] They provide higher speeds of up to 2 Mbps in a fixed or stationary wireless environment, and 384 kbps in a mobile environment. 3G UMTS networks aim to converge existing networks to a global network based on one international standard. If the user is under the coverage of an IEEE 802.11b WLAN network, his or her communication device can access high-bandwidth data service using that network. If IEEE 802.11b WLAN service is not available, the user may be handed over to the 3G UMTS network.[4] Mobile operators can significantly increase data traffic revenue and test new applications on the IEEE 802.11b WLAN network. High-demand traffic may be diverted from 3G UMTS network to the IEEE 802.11b network, preventing network congestion conditions. Mobile operators can provide improved in-building coverage by internetworking with the IEEE 802.11b WLAN network. One key capability in the next-generation wireless world will be Voice-over-IP (VoIP)[5] using IEEE 802.11b WLANs. The use of an

IEEE 802.11b WLAN to transmit voice is a great solution when people need to be constantly in contact with one another. IEEE 802.11b WLAN phones, which work just like cell phones when they are in the coverage of the WLAN, are very useful in places where workers are moving around. Corporate users will benefit significantly from IEEE 802.11b WLAN-3G internetworking. Enterprise-oriented internetworking solutions will provide secure mobile access for corporate users to connect to their office networks through 3G and various IEEE 802.11b WLANs such as office WLAN, home WLAN, and public WLAN.

This chapter describes 3G UMTS networks and internetworking between IEEE 802.11b WLAN and 3G UMTS networks. The section "3G Standardization and Development" presents 3G standardization efforts and describes strategies for 2G networks. The section "Evolution Towards 3G UMTS" describes key technologies (HSCSD, GRPS, and EDGE) for transitioning to 3G UMTS networks. The section "Universal Mobile Telecommunications Service" describes the 3G UMTS Core Network Architecture, reference points, and UMTS Terrestrial Radio Access Network (UTRAN) and protocol structure. The section "Internetworking with IEEE 802.11b" presents five possible network layer architectures for internetworking and handover between IEEE 802.11b WLAN and 3G UMTS networks without making any major changes to existing networks and technologies, especially at the lower layers such as the medium access control (MAC) and physical (PHY) layers. This will ensure that existing 3G networks will continue to function.

## 3G STANDARDIZATION AND DEPLOYMENT

2G wireless systems include Global System for Mobile Communications (GSM), IS-136, and IS-95 Code Division Multiple Access (CDMA).[6,7,8] GSM networks have the highest penetration worldwide. The International Telecommunications Union-Radio (ITU-R) developed the International Mobile Telephony-2000 (IMT-2000) specifications.[9] IMT-2000 is a set of standards for creating a global 3G network that includes terrestrial systems, satellite systems, and fixed access and mobile access networks. The international standardization effort for IMT-2000 involves the European Telecommunications Standards Institute (ETSI) Special Mobile Group, Research Institute of Telecommunications Transmission (RITT) in China, Association of Radio Industries and Businesses (ARIB) and Telecommunication Technology Committee (TTC) in Japan, Telecommunications Technology Association (TTA) in Korea, and the Telecommunications Industry Association (TIA) and T1P1 in the United States. ETSI SMG identified usage of Wideband CDMA (W-CDMA) for 3G networks.[10] China decided to deploy Synchronous Time Division CDMA (TD-SCDMA) for 3G networks,[11] and ARIB, Japan, decided to use W-CDMA for 3G networks. TTA, Korea, presented two schemes, one similar to W-CDMA and other close to the TIA cdma2000 approach. In the United States, the TIA presented several proposals for 3G UMTS, UWC-136[12] as an evolution of IS-136, cdma2000[13] as an evolution of IS-95, and W-CDMA for GSM networks.

GSM and IS-136, being time division multiple access (TDMA) systems, will evolve to code division multiple access (CDMA) systems in a number of steps. GSM networks must incorporate General Packet Radio Service (GPRS) for evolving to 3G UMTS network capabilities.[14] The next step for GSM networks will be adding enhanced data rates for GSM evolution (EDGE) capabilities.[15] EDGE allows GSM operators to use existing GSM radio bands while offering high bandwidth data services. It will offer 384 kbps data rates for pedestrian and low-speed environments. EDGE will offer 144 kbps for high-speed vehicular environments and 2 Mbps for the indoor office environment. UMTS is a new radio access network based on W-CDMA. It will offer 384 kbps in wide areas and up to 2 Mbps in local areas. GSM operators have two complementary options to upgrade their networks to 3G. The first option will be to use GPRS and EDGE in the existing radio spectrum. The second option will be to deploy UMTS in new 2-GHz bands. cdmaOne (TIA-IS-95) allows for channel aggregation to provide data rates in the range of 64 to 115 kbps. This simplifies migration of cdmaOne systems to cdma2000 systems.

## EVOLUTION TOWARD 3G UMTS — HSCSD, GPRS, AND EDGE

We describe key technologies for transitioning to 3G UMTS networks. High-Speed Circuit Switched Data (HSCSD) provides for high bandwidth data rates by co-allocation of multiple full rate traffic channels in GSM networks. GPRS is a packet data service using TCP/IP and X.25 to offer speeds up to 115 kbps. EDGE is essentially a radio interface improvement scheme. EDGE is an enhancement for both GSM circuit switched (HSCSD) and packet switched (GPRS) modes of operation. EDGE packet-switched mode of operation is known as Enhanced GPRS (EGPRS). EDGE circuit switched mode of operation is known as Enhanced CSD (ECSD).

### HSCSD

HSCSD provides for high bandwidth data rates by co-allocation of multiple full rate traffic channels (TCH/F) in GSM networks.[16] Baseline data rates have increased from 9.6 to 14.4 kbps due to reduction in error correction overhead of GSM Radio Link Protocol (RLP). Multiple 14.4 kbps time slots may be combined to offer access rates up to 57.6 kbps. Multiple time slots are dynamically allocated based on network operator policies and the user's data transfer needs. HSCSD requires software upgrades to the base station (BS) and the mobile switching center (MSC). The end-user experience will be similar to that of an Internet Service Provider (ISP) that offers fast secure dial-up using mobile equipment. HSCSD does not require any changes to existing mobility management procedures. Simultaneous handoff should take place for all time slots comprising the HSCSD connection. It should be noted that multiple time slots for HSCSD connections will probably be available in off-peak network usage times.

## GPRS

GPRS is a packet data service using TCP/IP and X.25 to offer speeds up to 115 kbps.[17] It provides short connection setup times and virtual connections, and users are charged for the actual data transmitted. Network resource and bandwidth are only used when data is actually transmitted. Bandwidth can be shared efficiently and simultaneously among several users. A GPRS core network is defined in parallel to the existing GSM core network. Two new types of nodes are introduced in GPRS: the Serving GPRS Support Node (SGSN) and the Gateway GPRS Support Node (GGSN). GGSN is responsible for the connection to other packet-switched networks and stores information about location of GPRS users. It encapsulates TCP/UDP packets and forwards them to the SGSN using GPRS Tunneling Protocol (GTP). GGSN may also offer packet filtering services. It is connected with SGSN via an IP-based GPRS backbone network. The Home Location Register (HLR) is enhanced to store GPRS subscription data such as the IP address of mobile users and routing information. HLR also maps each subscriber to one or more GGSN. SGSN and GGSN nodes interface with the HLR through Signaling System 7 (SS7) links. SGSN is responsible for authorization, authentication, admission control, charging, and mobility management of mobile users. SGSN encapsulates TCP/UDP packets and forwards them to the GGSN using GTP. SGSN is connected to the base station system by Frame Relay. Figure 23.1 illustrates a GPRS-Enhanced GSM network.

Mobile user packet data session is known as a Packet Data Protocol (PDP) context. At power-up, mobile users perform a GPRS attach. At GPRS attach, the mobile user's profile is downloaded from the HLR to SGSN. The mobile user must perform PDP context activation before it can send or receive IP packets. SGSN validates the PDP context activation request against the subscription information downloaded from the HLR during GRPS attach. The GGSN that should be used for TCP/UDP traffic routing is identified by a Domain Name Service (DNS) query of the access point name (the destination).[18] A GPRS tunnel is created using GTP between the SGSN and GGSN.

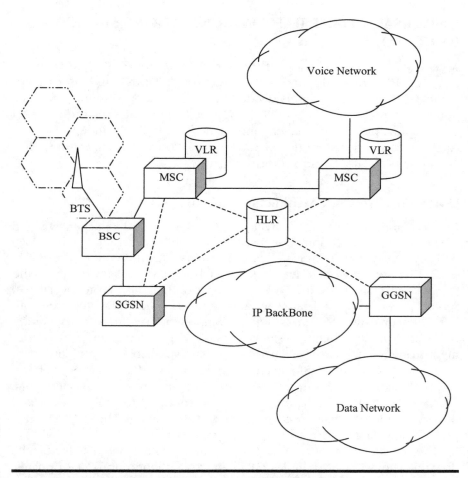

**Figure 23.1    General Packet Radio Service (GPRS)-enhanced GSM network.**

## EDGE

EDGE is designed to offer 3G services in existing spectrum bands. It is essentially a radio interface improvement scheme. The major enhancement in EDGE is the introduction of a new modulation system known as 8 PSK (Phase Shift Keying). 8 PSK will coexist with the existing GMSK (Gaussian Minimum Phase Shit Keying). It will provide higher data rates in a reduced coverage area. GSM/GPRS protocols are reused wherever possible. EDGE is an enhancement for both GSM circuit-switched (HSCSD) and packet-switched (GPRS) modes of operation. EDGE packet-switched mode of operation is known as enhanced GPRS (EGPRS). EDGE circuit-switched mode of operation is known as enhanced CSD (ECSD).

## UNIVERSAL MOBILE TELECOMMUNICATIONS SERVICE (W-CDMA)

A 3G UMTS network consists of three interacting domains: Core Network (CN), UMTS Terrestrial Radio Access Network (UTRAN), and User Equipment (UE). A UMTS system is divided into a set of domains and reference points that interconnect them. The 3G UMTS protocol structure is based on the principle that the layers and planes are logically independent of each other. We describe the 3G UMTS Core Network Architecture, reference points, UTRAN, and protocol structure.

### UMTS Core Network Architecture (UCN)

UCN is composed of a circuit-switched (CS) domain and a packet-switched (PS) domain. The CS domain consists of a mobile switching center (MSC) and a gateway MSC (GMSC). Figure 23.2 illustrates the entities present in a UMTS core network. The packet-switched (PS) domain contains the SGSN, the GGSN, DNS, the Dynamic Host Configuration Protocol (DHCP) server, packet charging gateway, and firewalls. The HLR interfaces with both domains over SS7 links. Other components required for operation of the UCN include billing systems, provisioning systems, and service/element management systems. The 3G-MSC is responsible for mobility management. It handles IMSI attach, authentication, HLR updates, Serving Radio Network Subsystem (SRNS) relocation, and intersystem handovers. The 3G-MSC handles call setup messages from and to mobile users and provides supplementary services such as call waiting. The 3G-MSC also provides CS data services for services such as fax.

The 3G-SGSN provides functionality similar to the 3G-MSC for the PS domain. The 3G-SGSN handles GPRS attach, authentication, VLR updates, SRNS relocation, and intersystem handover for user packet data session. It accepts session setup messages and enforces admission control. The 3G-SGSN is also responsible for tunneling user TCP/IP data to the 3G-GGSN using GTP. It collects statistics relating to mobile users' internal data usage, which may be used for charging. The 3G-GGSN provides internetworking with the external PS network, and also offers packet filtering services. The 3G-GGSN may allocate dynamic IP addresses to mobile users on PDP context activation, and is also responsible for tunneling user TCP/IP data to the 3G-SGSN using GTP. The 3G-GGSN collects statistics relating to mobile users' external data usage, which may be used for charging. The DNS server translates Access Point Names (APN) to the 3G-GGSN IP addresses. A DHCP server may be present to automatically allocate IP addresses for mobile users at PDP context activation. A packet-data firewall is present to protect the GPRS PS domain.

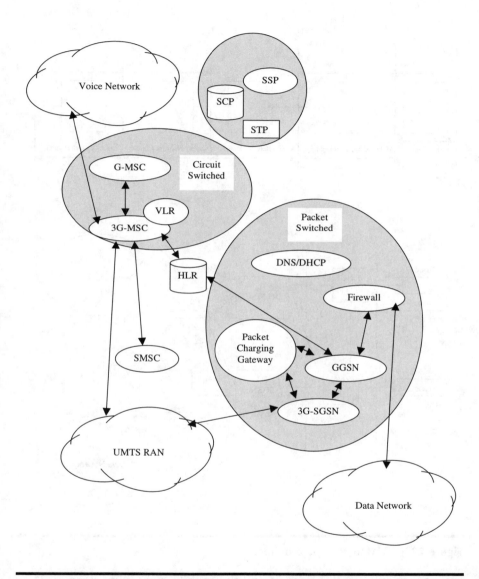

**Figure 23.2   UMTS core network architecture.**

## UMTS Network Reference Architecture

A set of reference points and domains have been defined for the UMTS network. $C_u$ is defined as the reference point between the UMTS subscriber

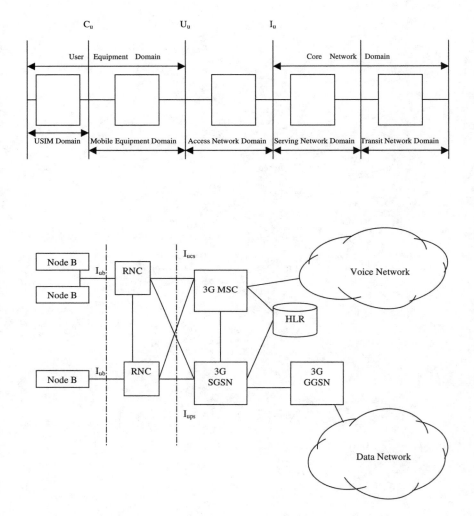

**Figure 23.3    UMTS reference points.**

identity module (USIM) domain and the mobile equipment domain. $U_u$ is defined as the reference point between the mobile equipment domain and the UMTS radio interface domain. $I_u$ is defined as the reference point between the UMTS radio interface domain and serving network domain. The $I_u$ reference point is split into $I_{ucs}$ and $I_{ups}$. $I_{ucs}$ connects the UMTS radio interface domain to the CS domain. $I_{ups}$ connects the UMTS radio interface domain to the PS domain. Figure 23.3 illustrates UMTS domains and reference points.

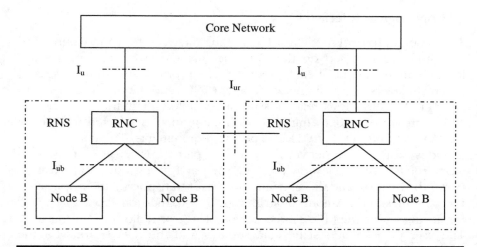

**Figure 23.4   UMTS architecture.**

## UMTS Terrestrial Radio Access Network (UTRAN)

UTRAN consists of a set of Radio Network Subsystems (RNSs) (Figure 23.4). An RNS is responsible for radio resources and coverage in a set of cells. It has two main elements: Node B and Radio Network Controller (RNC).

RNC enables autonomous Radio Resource Management (RRM) by UTRAN. It assists in soft handover of the user equipment (UE) when a mobile user moves from one cell to another. It combines and splits $I_{ub}$ data streams received from multiple Node Bs, and is also responsible for frame synchronization, outer loop power control, and SRNS relocation.

Node B is the physical unit of radio transmission and reception with cells. It can support both time division duplex (TDD) and frequency division duplex (FDD) modes and can be collocated with a GSM base transceiver system (BTS) to reduce implementation costs. It connects to the user equipment via the $U_u$ interface, and the RNC via the $I_{ub}$ interface. The main task of Node B is the conversion to and from the $U_u$ radio interface, including forward error correction (FEC), rate adaptation, W-CDMA spreading and despreading, and quadrature phase shift keying (QPSK) modulation on the air interface. It measures the quality and strength of the connection and determines the Frame Error Rate (FER), transmitting these data to the RNC as a measurement report. Node B also participates in power control, because it enables the user equipment to adjust its power using Down Link (DL) Transmission Power Control (TPC) commands via the inner-loop power control on the basis of Up Link (UL) TPC information. The predefined values for the inner-loop power control are derived from the RNC via the outer-loop power control.

## UTRAN Logical Interfaces

The general protocol model for UTRAN interfaces is shown in Figure 23.5. The structure is based on the principle that the layers and planes are logically independent of each other.[19-21] The protocol structure consists of two main layers: the radio network layer (RNL), and the transport network layer (TNL). All UTRAN-related issues are visible only in the RNL, and the TNL represents standard transport technology that is selected for use by UTRAN but without any UTRAN-specific requirements. The control plane includes radio access network application part (RANAP) at $I_u$, radio network subsystem application part (RNSAP) at $I_{ur}$, or Node B application part (NBAP) at $I_{ub}$, and the signaling bearer for transporting the Application Protocol messages. Among other things, the Application Protocol is used for setting up bearers (i.e., radio access bearer or radio link) in the RNL. The user plane includes the data streams and the data bearers for the data streams. The data streams are characterized by one or more frame protocols specified for that interface.

The Transport Network Control Plane (TNCP) does not include any RNL information, and is completely in the transport layer. It includes the Access Link Control Application Part (ALCAP) protocols that are needed to set up the transport bearers (data bearer) for the user plane. It also includes the appropriate signaling bearers needed for the ALCAP protocols. The TNCP is a plane that acts between the control plane and the user plane. The introduction of TNCP is performed in such a way that the Application Protocol in the radio network control plane is kept completely independent of the technology selected for data bearer in the user plane.

The UMTS $I_{ucs}$ logical interface interconnects the UTRAN to the UMTS circuit-switched core network. The circuit-switched protocol architecture on the $I_{ucs}$ interface is illustrated in Figure 23.6. The radio network layer control plane consists of Radio Access Network Application Part (RANAP). The transport network user plane consists of SS7 protocols. Signaling Connection Control Part (SCCP), Message Transfer Part (MTP3-B), and Signaling Asynchronous Transfer Mode (ATM) Adaptation Layer for Network-to-Network Interface (SAAL-NNI) is present in the transport network user plane. SAAL-NNI is divided into Service-Specific Coordination Function (SSCF), Service Specific Connection Oriented Protocol (SSCOP), and ATM Adaptation Layer 5 (AAL5). SSCF and SSCOP are designed for signaling transport in ATM networks. AAL5 is responsible for segmenting data into ATM cells.

The UMTS $I_{ups}$ logical interface interconnects the UTRAN to the UMTS packet-switched core network. The packet-switched protocol architecture on the $I_{ups}$ interface is illustrated in Figure 23.7. Signaling Connection Control Part (SCCP), Message Transfer Part User Adaptation Layer (M3UA), Simple Control Transmission Protocol (SCTP), and IP are present in the

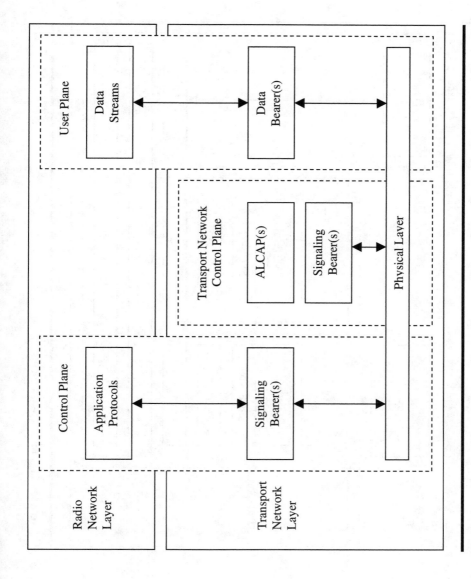

**Figure 23.5 UMTS protocol model.**

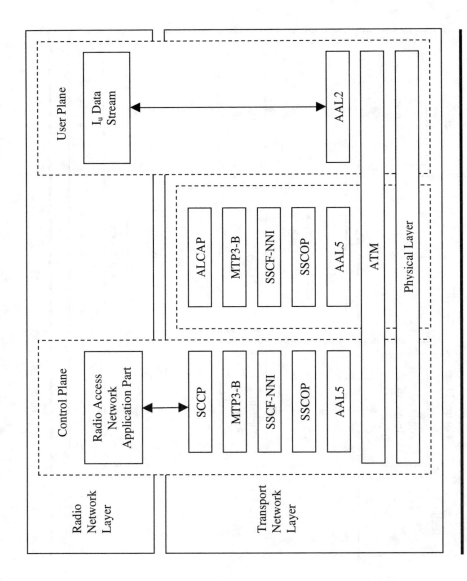

**Figure 23.6** I$_{ucs}$ interface protocols.

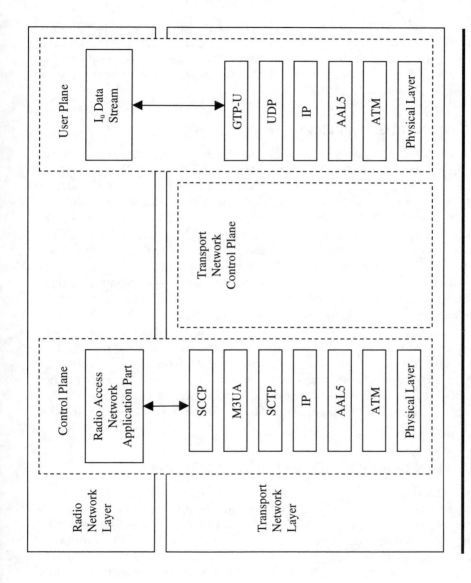

**Figure 23.7** $I_{ups}$ interface protocols.

transport network user plane. AAL5 is responsible for segmenting data into ATM cells.

## INTERNETWORKING WITH IEEE 802.11B

Interconnecting IEEE 802.11b networks and 3G UMTS networks will allow for seamless mobility. The interconnection scheme should have minimal to no impact on the MAC and PHY layers. We present five solutions that will allow 3G UMTS-IEEE 802.11b WLAN internetworking and have no impact on the MAC and PHY layers.

### 3G-WLAN Internetworking by Emulating RNC

The IEEE 802.11b network is connected to the 3G UMTS network at the $I_{ups}$ interface (Figure 23.8). An Internetworking Unit (IWU) is introduced at the $I_{ups}$ interface for interconnection of the 3G UMTS–IEEE 802.11b WLAN networks. The Internetworking Unit (IWU) emulates a Radio Network Controller (RNC). The Internetworking Unit (IWU) presents an IEEE 802.3 Local Area Interface to the IEEE 802.11b WLAN network. The Internetworking Unit (IWU) presents an $I_{ups}$ interface to the 3G UMTS network. A number of IEEE 802.11b cells are organized into a single distribution area. An IEEE 802.11b distribution area will appear as another routing area to the 3G-SGSN. IEEE 802.11b terminals are treated as 3G UMTS users. 3G UMTS mobility management schemes will keep track on the IEEE 802.11b users irrespective of the network they are connected to. The main advantage of this interconnection scheme is that mobility management, roaming, billing, and security are taken care of by existing 3G UMTS procedures. Minimum changes are required to the existing networks. The main drawback of this scheme is that IWU present as single point of failure and a potential bottleneck in the 3G UMTS–IEEE 802.11b WLAN interconnected network.

### 3G-WLAN Internetworking by Emulating 3G-SGSN

The IEEE 802.11b network is connected to the 3G UMTS network at the $G_n$ interface (Figure 23.9). An IWU is introduced at the $G_n$ interface for interconnection of the 3G UMTS–IEEE 802.11b WLAN networks. The IWU emulates a Serving GPRS Support Node (3G-SGSN). The IWU presents an IEEE 802.3 local area interface to the IEEE 802.11b WLAN network and a $G_n$ interface to the 3G UMTS network. IEEE 802.11b terminals are treated as 3G UMTS users. 3G UMTS mobility management schemes will keep track on the IEEE 802.11b users irrespective of the network they are connected to. The main advantage of this interconnection scheme is that

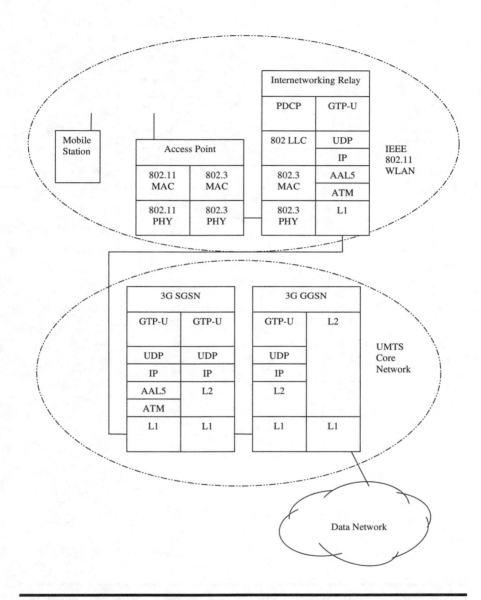

**Figure 23.8   3G UMTS-IEEE 802.11b internetworking by RNC emulation.**

mobility management, roaming, billing, and security are taken care of by existing 3G UMTS procedures, and only minimum changes are required to the existing networks. The main drawback of this scheme is that the IWU is present as single point of failure and a potential bottleneck in the 3G UMTS–IEEE 802.11b WLAN interconnected network.

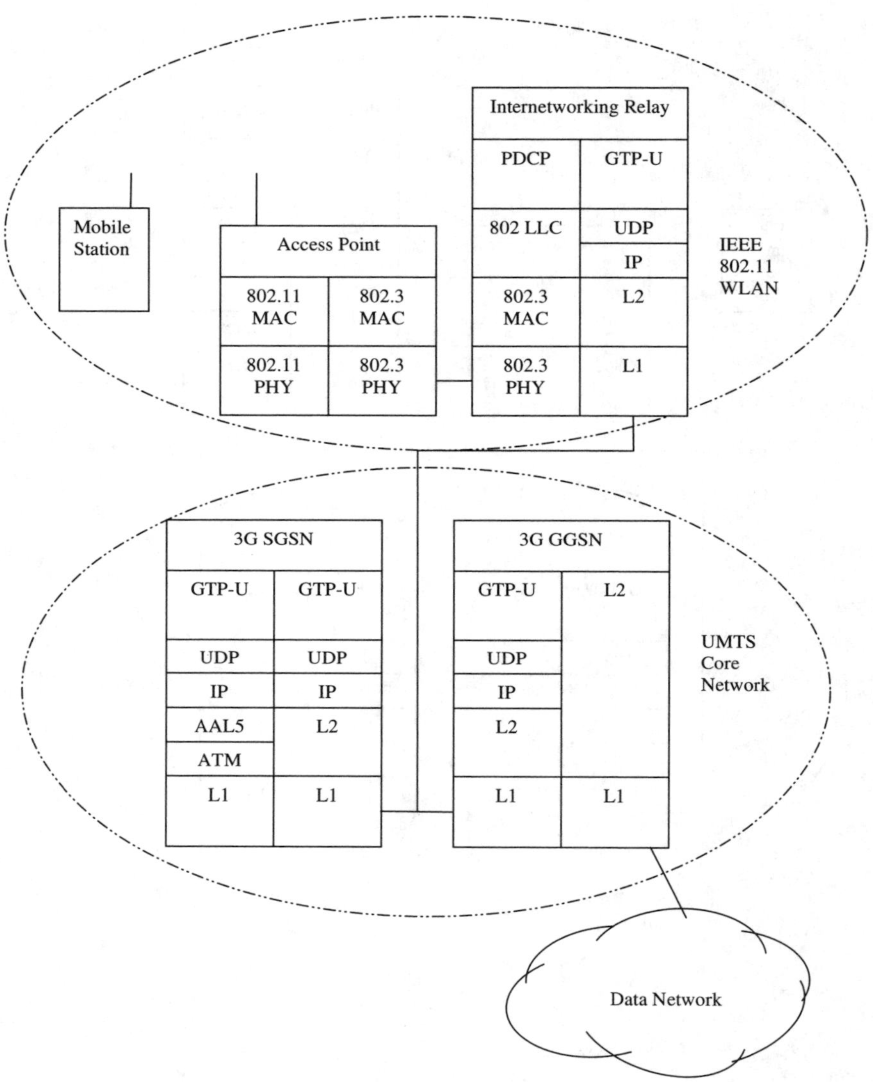

**Figure 23.9  3G UMTS-IEEE 802.11b internetworking by 3G SGSN emulation.**

## 3G-WLAN Internetworking by Emulating Virtual Access Point (VAP)

The 3G UMTS network is viewed by the IEEE 802.11b WLAN network as an access point (Figure 23.10). This interconnection method reverses the roles of the 3G UMTS and IEEE 802.11b WLAN networks as discussed in the RNC and 3G-SGSN emulation schemes. Mobility is managed by the

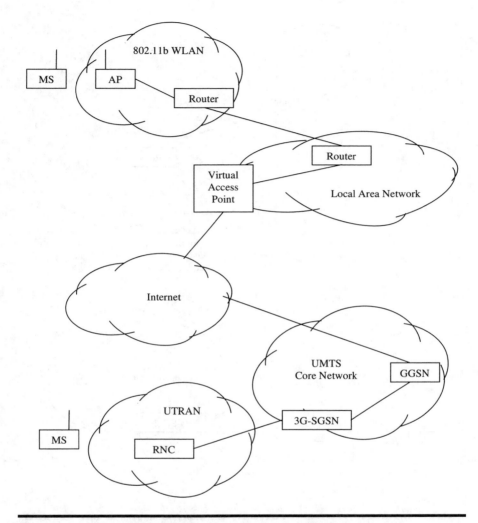

**Figure 23.10  3G UMTS-IEEE 802.11b internetworking by virtual access point scheme.**

IEEE 802.11b WLAN network according to the Inter Access Point Protocol (IAPP). The entire 3G UMTS network is treated as a Pico cell associated with a "virtual" access point. A VAP is introduced for interconnection of the 3G UMTS–IEEE 802.11b WLAN networks.

## 3G-WLAN Internetworking through the Mobility Gateway

A mobile "proxy" may be placed in either the 3G UMTS network or the IEEE 802.11b WLAN network (Figure 23.11). The mobility gateway will

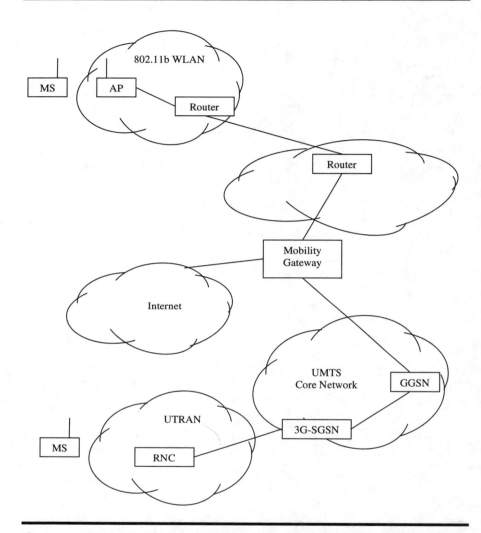

**Figure 23.11 UMTS-IEEE 802.11b internetworking by mobility gateway scheme.**

be responsible for routing of packets and mobility management. This proxy architecture is highly scalable because we could have a number of mobility gateways. Organizations already have IP proxies that could be upgraded to mobility gateways. This scheme suffers from lack of standardization of the proxy architecture and mobility management schemes.

## 3G-WLAN Internetworking by Mobile IP

Figure 23.12 illustrates 3G-WLAN Internetworking by Mobile IP.[22] Mobile IP is used for forwarding IP datagrams when a mobile user roams from

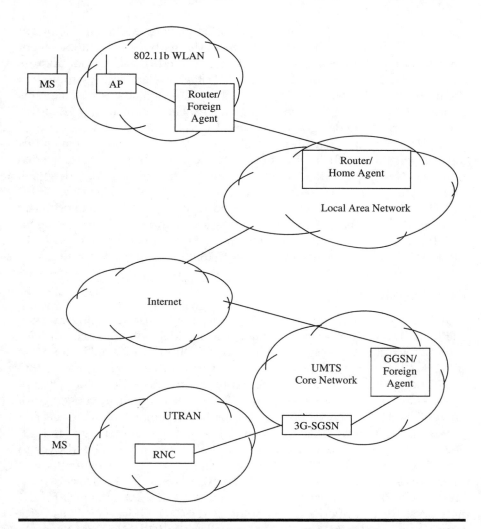

**Figure 23.12  3G UMTS-IEEE 802.11b internetworking by mobile IP.**

one network to another. Mobile users are identified by fixed IP addresses. Mobile devices are required to have a dual-mode 3G UMTS–IEEE 802.11b WLAN protocol stacks. The 3G UMTS network and IEEE 802.11b WLAN networks function as peer networks. Mobile users register with home agents in their "home" IEEE 802.11b network. A mobile device could initiate a handover when it moves out of the IEEE 802.11b WLAN network coverage and detects the presence of a 3G UMTS network. The 3G UMTS network is treated as "foreign" network by the mobile device. When a mobile user roams into a foreign network, it registers with a foreign agent to obtain a care-of address. The mobile device is allocated a care-of address by a foreign agent on the 3G UMTS network. Mobile users inform their home agents of their care-of address. The mobile device home agent on the "home" IEEE 802.11b network is informed of the care-of address by a mobile IP registration procedure. Datagrams are always routed to the mobile user by its home agent. The home agent receives all datagrams addressed to the mobile device and encapsulates them using "IP-in-IP". These encapsulated datagrams are tunneled to the mobile device's foreign agent. The foreign agent removes the IP-in-IP header of the datagrams and delivers them to the mobile device. While the mobile device is attached to the 3G UMTS network, it constantly searches for an IEEE 802.11b signal. The mobile device could initiate a handover when it detects an IEEE 802.11b WLAN network while connected to the 3G UMTS network. The Foreign agent in the 3G UMTS network is deactivated. The home agent is informed by the mobile device that it no longer requires IP-in-IP tunneling. This scheme suffers from a triangular routing overhead.

## CONCLUSION

Seamless wireless data and voice communication is fast becoming a reality. IEEE 802.11b WLAN networks have been widely deployed in offices, homes, and public hot spots such as coffee shops and hotels. However, IEEE 802.11b is limited by the small coverage area (100 to 300 ft). 3G UMTS networks aim at converging existing networks to a global network based on one international standard. If the user is under the coverage of an IEEE 802.11b WLAN network, his or her communication device can access high-bandwidth data service using that WLAN network. If IEEE 802.11b WLAN service is not available, the user may be handed over to the 3G UMTS network. This chapter described 3G UMTS networks and internetworking between IEEE 802.11b WLAN and 3G UMTS networks. We presented 3G standardization efforts and described strategies for 2G networks. We described key technologies (HSCSD, GRPS, and EDGE) for transitioning to 3G UMTS networks, as well as the 3G UMTS Core Network, reference points, and UMTS Terrestrial Radio Access Network (UTRAN) and protocol structure. We presented five possible network layer architectures for internetworking and handover between IEEE 802.11b WLAN and 3G-UMTS networks without making any major changes to existing networks and technologies, especially at the lower layers such as MAC and PHY layers. Mobile IP interconnection architecture provides an efficient method for internetworking of heterogeneous packet-oriented networks.

# REFERENCES

1. IEEE 802.11, 1999 Edition (ISO/IEC 8802-11: 1999) IEEE Standards for Information Technology — Telecommunications and Information Exchange between Systems — Local and Metropolitan Area Network — Specific Requirements — Part 11: Wireless LAN Medium Access Control (MAC) and Physical Layer (PHY) Specifications.
2. IEEE 802.3-2002, IEEE Standard for Information Technology — Telecommunications and information exchange between systems — Local and metropolitan Area Networks —Specific Requirements — Part 3: Carrier Sense Multiple Access with Collision Detection (CSMA/CD) Access Method and Physical Layer Specifications.
3. Collins, D. and Smith, C., *3G Wireless Networks*, McGraw-Hill Professional, New York, September, 2001.
4. Kaaranen, H., Naghian, S., Laitinen, L., Ahtiainen, A., and Niemi, V., *UMTS Networks: Architecture, Mobility and Service*, 1st ed., John Wiley & Sons, New York, August 2001.
5. Collins, D., *Carrier Grade Voice Over IP*, McGraw-Hill Professional, New York, September 2002.
6. Mouley, M. and Pautet, M.B., The GSM System for Mobile Communications, Palaiseau, France 1992.
7. Garg, V.K. and Wilkes, J.E., *Wireless and Personal Communications Systems*, Prentice Hall, Englewood Cliffs, NJ, 1996.
8. TIA/EIA IS-95, Mobile Station-Base Station Compatibility Standard for Dual-Mode Wideband Spread Spectrum Cellular System, PN-3422, 1994.
9. ETSI SMG, Proposal for a Consensus Decision on UTRA, ETSI SMG Tdoc 032/98.
10. Holma, H. and Toskala, A. (Eds.), *WCDMA for UMTS*, 2nd ed., John Wiley & Sons, New York, September 2002.
11. TD-SCDMA, http://www.tdscdma-forum.org/.
12. TIA TR45, Proposed RTT Submission (UWC 136), TR-45.3/98.03.03.19, March 1998.
13. TIA/EIA IS-2000-1 Introduction to cdma2000 Spread Spectrum Systems, November 1999.
14. ETSI Technical Specification GSM 02.60 GPRS Service Description — Stage 1 version 5.2.1, July 1998.
15. ETSI Tdoc SMG2 95/97, EDGE Feasibility Study, Work Item 184: Improved Data Rates through Optimized Modulation, version 0.3, December 1997.
16. Digital Cellular Telecommunication System (Phase 2+), High Speed Circuit Switched Data (HSCSD), Service Description, Stage 2, GSM 03.34.
17. RFC 793, Transmission Control Protocol, September 1981.
18. RFC 1034, Domain Names — Concepts and Facilities, November 1987.
19. 3GPP Technical Specification 25.410 UTRAN $I_u$ Interface: General Aspects and Principles.
20. 3GPP Technical Specification 25.420 UTRAN $I_{ur}$ Interface: General Aspects and Principles.
21. 3GPP Technical Specification 25.430 UTRAN $I_{ub}$ Interface: General Aspects and Principles.
22. RFC 3220, IP Mobility Support for IPv4, January 2002.

# 24

---

# SECURITY
# IN IEEE 802.11 WLANS

*Costas Lambrinoudakis and Stefanos Gritzalis*

## INTRODUCTION

Wireless communication networks offer to individuals and organizations several benefits such as portability, mobility, and flexibility. They serve as the transport mechanism among devices, enterprise networks, and the Internet, reducing installation costs while increasing everyday business productivity.

Wireless local area networks (WLANs), in particular, have been used in various environments, including businesses, homes, conference centers, and airports, to name just a few. WLAN devices allow users to move from place to place, avoiding cabling restrictions and without being disconnected from the network. A WLAN connects mobile clients to the network through a device called *access point*. All mobile clients can move freely within the access point cell, known as the *coverage area*. The access point cells can overlap each other, allowing roaming within a building or between nearby buildings.

Although wireless communication networks, and especially WLANs, eliminate some of the problems associated with traditional wired local area networks (LANs), new security risks, inherent in any wireless technology, are introduced. Security is a *sine qua non* condition in public communication infrastructures, whether fixed or mobile. It has a dominant role in ensuring that citizens can trust the capabilities and the service framework that such a new environment can offer within the modern information society. The willingness of the end users (businesses, governments, citizens) to conduct business and facilitate other activities in an advanced, wireless networked environment will be determined not only by the performance of the enabling technology but also by the deployment of an integrated trust framework that surrounds such activities.

Typical security requirements that must be addressed in a wireless network environment include: *access control* — access to specific information and services granted to authorized entities only; *confidentiality* — not disclosing information to nonauthorized entities; *authentication* — verifying the validity of the credentials (proof of identity) provided by a client; and *integrity* — nonauthorized entities cannot modify information.

This chapter addresses the aforementioned security requirements for the IEEE 802.11 WLANs. The section of this chapter titled "WLANs and Security Standards" provides an overview of the standardization activities in the WLAN arena. The section titled "WEP Characteristics" presents the design principles and the main operational characteristics of the Wired Equivalent Privacy (WEP) protocol and then highlights specific security problems with regard to the supported authentication, confidentiality protection, and data integrity check mechanisms. The section titled "The IEEE 802.11i Security Architecture" introduces the new security architecture specified in the IEEE 802.11i standard, focusing on the characteristics of the Temporal Key Integrity Protocol (TKIP) and the way in which WEP inefficiencies have been rectified.

## WLANS AND SECURITY STANDARDS

Working Group 11 of the IEEE 802 family "Local Area and Metropolitan Area Networks" has undertaken the task of producing technical standards for WLANs. The need was to describe the technical characteristics of the equipment and specify a standardized approach for wireless communication in such a way that interoperability between different products would be facilitated, performance would not be compromised, and an adequate level of security would be ensured. The first WLAN international standard, known as IEEE 802.11,[6] appeared in 1999. The IEEE 802.11 security-related provisions are known as the *Wired Equivalent Privacy (WEP)* protocol and are described in detail in the section titled "WEP Characteristics."

As happens with the majority of complex standards, IEEE 802.11 did not avoid certain ambiguities or problematic definitions, and several features were characterized as optional. It is therefore clear that because each manufacturer can interpret specific parts of the standard differently, or decide to implement or to ignore the optional features, interoperability is at risk. To overcome this problem, a group of major manufacturers, known as the "Wi-Fi Alliance," developed the Wi-Fi industry standard for products based on IEEE 802.11. The objective was for all manufacturers to reach a consensus on both the interpretation of the ambiguous IEEE

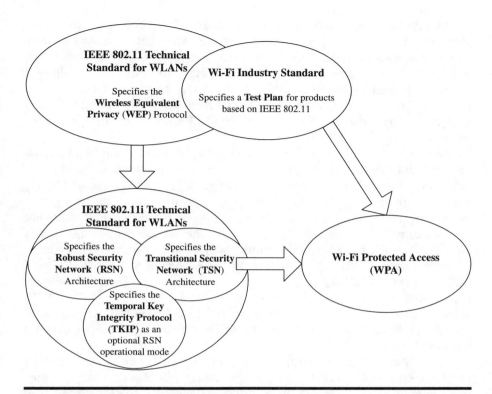

**Figure 24.1   WLAN security standards.**

802.11 specifications and make a decision on whether or not to implement specific options. As illustrated in Figure 24.1, Wi-Fi is a subset of IEEE 802.11, which means that there are specific parts of the standard that were left out, whereas there are additional specifications that were not included in IEEE 802.11. Compliance of a manufacturer's product with the Wi-Fi industry standard was verified through a test plan developed by the Wi-Fi Alliance. In this way, interoperability is ensured among the products passing the specified tests successfully.

Unfortunately, as soon as WLANs started penetrating the market in years 2000 and 2001, several scientific reports were published,[7] highlighting major security flaws in the WEP protocol and thus rendering WLANs insecure. As a response to this new scene, the IEEE Working Group 802.11 started revising the existing standard in an attempt to specify a new WLAN security architecture that would fulfill security requirements in a robust way — and, of course, would cure the identified security problems of the WEP protocol. This new standard is known as IEEE 802.11i, a draft of which was available in late 2003. The IEEE 802.11i envisages a new *Robust Security Network (RSN)*, accommodating mobile devices that offer more advanced encryption techniques and also exhibit several new hardware-supported capabilities. The problem, however, was that there was no way to upgrade existing WEP-based networks through software because RSN requires the replacement of all non-RSN-compatible devices. To allow a transition period for moving from WEP-based to RSN-compatible devices, IEEE 802.11i specifies another WLAN architecture, known as the *Transitional Security Network (TSN)*, in which WEP and RSN can operate in parallel.

Although RSN may, in theory, solve the WLAN security problems, what happens until the time RSN-compatible equipment is widely available and, on top of that, what happens with existing WEP-based systems? Should we throw them away or should we keep operating them, knowing that the security level is inadequate, thus putting at stake the confidentiality and integrity of our data ? There is only one answer: until the market switches to RSN, it is essential to find a way to improve the security level of WEP systems. This is exactly what the *Temporal Key Integrity Protocol (TKIP)*, specified in IEEE 802.11i as an alternative RSN operational mode, is all about. Because TKIP is described in detail in the section "The TKIP Security Protocol," here we will only say that it rectifies several WEP security problems by simply performing a software upgrade of existing wireless devices.

As happens in most cases, the manufacturers of wireless commercial products could not wait for the official release of the standard to start marketing their products. The Wi-Fi Alliance developed the *Wi-Fi Protected Access (WPA)* industry standard, which was based on the draft specification of TKIP, facilitating more secure and interoperable operation of WLANs by upgrading existing WEP-based systems.

## WEP CHARACTERISTICS

The main design principles and operational characteristics of the WEP protocol, in terms of the authentication, confidentiality protection, and integrity assurance mechanisms, are briefly described in the current section. However, taking into account that all cryptographic operations are solely based on the RC4 algorithm, we start with an overview of its operational principles.

### The RC4 Encryption Algorithm

RC4 (Rivest Cipher 4)[9,8] is a symmetric algorithm operating as stream cipher. In practical terms, this means that both the encryption and the decryption phases utilize the same secret key and also that the algorithm operates (encrypts or decrypts) on small units of plaintext rather than on large blocks of data (block ciphers). Therefore, for each byte of plaintext (or encrypted text) going in, an encrypted byte (or plaintext byte) is produced until all the bytes are processed.

As happens with the majority of stream ciphers, RC4 generates a *keystream* that, during encryption (see Figure 24.2), is combined with the plaintext through a bitwise XOR operation. Therefore, the encryption process can be described by the following simple formula:

$$Ciphertext = Keystream \oplus Plaintext$$

Considering the main property of the bitwise XOR operation, which is that if you apply (see Figure 24.2) the same value twice you get the original value, the decryption process is given by:

$$Plaintext = Keystream \oplus Ciphertext$$

Ideally, keystream should be a random sequence of bytes. In practice, however, the side decrypting the data must have the means to reproduce the keystream used during the encryption process, yielding what is known as *pseudorandom keystream*. Without going into the details, the keystream generation by RC4 algorithm is performed through a series of permutations of the elements of a 256-byte array, initialized to the values 0 up to 255, based on the value of the secret encryption/decryption key. It is, therefore, clear that the only way to reproduce the same sequence of bytes (keystream) is to know the key.

### RC4 Keys in the WEP Protocol

The size of the WEP (RC4) keys specified in IEEE 802.11 is 40 bits. However, most Wi-Fi manufacturers, knowing that a key of that size can be easily compromised with a brute force attack, offer products featuring 104-bit and, in some cases, even 128-bit keys. An additional problem is that in a WLAN environment, the secret key is statically shared between several mobile devices and access points, mainly due to the lack of mechanisms for automatically updating the key in all WLAN devices. Consequently, the encrypted text (ciphertext) is always the same if the plaintext remains constant which, in turn, means that if an attacker notes that the mobile device transits the same encrypted bytes, he knows that the plaintext is repeated. To overcome this problem, WEP has introduced the notion of a 24-bit *Initialization Vector (IV)*.

As illustrated in Figure 24.3, the side encrypting the data (for example, the mobile device) generates a random 24-bit IV, which is concatenated to the 104-bit secret key, forming a 128-bit RC4 key. This 128-bit sequence

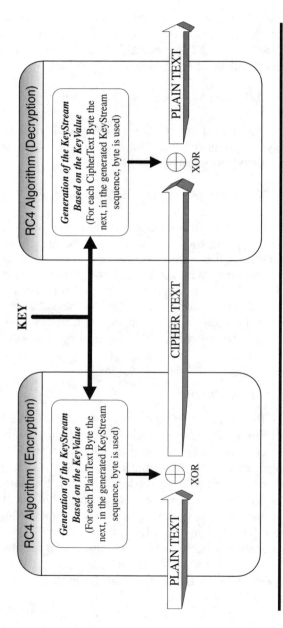

**Figure 24.2 The RC4 encryption algorithm.**

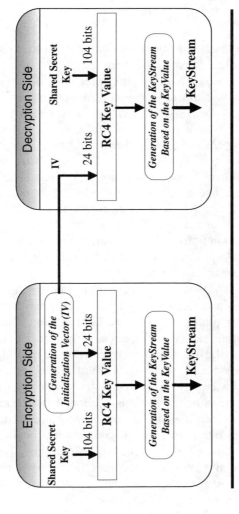

Figure 24.3 RC4 key and initialization vector in the WEP protocol.

is used for generation of the RC4 keystream, as described in the previous section. The IV value is transmitted, together with the encrypted data, to the decrypting side (for example, the access point) to concatenate it with the 104-bit shared secret key and thus generate the same 128-bit RC4 key and consequently the same RC4 keystream that will facilitate decryption of the data.

## WEP Authentication

The purpose of an authentication mechanism is to ensure that only authorized devices gain access to the network. WEP supports two distinct authentication mechanisms, namely the *open system authentication* and the *shared-key authentication*. During open system authentication, the mobile device transmits its medium access control (MAC) address to the access point, together with a request for authentication. The access point maintains a list of all MAC addresses that have been approved to join the network, and if the received MAC address is in this list, it responds by transmitting an Accept message to the mobile device. Otherwise, it transmits a Reject message. The protection offered by the open system authentication mechanism is against accidental connection to the wrong WLAN or against very simple attacks. In practice, any attacker can gain access to the network by transmitting, during authentication, instead of its own unauthorized MAC address, a MAC address that has been included in the list maintained by the access point.

As far as the shared-key authentication mechanism is concerned, although it establishes a better security level as compared to open authentication, it is still inadequate, as explained later in this section. Figure 24.4 illustrates the messages exchanged between the mobile device and the access point. Initially, the mobile device transmits a *request for authentication* message. In response, the access point transmits a *challenge text*, which the mobile device encrypts, using the RC4 algorithm in a way similar to that presented in the previous section. Upon completion of the encryption process, the mobile device transmits back to the access point the challenge text in encrypted form, together with the IV value that has been concatenated to the shared secret key to generate the RC4 key. The access point, knowing the IV value, decrypts the received cipher. Provided that the decrypted text matches the challenge text that was initially transmitted by the access point, the authentication is successful and an Accept message is transmitted to the mobile device. Otherwise, a Reject message is transmitted. However, it has been proved that even the shared-key authentication mechanism cannot ensure that unauthorized devices will not gain access to a WLAN network, mainly due to the problems discussed in the following text.

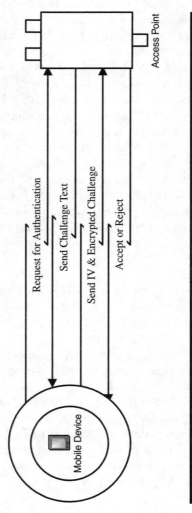

**Figure 24.4  WEP authentication.**

## WEP Authentication Security Problems

Normally, an authentication process facilitates an authenticating device to validate the identity of a requesting device and vice versa. In the case of WEP authentication, the mobile device cannot validate the identity of the access point as any access point can respond to an authentication request by either transmitting an Accept or Reject message without providing any proof that it possesses the shared secret key and that it actually checked the validity of the encrypted challenge text. Furthermore, during WEP authentication, the mobile device is not supplied with any temporary token (session key) that can be utilized for authenticating subsequent transactions. In other words, after a mobile device joins the WLAN, following a successful authentication, identity cannot be revalidated.

Maybe the most severe WEP authentication security flaw originates from the fact that a mobile station transmits, together with the ciphertext (encrypted challenge text), the IV utilized during the encryption process. Assuming that an attacker monitors the communication between a legitimate mobile device, requesting authentication, and the access point, he can record the following data (refer to Figure 24.5): The challenge text A (CH-A) transmitted by the access point, the initialization vector A (IV-A), and the encrypted challenge text A (ECH-A) transmitted by the mobile station A. Based on the RC4 operation, the following relationships hold true:

$$ECH\text{-}A = KeyStream\text{-}A \oplus CH\text{-}A \; and \; thus$$

$$KeyStream\text{-}A = ECH\text{-}A \oplus CH\text{-}A$$

Consequently, by performing a bitwise XOR operation between ECH-A and CH-A, the attacker regenerates the RC4 KeyStream A, which corresponds to the initialization vector IV-A.

As illustrated in Figure 24.5, following an attacker's request for authentication, a bitwise XOR operation is performed between the challenge text B (transmitted by the access point) and *KeyStream-A*. Then the encrypted challenge B is transmitted back to the access point, together with initialization vector *IV-A*. The access point will authenticate the attacker, even if he does not know the shared secret key.

## Confidentiality Protection

Protection of the confidentiality property implies that there are mechanisms ensuring that only authorized persons can access the information transmitted over a WLAN. It is apparent that due to the nature of wireless

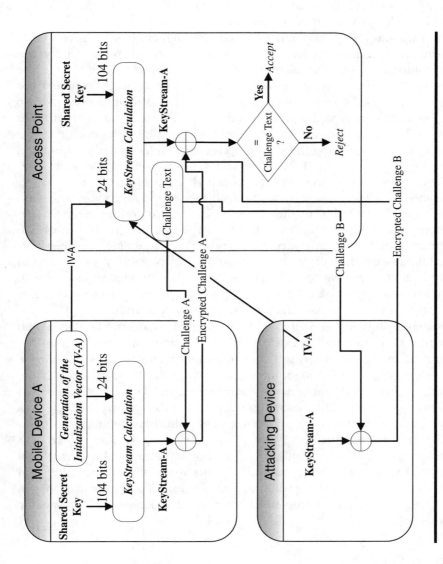

**Figure 24.5  Violation of the WEP authentication mechanism.**

technology, signals may travel outside the physical perimeter of an installation, turning confidentiality into a very difficult requirement to meet. WEP protects confidentiality by encrypting the data prior to its transmission. The encryption algorithm utilized is again RC4. Unfortunately, several security flaws have been identified yielding WEP as insecure, as far as confidentiality protection is concerned.

An important WEP vulnerability is the employment of static secret keys. If a mobile station such as a laptop is stolen, the secret key becomes compromised, along with all remaining mobile devices sharing the same secret key. Furthermore, assuming that specific parts of the plaintext are repeated, the use of shared static keys results in the repetition of specific patterns in the encrypted text. Such information is valuable to an attacker who wishes to initiate an analytic attack. The WEP protocol utilizes a random IV for modifying the shared secret keys (refer to the section titled "RC4 Keys in the WEP Protocol") and thus avoiding their static use. However, once again, this WEP approach is not problem free.

A mobile device concatenates IV with the shared secret key and then broadcasts the IV value to the access point. It is therefore clear that an attacker monitoring the network traffic can easily identify 24 out of 128 key bits. Combining this fact with known vulnerabilities in the key-scheduling algorithm of RC4 (*weak keys*)[4] the possibility of successful analytic attacks aiming to recover the secret key are increased. In fact, there are publicly available attack scripts that do recover the secret key after recording only a moderate amount of network traffic.

Another problem associated with IV use is that it cannot be considered as a true random value because IEEE 802.11 does not specify how IVs should be initialized or changed, meaning that manufacturers can choose their own IV generation algorithms and also that all their products could generate identical IV sequences. Furthermore, the limited IV length (24 bits) guarantees that, after some time, the same IV values will be repeated. It has been proved that a busy data access point transmitting 3000 bytes at 11 Mbps will exhaust the 24-bit IV space in approximately 10 hr. It is therefore possible for an attacker to record ciphertexts that have been produced with the same IV value and consequently with the same RC4 keystream. As explained earlier in this section, such information can assist analytic attacks aiming to identify the RC4 keystream for specific IV values.

## Integrity Protection

The integrity property represents one of the most important security requirements: only authorized persons should be allowed to modify the data. A well-known method for checking the integrity of a message, transmitted over some communication lines, is the *Cyclic Redundancy Check (CRC)*. The CRC is computed, by both the transmitting and receiving devices, as a function of all bytes in the message and is then transmitted together with the rest of the data. Even if a single bit in one of the message bytes is modified during transmission, the CRC computed at the receiving end will not match the one transmitted by the message sender, causing the rejection of the message.

The mechanism employed by the WEP protocol for checking the integrity of transmitted packets is based on a double CRC computation. The first CRC, known as the *Integrity Check Value (ICV)*, is computed prior to data encryption and is then appended to the plain data stream. The second CRC is computed with the encrypted data bytes prior to message transmission. It is stressed that the computation of a CRC on the encrypted bytes does not provide any protection from intentional alterations of the data as an attacker can easily recompute the CRC after altering the data. It only serves to detect accidental transmission errors. Theoretically, the same cannot happen with the ICV because it has been encrypted together with the rest of the data.

However, the combination of CRCs with the RC4 stream cipher introduces a serious vulnerability. Borisov and colleagues[2] have demonstrated that the CRC computation method can predict the CRC bits that will change if a specific data bit is modified. Recalling that the RC4 encryption is only based on bitwise XOR operations, an attacker can modify the (encrypted) data and then *fix* the (encrypted) ICV by simply flipping the appropriate bits.

## THE IEEE 802.11I SECURITY ARCHITECTURE

It has been demonstrated that the security provisions of the IEEE 802.11 standard, known as WEP, are not satisfactory because they cannot even ensure fulfillment of basic security requirements such as confidentiality and integrity. The identified WEP security problems are mainly because:

■ All WEP security mechanisms (i.e., for authentication, encryption, and integrity checking) utilize a single preshared secret key.
■ The initialization vector appended in the secret WEP key, aiming to avoid static use of the key, is too short, not protected from reuse, and is transmitted in clear text.
■ There is no provision to avoid the use of RC4 weak keys.

As explained in the section "WLANs and Security Standards," in an attempt to rectify WEP problems, the IEEE 802.11i standard has specified a new WLAN security architecture known as RSN, which supports much more advanced encryption techniques and exhibits new capabilities. The WPA industry standard has also adopted the IEEE 802.11i security architecture.

The main principle of the new security architecture is that the protection of data confidentiality and integrity must be performed on the basis of *temporal* (session) *keys*, providing the *prior mutual authentication* between the mobile device and the access point. The authentication procedure should be based on existing well protected keys, whereas the session keys should be dynamically computed from the authentication key and, as their name denotes, be different every time the mobile device connects to an access point.

As illustrated in Figure 24.6, IEEE 802.11i specifies three distinct security layers. At the *authentication layer,* clients interested in joining the network submit their credentials to the authentication server who, in turn, verifies their validity. The IEEE 802.11i standard does not provide the details of the authentication method, as it allows the employment of any existing authentication protocol of proven security, such as TSL, Kerberos, etc. Provided that the authentication was successful, the *access control layer* is called to decide if the authenticated user should be allowed to access the network and with what privileges. As specified in IEEE 802.11i, the implementation of the access control mechanism is based on the provisions of the IEEE 802.1X standard. Finally, there is the *WLAN layer* that protects, in accordance to the IEEE 802.11i specifications, the confidentiality and integrity of the communicated data.

## Authentication and Access Control

The IEEE 802.1X standard defines port-based network access control for Ethernet networks. More specifically, it uses the physical characteristics of the switched LAN infrastructure for authenticating devices connected to a LAN port. However, as indicated in Reference 1, 802.1X can be applied to a wireless LAN, in the sense that for each device requesting authentication, the access point must create a distinct logical port (analogous to the ports of a LAN switch).

Following the IEEE 802.1X terminology, the entities involved in the authentication process (refer to Figure 24.6) are: the *supplicant* (mobile device), which is the entity requesting authentication; the *authenticator* (access point), which is the entity that allows access to the network and grants privileges; and the *authentication server*, which implements the authentication services on behalf of the authenticator and thus grants or denies access of the supplicant to the authenticator services. During the 802.1X authentication procedure, the *Extended Authentication Protocol over LAN* (EAPoL) is employed for exchanging messages between the

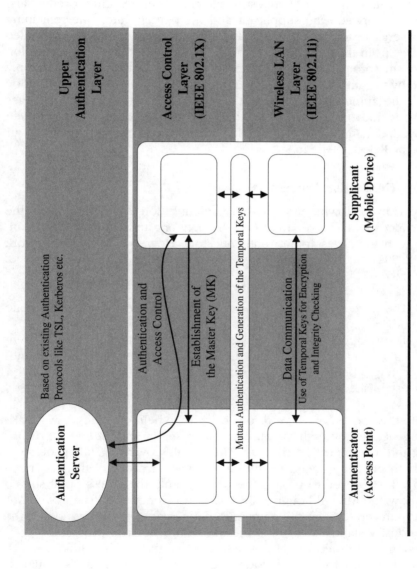

**Figure 24.6  The IEEE 802.11i security architecture.**

supplicant (mobile device), the authenticator (access point), and the authentication server. The typical sequence of the EAPOL messages exchanged is illustrated in Figure 24.7.

It should be stressed that, with the exception of the initial messages (Request Authentication, Request Credentials, and Submit Credentials) exchanged between the supplicant and the authenticator, the remaining ones (a series of requests from the authentication server and the associated responses from the supplicant) depend on the upper-layer authentication mechanism (protocol) that has been adopted, and that they are simply relayed by the authenticator either to the supplicant or to the authentication server. The authenticator does not understand the exchanged messages because it does not even know which authentication protocol has been employed. It simply waits until the authentication server transmits an Accept or Reject message.

## The Key Generation Process

Theoretically, following a successful authentication, a mobile device (the supplicant) should be ready to join the network and establish communication with the access point (authenticator). However, in the framework of IEEE 802.11i, there are still two open issues that must be resolved. Specifically:

- The access point has not yet been authenticated by the mobile device, and
- The temporal (session) keys, utilized for protecting the confidentiality and integrity of the data exchanged between the authenticated mobile device and the access point, have not yet been established.

Both the above issues are resolved with the help of a *master key* (MK) that is generated by both the authentication server and the mobile device during the authentication process. The details governing MK generation from both entities depend on the authentication protocol that has been employed. However, to protect the wireless link, the access point should also know the MK. Although IEEE 802.11i does not specify how the MK can be transferred from the authentication server to the access point, the WPA industry standard suggests the use of the RADIUS protocol.[5]

Having established the MK at both ends of the wireless link, a *handshaking* between the mobile device and the access point is initiated. Through this handshake, the mobile device authenticates the access point, and the temporal keys are computed. Before describing this process in more detail, let us introduce the temporal (session) keys required by the

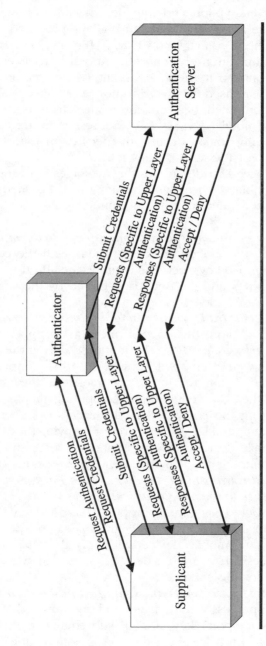

**Figure 24.7   IEEE 802.1X authentication through EAPOL messages.**

RSN and TKIP/WPA cipher methods. One key is required for data encryption (TKey-En), a second one for checking data integrity (TKey-In), and two more keys for encryption and integrity checking of the messages exchanged during the aforementioned handshaking process as well as for some additional notifications that follow (TKey-HEn and TKey-HIn, respectively). To avoid computation of identical temporal keys, every time that the mobile device connects to the access point, two big random numbers (different each time) are used for diversification. More specifically, all four temporal keys are computed as a function of the MK, the access point's MAC addresses, the mobile device's MAC address, and the two random numbers. The key computation details are specified in IEEE 802.11i and are slightly different for RSN and TKIP/WPA.

We can now proceed to describing the messages exchanged during the access point–mobile device handshaking (it is assumed that each device has already generated its random number):

- The *access point transmits its random number to the mobile device.* Upon receipt, the mobile device proceeds with the computation of all four temporal keys, because it knows the MK, the two random numbers (its own and the one just received by the access point), and the two MAC addresses.

- The *mobile device transmits its random number to the access point.* Prior to transmission, this specific message is appended with a *message integrity code (MIC).* The key used during the MIC calculation is one of the temporal keys (TKey-HIn) computed by the mobile device. Upon receipt of the message, the access point acquires all the data required for computing the four temporal keys. Following the computation, the access point employs the appropriate temporal key (TKey-HIn) for verifying the MIC received from the mobile device. If the verification is successful, the access point has confirmed that the mobile device possesses the genuine MK.

- *The access point transmits a "Ready to Start Encrypted Communication" message to the mobile device.* Similar to the previous step, a MIC is computed using the appropriate temporal key (TKey-HIn), and then transmitted together with the message. Upon receipt of the message, the mobile device will verify the MIC. If the verification is successful, the mobile device has confirmed that the access point possesses the genuine MK.

- *The mobile device transmits an "End of Handshake" message.* The handshaking process is terminated, and the mobile device is ready to utilize its temporal keys for all subsequent communications. Upon receipt of this message, the access point enables the use of its temporal keys, thus synchronizing with the mobile device.

## The TKIP Security Protocol

As described in the section titled "WLANs and Security Standards," the IEEE 802.11i standard specifies the RSN architecture, which accommodates mobile devices, offering more advanced encryption techniques and exhibiting several new hardware-supported capabilities. Considering, however, the fact that existing Wi-Fi devices cannot be upgraded to comply with RSN, an optional operational mode of RSN known as TKIP was also specified to cover this particular need. Generally speaking, RSN and TKIP follow the IEEE 802.11i security architecture presented in Figure 24.6, and thus the authentication, access control, and key generation mechanisms are implemented as described in the previous sections. The main difference lies in the WLAN layer of the security architecture and, specifically, in the encryption techniques employed for the protection of data confidentiality and integrity. RSN exploits the AES encryption algorithm, whereas, for compatibility purposes, TKIP continues to use RC4. To facilitate a comparison with the WEP protocol and thus highlight how the WEP security problems have been alleviated, the TKIP mechanisms for data encryption and data integrity checks are presented in the following text.

### Data Integrity Protection

The approach adopted by TKIP for checking the integrity of transmitted messages is based on the computation of a message authentication code, which, in the framework of the IEEE 802.11i standard, is referred to as MIC. More specifically, the bytes of a message, together with a secret key, are processed with some nonreversible algorithm and the resulting stream of bytes is the MIC.

The algorithm employed by TKIP for the MIC calculation is called Michael[3] and has been proposed by the cryptographer N. Ferguson. The choice of the specific algorithm has been based on the fact that it does not involve any processing power-intensive operations (such as multiplication), which was an essential requirement for both the mobile devices and the access points. Assuming that a mobile device needs to transmit a message to the associated access point, it calculates the MIC, engaging the temporal key TKey-In, and then appends it to the message prior to encryption. Upon receipt, the access point decrypts the message and then recalculates the MIC value, engaging its own copy of the TKey-In key. Provided that the calculated MIC matches the one transmitted by the mobile device, it can be safely assumed that the message was not modified (accidentally or intentionally) during transmission.

### Data Confidentiality Protection

Keeping in mind the WEP security problems as far as confidentiality protection is concerned (refer to the section titled "Confidentiality Protection"), the main TKIP design criteria were focused on the selection and use of the initialization vector (IV) to ensure that the RC4 encryption key is different for every packet, the key structure and length match the WEP specifications (for compatibility purposes), RC4 weak keys are avoided, and finally that replay attacks can be identified.

The IV length, specified by TKIP, is 48 bits. However, to maintain compatibility with the WEP protocol, the key generation mechanism has become much more complex, as clearly illustrated in Figure 24.8.

The choice of a 48-bit IV effectively eliminates the possibility of exhausting the entire space of different IV values, even in cases of extremely busy networks. Consequently, it can be safely assumed that for every packet, a different (not repeated) IV value is engaged. Furthermore, the inclusion of the device MAC address in the RC4 key generation process guarantees that even if two devices are communicating, sharing the same temporal key and using the same IV value, the RC4 encryption key computed by each device will be different. Finally, the RC4 weak keys are avoided by *forcing* an appropriate value to the 8-bit space provided in the RC4 encryption key structure (shaded area in Figure 24.8).

The last issue is how TKIP can identify *replay attacks*. The specified approach is again related to the IV characteristics. More specifically, the IV value always starts from zero and increments by one for each packet transmitted. Because the IV is guaranteed not to repeat for a given temporal key, any packet received with an IV value that is not greater than the IV value of the previous packet will be rejected.

## CONCLUSIONS

WLANs facilitate the deployment of new services that traditional wired LANs cannot provide. On the other hand, they introduce new security risks. It has been demonstrated that the WEP protocol, being the result of the first technical standard in the area, namely the IEEE 802.11, cannot be considered as a complete and robust security solution. However, the right combination of existing standards such as 802.1X with appropriate technologies can mitigate the problems. This has been demonstrated by the IEEE 802.11i standard that introduced a new WLAN security architecture (RSN) while maintaining, to the greatest possible degree, compatibility with existing WLAN equipment (TKIP, WPA). It is certain, however, that to achieve the goal of security and privacy in future mobile communication networks, further research and technology development are necessary.

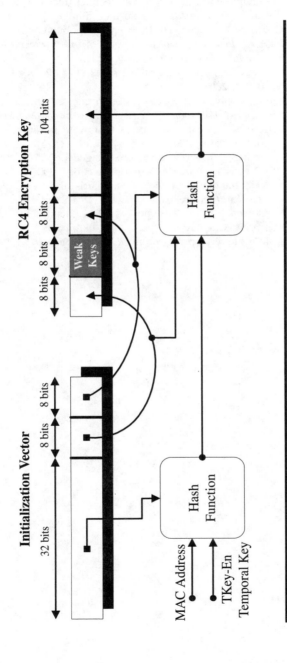

**Figure 24.8  TKIP calculation of the RC4 encryption key.**

# REFERENCES

1. 802.11 Security, http://www.microsoft.com/technet/treeview/default.asp?url=/technet/prodtechnol/winxpprp/reskit/prdc_mcc_hmwd.asp.
2. Borisov N., Goldberg I., Wagner D., Intercepting mobile communications: the insecurity of 802.11, *Proceedings of the Seventh Annual International Conference on Mobile Computing and Networking*, 2001.
3. Ferguson Michael N., An improved MIC for 802.11 WEP, Document Number IEEE 802.11-02/020r0, 2002.
4. Fluhrer S., Mantin, I., Shamir A., Weaknesses in the key scheduling algorithm of RC4, *Eighth Annual Workshop on Selected Areas in Cryptography*, 2001.
5. Hassel J., RADIUS: Securing Public Access to Private Resources, Cambridge, MA: O'Reilly and Associates, 2003.
6. IEEE 802.11 standards, http://standards.ieee.org/getieee802.
7. Karygiannis T., Owens L., Wireless Network Security: 802.11, Bluetooth and Handheld Devices, National Institute of Standards and Technologies, 2002. http://stabdards.ieee.org/getieee802.
8. Menezes A.J., Oorschot P.C., Vanstone S.A., Boca Raton, FL: *Handbook of Applied Cryptography*, CRC Press, 1996.
9. Rivest R., The RC4 Encryption Algorithm, *RSA Data Security,* 1992.

# 25

DISTRIBUTED CONTENTION IN THE MEDIUM ACCESS CONTROL OF IEEE 802.11 WLANS: A SPACE AND TIME PERSPECTIVE

*Luciano Bononi and Lorenzo Donatiello*

## INTRODUCTION

In recent years, the explosive proliferation of portable and laptop computers has led to local area network (LAN) technologies being required to support wireless connectivity for mobile hosts, i.e., the wireless LAN (WLAN) concept.[14,16,27] Mobile and wireless solutions for communication have been studied for many years to make it possible for mobile users to access information anywhere and at anytime.[27] Common and widely popular Internet services (e.g., World Wide Web and e-mail) are considered the most promising killer applications for wireless technologies, services, and infrastructures deployment by network service providers and private customers. The evolution of advanced mobile and wireless Internet services, such as real-time data, voice, and video communication services, demands the introduction of quality-of-service (QoS) support and management at every layer of the network protocol stack. The integration of Internet communication with innovative and challenging last-mile wireless connectivity will be required to support full services and protocol integration between the two worlds. On the other hand, the new wireless scenarios introduce challenging new problems and system constraints requiring new protocols and solutions to be investigated and new standards to be proposed, possibly by seamlessly integrating and supporting legacy technology, solutions, and services.

Given these developments, many research activities and projects are currently dedicated to identifying, proposing, evaluating, and tuning new solutions, and to redesigning protocols that have become outdated in the wired networking community. One of the most challenging problems requiring solutions due to the advent of WLAN technology is the Medium Access Control (MAC) layer design. Briefly stated, the MAC layer is responsible for the host's ability to share in a mutually exclusive way (i.e., one at a time) a broadcast communication channel to support multiple transmissions. Most of the MAC protocols and technologies for wired LANs (such as Ethernet[32] and Token Ring) cannot be implemented in wireless scenarios because the system assumptions that drove their design and tuning have changed to a new set of characteristics and constraints for WLAN systems and mobile devices. As an example, this chapter will illustrate the new space and time perspective of the contention problem for the class of distributed random-access MAC protocols. This class of MAC protocol is fundamental because it is the basic access scheme in today's Standard IEEE 802.11 MAC, the most widely deployed WLAN technology, and also the basis for new protocol enhancements aimed at obtaining higher performance (such as IEEE 802.11b and IEEE 802.11a[25]) and QoS support (such as IEEE 802.11e[26]). Many solutions and guidelines that have been inherited from the wired scenarios, and the preliminary wireless solutions proposed to overcome the distributed contention problem in both the time and space domains of wireless channels, will be illustrated in this chapter. Additionally, the resource-oriented design of the distributed contention-based MAC protocols will be emphasized. This is required, given the resource bottlenecks of wireless systems (for example, limited channel bandwidth and battery energy), and given the challenging resource requirements of user applications that would make no concessions to the wireless Internet.

A newly identified characteristic that has revealed its potential in the design of wireless protocols in recent years is their adaptive behavior with respect to additional information gathered as system feedback. The design of many wireless protocols has been focused on identifying, implementing, and tuning adaptive behavior, with the goal of optimizing trade-offs in system performance and resource utilization. Already, attention is shifting to consideration of these trade-offs in view of the QoS achievable for supporting user applications.

## MAC PROTOCOLS FOR WLANS: REINVENTING THE WHEEL?

Wireless medium problems and resource restrictions in wireless systems made the MAC goal difficult to fulfill in WLANs. The wireless MAC protocol design plays an important role: MAC protocols define the way resources are used and shared, and affect system performance. MAC protocols determine which services the system can concurrently support, and the QoS guarantees for user applications. Considerable investigation is required into the role of protocols and, specifically, of the class of distributed algorithms for network systems management. The new constraints of the wireless scenarios have caused many problems in realizing optimal tuning of the protocols and algorithms derived from those adopted in wired networks. In this section, we will sketch some of the new assumptions and characteristics that have been introduced by the WLAN scenarios, requiring new designs, tuning, and implementation of MAC protocols. The need for this underlines why the most successful solutions for the wired scenarios have become unpractical for wireless scenarios.

### Broadcast Nature, Space Propagation, Transmission Power, and Receiver Sensitivity

From a physical viewpoint, wireless signals are electromagnetic waves that extend outward from their sources into open space (i.e., light from a lamp). This physical characteristic is usually described as the "broadcast nature" of wireless transmissions. Signals cannot be restricted to a wire but diffuse around the transmitter. This new assumption is fundamental in the channel definition and in the MAC design, which is devoted to manage the wireless channel access. In any case, signal propagation is limited within a variable local range around the transmitter and can be described in a simple way with propagation models.[7,34] Propagation models describe the combination of effects of the medium characteristics, the environment obstacles, and the transmission power of the signal source (i.e., the wireless transmitter). A propagation law defines the relationship between transmission power (energy) and the space that transmission signals occupy in the wireless scenario. In the wireless medium, the transmission power of wireless signals ($Ptx$) is subject to a natural decay; the more the distance $d$ from the transmitter, the lower the residual power ($Prx$) for the signal being detected by a receiver (see Figure 25.1). If the residual signal power perceived by the receiving network interface is above the reception power threshold $Rth$, then data communication is possible between sender and receiver. Otherwise, to allow a communication (i.e., link establishment) between sender and receiver, it would be

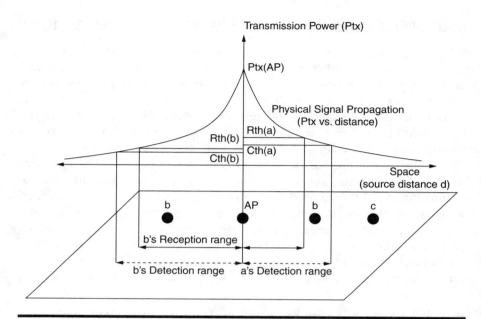

**Figure 25.1 Space propagation of wireless signals and device sensitivity.**

necessary to increase the transmission power of the sender. In general, network interfaces can be managed to transmit signals with variable transmission power *Ptx*. In the receiving phase, network interfaces can be summarized in terms of their own receiving threshold *Rth* and carrier-sensing threshold *Cth*, which characterize the receiver sensitivity (see Figure 25.1). For every transmitter–receiver pair, if the receiving power perceived for the ongoing transmission is greater than *Rth*, then it would be sufficient for reception (e.g., host *b* in Figure 25.1); if it is greater than *Cth*, it would be sufficient for carrier sensing only (i.e., to detect the channel as busy; see host *a* in Figure 25.1). Otherwise, it would be just interference (see host *c*).

Reception and carrier-sensing capabilities of wireless devices can be exploited by MAC protocols to locally gather system feedback events useful in managing the transmissions. In the detection area (see Figure 25.1) of a wireless transmission-device (i.e., the area where $Cth \leq Prx$), a receiver can sense that the wireless medium is busy. This makes possible another relevant assumption for the design of the MAC. A carrier-sensing (CS) mechanism is succinctly described as the physical capability of a network interface (NI), within the detection area of a transmitter, to detect and pass to the MAC layer the event indication: "This channel is busy due to ongoing transmissions." The CS mechanism, the heterogeneous sensitivity of network interfaces (such as hosts *a* and *b*), and the wireless signal propagation may generate suboptimal behavior and space contention problems such as irregular or spotted coverage areas, exposed terminals, hidden terminals, and capture effects (described in the following subsections).

## Half-Duplex Channels and Directional Links

To complicate the adoption of CS, a single wireless communication device, i.e., a wireless NI, can transmit and receive separately but cannot transmit and receive at the same time on the same wireless channel. In other words, a single network interface cannot "listen" to incoming communications while it is transmitting on the same channel. The dramatic effect of the half-duplex characteristic of wireless channels is that early Collision Detection (CD) techniques adopted in wired LAN MAC protocols (e.g., IEEE 802.3 and Ethernet) cannot be easily implemented on a single wireless channel.[30,36] A bidirectional full-duplex channel can be obtained by adopting duplexing techniques such as time division duplex (TDD), or frequency division duplex (FDD).[34] In most WLANs, (logical) bidirectional links are commonly defined as TDD channels; this usually requires strict time-synchronization in the system. All data transmissions and receptions have to be in the same channel as there are no "bridge" nodes (with the possible exception of base stations) to translate the transmissions from one physical channel to another.

On the other hand, to obtain a bidirectional link, communication must be possible in both directions — from sender to receiver and vice versa. Unfortunately, this assumption may be critical in the wireless scenario; for example, when two hosts have different transmission powers or different receiver sensitivities. The bidirectional nature of wireless links is another assumption that plays a fundamental role in the design of the MAC (and logical link control) layer — for example, if the MAC protocol requires immediate acknowledgments (ACKs) to ascertain a successful transmission. The receiver only would know the transmission result; hence, immediate acknowledgments are commonly used to give the sender-side feedback on the transmission results.

## Signal Vulnerability to Collisions: Space–Time Contention and Collision Domains

A *signal collision* is a receiver-centric event and occurs on a receiver when two or more portions of concurrent transmissions overlap each other, both in time and in the receiver's space (see Figure 25.3A). Given the limited signal propagation, if two time-concurrent transmissions are sufficiently separated by enough space, then the intended receivers will not experience any collision. In presence of a collision, if the received signal power of one of the colliding signals is much greater than the others, it may happen that the receiver is able to capture a transmission anyway. In this case, a *capture effect* of signals can be obtained in spite of the collision. A tagged receiver's *collision domain* (on a single shared communication channel) is a *space domain*, defined as the union of the intersections of the coverage areas of any pair of neighbor nodes (i.e., potential transmitters) containing the tagged receiver. In other words, a collision domain is the space domain where at least two nodes risk generating a collision of their respective time-concurrent signals on a tagged receiver. The larger the transmission power $Ptx$, the larger the collision domains, and hence the larger the number of potential colliding transmitters and the space vulnerability of any ongoing transmission.

If there is no capture effect, any collision of concurrent signals transmitted on the same collision domain may cause destructive interference on the receiver. The main goal of a MAC protocol policy is to avoid destructive collisions, and to adapt the space density of transmissions, i.e., the space contention, in an adaptive way. The space contention (or space vulnerability) can be thought of as the risk of collisions due to the density of transmitters defining the collision domain of the receiver. To reduce this risk, an ideal MAC protocol should allow only one transmission at a time in the receiver's collision domain. On the other hand, the space density of concurrent transmissions over the channel should be maximized to allow the maximum channel space reuse. From the resource optimization and space contention viewpoint, the ideal MAC protocol would allow only one transmission at a time in a receiver's collision domain and the maximum number of concurrent transmissions in the system area provided that all the receivers belong to nonoverlapping collision domains. This condition can be satisfied by reducing the collision domains through an adaptive reduction of transmission power. Unfortunately, transmitters do not have a suitable knowledge of collision domains. Moreover, factors such as node mobility, variable wireless propagation properties and obstacles, and heterogeneous device sensitivity complicate the dynamic management of variable transmission power. In the following sections, some practical solutions for controlling MAC space-contention adaptation will be presented.

By focusing on a single-collision domain (for example, by assuming that each node can receive from any other), one complementary solution, at the MAC layer, is to separate the transmissions in the time domain by reducing time contention instead of space contention. Time contention requires control of the access densities in time and can be thought of as a random time-spreading of the transmissions. If the accesses are performed with low time density, the risk that two transmitters would collide (i.e., the vulnerability time of transmissions) would be low. The resource optimization viewpoint in the time domain would require the transmission density to be maximized to allow the maximum degree of channel utilization and channel throughput, and the minimum average access delay for transmissions.

In summary, three important factors determine the MAC protocol efficiency in managing distributed accesses to the shared channel: the time contention (related to spreading the access time and reducing the transmission vulnerability), space contention (aimed at realizing multiple, concurrent, and nonoverlapping collision domains as mutually exclusive areas for transmissions), and transmission power (i.e., transmission energy which, together with receiver sensitivity, determines space and propagation factors). Any centralized or distributed MAC protocol should manage the time schedule of transmissions, depending on the variable traffic load requirements, to avoid collisions on receivers, to exploit the maximum degree of spatial reuse and time utilization of the limited channel resource, and to control the energy consumed for transmission and reception phases. Moreover, the effect of user mobility must be considered, resulting in highly dynamic and unpredictable collision domains and contention levels.

## IEEE 802.11 WLANS INFRASTRUCTURE

Major factors in the extensive deployment of WLANs have been the availability of appropriate networking standards, good performance, and low cost. The IEEE 802.11 Standard (Wi-Fi) technology has been widely adopted as the MAC layer in WLANs. This led research in the field of WLANs to be mainly focused on IEEE 802.11-based MAC solutions.

The IEEE 802.11–WLAN is a single-hop infrastructure based on access points (APs) serving and managing local (mobile) nodes in a basic service set (BSS). If nodes leave the WLAN area, they should register with a new AP, if one exists. The limited range of communication around APs could limit the mobility of users. On the other hand, the integration of WLANs and Mobile Ad hoc Networks (MANETs) could help in extending the range around APs, enabling support for wide hot-spot connectivity.

A distributed MAC for IEEE 802.11–WLANs has been identified as the suitable basis for the implementation of the IEEE 802.11–MAC, both for WLANs and MANETs.[25] The centralized support for communication, called Point Coordination Function (PCF), was designed on top of the distributed access scheme, known as Distributed Coordination Function (DCF). The centralized PCF is an optional (and often not implemented) management service, in charge of the AP, and can support QoS, priority schemes, and asymmetric channel scheduling among coordinated nodes, but is influenced by system factors such as load changes, weak synchronization and clock drifts, interference among neighbor APs, and host mobility.

The DCF distributed MAC protocol enables less critical and easier implementation and has been defined to work also under peer-to-peer management, with no need for coordinating nodes. It is expected to work (in a best effort, i.e., suboptimal way) in spite of any unpredictable host behavior, interference, and weak synchronization and faulty node distribution in wireless scenarios. For these reasons, WLANs and MANETs are commonly based on the DCF distributed MAC protocol. On the other hand, distributed random-access MAC protocols have been demonstrated to suffer from scalability, efficiency, and QoS problems under high loads.

The idea behind distributed, contention-based MAC protocols is to define distributed protocols as event-based algorithms, randomly spreading the accesses of nodes in an effort to attain system stability and acceptable resource utilization and performance from the aggregate behavior of nodes. The distributed contention-based MAC protocols are driven by the limited feedback information and events perceived by the network interface of every node. In the following sections, we will cover the evolution of proposals based on different assumptions about feedback information that could be exploited by wireless nodes.

In the remaining sections of this chapter, we will also provide the reader with an overview of solutions and milestones that have been proposed for the class of the distributed contention-based MAC protocols for WLANs. The overview will illustrate the collision-avoidance and contention-control issues of the MAC design leading to the IEEE 802.11 MAC protocol for WLANs and MANETs. This design can be considered the wireless counterpart of Ethernet technology, i.e., "Ethernet on the air." We will also analyze solutions for increasing both the MAC protocol's efficiency and its ability to react to time and space congestion. In addition, we will investigate the protocol's robustness to time and space vulnerabilities (hidden/exposed terminals and contention control).

## EVOLUTION OF DISTRIBUTED CONTENTION-BASED MAC: FROM ALOHA TO IEEE 802.11

This section describes the evolution of proposals in the field of distributed contention-based multiple-access MAC protocols for the wireless environment. The list of proposals is not exhaustive due to reasons of space, but it is an illustration of the milestones in the evolution of the protocols leading to the definition of the IEEE 802.11 MAC. Specifically, given the focus of the chapter, this perspective is oriented toward contention control, i.e., the reduction of time and space vulnerability of transmissions in the distributed contention-based MAC protocol design.

## The Time Vulnerability of Wireless Transmissions

The first MAC protocol defined for distributed, multiple-access wireless transmission of data frames was the ALOHA protocol.[1] The MAC protocol policy was straightforward: every node transmits any data in the buffer queues immediately, whenever it is ready. During the transmission, CD is not possible, and the transmission is attempted up to the end of the data frame over half-duplex channels. After the transmission, an acknowledgment (ACK) is provided on a separate channel to ascertain the success of the transmission. The transmitter waits for the ACK up to a maximum amount of time (ACK time-out). In case of unsuccessful transmission (i.e., missing acknowledgment), a new transmission attempt is required. The simple bidirectional (Data + ACK) structure of the frame transmission implements the prototype definition of a reliable logical link layer. To avoid synchronization of retransmissions among multiple contending nodes, resulting in a deterministic sequence of collisions, every retransmission is scheduled after a pseudorandom waiting period. In this protocol, the carrier-sensing (CS) concept was not considered, i.e., every node was not assumed to "listen" to the channel before transmitting. The time vulnerability of a frame being transmitted is defined as the maximum time window during which another frame transmission may be originating a collision on the receiver's collision domain. In Reference 29, it was demonstrated that the vulnerability time for each frame (by assuming a constant frame size) in the ALOHA access scheme is twice the average frame size expressed in time units. By assuming independent Poisson-distributed arrivals of frame transmissions (with constant size) and collisions as the only source of transmission errors, the expected channel utilization was upper bounded by only 18 percent of the channel capacity (i.e., the maximum channel throughput). In other words, by increasing the load offered by independent nodes (load G defined as the frame size multiplied by the Poisson interarrival rate), the probability of a collision increases and the MAC policy would not be able to support average channel utilization greater than 18 percent (see Figure 25.2). This is a theoretical result that well describes the scalability limits of this MAC policy from the contention-control viewpoint.

The Slotted ALOHA protocol was introduced to reduce the time vulnerability of each frame. The time is divided into frame slots (with fixed frame size) with each slot capable of containing a frame transmission. This protocol is similar to ALOHA, but a synchronization of nodes is introduced such that every transmission starts only at the beginning of a frame slot. In this way, the advantage is that the time vulnerability is limited by the single frame slot in which the transmission is performed. Analytical models demonstrate that the expected average channel utilization was upper-bounded by 36 percent of the channel capacity, i.e., twice

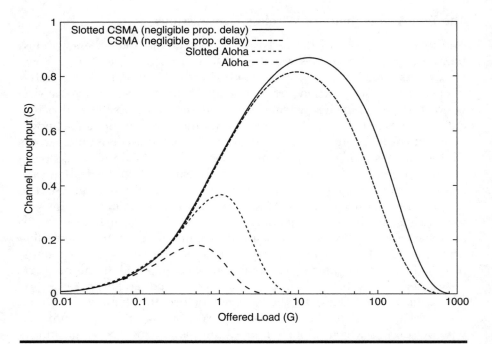

**Figure 25.2 A comparison of analytical channel throughput of distributed MAC protocols.**

the ALOHA value (see Figure 25.2). This theoretical result shows that the scalability of this MAC policy is better than ALOHA, but is still far from optimal.

Previous studies gave motivations for designing and introducing the carrier sense multiple access (CSMA) concept.[29] In CSMA MAC, every node "listens" to the channel before transmitting, and if the channel is found to be busy, it defers the transmission to a later time (i.e., nonpersistent carrier sensing); otherwise, it transmits immediately. The advantage of carrier sensing is that ongoing transmissions can be detected, and the next candidate transmitter would nicely avoid colliding with them. Unfortunately, if the propagation delays are significant with respect to the frame size, the performance of CSMA would be negatively affected. The propagation delay is due to the orthogonal time and space domains, which temporarily hide ongoing transmission from other potential transmitters and may cause a collision on the intended receivers. In CSMA, the vulnerability of the frame was demonstrated to be reduced to only twice the maximum propagation delay of wireless signals ($2*\tau$) of the most distant transmitters. It is usually assumed that the propagation delay $\tau$ is orders of magnitude lower than the size of typical frames.[29] This is specifically more probable for common WLAN scenarios. The CSMA throughput was analytically modeled and was defined as high as 80 percent of the channel capacity (see Figure 25.2).

By applying the slot-based concept to CSMA, the Slotted CSMA protocol was proposed as a further enhancement of CSMA.[29] A minislot is defined as the upper bound of the propagation delay $\tau$ between different transmitters in the system. In the Slotted CSMA protocol, every node with a frame to transmit "listens" to the channel at the beginning of a minislot, and if the channel was found to be busy, it would defer the transmission to a later time (i.e., nonpersistent carrier sensing); otherwise, it would transmit immediately at the beginning of the next minislot. The advantage of this policy is that the beginning of possible transmissions are synchronized up to a minislot time. Hence, transmissions are guaranteed to be detected before the beginning of the next minislot by all the transmitters within each other's detection range. In such a way, the time vulnerability of the frame was demonstrated to be reduced to the minislot time ($\tau$). Again, if the propagation delays are significant with respect to the frame size, the performance of Slotted CSMA would degrade. Slotted CSMA throughput was modeled and gave better values than the CSMA channel capacity for the same scenario assumptions (see Figure 25.2). All previous MAC policies were based on the assumption that an acknowledgment-based (ACK) feedback indication of successful transmission would be received from the intended receiver. The ACK is usually provided as a short reply frame, and can be exploited to realize the link layer concept of a reliable link between transmitter and receiver. In some systems, and in early wireless MAC proposals, ACKs were sent on separate control channels. Nowadays, the ACK transmission is usually piggybacked by the data receiver immediately after the end of the data reception on the single, shared, time division, half-duplex communication channel. In this way, at the MAC/LLC layer, the receiver could immediately exploit the contention won by the transmitter for sending the ACK frame (i.e., a new contention is not required as the shared channel has been successfully captured by the sender).

Different policies and definitions of the MAC and LLC layers can be defined by assuming the explicit indication of the cause of unsuccessful transmissions (e.g., if a frame is received with errors, if it is subject to collision, etc.). In any case, the LLC layer is beyond the scope of this chapter. The need for understanding the reasons behind unsuccessful transmissions has been considered in the literature, and analysis has shown that the more information feedback is provided regarding the cause of a contention failure (i.e., collision, number of colliding nodes, bit error due to interference, etc.), the greater the performance and adaptive behavior that can be obtained by the MAC protocol.[35] Unfortunately, in most scenarios, the only feedback information provided after a transmission attempt is the existence of ACK frames within a time-out period.

In early wired LANs, researchers considered solutions such as Ethernet based on CSMA techniques and transmitters with CD capabilities,[32] i.e., the nodes were listening to the channel while they were transmitting. As mentioned before, in wireless systems, CSMA and slotting techniques can be exploited to reduce the vulnerability time of frames being transmitted. In any case, when the frame transmission starts, there is no way to detect beforehand if a collision is occurring at the receiver node. With this assumption, CSMA/CD schemes such as Ethernet and IEEE 802.3 cannot be exploited in wireless MAC protocols. The implementation of collision detection in wireless scenarios has been investigated in some pioneering research works, e.g., see Reference 30, Reference 36, and Reference 37. Given the characteristics of wireless systems, the only practical way to obtain something equivalent to CD is the adoption of separate signaling channels and multiple NIs. This would require twice the channel bandwidth, power, and number of network interfaces than other mechanisms. We will see in the following text how researchers defined new MAC policies for wireless systems that can be considered approximately equivalent to the collision detection schemes, from the channel reservation and channel utilization viewpoint.

## Collision Avoidance and Space Vulnerability

The ALOHA and CSMA MAC protocols illustrated in the previous subsection were conceived to reduce the vulnerability period of contention-based frame transmissions. The Collision Avoidance (CA) techniques have been designed to create in advance the same conditions provided by CD, by using a single shared channel and a single NI. If contention-based transmission evolves into a collision with the intended receiver, then the energy and channel occupancy wasted due to the colliding transmissions should be as limited as possible, similar to the CD mechanism.

Collision avoidance is obtained on top of a time contention, and can be thought of as a preliminary *spatial reservation* of the collision domains around both the sender and the receiver, to preserve both the data and ACK transmissions. The spatial reservation can be performed by solving the space contention for the channel among multiple transmitters in the neighborhood of both the sender and the receiver. Before illustrating the proposals for collision avoidance solutions, the most representative problems related to space vulnerability at the MAC layer are presented.

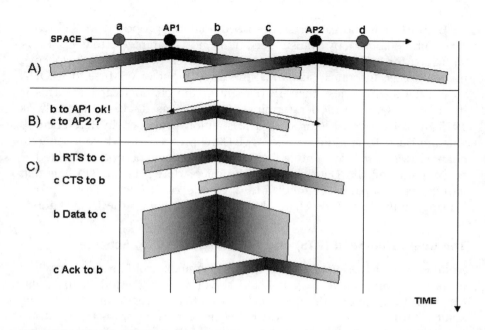

**Figure 25.3** (A) A collision by two hidden terminals, AP1 and AP2; (B) an exposed terminal *c* cannot transmit due to *b*'s transmission to AP1; (C) the RTS/CTS mechanism to seize the collision domain of *b* and *c* before the long Data + ACK transmission.

## The Hidden and Exposed Terminals

The heterogeneous sensitivity and mobility of hosts, the adoption of shared channels for transmissions, the space propagation properties of wireless signals and a carrier sensing-based policy for MAC protocol implementation may cause some nodes to experience the *hidden-terminal* and the *exposed-terminal* problems. The collision due to the hidden-terminal problem may happen to nodes (e.g., *b* and *c*) receiving the concurrent transmissions of at least two transmitting hosts (AP1 and AP2), which cannot sense and detect each other (see Figure 25.3A). In this scenario, if the two hidden transmitters belong to the receiver's collision domain, any space and time overlapping of concurrent transmissions may result in a destructive collision. Exposed terminals are terminals (e.g., host *c* in Figure 25.3B) whose transmission is not allowed (e.g., by the carrier-sensing mechanism) due to exposure to irrelevant transmissions (e.g., from *b* to AP1). The exposed-terminal problem is one reason for the spatial channel reuse limit. A discussion of details for hidden and exposed terminals and capture effect can be found in Reference 10 and Reference 33.

Because it is the carrier sense-based solution proposed in Reference 29 is implemented by transmitters, it is not guaranteed that CSMA is sufficient for transmitters to detect each other before their respective transmissions. In critical scenarios, the throughput of CSMA and ALOHA would fall again to less than 18 percent of the channel capacity. This is the reason why the hidden-terminal problem was discussed as early as 1975 by Tobagi and Kleinrock.[39] The solution proposed in Reference 39 was the Busy Tone Multiple Access (BTMA) protocol, in which a separate channel was used to send a busy tone whenever a node detected a transmission on the Data channel. The major drawback of BTMA is that it requires a separate channel and a couple of NIs to perform carrier sensing on the busy-tone channel while transmitting on the Data channel.

## The Request-to-Send (RTS) and Clear-to-Send (CTS) Scheme

Another way to oppose to the hidden-terminal problem among wired terminals connected to a centralized server was suggested in the Split-Channel Reservation Multiple Access (SRMA) protocol in 1976.[40] The solution was based on the handshake of short request-to-send/clear-to-send (RTS/CTS) messages between senders (i.e., terminals) and receivers (i.e., the server) over three separate channels (RTS, CTS, and Data channels). The RTS/CTS scheme was originally designed to manage efficient transmission scheduling between senders and receivers, without causing interference among hidden terminals on the server.[40] Some time later, the RTS/CTS scheme was interpreted and adopted in quite a different way with respect to SRMA, and it was performed on a single transmission channel as originally proposed for wired LANs in the Carrier Sense Multiple Access with Collision Avoidance (CSMA/CA) scheme in 1983.[13] The idea was to send a short request-to-send (RTS) frame (by applying the CSMA MAC channel access scheme) to the intended receiver before a data transmission. If the receiver correctly receives the RTS, then it immediately responds with a clear-to-send (CTS) back to the transmitter. In this way the double successful transmission of both RTS and CTS should reserve the channel space (i.e., the collision domain) and should alleviate the space vulnerability of sender and receiver to hidden transmitters.

The principle of this solution, modified in an opportune way, was, over the years, successively adopted in many protocols and standards. Collision avoidance based on RTS/CTS is an optional function included in current IEEE 802.11 DCF implementation.

The first introduction of the RTS/CTS in wireless systems was in the Multiple Access Collision Avoidance (MACA) protocol.[28] MACA is a random-access MAC protocol that tried to deal with the hidden-terminal problem by taking a step back with respect to the CSMA approach. In MACA, it was observed that the relevant contention is at the receiver side, not at the sender's. The carrier-sensing approach at the transmitter is not fully appropriate for the purpose of collision avoidance. Carrier sensing at the sender would provide information about potential collisions at the sender but not at the receiver. Because the carrier-sensing mechanism implemented on the transmitter side cannot ensure protection against hidden terminals and leads to exposed terminals, the radical proposal is to ignore carrier sensing before transmissions. The idea is to rely on the contention of two really short frames, RTS and CTS, to reserve the collision domain of both sender and receiver before sending the Data and ACK frames (see Figure 25.3C). Both RTS and CTS are 30 bytes long and contain information about the expected duration of the channel occupancy for the subsequent data transmission. The main critical assumption in this definition is the perfect symmetry of links, i.e., (A receives B) $\Leftrightarrow$ (B receives A). The transmitting node $b$ sends the RTS to the receiver $c$ as a broadcast message (see Figure 25.3C). If the receiver $c$ receives the RTS, and it is not deferring due to a previous reservation, it immediately responds with the CTS, which also contains information about the expected duration of the channel occupancy. If a node different from the sender (e.g., AP2) receives the CTS (the node is in the collision domain of the receiver), then it would be informed about the transmission duration, and it would enter the deferring phase for that interval. If the sender $b$ receives the CTS, it knows that the receiver $c$ is within the transmission range, and the channel space should have been reserved successfully. This indicates that the data transmission will likely be successful.

All nodes receiving the RTS that differs from the intended receiver (e.g., AP1) could hear the CTS reply. If they are not receiving the CTS, they could assume they are not in the collision domain of the receiver (i.e., the receiver is not exposed); hence, they could start their own transmissions, increasing in this way the spatial reuse of the channel (Note that the final ACK after the data transmission in Figure 25.3C is not expected by the sender with MACA.) This could be considered a first solution for terminals "exposed" to the sender's RTS. In this scheme, the confirmation (CTS) of channel capture success is based on a lower risk of wasting channel bandwidth than the risk of collisions of two or more long Data frames. If CTS is not received within a time-out period, the sender assumes that a collision occurred and reschedules a new RTS transmission by adopting a time-contention control scheme (Backoff Protocol; see the following subsection) to select the slot time for the new RTS transmission. RTS and CTS are 30 bytes long, and their time duration defines the slot time for quantizing transmissions.

Other solutions for CA are based on the reversing of the RTS/CTS handshake. In MACA By Invitation (MACA-BI)[38] and in many receiver-initiated multiple access (RIMA) versions,[42] the RTS/CTS scheme is initiated by candidate receivers sending ready-to-receive (RTR) frames to the neighbors. In this way the CA scheme reduces the overhead required to reserve collision domains.

## Adaptive Contention Control and Time Vulnerability: Backoff Protocols

The Backoff Protocol in MACA is a *time-contention control protocol* that is frequently associated with contention-based, CA, slotted-access schemes. Whenever a collision is detected by the sender (e.g., missing ACK or missing CTS), this event can be considered an indication of a high contention level in the channel. A time-spreading randomized retransmission is required to reduce contention and avoid new collisions due to the choice of the same slot. The backoff scheme realizes adaptive contention reduction based on the experience of collisions for a frame transmission. For every new frame transmission, the backoff scheme is restarted. The first transmission attempt is performed in one of the next slots selected with pseudorandom uniform distribution in the interval (contention window), $CW\_Size = [0, CW\_min-1]$, where CW_min is an integer. $CW\_Size$ is increased after each collision, up to a maximum value, $CW\_MAX$, and reduced to the minimum, $CW\_min$, after a successful transmission. In the Backoff Protocol adopted in MACA, $CW\_Size$ is doubled after every collision (i.e., a Binary Exponential Backoff [BEB] protocol), $CW\_min = 2$ and $CW\_MAX = 64$.

MACA for Wireless (MACAW)[2] is a modified version of MACA, where the new wireless scenario assumptions still play an important role. Note that MAC protocols should deliver high network utilization together with fair access to every node (i.e., there is no "one node takes all" solution). In Reference 2, the unfairness problem of the BEB protocol was first described: the local collisions from one sender could result in high values of $CW\_Size$, while other senders could keep winning contention within $CW\_min$ slots. The suggested solution to this problem was to insert $CW\_Size$ information in the frame header. Every node receiving the frame would copy the $CW\_Size$ value locally, thereby obtaining a more fair access distribution. A multiplicative increase and linear decrease (MILD) algorithm was recommended to avoid wild fluctuations and so was applied to $CW\_Size$. Also, the concept of "stream-oriented" fairness instead of station-oriented fairness, later considered in IEEE 802.11e, was taken into account in Reference 2. The idea was to adopt one backoff queue per stream with local scheduling and the resolution of virtual frame collisions within the local station. In this way, the density of accesses on the channel would not be the same for all the stations but a function of virtual contention among flows inside each station. Recently, this idea has been reconsidered in the IEEE 802.11e design, leading to a distributed implementation of differentiated accesses (i.e., transmission opportunities) for flows with different priority classes.

New special frames were defined in MACAW as proposed solutions for synchronization problems and to make it possible for the receiver to contend for bandwidth, even in the presence of congestion.[2] MACAW introduced for the first time the assumption that channel contention in wireless scenarios is location dependent, and some kind of collective enterprise should be adopted to allocate media accesses fairly. Finally, the MAC protocol should propagate synchronization information about contention periods to allow every device to contend in an effective way, e.g., by exploiting contention initiated on the receiver side (RRTS).

In MACAW, following the ideas of Tobagi and Kleinrock,[41] and Appletalk[37] (later reconsidered in the early IEEE 802.11 working groups), immediate acknowledgment is introduced after the RTS-CTS-Data exchange of information at the MAC-logical link control sublayer (see Figure 25.3C). In this way, if RTS-CTS-Data-ACK fails, immediate retransmission at the link layer can be performed, if for some reason the frame was not correctly received. This condition is assumed by the sender if the ACK is not received, even if the CTS was received. This improves many network and user performance indices with respect to transport layer retransmission management, due to the characteristics of the wireless scenario (mainly the high risk of bit error and interference). Unfortunately, the immediate ACK from the receiver to complete the transmission sequence makes the sender acting as a receiver in the RTS-CTS-Data-ACK transmission scheme. The solution proposed in MACA for alleviating exposed terminals is now a risky drawback of MACAW because concurrent transmitters could interfere with the reception of ACKs. This limits the spatial reuse of the channel that was obtained by the RTS/CTS policy in MACA. MACA and MACAW are not based on the carrier-sensing activity at the transmitter before the transmission of the RTS. Also, at least a double propagation-delay time of idle-channel space should be required between the channel becoming idle and the RTS transmission, to allow for the full reception of ACKs.[2]

Floor Acquisition Multiple Access (FAMA)[17,18] is a refinement of MACA and MACAW with the introduction of (1) carrier sensing on both senders and receivers, before and after every transmission, to acquire additional information on the channel capture, (2) nonpersistence in the CSMA access scheme (if the channel is found to be busy, a random wait is performed before a new carrier sensing), (3) lower bound of the size for RTSs and CTSs based on worst-case assumptions regarding propagation delays and processing time, and (4) RTS size shorter than CTS (CTS dominance) to avoid hidden collisions among RTS and CTS. It is worth noting that, from MACAW onward, the frame transmission is considered complete when the RTS-CTS-Data-ACK is completed. The need for ACK reception on the sender to complete the handshake implies that both receiver and sender must be protected against hidden terminals. (As mentioned before, the main drawback of hidden terminals is the collision that may happen on a terminal acting as a receiver.)

Sender and receiver aim to reserve the "floor" around them, i.e., their collision domains, to protect the Data reception on the receiver and the ACK reception on the sender against their respective hidden terminals. This conservative approach may reduce long collisions and link layer transmission delays, and hence result in a better utilization of scarce resources such as channel bandwidth and battery energy. In Reference 17, sufficient conditions to lead RTS/CTS exchange a floor acquisition strategy are discussed (with and without carrier sensing).

## Analysis of CA Schemes

The RTS/CTS mechanism has many interesting features and a couple of drawbacks. The adoption of RTS/CTS guarantees in most cases that the transmission will be effective because a successful RTS/CTS handshake ensures that (1) the sender successfully captures the channel in its local range of connectivity, (2) the receiver is active, (3) the receiver reserves the channel in its local range of connectivity (not necessarily the same area of the sender) and it is ready and able to receive data from the sender, and (4) the RTS/CTS exchange would allow the sender and receiver to tune their transmission power in an adaptive way. The latter point would permit saving energy, reducing interference, limiting spatial contention, and enhancing channel reuse by reducing the collision domains. Recent studies have shown that the hidden terminal problem is not as frequent as the amount of research on this topic might make people think.[45] On the other hand, as a conservative approach, the RTS/CTS solution is the basis for many research proposals. Ongoing activities are based on the adoption of directional antennas to implement directional CA schemes. The idea is to adopt directional antenna beams to reserve the channel over small area sectors between sender and receiver. In this way, less energy can be used and more spatial reuse of the channel can be obtained. Directional MAC protocols and directional CA schemes are ongoing research activities.

The first drawback of RTS/CTS is that in ideal conditions (i.e., when contention and interference are not major factors), the additional transmissions of RTS and CTS frames would require more bandwidth and energy than the strictly sufficient amount. One possible solution to this drawback, adopted in IEEE 802.11 networks, is to set a *RTS/CTS_threshold* defining the lower size of frames that require the adoption of RTS/CTS exchange. With this scheme, the RTS/CTS goal is twofold: (1) a channel reservation is performed to oppose to hidden terminals, and (2) long collisions can be avoided.

The second drawback is illustrated by a set of worst-case scenarios in which the adoption of RTS/CTS would not guarantee successful transmission due to collisions among RTSs and CTSs, and due to the characteristics of interference and propagation of wireless signals.[2,18,45] The proposed solution to enhance the RTS/CTS scheme was the Conservative CTS-Reply (CCR) scheme, a quite simple modification of the standard RTS/CTS solution. A conservative *RTS_threshold* power level is defined that should be reached by the RTS signal on the receiver side to allow the receiver to send the CTS back.[45] Under this assumption, data exchange is activated only if the transmitter is received with high power, and a capture effect is probable, despite interference.

Although the CA scheme reduces the risk of a collision caused by space contention (e.g., hidden terminals), the contention-control schemes (which may be considered as secondary components of collision avoidance) are defined to adaptively reduce the risk of a new collision following a previous one.

In the following text, we will consider only collisions caused by the selection of the same transmission slot on behalf of more than one transmitter in the collision domain of the receiver. We assume that collision avoidance (RTS-CTS) is performed in the background. Collisions become more probable if the number of users waiting for transmission in a given collision domain is high, i.e., if the channel contention is high.

The *collision resolution protocols* can be defined as an alternative to the contention-control protocols and have not been considered as a practical choice in CSMA/CA access schemes. A good description and comparison of collision avoidance and collision resolution schemes such as ICRMA, RAMA, TRAMA, DQRUMA, DQRAP, and CARMA can be found in Reference 20.

## IEEE 802.11 DISTRIBUTED FOUNDATION WIRELESS MAC

IEEE Standard 802.11-1997 and 1999, as well as successive releases, specify a single MAC sublayer and three physical layer specifications.[25] In this section, we will provide an overview of distributed contention-control management in IEEE 802.11 networks (specifically DCF) and some proposed enhancements.

In the IEEE 802.11 systems, the Distributed Foundation Wireless Medium Access Control protocol (DFWMAC) includes the definition of two access schemes coexisting in a time-interleaved superframe structure.[25]

DCF is the basic access scheme, and it is a distributed, contention-based, random-access MAC protocol for asynchronous accesses to the channel. On top of DCF, an optional PCF is defined as an access scheme to support infrastructure-based systems with a central coordinator (i.e., AP) for centralized, contention-free, polling-based accesses. Stations can operate in both configurations, based on the active coordination function. In the following text, for reasons of space, we have omitted all details that can be found in Reference 25. How PCF and DCF coexist and may interleave in controlling the channel is based on the CS mechanism and on the definition of variable interframe spaces (IFSs). IFSs are idle times that must be sensed on the channel before transmission is allowed, and can be exploited to implement distributed priorities to access the channel. The shorter the IFS, the higher the priority obtained.

In increasing order of duration, Short IFSs (SIFSs) are granted only to hosts allowed to access the channel by the context of the communication when the channel contention is already solved, and Point Coordination Function iFS (PIFS) are allowed only for central coordinators of PCF (if any), and Distributed Coordination iFS (DIFS) are required for all contending stations in the distributed DCF access scheme. In DCF, all stations must wait for an idle DIFS after every transmission, and then contend for the channel by implementing contention control based on the BEB protocol, whose parameters are $CW\_min = 32$, $CW\_MAX = 1024$.[22,24,25] DCF contention-based access is performed by adopting collision avoidance and a BEB contention-control scheme, which may be affected in a significant way by the time and space contention problems due to the distributed random-access characteristics discussed in previous sections. PCF may be affected by the contention problem as well, in an indirect way. The transmission requests from the Mobile Hosts to the AP are sent in DCF frames and are subject to contention-based accesses. In other words, the DCF access scheme is considered the basic access scheme in IEEE 802.11 networks, and hence its optimization is a relevant research activity for both DCF and PCF access schemes.

When a collision occurs, it is considered as feedback indicating a high level of contention for channel access in backoff protocols, and causes a further time-spreading of the scheduling of the following accesses. Hence, a contention-based MAC protocol is subject to channel wastage caused both by collisions and the idle periods introduced by the time-spreading of accesses (i.e., idle slots). Because the reduction of idle periods generally produces an increase in the number of collisions, the MAC protocol should balance these two conflicting costs[4,5,6,9,19] to maximize channel and energy utilization. Because these costs change dynamically depending on the network load, and on the number of contending hosts, the MAC protocol should be made adaptive to the contention level of the collision domain.[11,19,24] It is widely recognized that the dynamic nature of the wireless link demands fast and low-cost adaptation on the part of MAC protocols.[3,8,9,16] Therefore, the study of tuning factors of adaptive protocols is an important issue already in protocol design. It is also necessary to understand the problems one might encounter with adaptive protocols, such as excessive overheads, stability, and fairness problems.

The objectives of the backoff scheme are (1) distribution (as uniform as possible) of the transmission attempts over a variable-sized time window, and (2) a small access delay under light load conditions. Analytical investigation of stability and characteristics of various backoff schemes have been presented in Reference 19, Reference 21, Reference 22, and Reference 24.

Several authors have investigated the enhancement of the IEEE 802.11 DCF MAC protocol to increase its performance when used in WLANs (i.e., a typical single-collision domain). Unfortunately, in a real-world scenario, a station does not have exact knowledge of the network and load configurations but can, at best, only estimate it. In Reference 12 and Reference 15, through a performance analysis, the tuning of the parameters of the standard has been studied. In Reference 3 and Reference 44, solutions have been proposed for achieving a more uniform distribution of accesses in the BEB scheme. The most promising direction for improving backoff protocols is to obtain the network status through channel observation.[4,23] A great deal of work has been done to study the information that can be obtained by observing system parameters.[21,35,43] For the IEEE 802.11 MAC protocol, some authors have proposed adaptive control of network congestion by investigating the number of users in the system.[3,8,9] This investigation would be time consuming, and hence difficult to obtain and subject to significant errors, especially in high-contention situations.[8] In Reference 4, a simple mechanism named Distributed Contention Control (DCC) was proposed to exploit the information obtained by the carrier-sensing mechanism as a preliminary contention-level estimation for adoption in the contention-control mechanism. DCC allows the implementation of a priority mechanism based on local priority parameters (with no need for priority negotiations), which could be considered a complementary scheme with respect to prority schemes defined in IEEE 802.11e.[26,31] The implementation details of DCC, stability analysis, and performance results can be found in Reference 4. The Asymptotically Optimal Backoff (AOB) mechanism proposed in Reference 6 tunes the backoff parameters to the network contention level by using two simple and low-cost estimates obtained through the information provided by the carrier-sensing mechanism: the slot utilization, and the average size of transmitted frames. AOB guarantees that the channel utilization converges to the optimal value when the network is congested, and that no overheads are introduced in a low-contention scenario.

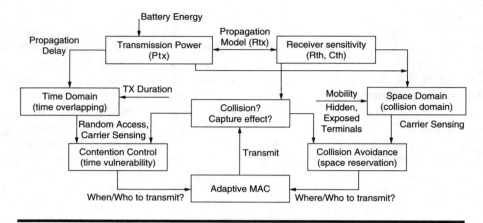

**Figure 25.4 A roadmap of the factors and their chief interrelationships as discussed in this chapter.**

## CONCLUSIONS

In WLANs, the MAC protocol is the candidate to manage the limited shared channel among mobile hosts in a highly dynamic scenario. MAC protocols also influence utilization of scarce resources such as channel bandwidth and battery energy. In this chapter, we illustrated the motivations leading to a new design and tuning of existing and new MAC protocols, based on assumptions regarding the new wireless systems. Some assumptions, problems, and limiting constraints of the wireless communication channels have been provided as background information.

The perspective of distributed random-access MAC protocols has been presented to illustrate in an incremental way the problems considered and solutions proposed that led to the current IEEE 802.11 definition. A sketched roadmap of the top-level topics considered and their chief interrelationships are summarized in Figure 25.4.

## NOTE:

This work has been partially supported by Italian MIUR funds.

## REFERENCES

1. Abramson, N., The ALOHA system — another alternative for computer communications, *Proceedings of the Fall Joint Computing Conference, AFIPS,* 1970.
2. Bharghavan, V., Demers, A., Shenker, S., and Zhang, L., MACAW: A media access protocol for wireless LANs, *Proceedings of the ACM SIGCOMM '94,* pp. 212–225, London, 1994.

3. Bianchi, G., Fratta, L., and Olivieri, M., Performance evaluation and enhancement of the CSMA/CA MAC protocol for 802.11 Wireless LANs, *Proceedings of PIMRC 1996*, Oct. 1996, Taipei, Taiwan, pp. 392–396.

4. Bononi, L., Conti, M., and Donatiello, L., Design and performance evaluation of a distributed contention control (DCC) mechanism for IEEE 802.11 Wireless Local Area Networks, *Journal of Parallel and Distributed Computing (JPDC)*, Vol. 60, N4, April 2000, pp. 407–430.

5. Bononi, L., Conti, M., and Donatiello, L., A distributed mechanism for power saving in IEEE 802.11 WLANs, *Mobile Networks and Applications (MONET)* 6, Kluwer/Academic Press, pp. 211–222, June 2001.

6. Bononi, L., Conti, M., and Gregori, E., Run-Time optimization of IEEE 802.11 Wireless LANs performance, *IEEE Transactions on Parallel and Distributed Systems (IEEE TPDS)*, Vol.15, No.1, Jan 2004, pp. 66–80.

7. Boukerche, A. and Bononi, L., Simulation and Modeling of Wireless Mobile and Ad Hoc Networks, *Mobile Ad Hoc Networking*, Chapter 14, IEEE/Wiley, August 2004.

8. Cali, F., Conti, M., and Gregori, E., Dynamic IEEE 802.11: Design, modeling and performance evaluation, *IEEE Journal on Selected Areas in Communications*, 18(9), September 2000. pp. 1774–1786.

9. Cali, F., Conti, M., and Gregori, E., Dynamic Tuning of the IEEE 802.11 Protocol to Achieve a Theoretical Throughput Limit, *IEEE/ACM Transactions on Networking*, Volume 8, No. 6 (Dec. 2000), pp. 785–799.

10. Chandra, A., Gumalla, V., and Limb, J.O., Wireless Medium Access Control Protocols, *IEEE Comm. Surveys*, Second Quarter 2000.

11. Chen, K.C., Medium access control of wireless LANs for mobile computing, *IEEE Networks*, 9-10/1994.

12. Chhaya, H.S., Performance evaluation of the IEEE 802.11 MAC Protocol for Wireless LANs, Master's Thesis, Graduate School of Illinois Institute of Technology, May 1996.

13. Colvin, A., CSMA with collision avoidance, *Computer Communications*, 6(5):227–235, 1983.

14. Corson, M.S., Maker, J.P., and Cerincione, J.H., Internet-based mobile ad hoc networking, *Internet Computing*, July–August 1999, pp. 63–70.

15. Crow, B.P., Performance Evaluation of the IEEE 802.11 Wireless Local Area Network Protocol, master's thesis. University of Arizona, 1996.

16. Forman, G.H. and Zahorjan, J., The challenges of mobile computing, *IEEE Computer*, April 94, pp. 38-47.

17. Fullmer, C.L. and Garcia-Luna-Aceves, J.J., Floor acquisition multiple access (FAMA) for packet radio networks, *Proceedings of ACM Sigcomm '95*, Cambridge, MA, 1995.

18. Fullmer, C.L., Collision Avoidance Techniques for Packet Radio Networks, Ph.D. thesis, University of California-Santa Cruz, 1998.

19. Gallagher, R.G., A Perspective on Multiaccess Channels, *IEEE Transactions on Information Theory*, Vol. IT-31, No.2, 3/1985, pp. 124–142.

20. Garces, R., Collision Avoidance and Resolution Multiple Access, Ph.D. thesis, University of California-Santa Cruz, 1999.

21. Georgiadis, L. and Papantoni-Kazakos P., Limited Feedback Sensing Algorithms for the Packet Broadcast Channel, *IEEE Transactions on Information Theory*, Vol. IT-31, No. 2, 3/1985, pp. 280–294.

22. Goodman, J., Greenberg, A.G., Madras, N., and March, P., Stability of binary exponential backoff, *Journal of the ACM*, Vol. 35, Issue 3, pp. 579–602, 1988.

23. Hajek, B. and Van Loon, T., Decentralized dynamic control of a multiaccess broadcast channel, *IEEE Transactions on Automatic Control*, Vol. 27, 1982, pp. 559–569.

24. Hastad, J., Leighton, T., and Rogoff, B., Analysis of backoff protocols for multiple access channels, *Siam Journal of Computing*, Vol. 25, No. 4, 8/1996, pp. 740–774.

25. IEEE 802.11 WG, IEEE Std. 802.11, 1999 edition, Part II: Wireless LAN Medium Access Control (MAC) and Physical Layer (PHY) Specifications, 1999.

26. IEEE 802.11 WG, Draft supplement for IEEE Std. 802.11, 1999 edition, Part II: Wireless LAN Medium Access Control (MAC) and Physical Layer (PHY) Specifications: Medium Access Control Enhancements for Quality of Service (QoS), IEEE 802.11e/D2.0, November 2001.

27. Imielinsky, T. and Badrinath, B.R., Mobile wireless computing: solutions and challenges in data management, *Communications of the Association of Computer Machinery*, Vol. 37, No. 10, October 1994.

28. Karn, P., MACA — A new channel access method for packet radio, *Proceedings of the 9th Computer Networking Conference*, 1990.

29. Kleinrock, L. and Tobagi, F.A., Packet Switching in Radio Channels: Part I — Carrier sense multiple-access modes and their throughput-delay characteristics, *IEEE Transactions on Communications*, Vol. Com-23, No. 12, pp. 1400–1416, 1975.

30. Lo, W.F. and Mouftah, H.T., Carrier sense multiple access with collision detection for radio channels, Proceedings of the *IEEE 13th International Communication and Energy Conference*, pp. 244–247, 1984.

31. Mangold, S., Choi, S., May, P., Klein, O., Hierz, G., and Stibor, L., IEEE 802.11e wireless LAN for quality of service, *Proceedings of the European Wireless Conference 2002*, Florence, Italy, February 2002, pp. 32–39.

32. Metcalfe, R.M. and Boggs, D.R., Ethernet: Distributed packet switching for local computer networks, *Communications of the Association of Computer Machinery*, Vol. 19, 1976.

33. Ramamurthi, B., Goodman, D.J., and Saleh, A., Perfect capture for local radio communications, *IEEE Journal on Selected Areas in Communications*, Vol. SAC-5, No. 5, June 1987.

34. Rappaport, T.S., *Wireless Communications: Principles and Practice*, 2nd edition, Prentice Hall, Upper Saddle River, NJ, 2002.

35. Rivest, R.L., Network Control by Bayesian Broadcast, *IEEE Transactions on Information Theory*, Vol. IT-33, No.3, 5/1987, pp. 323–328.

36. Rom, R., Collision detection in radio channels, *Local Area and Mutiple Access Networks*, pp. 235–249, CS Press, 1986.

37. Sidhu, G., Andrews, R., and Oppenheimer, A., *Inside Appletalk*, Addison-Wesley, 1989.

38. Talucci, F. and Gerla, M., MACA-BI (MACA by Invitation): A Wireless MAC Protocol for High Speed Ad Hoc Networking, *Proceedings IEEE ICUPC '97*, Vol. 2, pp. 913–17, San Diego, CA, October 1997.

39. Tobagi, F.A. and Kleinrock, L., Packet switching in radio channels: Part II — the hidden terminal problem in carrier sensing multiple access and busy tone solution, *IEEE Transactions on Communications*, Vol. Com-23, No. 12, pp. 1417–1433, 1975.

40. Tobagi, F.A. and Kleinrock, L., Packet Switching in Radio Channels: Part III — Polling and (dynamic) split channel reservation multiple access, *IEEE Transactions on Computing*, Vol. 24, pp. 832–845, 1976.

41. Tobagi, F.A. and Kleinrock, L., The effect of acknowledgment traffic on the capacity of packet switched radio channels, *IEEE Transactions on Communications*, Vol. Com-26, No. 6, pp. 815–826, 1978.

42. Tzamaloukas, A., Sender- and Receiver-Initiated Multiple Access Protocols for Ad Hoc Networks, Ph.D. Thesis, University of California-Santa Cruz, 2000.

43. Tsitsiklis, J.N., Analysis on a multiaccess control scheme, *IEEE Transactions on Automatic Control*, Vol. AC-32, No. 11, 11/1987, pp. 1017–1020.

44. Weinmiller, J., Woesner, H., Ebert, J.P., and Wolisz, A., Analyzing and tuning the distributed coordination function in the IEEE 802.11 DFWMAC Draft Standard, *Proceedings of the International Workshop on Modeling, MASCOT 96*, San Jose, CA.

45. Xu, K., Gerla, M., and Bae, S., How effective is the IEEE 802.11 RTS/CTS handshake in ad hoc networks?, *Proceedings of GLOBECOM 2002*, Vol. 1, November 2002, pp. 72–76.

# INDEX

# INDEX